Continuum Mechanics and Thermodynamics

Continuum mechanics and thermodynamics are foundational theories of many fields of science and engineering. This book presents a fresh perspective on these important subjects, exploring their fundamentals and connecting them with micro- and nanoscopic theories.

Providing clear, in-depth coverage, the book gives a self-contained treatment of topics directly related to nonlinear materials modeling with an emphasis on the thermo-mechanical behavior of solid-state systems. It starts with vectors and tensors, finite deformation kinematics, the fundamental balance and conservation laws, and classical thermodynamics. It then discusses the principles of constitutive theory and examples of constitutive models, presents a foundational treatment of energy principles and stability theory, and concludes with example closed-form solutions and the essentials of finite elements.

Together with its companion book, *Modeling Materials* (Cambridge University Press, 2011), this work presents the fundamentals of multiscale materials modeling for graduate students and researchers in physics, materials science, chemistry, and engineering.

A solutions manual is available at www.cambridge.org/9781107008267, along with a link to the authors' website which provides a variety of supplementary material for both this book and *Modeling Materials*.

Ellad B. Tadmor is Professor of Aerospace Engineering and Mechanics, University of Minnesota. His research focuses on multiscale method development and the microscopic foundations of continuum mechanics.

Ronald E. Miller is Professor of Mechanical and Aerospace Engineering, Carleton University. He has worked in the area of multiscale materials modeling for over 15 years.

Ryan S. Elliott is Associate Professor of Aerospace Engineering and Mechanics, University of Minnesota. An expert in stability of continuum and atomistic systems, he has received many awards for his work.

Continuum Mechanics and Thermodynamics

From Fundamental Concepts to Governing Equations

ELLAD B. TADMOR
University of Minnesota, USA

RONALD E. MILLER
Carleton University, Canada

RYAN S. ELLIOTT
University of Minnesota, USA

CAMBRIDGE
UNIVERSITY PRESS

CAMBRIDGE
UNIVERSITY PRESS

University Printing House, Cambridge CB2 8BS, United Kingdom

One Liberty Plaza, 20th Floor, New York, NY 10006, USA

477 Williamstown Road, Port Melbourne, VIC 3207, Australia

4843/24, 2nd Floor, Ansari Road, Daryaganj, Delhi - 110002, India

79 Anson Road, #06-04/06, Singapore 079906

Cambridge University Press is part of the University of Cambridge.

It furthers the University's mission by disseminating knowledge in the pursuit of education, learning and research at the highest international levels of excellence.

www.cambridge.org
Information on this title: www.cambridge.org/9781107008267

First published 2012
Reprinted with corrections 2013
5th printing 2016

A catalogue record for this publication is available from the British Library

Library of Congress Cataloging in Publication data
Tadmor, Ellad B., 1965–
Continuum mechanics and thermodynamics : from fundamental concepts to governing equations / Ellad B. Tadmor, Ronald E. Miller, Ryan S. Elliott.
p. cm.
Includes bibliographical references and index.
ISBN 978-1-107-00826-7
1. Continuum mechanics. 2. Thermodynamics – Mathematics. I. Miller, Ronald E. (Ronald Earle) II. Elliott, Ryan S. III. Title.
QA808.2.T33 2012
531 – dc23 2011040410

ISBN 978-1-107-00826-7 Hardback

Additional resources for this publication at www.cambridge.org/9781107008267

Contents

Preface

This book on *Continuum Mechanics and Thermodynamics* (CMT) (together with the companion book, by Tadmor and Miller, on *Modeling Materials* (MM) [TM11]) is a comprehensive framework for understanding modern attempts at modeling materials phenomena from first principles. This is a challenging problem because material behavior is dictated by many different processes, occurring on vastly different length and time scales, that interact in complex ways to give the overall material response. Further, these processes have traditionally been studied by different researchers, from different fields, using different theories and tools. For example, the bonding between individual atoms making up a material is studied by physicists using quantum mechanics, while the macroscopic deformation of materials falls within the domain of engineers who use continuum mechanics. In the end a multiscale modeling approach – capable of predicting the behavior of materials at the macroscopic scale but built on the quantum foundations of atomic bonding – requires a deep understanding of topics from a broad range of disciplines and the connections between them. These include quantum mechanics, statistical mechanics and materials science, as well as continuum mechanics and thermodynamics, which are the focus of this book.

Together, continuum mechanics and thermodynamics form the fundamental theory lying at the heart of many disciplines in science and engineering. This is a nonlinear theory dealing with the macroscopic response of material bodies to mechanical and thermal loading. There are many books on continuum mechanics, but we believe that several factors set our book apart. First, is our emphasis on fundamental concepts. Rather than just presenting equations, we attempt to explain where the equations come from and what are the underlying assumptions. This is important for those seeking to integrate continuum mechanics within a multiscale paradigm, but is also of great value for those who seek to master continuum mechanics on its own, and even for experts who wish to reflect further upon the basis of their field and its limitations. To this end, we have adopted a careful expository style, developing the subject in a step-by-step fashion, building up from fundamental ideas and concepts to more complex principles. We have taken pains to carefully and clearly discuss many of the subtle points of the subject which are often glossed over in other books.

A second difference setting our CMT apart from other books on the subject is the integration of thermodynamics into the discussion of continuum mechanics. Thermodynamics is a difficult subject which is normally taught using the language of heat engines and Carnot cycles. It is very difficult for most students to see how these concepts are related to continuum mechanics. Yet thermodynamics plays a vital role at the foundation of continuum mechanics. In fact, we think of continuum mechanics and thermodynamics as a single unified subject. It is simply impossible to discuss thermomechanical processes in materials without including thermodynamics. In addition, thermodynamics introduces key constraints on allowable forms of constitutive relations, the fundamental equations describing material response, that form the gateway to the underlying microscopic structure of the material.

The third difference is that we have written CMT with an eye to making it accessible to a broad readership. Without oversimplifying any of the concepts, we endeavor to explain everything in clear terms with as little jargon as possible. We do not assume prior knowledge of the subject matter. Thus, a reader from any field with an undergraduate education in engineering or science should be able to follow the presentation. We feel that this is particularly important as it makes this vital subject accessible to researchers and students from physics, chemistry and materials science who traditionally have less exposure to continuum mechanics.

The philosophy underlying CMT and its form provide it with a dual role. On its own, it is suitable as a first introduction to continuum mechanics and thermodynamics for graduate students or researchers in science and engineering. Together with MM, it provides a comprehensive and integrated framework for modern predictive materials modeling. With this latter goal in mind, CMT is written using a similar style, notation and terminology to that of MM, making it easy to use the two books together.

Acknowledgments

As we explained in the preface, this book is really one part of a two-volume project covering many topics in materials modeling beyond continuum mechanics and thermodynamics (CMT). In the following few pages, we choose to express our thanks to everyone involved in the *entire* project, whether their contribution directly affected the words on these pages or only the words in the companion volume (*Modeling Materials* or MM for short). We mention this by way of explanation, in case a careful reader is wondering why we thank people for helping us with topics that clearly do not appear in the table of contents. The people thanked below most certainly helped shape our understanding of materials modeling in general, even if not with respect to CMT specifically.

Our greatest debt goes to our wives, Jennifer, Granda and Sheila, and to our children: Maya, Lea, Persephone and Max. They have suffered more than anyone during the long course of this project, as their preoccupied husbands and fathers stole too much time from so many other things. They need to be thanked for such a long list of reasons that we would likely have to split these two books into three if we were thorough with the details. Thanks, all of you, for your patience and support. We must also thank our own parents Zehev and Ciporah, Don and Linda, and Robert and Mary for giving us the impression – perhaps mistaken – that everybody will appreciate what we have to say as much as they do.

The writing of a book is always a collaborative effort with so many people whose names do not appear on the cover. These include students in courses, colleagues in the corridors and offices of our universities and unlucky friends cornered at conferences. The list of people that offered a little piece of advice here, a correction there or a word of encouragement somewhere else is indeed too long to include, but there are a few people in particular that deserve special mention.

Some colleagues generously did calculations for us, verified results or provided other contributions from their own work. We thank Quiying Chen at the NRC Institute for Aerospace Research in Ottawa for his time in calculating UBER curves with density functional theory. Tsveta Sendova, a postdoctoral fellow at the University of Minnesota (UMN), coded and ran the simulations for the two-dimensional NEB example we present. Another postdoctoral fellow at UMN, Woo Kyun Kim, performed the indentation and thermal expansion simulations used to illustrate the hot-QC method. We thank Yuri Mishin (George Mason University) for providing figures, and Christoph Ortner (Oxford University) for providing many insights into the problem of full versus sequential minimization of multivariate functions, including the example we provide in the MM book. The hot-QC project has greatly benefited from the work of Laurent Dupuy (SEA Saclay) and Frederic Legoll (École Nationale des Ponts et Chaussées). Their help in preparing a journal paper on the subject has also proven extremely useful in preparing the chapter on dynamic multiscale methods. Furio Ercolessi must be thanked in general for his fantastic web-based notes on so many important subjects discussed herein, and specifically for providing us with his molecular dynamics code as a teaching tool to provide with MM.

Other colleagues patiently taught us the many subjects in these books about which we are decidedly *not* experts. Dong Qian at the University of Cincinnati and Michael Parks at Sandia National Laboratories very patiently and repeatedly explained the nuances of various multiscale methods to us. Similarly, we would like to thank Catalin Picu at the Rensselaer Polytechnic Institute for explaining CACM, and Leo Shilkrot for his frank conversations about CADD and the BSM. Noam Bernstein at the Navy Research Laboratories (NRL) was invaluable in explaining DFT in a way that an engineer could understand, and Peter Watson at Carleton University was instrumental in our eventual understanding of quantum mechanics. Roger Fosdick (UMN) discussed, at length, many topics related to continuum mechanics including tensor notation, material frame-indifference, Reynolds transport theorem and the principle of action and reaction. He also took the time to read and comment on our take on material frame-indifference.

We are especially indebted to those colleagues that were willing to take the time to carefully read and comment on drafts of various sections of the books – a thankless and delicate task. James Sethna (Cornell University) and Dionisios Margetis (University of Maryland) read and commented on the statistical mechanics chapter. Noam Bernstein (NRL) must be thanked more than once, for reading and commenting on both the quantum mechanics chapter and the sections on cluster expansions. Nikhil Admal, a graduate student working with Ellad at UMN, contributed significantly to our understanding of stress and read and commented on various continuum mechanics topics, Marcel Arndt helped by translating an important paper on stress by Walter Noll from German to English and worked with Ellad on developing algorithms for lattice calculations, while Gang Lu at the California State University (Northridge) set us straight on several points about density functional theory. Other patient readers to whom we say "thank you" include Mitch Luskin from UMN (numerical analysis of multiscale methods and quantum mechanics), Bill Curtin from Brown University (static multiscale methods), Dick James from UMN (restricted ensembles and the definition of stress) and Leonid Berlyand from Pennsylvania State University (thermodynamics).

There are a great many colleagues who were willing to talk to us at length about various subjects in these books. We hope that we did not overstay our welcome in their offices too often, and that they do not sigh too deeply anymore when they see a message from us in their inbox. Most importantly, we thank them very much for their time. In addition to those already mentioned above, we thank David Rodney (Institut National Polytechnique de Grenoble), Perry Leo and Tom Shield (UMN), Miles Rubin and Eli Altus (Technion), Joel Lebowitz, Sheldon Goldstein and Michael Kiessling (Rutgers)[1] and Andy Ruina (Cornell). We would also be remiss if we did not take the time to thank Art Voter (Los Alamos National

[1] Ellad would particularly like to thank the Rutgers trio for letting him join them on one of their lunches to discuss the foundations of statistical mechanics – a topic which is apparently standard lunch fare for them along with the foundations of quantum mechanics.

Laboratory), John Moriarty (Lawrence Livermore National Laboratory) and Mike Baskes (Sandia National Laboratories) for many insightful discussions and suggestions of valuable references.

There are some things in these books that are so far outside our area of expertise that we have even had to look beyond the offices of professors and researchers. Elissa Gutterman, an expert in linguistics, provided phonetic pronunciation of French and German names. As none of us are experimentalists, our brief foray into pocket watch "testing" would not have been very successful without the help of Steve Truttman and Stan Conley in the structures laboratories at Carleton University. The story of our cover images involves so many people, it deserves its own paragraph.

As the reader will see in the introduction to both books, we are fond of the symbolic connection between pocket watches and the topics we discuss herein. There are many beautiful images of pocket watches out there, but obtaining one of sufficient resolution, and getting permission to *use* it, is surprisingly difficult. As such, we owe a great debt to Mr. Hans Holzach, a watchmaker and amateur photographer at Beyer Chronometrie AG in Zurich. Not only did he generously agree to let us use his images, he took over the entire enterprise of retaking the photos when we found out that his images did not have sufficient resolution! This required Hans to coordinate with many people that we also thank for helping make the strikingly beautiful cover images possible. These include the photographer, Dany Schulthess (www.fotos.ch), Mr. René Beyer, the owner of Beyer Chronometrie AG in Zurich, who compensated the photographer and permitted photos to be taken at his shop, and also to Dr. Randall E. Morris, the owner of the pocket watch, who escorted it from California to Switzerland (!) in time for the photo shoot. The fact that total strangers would go to such lengths in response to an unsolicited e-mail contact is a testament to their kind spirits and, no doubt, to their proud love of the beauty of pocket watches.

We cannot forget our students. Many continue to teach us things every day just by bringing us their questions and ideas. Others were directly used as guinea pigs with early drafts of parts of these books.[2] Ellad would like to thank his graduate students and postdoctoral fellows over the last five years who have been fighting with this project for attention, specifically Nikhil Admal, Yera Hakobian, Hezi Hizkiahu, Dan Karls, Woo Kyun Kim, Leonid Kucherov, Amit Singh, Tsvetanka Sendova, Valeriu Smiricinschi, Slava Sorkin and Steve Whalen. Ron would likewise like to thank Ishraq Shabib, Behrouz Shiari and Denis Saraev, whose work helped shape his ideas about atomistic modeling. Ryan would like to thank Kaushik Dayal, Dan Gerbig, Dipta Ghosh, Venkata Suresh Guthikonda, Vincent Jusuf, Dan Karls, Tsvetanka Sendova, Valeriu Smirichinski and Viacheslav (Slava) Sorkin. Harley Johnson and his 2008–2009 and 2010–2011 graduate classes at the University of Illinois (Urbana-Champaign) who used the books extensively provided great feedback to improve the manuscripts, as did Bill Curtin's class at Brown in 2009–2010. The 2009 and 2010 classes of Ron's "Microstructure and Properties of Engineering Materials" class caught many initial errors in the chapters on crystal structures and molecular statics and

[2] Test subjects were always treated humanely and no students were irreparably harmed during the preparation of these books.

dynamics. Some students of Ellad's Continuum Mechanics course are especially noted for their significant contributions: Yilmaz Bayazit (2008), Pietro Ferrero (2009), Zhuang Houlong (2008), Jenny Hwang (2009), Karl Johnson (2008), Dan Karls (2008), Minsu Kim (2009), Nathan Nasgovitz (2008), Yintao Song (2008) and Chonglin Zhang (2008).

Of course, we should also thank our own teachers. Course notes from Michael Ortiz, Janet Blume, Jerry Weiner, Nicolas Triantafyllidis and Tom Shield were invaluable to us in preparing our own notes and this book. Thanks also to Ellad and Ron's former advisors at Brown University, Michael Ortiz and Rob Phillips (both currently at Caltech) and Ryan's former advisors Nicolas Triantafyllidis and John A. Shaw at the University of Michigan (Nick is currently at the École Polytechnique, France), whose irresistible enthusiasm, curiosity and encouragement pulled us down this most rewarding of scientific paths.

Ryan would like to thank the University of Minnesota and the McKnight Foundation whose *McKnight Land-Grant Professorship* helped support his effort in writing this book. Further, he would like to sincerely thank Patrick Le Tallec, Nicolas Triantafyllidis, Renata Zwiers, Kostas Danas, Charis Iordanou and everyone at the *Laboratoire de Mécanique des Solides* (LMS), the École Polytechnique, France for their generous support, hosting and friendship during Ryan and Sheila's "Paris adventure" of 2010. Finally, Ryan would like to acknowledge the support of the National Science Foundation.

We note that many figures in these books were prepared with the drawing package Asymptote (see http://asymptote.sourceforge.net/), an open-source effort that we think deserves to be promoted here. Finally, we thank our editor Simon Capelin and the entire team at Cambridge, for their advice, assistance and truly astounding patience.

Notation

This book is devoted to the subject of continuum mechanics and thermodynamics. However, together with the companion book by Tadmor and Miller, *Modeling Materials* (MM) [TM11], it is part of a greater effort to create a unified theoretical foundation for multiscale modeling of material behavior. Such a theory includes contributions from a large number of fields including those covered in this book, but also quantum mechanics, statistical mechanics and materials science. We have attempted as much as possible to use the most common and familiar notation from within each field as long as this does not lead to confusion. To keep the amount of notation to a minimum, we generally prefer to append qualifiers to symbols rather than introducing new symbols. For example, $\boldsymbol{f}$ is force, which if relevant can be divided into internal, $\boldsymbol{f}^{\text{int}}$, and external, $\boldsymbol{f}^{\text{ext}}$, parts.

We use the following general conventions:

- Descriptive qualifiers generally appear as superscripts and are typeset using a Roman (as opposed to Greek) nonitalic font.
- The weight and style of the font used to render a variable indicates its type. Scalar variables are denoted using an italic font. For example, T is temperature. Array variables are denoted using a sans serif font, such as A for the matrix A. Vectors and tensors (in the mathematical sense of the word) are rendered in a boldface font. For example, $\boldsymbol{\sigma}$ is the stress tensor.
- Variables often have subscript and superscript indices. Indices referring to the components of a matrix, vector or tensor appear as subscripts in italic Roman font. For example, v_i is the ith component of the velocity vector. Superscripts will be used as counters of variables. For example, $\boldsymbol{F}^e$ is the deformation gradient in element e. Iteration counters appear in parentheses, for example $\boldsymbol{f}^{(i)}$ is the force in iteration i.
- The Einstein summation convention will be followed on repeated indices (e.g. $v_i v_i = v_1^2 + v_2^2 + v_3^2$), unless otherwise clear from the context. (See Section 2.2.2 for more details.)
- A subscript is used to refer to multiple equations on a single line, for example, "Eqn. $(3.32)_2$" refers to the second equation in Eqn. (3.32) ("$a_i(\boldsymbol{x}, t) \equiv \dots$").
- Important equations are emphasized by placing them in a shaded box.

Below, we describe the main notation and symbols used in the book, and indicate the page on which each is first defined.

Mathematical notation

Notation	Description	Page
$\equiv$	equal to by definition	22
$:=$	variable on the left is assigned the value on the right	283
$\forall$	for all	22
$\in$	contained in	22
$\subset$	a subset of	107
iff	if and only if	22
$O(n)$	orthogonal group of degree n	32
$SL(n)$	proper unimodular (special linear) group of degree n	217
$SO(n)$	proper orthogonal (special orthogonal) group of degree n	32
$\mathbb{R}$	set of all real numbers	22
$\mathbb{R}^n$	real coordinate space (n-tuples of real numbers)	25
$\lvert \bullet \rvert$	absolute value of a real number	25
$\lVert \bullet \rVert$	norm of a vector	25
$\langle \bullet, \bullet \rangle$	inner product of two vectors	25
$\langle \mathcal{D}_{\boldsymbol{x}} \bullet; \boldsymbol{u} \rangle$	nonnormalized directional derivative with respect to $\boldsymbol{x}$ in the direction $\boldsymbol{u}$	57
$f[\bullet]$	square brackets indicate f is a linear function of its arguments	24
$\boldsymbol{A}^T$	transpose of a second-order tensor or matrix: $[\boldsymbol{A}^T]_{ij} = A_{ji}$	19
$\boldsymbol{A}^{-T}$	transpose of the inverse of $\boldsymbol{A}$: $\boldsymbol{A}^{-T} \equiv (\boldsymbol{A}^{-1})^T$	43
$\boldsymbol{a} \cdot \boldsymbol{b}$	dot product (vectors): $\boldsymbol{a} \cdot \boldsymbol{b} = a_i b_i$	25
$\boldsymbol{a} \times \boldsymbol{b}$	cross product (vectors): $[\boldsymbol{a} \times \boldsymbol{b}]_k = \epsilon_{ijk} a_i b_j$	29
$\boldsymbol{a} \otimes \boldsymbol{b}$	tensor product (vectors): $[\boldsymbol{a} \otimes \boldsymbol{b}]_{ij} = a_i b_j$	39
$\boldsymbol{A} : \boldsymbol{B}$	contraction (second-order tensors): $\boldsymbol{A} : \boldsymbol{B} = A_{ij} B_{ij}$	44
$\boldsymbol{A} \cdot\cdot \boldsymbol{B}$	transposed contraction (second-order tensors): $\boldsymbol{A} \cdot\cdot \boldsymbol{B} = A_{ij} B_{ji}$	44
$A_{(ij)}$	symmetric part of a second-order tensor: $A_{(ij)} = \frac{1}{2}(A_{ij} + A_{ji})$	48
$A_{[ij]}$	antisymmetric part: $A_{[ij]} = \frac{1}{2}(A_{ij} - A_{ji})$	48
$\lambda_\alpha^{\boldsymbol{A}}, \boldsymbol{\Lambda}_\alpha^{\boldsymbol{A}}$	αth eigenvalue and eigenvector of the second-order tensor $\boldsymbol{A}$	49
$I_k^{\boldsymbol{A}}$	kth principal invariant of the second-order tensor $\boldsymbol{A}$	49
$đ$	inexact differential	159
$\det \boldsymbol{A}$	determinant of a matrix or a second-order tensor	21
$\operatorname{tr} \boldsymbol{A}$	trace of a matrix or a second-order tensor: $\operatorname{tr} \boldsymbol{A} = A_{ii}$	19
$\nabla \bullet$, grad $\bullet$	gradient of a tensor (deformed configuration)	57
$\nabla_0 \bullet$, Grad $\bullet$	gradient of a tensor (reference configuration)	77
curl $\bullet$	curl of a tensor (deformed configuration)	58
Curl $\bullet$	curl of a tensor (reference configuration)	77
div $\bullet$	divergence of a tensor (deformed configuration)	59
Div $\bullet$	divergence of a tensor (reference configuration)	77
$\nabla^2 \bullet$	Laplacian of a tensor (deformed configuration)	60
$\vec{\alpha}_e$	local node number on element e for global node number α	294

General symbols – Greek

Symbol	Description	Page
α	stretch parameter	78
$\boldsymbol{\Gamma}, \Gamma_i$	set of extensive state variables	135
$\boldsymbol{\Gamma}^{\mathrm{i}}, \Gamma^{\mathrm{i}}_i$	set of intensive state variables obtained from $\boldsymbol{\Gamma}$	173
$\boldsymbol{\gamma}, \gamma_i$	set of intensive state variables work conjugate with $\boldsymbol{\Gamma}$	157
δ_{ij}	Kronecker delta	19
ϵ_{ijk}	permutation symbol	20
$\boldsymbol{\epsilon}, \epsilon_{ij}$	small strain tensor	93
θ	polar coordinate in polar cylindrical system	61
θ	zenith angle in spherical system	62
θ^i	curvilinear coordinates in a general coordinate system	60
κ	bulk viscosity (fluid)	220
λ	Lamé constant	235
μ	shear viscosity (fluid)	220
μ	shear modulus (solid)	235
ν	Poisson's ratio	235
$\boldsymbol{\xi}, \xi_I$	parent space for a finite element	294
Π	total potential energy of a system and the applied loads	247
ρ	mass density (deformed configuration)	106
ρ_0	mass density (reference configuration)	107
$\boldsymbol{\sigma}, \sigma_{ij}$	Cauchy stress tensor	116
$\boldsymbol{\tau}, \tau_{ij}$	Kirchhoff stress tensor	123
ϕ	azimuthal angle in spherical system	63
$\boldsymbol{\varphi}, \varphi_i$	deformation mapping	72
ψ	specific Helmholtz free energy	193
$\boldsymbol{\psi}, \psi_i$	spin axial vector	98

General symbols – Roman

Symbol	Description	Page
$\breve{\boldsymbol{a}}, \breve{a}_i$	acceleration vector (material description)	94
$\boldsymbol{a}, a_i$	acceleration vector (spatial description)	94
B	bulk modulus	241
$\boldsymbol{B}, B_{ij}$	left Cauchy–Green deformation tensor	85
B	matrix of finite element shape function derivatives	302
$\breve{\boldsymbol{b}}, \breve{b}_i$	body force (material description)	122
$\boldsymbol{b}, b_i$	body force (spatial description)	112
C_v	molar heat capacity at constant volume	144
$\boldsymbol{C}, C_{IJ}$	right Cauchy–Green deformation tensor	79

$\boldsymbol{C}, C_{IJKL}$	referential elasticity tensor	226
c_v	specific heat capacity at constant volume	320
$\boldsymbol{c}, c_{ijkl}$	spatial (or small strain) elasticity tensor	228
$\mathsf{c}, \mathsf{c}_{mn}$	elasticity matrix (in Voigt notation)	230
$\boldsymbol{D}, D_{iJkL}$	mixed elasticity tensor	227
D	matrix representation of the mixed elasticity tensor	303
$\boldsymbol{d}, d_{ij}$	rate of deformation tensor	96
$\mathcal{E}$	total energy of a thermodynamic system	141
E	Young's modulus	235
$\boldsymbol{E}, E_{IJ}$	Lagrangian strain tensor	87
E	finite element strain operator matrix	302
$\boldsymbol{e}_i$	orthonormal basis vectors	23
$\boldsymbol{e}, e_{ij}$	Euler–Almansi strain tensor	90
$\mathcal{F}$	frame of reference	196
$\boldsymbol{F}^{\text{ext}}, F_i^{\text{ext}}$	total external force acting on a system	10
$\boldsymbol{F}, F_{iJ}$	deformation gradient	78
F	matrix representation of the deformation gradient	301
f	column matrix of finite element nodal forces	281
G	material symmetry group	216
g	specific Gibbs free energy	195
$\boldsymbol{g}^i, \boldsymbol{g}_i$	contravariant and covariant basis vectors, respectively	28
$\boldsymbol{H}_0, H_{0i}$	angular momentum about the origin	120
h	outward heat flux across a body surface	173
h	specific enthalpy	194
$\boldsymbol{I}$	identity tensor	41
I	identity matrix	20
J	Jacobian of the deformation gradient	79
$\hat{J}$	Jacobian of the finite element parent space mapping	295
J	affine mapping from the parent element space to physical space	295
$\mathcal{K}$	macroscopic (continuum) kinetic energy	140
K	finite element stiffness matrix	287
k	thermal conductivity	210
$\boldsymbol{L}, L_i$	linear momentum	110
$\boldsymbol{l}, l_{ij}$	spatial gradient of the velocity field	95
$\boldsymbol{M}_0^{\text{ext}}, M_{0i}^{\text{ext}}$	total external moment about the origin acting on a system	120
N	number of particles or atoms	110
n_{d}	dimensionality of space	16
n	number of moles of a gas	144
$\mathcal{P}^{\text{def}}$	deformation power	172
$\mathcal{P}^{\text{ext}}$	external power	170
$\boldsymbol{P}, P_{iJ}$	first Piola–Kirchhoff stress tensor	122
P	matrix representation of the first Piola–Kirchhoff stress	301
p	pressure (or hydrostatic stress)	119

$\Delta\mathcal{Q}$	heat transferred to a system during a process	140
$\mathcal{Q}_t$	orthogonal transformation between frames of reference	197
$\mathbf{Q}$,$\mathsf{Q}_{\alpha i}$	orthogonal transformation matrix	31
$\boldsymbol{q}$, q_i	spatial heat flux vector	174
$\boldsymbol{q}_0$, q_{0I}	reference heat flux vector	175
$\mathcal{R}$	rate of heat transfer	170
$\boldsymbol{R}$, R_{iJ}	finite rotation (polar decomposition)	83
r	radial coordinate in polar cylindrical (and spherical) system	61
r	spatial strength of a distributed heat source	173
r_0	reference strength of a distributed heat source	175
$\mathcal{S}$	entropy	150
$\dot{\mathcal{S}}^{\text{ext}}$	external entropy input rate	176
$\dot{\mathcal{S}}^{\text{int}}$	internal entropy production rate	176
S^α	shape function for finite element node α (physical space)	280
$\boldsymbol{S}$, S_{IJ}	second Piola–Kirchhoff stress tensor	125
S	matrix of finite element shape functions	279
s	specific entropy	175
$\dot{s}^{\text{ext}}$	specific external entropy input rate	176
$\dot{s}^{\text{int}}$	specific internal entropy production rate	176
$\boldsymbol{s}$, s_{ijkl}	spatial (or small strain) compliance tensor	230
s, s_{mn}	compliance matrix (in Voigt notation)	231
$\mathfrak{s}^\alpha$	shape function for finite element node α (parent space)	294
T	temperature	137
$\boldsymbol{T}$, T_i	nominal traction (stress vector)	123
$\boldsymbol{t}$, t_i	true traction (stress vector)	113
$\bar{\boldsymbol{t}}$, $\bar{t}_i$	true external traction (stress vector)	112
$\mathcal{U}$	internal energy	140
$\boldsymbol{U}$, U_{IJ}	right stretch tensor	83
u	spatial specific internal energy	170
u_0	reference specific internal energy	175
$\boldsymbol{u}$, u_i	displacement vector	91
$\widetilde{\boldsymbol{u}}$, $\widetilde{u}_i$	finite element approximation to the displacement field	279
u	column matrix of finite element nodal displacements	278
V_0	volume (reference configuration)	79
V	volume (deformed configuration)	79
$\boldsymbol{V}$, V_{ij}	left stretch tensor	83
v	specific volume	132
$\breve{\boldsymbol{v}}$, $\breve{v}_i$	velocity vector (material description)	94
$\boldsymbol{v}$, v_i	velocity vector (spatial description)	94
$\Delta\mathcal{W}$	work performed on a system during a process	140
W	strain energy density function	194
$\boldsymbol{w}$, w_{ij}	spin tensor	97

$\boldsymbol{X}$, X_I	position of a continuum particle (reference configuration)	72
$\boldsymbol{x}$, x_i	position of a continuum particle (deformed configuration)	72
X	column matrix of finite element nodal coordinates	278
z	axial coordinate in polar cylindrical system	61

1 Introduction

A solid material subjected to mechanical and thermal loading will change its shape and develop internal stress and temperature variations. What is the best way to describe this behavior? In principle, the response of a material (neglecting relativistic effects) is dictated by that of its atoms, which are governed by quantum mechanics. Therefore, if we could solve Schrödinger's equation for all of the atoms in the material (there are about 10^{22}=10 000 000 000 000 000 000 000 atoms in a gram of copper) and evolve the dynamics of the electrons and nuclei over "macroscopic times" (i.e. seconds, hours and days), we would be able to predict the material behavior. Of course, when we say "material," we are already referring to a very complex system. In order to predict the response of the material we would first have to construct the material structure in the computer, which would require us to use Schrödinger's equation to simulate the process by which the material was manufactured. Conceptually, it may be useful to think of materials in this way, but we can quickly see the futility of the approach: the state of the art of quantum calculations involves just hundreds of atoms over a time of nanoseconds.

Fortunately, in many cases it is not necessary to keep track of all the atoms in a material to describe its behavior. Rather, the overall response of such a collection of atoms is often much more readily amenable to an elegant, mathematical description. Like the pocket watch on the cover of this book, the complex and intricate inner workings of a material are often not of interest. It is the outer expression of these inner workings – the regular motion of the watch hands or macroscopic material response – that is of primary concern. To this end, lying at the opposite extreme to quantum mechanics, we find continuum mechanics and thermodynamics (CMT). The CMT disciplines completely ignore the discreteness of the world, treating it in terms of "macroscopic observables" – time and space averages over the underlying swirling hosts of electrons and atomic nuclei. This leads to a theory couched in terms of continuously varying fields. Using clear thinking inspired by our understanding of the basic laws of nature (which have been validated by experiments) it is possible to construct a remarkably coherent and predictive framework for material behavior. In fact, CMT have been so successful that with the exception of electromagnetic phenomena, almost all of the courses in an engineering curriculum from aerodynamics to solid mechanics are simply an application of simplified versions of the general CMT theory to situations of special interest. Clearly there is something to this macroscopically averaged view of the world. Of course, the continuum picture becomes fuzzy and eventually breaks down when we attempt to apply it to phenomena governed by small length and time scales.[1] Those are

[1] Having said that, it is important to note that continuum mechanics works remarkably well down to extremely small scales. Micro electro mechanical systems (MEMS) devices, which are fully functioning microscopic

exactly the "multiscale" situations that we explore in depth in the companion book to this one titled *Modeling Materials: Continuum, Atomistic and Multiscale Techniques* (MM) [TM11]. Here, we focus on CMT.

Continuum mechanics involves the application of the principles of classical mechanics to material bodies approximated as continuous media. Classical mechanics itself has a long and distinguished history. As Clifford Truesdell, one of the fathers of modern continuum mechanics, states in the introduction to his lectures on the subject [Tru66a]:

> The classical nature of mechanics reflects its greatness: Ever old and ever new, it continues to pour out for us understanding and application, linking a changing world to unchanged law.

The unchanged laws that Truesdell refers to are the balance principles of mechanics: conservation of mass and the balance of linear and angular momentum. Together with the first law of thermodynamics (conservation of energy), these principles lead to a set of coupled differential equations governing the evolution of material systems.[2] The resulting general theory of continuum mechanics and thermodynamics is applicable to arbitrary materials undergoing arbitrarily large deformations. We develop this theory and explore its applications in two main parts. Part I on *theory* focuses on the basic theory underlying CMT, going from abstract mathematical ideas to the response of real materials. Part II on *solutions* focuses on the application of the theory to solve actual problems.

Part I begins with Chapter 2 on scalars, vectors and tensors and the associated notation used throughout the book. This chapter deals with basic physical and mathematical concepts that must be understood before we can discuss the mechanics of continuum bodies. First and foremost we must provide basic definitions for *space* and *time*. Without such definitions it is meaningless to speak of the positions of physical objects and their time evolution. Newton was well aware of this and begins his *Principia* [New62] with a preface called the *Scholium* devoted to definitions. In many ways Newton's greatness lies not in his famous laws (which are based on earlier work) but in his ability to create a unified framework out of the confusion that preceded him by defining his terms.[3] Once space and time are agreed upon, the next step is to identify suitable mathematical objects for describing physical variables. We seek to define such things as the positions of particles, their velocities and more complex quantities like the stress state at a point in a solid. A key property of all such variables is that they should exist independently of the particular coordinate system in which they are represented. Variables that have this property are called *tensors* or *tensor fields*. Anyone with a mathematical or scientific background will have come across the term "tensor," but few really understand what a tensor is. This is because tensors are often

machines smaller than the diameter of a human hair (~100 microns), are for the most part described quite adequately by continuum mechanics. Even on the nanoscale where the discrete nature of materials is apparent, continuum mechanics is remarkably accurate to within a few atomic spacings of localized defects in the atomic arrangement.

[2] The second law of thermodynamics also plays an essential role. However, in the (standard) presentation of the theory developed here it does not explicitly enter as a governing equation of the material. Rather, it serves to restrict the possible response to external stimuli of a material (see Chapter 6).

[3] Amazingly, more than 300 years after Newton published *Principia*, the appropriate definitions for space and time in classical mechanics remain controversial. We discuss this in Section 2.1.

defined with a purely rules-based approach, i.e. a recipe is given for checking whether a given quantity is or is not a tensor. This is fine as far is it goes, but it does not lead to greater insight. The problem is that the idea of a tensor field is complex and to gain a true and full understanding one must immerse oneself in the rarefied atmosphere of differential geometry. We have placed ourselves squarely between these two extremes and have attempted to provide a more nuanced fundamental description of tensors while keeping the discussion as accessible as possible. For this reason we mostly adopt the Cartesian coordinate system in our discussions, introducing the more general covariant and contravariant notation of curvilinear coordinates only where necessary.

Our next step takes us away from the abstract world of tensor algebra and calculus to the description of physical bodies. As noted above, we know that in reality bodies are made of material and material is made of atoms which themselves are made of more fundamental particles and – who knows – perhaps those are made of strings or membranes existing in a higher-dimensional universe. Continuum mechanics ignores this underlying discrete structure and provides a *model* for the world in which a material is infinitely divisible. Cut a piece of copper in two and you get two pieces of copper, and so on ad infinitum. The downside of this simplification is that it actually becomes *more* complicated to describe the shape and evolution of bodies. For a discrete set of particles all we need to know is the positions of the particles and their velocities. In contrast, how can we describe the "position" that an evolving blob of material occupies in space? This broadly falls under the topic of *kinematics of deformation* covered in Chapter 3. The study of kinematics is concerned exclusively with the abstract motion of bodies, taking no consideration of the forces that may be required to impart such a motion. As a result, kinematics is purely the geometric, descriptive aspect of mechanics, phrased in the language of *configurations* that a blob of material can adopt. In a sense one can think of a configuration being the "sheet music" of mechanics. The external mechanical and thermal loading are what ultimately realize this configuration, just as the musicians and their instruments ultimately bring a symphony to life.

A continuum body can take on an infinity of possible configurations. It is convenient to identify one of these as a *reference configuration* and to refer all other configurations to this one. Once a reference configuration is selected, it is possible to define the concept of *strain* (or more generally "local deformation"). This is the change in shape experienced by the infinitesimal environment of a point in a continuum body relative to its shape in the reference configuration. Since it is shape change (as opposed to rigid motion) that material bodies resist, strain becomes a key variable in a continuum theory. An important aspect of continuum mechanics is that shape change can be of arbitrary magnitude. This is referred to somewhat confusingly as "finite strain" as if contrasting the theory with another one dealing with "infinite strain." Really the distinction is with theories of "infinitesimal strain" (like the theories of strength of materials and linear elasticity taught as part of an engineering curriculum). This makes continuum mechanics a nonlinear theory – very general in the sort of problems it can handle, but also more difficult to solve.

Having laid out the geometry of deformation, we must next turn to the laws of nature to determine how a body will respond to applied loading. This topic naturally divides into two parts. Chapter 4 focuses on this question from a purely mechanical perspective. This means that we ignore temperature and think only of masses and the mechanical forces acting on them. At the heart of this description are three laws taken to be fundamental principles in classical mechanics: conservation of mass and the balance of linear and of angular momentum. Easily stated for a system of particles, the extension of these laws to continuous media leads to some interesting results. The big name here is Cauchy, who through some clever thought experiments was able to infer the existence of the stress tensor and its properties. Cauchy was concerned with what we today would call the "true stress" or for obvious reasons the "Cauchy stress." This is the force per unit area experienced by a point in a continuum when cut along some plane passing through that point. The notion of configurations introduced above means that the stress tensor can be recast in a variety of forms that, although lacking the clear physical interpretation of Cauchy's stress, have certain mathematical advantages. In particular, the first and second Piola–Kirchhoff stress tensors represent the stress relative to the reference configuration mentioned above.

The second set of the laws of nature that must be considered to fully characterize a continuum mechanics problem are those having to do with temperature, i.e. the laws of thermodynamics discussed in Chapter 5. In reality, a material is not just subjected to mechanical loading which leads to stresses and strains in the body; it also experiences thermal loading which can lead to an internally varying temperature field. Furthermore, the mechanical and thermal effects are intimately coupled into what can only be described as thermomechanical behavior. Thermodynamics is for most people a more difficult subject to understand than pure mechanics. This is another consequence of the "simplification" afforded by the continuum approximation. Concepts like temperature and entropy that have a clear physical meaning when studied at the level of discrete particles become far more abstract at the macroscopic level where their existence must be cleverly inferred from experiments.[4] The three laws of thermodynamics (numbered in a way to make C programmers happy) are the *zeroth law*, which deals with thermal equilibrium and leads to the concept of temperature, the *first law*, which expresses the conservation of energy and defines energy, and the *second law*, which deals with the concept of entropy and the direction of time (i.e. why we have a past and a future). Unlike a traditional book on thermodynamics, we develop these concepts with an eye to continuum mechanics. We do not talk about steam engines, but rather show how thermodynamics contributes a conservation law to the field equations of continuum mechanics, and how restrictions related to the second law impact the possible models for material behavior – the so-called "constitutive relations" described next.

The theory we have summarized so far appears wonderfully economical. Using a handful of conservation laws inferred from experiments, a very general theoretical formulation is established which (within a classical framework) fully describes the behavior of materials subjected to arbitrary mechanical and thermal loading. Unfortunately, this theory is not closed. By this we mean that the theoretical formulation of continuum mechanics and thermodynamics possesses more unknowns than equations to solve for them. If one thinks about this for a minute, it is not surprising – we have not yet introduced the particular nature

[4] A student wishing to truly understand thermodynamics is strongly encouraged to also explore this subject from the perspective of statistical mechanics as is done in Chapter 7 of [TM11].

of the material into the discussion. Clearly the response of a block of butter will be different than that of steel when subjected to mechanical and thermal loading. The equations relating the response of a material to the loading applied to it are called *constitutive relations* and are discussed in Chapter 6. Since we are dealing with a general framework which allows for arbitrary "finite" deformation, the constitutive relations are generally nonlinear. Continuum mechanics cannot predict the particular form of the constitutive relations for a given material – these are obtained either empirically through experimentation or more recently using multiscale modeling approaches as described in MM [TM11]. However, continuum mechanics *can* place constraints on the allowable forms for these relations. This is very important, since it dramatically reduces the set of possible functions that can be used for interpreting experiments or multiscale simulations. One constraint already mentioned above is the restrictions due to the second law of thermodynamics. For example, it is not possible to have a material in which heat flows from cold to hot.[5] Another fundamental restriction is related to the *principle of material frame-indifference* (or "objectivity"). Material frame-indifference is a difficult and controversial subject with different, apparently irreconcilable, schools of thought. Most students of continuum mechanics – even very advanced "students" – find this subject quite difficult to grasp. We provide a new presentation of material frame-indifference that we feel clarifies much of the confusion and demonstrates how the different approaches mentioned above are related and are in fact consistent with each other. A third restriction on the form of constitutive relations is tied to the symmetry properties of the material. This leads to vastly simplified forms for special cases such as isotropic materials whose response is independent of direction. Even simpler forms are obtained when the equations are linearized, which in the end leads to the venerable (generalized) Hooke's law – a linear relation between the Cauchy stress and the infinitesimal strain tensor.

The addition of constitutive relations to the conservation and balance laws derived before closes the theory. It is now possible to write down a system of coupled, nonlinear partial differential equations that fully characterize a thermomechanical system. Together with appropriate boundary conditions (and initial conditions for a dynamical problem) a well-defined *(initial) boundary-value problem* can be constructed. This is described in Chapter 7. Special emphasis is placed in this chapter on purely mechanical static problems. In this case, the boundary-value problem can be conveniently recast as a variational problem, i.e. a problem where instead of solving a complicated system of nonlinear differential equations, a single scalar energy functional has to be minimized. This variational principle, referred to as the *principle of minimum potential energy* (PMPE), is of great importance in continuum mechanics as well as more general multiscale theories such as those discussed in MM [TM11]. A key component of the derivation of the PMPE is the theory of stability, which is concerned with the conditions under which a mechanical system is in *stable* equilibrium as opposed to *unstable* equilibrium. (Think of a pencil lying on a table as opposed to one balanced on its end.) We only give a flavor of this rich and complex theory, sufficient for our purposes of elucidating the derivation of PMPE.

[5] This is true for thermomechanical systems. However, if electromagnetic effects are considered, the application of an appropriate electric potential to certain materials can lead to heat flow in the "wrong" direction without violating the second law.

The discussion of stability and PMPE concludes the first part of the book. At this stage, we are able to write down a complete description of any problem in continuum mechanics and we have a clear understanding of the origins of all of the equations that appear in the problem formulation. Unfortunately, the complete generality of the continuum mechanics framework, with its attendant geometric and material nonlinearity, means that it is almost always impossible to obtain closed-form analytical solutions for a given problem. So how do we proceed? There are, in fact, three possible courses of action, which are described in Part II on *Solutions*. First, in certain cases it *is* possible to obtain closed-form solutions. Even more remarkably, some of these solutions are *universal* in that they apply to all materials (in a given class) regardless of the form of the constitutive relations. In addition to their academic interest, these solutions have important practical implications for the design of experiments that measure the nonlinear constitutive relations for materials. The known universal solutions are described in Chapter 8.

The second option for solving a continuum problem (assuming the analytical solution is unknown or, more likely, unobtainable) is to adopt a numerical approach. In this case, the continuum equations are solved approximately on a computer. The most popular numerical approach is the *finite element method* (FEM) described in Chapter 9. In FEM the continuum body is discretized into a finite set of domains, referred to as "elements," bounded by "nodes" whose positions and temperatures constitute the unknowns of the problem.[6] When substituting this representation into the continuum field equations, the result is a set of coupled nonlinear algebraic equations for the unknowns. Entire books are written on FEM and our intention is not to compete with those. We do, however, offer a derivation of the key equations that is different from most texts. We focus on static boundary-value problems and approach the problem from the perspective of the PMPE. In this setting, the FEM solution to a general nonlinear continuum problem corresponds to the minimization of the energy of the system with respect to the nodal degrees of freedom. This is a convenient approach which naturally extends to multiscale methods (like those described in Chapter 12 of [TM11]) where continuum domains and atomistic domains coexist.

The third and final option for solving continuum problems is to simplify the equations by linearizing the kinematics and/or the constitutive relations. This approach is discussed in Chapter 10. As noted at the start of this introduction, this procedure leads to almost all of the theories studied as independent subjects in an engineering curriculum. For example, few students understand the connection between heat transfer and elasticity theory. The ability of continuum mechanics to provide a unified framework for all of these subjects is one of the reasons that this is such an important theory. Most students who take a continuum mechanics course leave with a much deeper understanding of engineering science (once they have recovered from the shell shock). We conclude in Chapter 11 with some suggested *further reading* for readers wishing to expand their understanding of the topics covered in this book.

[6] It is amusing that the continuum model is introduced as an approximation for the real discrete material, but that to solve the continuum problem one must revert back to a discrete (albeit far coarser) representation.

PART I

THEORY

2 Scalars, vectors and tensors

Continuum mechanics seeks to provide a fundamental model for material response. It is sensible to require that the predictions of such a theory should not depend on the irrelevant details of a particular coordinate system. The key is to write the theory in terms of variables that are unaffected by such changes; *tensors*[1] (or *tensor fields*) are the measures that have this property. Tensors come in different flavors depending on the number of spatial directions that they couple. The simplest tensor has no directional dependence and is called a *scalar invariant* to distinguish it from a simple scalar. A *vector* has one direction. For two directions and higher the general term *tensor* is used.

Tensors are tricky things to define. Many books define tensors in a technical manner in terms of the rules that tensor components must satisfy under coordinate system transformations.[2] While certainly correct, we find such definitions unilluminating when trying to answer the basic question of "what is a tensor?". In this chapter, we provide an introduction to tensors from the perspective of linear algebra. This approach may appear rather mathematical at first, but in the end it provides a far deeper insight into the nature of tensors.

Before we can begin the discussion of the definition of tensors, we must start by defining "space" and "time" and the related concept of a "frame of reference," which underlie the description of all physical objects. The notions of space and time were first tackled by Newton in the formulation of his laws of mechanics.

2.1 Frames of reference and Newton's laws

In 1687, Isaac Newton published his *Philosophiae Naturalis Principia Mathematica* or simply *Principia*, in which a unified theory of mechanics was presented for the first time. According to this theory, the motion of material objects is governed by three laws. Translated from the Latin, these laws state [Mar90]:

[1] The term "tensor" was coined by William Hamilton in 1854 to describe the norm of a polynome in his theory of quaternions. It was first used in its modern sense by Woldemar Voigt in 1898.

[2] More correctly, tensors are defined in terms of the rules that their components must satisfy under a *change of basis*. A rectilinear "coordinate system" consists of an origin and a basis. The distinction between a basis and a coordinate system is discussed further below. However, we will often use the terms interchangeably.

I *Every body remains in a state, resting or moving uniformly in a straight line, except insofar as forces on it compel it to change its state.*
II *The [rate of] change of momentum is proportional to the motive force impressed, and is made in the direction of the straight line in which the force is impressed.*
III *To every action there is always opposed an equal reaction.*

Mathematically, Newton's second law (also called the *balance of linear momentum*) is

$$\boldsymbol{F}^{\text{ext}} = \frac{d}{dt}(m\boldsymbol{v}), \tag{2.1}$$

where $\boldsymbol{F}^{\text{ext}}$ is the total external force acting on a system, m is its mass and $\boldsymbol{v}$ is the velocity of the center of mass. For a body with constant mass, Eqn. (2.1) reduces to the famous equation, $\boldsymbol{F}^{\text{ext}} = m\boldsymbol{a}$, where $\boldsymbol{a}$ is acceleration. (The case of variable mass systems is discussed further on page 13.)

Less well known than Newton's laws of motion is the set of definitions that Newton provided for the fundamental variables appearing in his theory (force, mass, space, time, motion and so on). These appear in the *Scholium* to the *Principia* (a chapter with explanatory comments and clarifications). Newton's definitions of space and time are particularly eloquent [New62]:

Space "Absolute space, in its own nature, without reference to anything external, remains always similar and unmovable."

Time "Time exists in and of itself and flows equably without reference to anything external."

These definitions were controversial in Newton's time and continue to be a source of active debate even today. They were necessary to Newton, since otherwise his three laws were meaningless. The first law refers to the velocity of objects and the second law to the rate of change of velocity (acceleration). But velocity and acceleration relative to what? Newton was convinced that the answer was *absolute space* and *absolute time*. This view was strongly contested by the *relationists* led by Gottfried Leibniz, who as a point of philosophy believed that only relative quantities were important and that space was simply an abstraction resulting from the geometric relations between bodies [DiS02].

Newton's bucket The argument was settled (at least temporarily) by a simple thought experiment that Newton described in the *Principia*.[3] Take a bucket half filled with water and suspend it from the ceiling with rope. Twist the rope by rotating the bucket as far as possible. Wait until the water settles and then let go. The unwinding rope will cause the bucket to begin spinning. Initially, the water will remain still even though the bucket is spinning, but then slowly due to the friction between the walls of the bucket and the water, the water will begin to spin as well until it is rotating in unison with the bucket. When the

[3] The story of this experiment and how it inspired later thinkers such as Ernst Mach and Albert Einstein is eloquently told in Brian Greene's popular science book on modern physics [Gre04].

water is spinning its surface will assume a concave profile, higher near the bucket walls than in the center. The rotation of the bucket and water will continue as the rope unwinds and begins to wind itself up in the opposite direction. Eventually, the bucket will slow to a stop, but the water will continue spinning for a while, before the entire process is repeated in the opposite direction. Not an experiment for the cover of *Nature*, but quite illuminating as we shall see.

The key point is the fact that the surface of the water assumes a concave profile. The reason for this appears obvious. When the water is spinning it is accelerating outward (in the same way that a passenger in a turning vehicle is pushed out to the side) and since there is nowhere for the water to go but up, it climbs up the walls of the bucket. This is certainly correct; however, it depends on the definition of *spinning*. Spinning relative to what? It cannot be the bucket itself, because when the experiment starts and the water appears still while the bucket is spinning, one can say that the water is spinning in the opposite direction relative to a stationary bucket – and yet the surface of the water is flat. Later when both the bucket and water are spinning together, so that the relative spin is zero, the water is concave. At the end when the bucket has stopped and the water is still spinning relative to it, the surface of the water is still concave. Clearly, the shape of the water surface cannot be explained in terms of the relative motion of the bucket and water. So what is the water spinning relative to? You might say the earth or the "fixed stars,"[4] but Newton countered with a thought experiment. Imagine that the experiment was done in otherwise empty space. Since the experiment with the bucket requires gravity, imagine instead two "globes" tied together with a rope. There is nothing in the universe except for the two globes and the rope: "an immense vacuum, where there was nothing external or sensible with which the globes could be compared" [New62]. If the rope is made to rotate about an axis passing through its center and perpendicular to it, we expect a tension to be built up in the rope due to the outward acceleration of the globes – exactly as in the bucket experiment. But now there is clearly nothing to relate the spinning of the rope and globes to *except* absolute space itself. QED as far as Newton was concerned.[5] Absolute space and time lie at the heart of Newton's theory. It is not surprising, therefore, that Newton considered his discovery of these concepts to be his most important achievement [Gre04].

Frame of reference In practice, Newton recognized that it is not possible to work directly with absolute space and time since they cannot be detected, and so he introduced the concepts of *relative space* and *relative time* [New62]:

[4] Recall that the word *planet* comes from the Greek "*planetai*" meaning "wanderers," because the planets appear to move relative to the fixed backdrop of the stars.

[5] Even Leibniz had to accept Newton's argument, although he remained unconvinced about the reality of absolute space: "I find nothing in . . . the Scholium . . . that proves, or can prove, the reality of space in itself. However, I grant there is a difference between an absolute true motion of a body, and a mere relative change of its situation with respect to another body" [Ale56]. Two hundred years later Ernst Mach challenged Newton's assertion by claiming that the water in the bucket is spinning relative to all other mass in the universe. Mach argued that if it were possible to perform Newton's experiment with the globes in an empty universe, then there would be no tension in the rope because there would be no other mass relative to which it was spinning. Albert Einstein was intrigued by Mach's thinking, but the conclusion to emerge from the special theory of relativity was that in fact there would be tension in the rope even in an empty universe [Gre04, p. 51].

> Relative space is some movable dimension or measure of the absolute spaces, which our senses determine by its position to bodies and which is commonly taken for immovable space; such is the dimension of a subterraneous, an aerial, or celestial space, determined by its position in respect of the earth.
>
> Relative, apparent, and common time is some sensible and external (whether accurate or unequable) measure of duration by the means of motion, which is commonly used instead of true time, such as an hour, a day, a month, a year.

Today, we refer to this combination of relative space and relative time as a *frame of reference*. A modern definition is that a frame of reference is a rigid physical object, such as the earth, the laboratory or the "fixed stars," relative to which positions are measured, and a clock to measure time.

Inertial frames of reference With the definition of absolute space and absolute time, Newton's laws of motion were made explicit. However, it turns out that Newton's equations also hold relative to an infinite set of alternative frames of reference that are moving uniformly relative to the absolute frame. These are called *inertial frames of reference*.[6]

Consider an inertial frame of reference that is moving at a constant velocity $\bar{v}$ relative to absolute space. Say that the position of some object is (x_1, x_2, x_3) in the absolute frame and (x'_1, x'_2, x'_3) in the inertial frame.[7] Assume the frames' origins coincide at time $t = 0$. The positions of the object and measured times in both frames are related through

$$x'_1 = x_1 - \bar{v}t, \qquad x'_2 = x_2, \qquad x'_3 = x_3, \qquad t' = t,$$

where, without loss of generality, the coordinate systems associated with the two frames have been aligned so that the relative motion is along the 1-direction. A mapping of this type is called a *Galilean transformation*. Note that the velocities of the object along the 1-direction measured in the two frames are related through

$$v'_1 = \frac{dx'_1}{dt} = v_1 - \bar{v}.$$

It is straightforward to show that Newton's laws of motion hold in the inertial frame. The first law is clearly still valid since an object moving uniformly relative to absolute space also moves uniformly relative to the inertial frame. The third law also holds under the assumption that force is invariant with respect to uniform motion. (This property of force, called *objectivity*, is revisited in Section 6.3.3.) The fact that the second law holds in all inertial frames requires more careful thought. The law is clearly satisfied for the case where the mass of the system is constant. In this case, $\boldsymbol{F}^{\text{ext}} = m\boldsymbol{a}$, which holds in all inertial frames since the acceleration is the same:

$$a'_1 = \frac{dv'_1}{dt} = \frac{d(v_1 - \bar{v})}{dt} = \frac{dv_1}{dt} = a_1,$$

[6] See also Section 6.3, where the relationship between inertial frames and the transformation between frames of reference and objectivity is discussed.

[7] Locating objects relative to a frame of reference requires the introduction of a coordinate system (see Section 2.3.2). Here a Cartesian coordinate system is used.

where the fact that $\bar{v}$ is constant was used. What about the case where the mass of the system is variable, for example, a rocket which burns its fuel as it is flying or a rolling cart containing sand which is being blown off as the cart moves? In these cases, a direct application of Newton's second law would appear to show a dependence on the motion of the frame, since

$$\frac{d(mv_1')}{dt} = \frac{dm}{dt}v_1' + m\frac{dv_1'}{dt} = \frac{dm}{dt}(v_1 - \bar{v}) + m\frac{d(v_1 - \bar{v})}{dt} = \frac{d(mv_1)}{dt} - \bar{v}\frac{dm}{dt}. \quad (2.2)$$

This result suggests that the rate of change of momentum for variable mass systems is *not* the same in all inertial frames since it directly depends on the motion of the frame $\bar{v}$. The answer to this apparent contradiction is that there is another principle at work which is not normally stated but is assumed to be true. This is the principle of *conservation of mass*.[8] Newton's second law is expressed for a *system*, a "body" in Newton's language, and the mass of this body in a classical system is conserved. This appears to suggest that variable mass systems are impossible, since m = constant. However, consider the case where the system consists of two bodies, A and B, with masses m_A and m_B. The bodies can exchange mass between them, so that $m_A = m_A(t)$ and $m_B = m_B(t)$, but their sum is conserved, $m_A + m_B = m =$ constant. In this case, the rate of change of momentum is indeed the same in all inertial frames, since dm/dt in Eqn. (2.2) is zero and therefore, $d(mv_1')/dt = d(mv_1)/dt$. If one wants to apply Newton's second law to a *subsystem* which is losing or gaining mass, say only body A in the above example, then one must explicitly account for the momentum transferred in and out of the subsystem by mass transfer. One can view this additional term as belonging to the force which is applied to the subsystem. This is the principle behind the operation of a rocket (see Exercise 2.1) or the recoil of a gun when a bullet is fired.[9]

We have established that Newton's laws of motion (with the added assumption of conservation of mass) hold in all inertial frames of reference. This fact was understood by Newton, who stated in Corollary V to his equations of motion [New62]:

> When bodies are enclosed in a given space, their motions in relation to one another are the same whether the space is at rest or whether moving uniformly straight forward without circular motion.

Once one inertial frame is known, an infinite number of other inertial frames can be constructed through a Galilean transformation. The practical problem with this way of defining inertial frames is that it is not possible to know whether a frame of reference is moving uniformly relative to absolute space, since it is not possible to detect absolute space. For this reason the modern definition of inertial frames does not refer to absolute space, but instead relies on Thomson's law of inertia, which is described shortly.

[8] Many books on mechanics take the view that Newton's laws only hold for systems of *point particles* that by definition have constant mass. In this case, conservation of mass is trivially satisfied and need not be mentioned. The view presented here is more general and consistent with the generalization of Newton's laws to continuum systems which is adopted in the later chapters.

[9] Interestingly, the correct treatment of variable mass systems is not uniformly understood even by researchers working in the field. See, for example, the discussion in [PM92].

Problems with absolute space Despite the apparent acceptance of absolute space when it was introduced, it continued (and continues) to trouble many people. Two main criticisms are raised against it.

1. Metaphysical nature of absolute space
   The absolute space which Newton introduced is an undetectable, invisible, all filling, fixed scaffolding relative to which positions are measured. A sort of universal global positioning system with a capital "G." Regardless of one's religious views, one wants to say God's frame of reference, and that is in some sense how Newton viewed it. The almost spiritual nature of this medium is apparent. Here we have an invisible thing that cannot be seen or sensed in any way and yet it has a profound effect on our every day experiences since it determines the acceleration upon which the physical laws of motion depend. Newton was strongly criticized for this aspect of his work by philosophers of science. For example, Ernst Mach stated: "With respect to the monstrous conceptions of absolute space and absolute time I can retract nothing. Here I have only shown more clearly than hitherto that Newton indeed spoke much about these things, but throughout made no serious application of them" [Mac60]; or according to Hans Reichenbach: "Newton begins with precisely formulated empirical statements, but adds a mystical philosophical superstructure . . . his theory of mechanics arrested the analysis of the problems of space and time for more than two centuries, despite the fact that Leibniz, who was his contemporary, had a much deeper understanding of the nature of space and time" [Rei59]. These claims have more recently been debunked as stemming from a misunderstanding of the role that absolute space plays in Newton's theory, a misunderstanding of Leibniz's theoretical shortcomings and a misunderstanding of Einstein's theory of relativity in which spacetime plays a similar role to that of Newton's definitions [Ear70, Art95, DiS06].
2. Equivalence of inertial frames
   The second complaint raised against Newton is that since all inertial frames are equivalent from the perspective of Newtonian dynamics and there is no way to tell them apart, it is not sensible to single out one of them, *absolute space*, as being special. Instead, one must think of all inertial frames as inherently equivalent. The definition of an inertial frame must therefore change since it can no longer be defined as a frame of reference in uniform motion relative to absolute space. A solution was proposed by James Thomson in 1884, which he called the *law of inertia*. It is paraphrased as follows [DiS91]:[10]

   > For any system of interacting bodies, it is possible to construct a reference-frame and time scale with respect to which all accelerations are proportional to, and in the direction of, impressed forces.

   This is meant to be added to Newton's laws of motion as a fourth law on equal standing with the rest. In this way inertial frames are *defined* as frames in which Newton's second law holds without reference to absolute space. The conclusion from this is that the often asked question regarding why the laws of motion hold only relative to inertial frames is

[10] Thomson's law is revisited from the perspective of material frame-indifference (objectivity) in Section 6.3.3.

ill-posed. The laws of motion do not hold relative to inertial frames, they *define* them [DiS91]. This view on inertial frames is often the one expressed in modern books on mechanics. With this interpretation, an inertial frame is defined as a frame of reference in which Newton's laws of motion are valid.

Relativistic spacetime Thomson's definition of the law of inertia is not the end of the story, of course. Just as the Newtonian picture was falling into place, James Clerk Maxwell was developing the theory of electromagnetism. One of the uncomfortable conclusions to emerge from Maxwell's theory was that electromagnetic waves travel at a constant speed, $c = 299\,792.458$ km/s, relative to *all* frames of reference, a fact that was confirmed experimentally for light. This conclusion makes no sense in the Newtonian picture. How can something travel at the same speed relative to two frames of reference that are in relative motion?

Surprisingly, a hint to the answer is already there in Newton's words: *"time exists in and of itself and flows equably without reference to anything external."* Einstein showed that this was entirely incorrect. Time does not exist "in and of itself." It is intimately tied with space and is affected by the motion of observers. The result is relativistic spacetime, which is beyond the scope of this book. It is, however, interesting to point out that Einstein's spacetime, like Newton's absolute space is *something*. In the absence of gravity, in the special theory of relativity, Einstein speaks of an "absolute spacetime" not much different philosophically from Newton's absolute space [DiS06]. In general relativity, spacetime "comes alive" [Gre04] and interacts with physical objects. In this way, the criticism that space and time are metaphysical is removed.

Within this context, it may be possible to regard Newton's absolute space as a legitimate concept that can be considered a limiting case of relativistic spacetime. If this is true, then perhaps the original definition of inertial frames in terms of absolute space is tenable, removing the need for Thomson's law of inertia. Philosophers of science are still arguing about this point.

2.2 Tensor notation

Having introduced the concepts of space, time and frame of reference, we now turn to a "nuts and bolts" discussion regarding the notation of tensor algebra. In the process of doing so we will introduce important operations between tensors. It may seem a bit strange to start discussing a notation for something that we have not defined yet. Think of it as the introduction of a syntax for a new language that we are about to learn. It will be useful for us later, when we learn the words of this language, to have a common structure in which to explain the concepts that emerge. Walter Jaunzemis, in his entertaining introduction to continuum mechanics, put it very nicely: "Continuum mechanics may appear as a fortress surrounded by the walls of tensor notation" [Jau67]. We begin therefore at the walls.

2.2.1 Direct versus indicial notation

Tensors represent physical properties such as mass, velocity and stress that do not depend on the coordinate system. It should therefore be possible to represent tensors and the operations on them and between them without reference to a particular coordinate system. Such a notation exists and is called *direct notation* (or *invariant notation*). Direct notation provides a symbolic representation for tensor operations but it does not specify how these operations are actually performed. In practice, in order to perform operations on tensors they must always be projected onto a particular coordinate system where they are represented by a set of components. The explicit representation of tensor operations in terms of their components is called *indicial notation*. This is the notation that has to be used when tensor operations involving numerical values are performed.

The number of spatial directions associated with a tensor is called its *rank* or *order*. We will use these terminologies interchangeably. A scalar invariant, such as mass, is not associated with direction at all, i.e. a body does not have a different mass in different directions. Therefore, a scalar invariant is a rank 0 tensor or alternatively a zeroth-order tensor. A vector, such as velocity, is associated with one spatial direction and is therefore a rank 1 or first-order tensor. Stress involves two spatial directions, the orientation of a plane sectioning a body and a direction in space along which the stress is evaluated. It is therefore a rank 2 or second-order tensor. Tensors of any order are possible. In practice, we will only be dealing with tensors up to fourth order.

In both indicial and direct notations, tensors are represented by a symbol, e.g. m for mass, v for velocity and σ for stress. In indicial notation, the tensor's spatial directions are denoted by indices attached to the symbol. Mass has no direction so it has no indices, velocity has one index, stress two, and so on: m, v_i, σ_{ij}. The number of indices is equal to the rank of the tensor and the range of an index $[1, 2, \ldots, n_\mathrm{d}]$ is determined by the dimensionality of space.[11] We will be dealing mostly with three-dimensional space ($n_\mathrm{d} = 3$); however, the notation we develop applies to any value of n_d. The tensor symbol with its numerical indices represents the components of the tensor, e.g. v_1, v_2 and v_3 are the components of the velocity vector. A set of simple rules for the interaction of indices provides a mechanism for describing all of the tensor operations that we will require. In fact, what makes this notation particularly useful is that *any operation defined by indicial notation has the property that if its arguments are tensors the result will also be a tensor.* We discuss this further at the end of Section 2.3, but for now we state it without proof.

In direct notation, no indices are attached to the tensor symbol. The rank of the tensor is represented by the typeface used to display the symbol. Scalar invariants are displayed in a regular font while first-order tensors and higher are displayed in a bold font (or with an underline when written by hand): m, $\boldsymbol{v}$, $\boldsymbol{\sigma}$ (or m, $\underline{v}$, $\underline{\sigma}$ by hand). As noted above, the advantage of direct notation is that it emphasizes the fact that tensors are independent of the choice of a coordinate system (whereas indices are always tied to a particular selection). Direct notation is also more compact and therefore easier to read. However, the lack of indices means that special notation must be introduced for different operations between

[11] See the discussion on finite-dimensional spaces in Section 2.3.

tensors. Many symbols in this notation are not universally accepted and direct notation is not available for all operations. We will discuss direct notation in Section 2.4, where tensor operations are defined.

In some cases, the operations defined by indicial notation can also be written using the matrix notation familiar from linear algebra. Here vectors and second-order tensors are represented as column and rectangular matrices of their components, for example

$$[\boldsymbol{v}] = \begin{bmatrix} v_1 \\ v_2 \\ v_3 \end{bmatrix}, \quad [\boldsymbol{\sigma}] = \begin{bmatrix} \sigma_{11} & \sigma_{12} & \sigma_{13} \\ \sigma_{21} & \sigma_{22} & \sigma_{23} \\ \sigma_{31} & \sigma_{32} & \sigma_{33} \end{bmatrix}.$$

The notation $[\boldsymbol{v}]$ and $[\boldsymbol{\sigma}]$ is a shorthand representation for the column matrix and rectangular matrix, respectively, formed by the components of the vector $\boldsymbol{v}$ and the second-order tensor $\boldsymbol{\sigma}$. This notation will sometimes be used when tensor operations can be represented by matrix multiplication and other matrix operations on tensor components.

Before proceeding to the definition of tensors, we begin by introducing the basic rules of indicial notation, starting with the most basic rule: the summation convention.

2.2.2 Summation and dummy indices

Consider the following sum:[12]

$$S = a_1 x_1 + a_2 x_2 + \cdots + a_{n_d} x_{n_d}.$$

We can write this expression using the summation symbol Σ:

$$S = \sum_{i=1}^{n_d} a_i x_i = \sum_{j=1}^{n_d} a_j x_j = \sum_{m=1}^{n_d} a_m x_m.$$

Clearly, the particular choice for the letter we use for the summation, i, j or m, is irrelevant since the sum is independent of the choice. Indices with this property are called *dummy indices*. Because summation of products, such as $a_i x_i$, appears frequently in tensor operations, a simplified notation is adopted where the Σ symbol is dropped and any index appearing twice in a product of variables is taken to be a dummy index, over which a sum is implied. For example,

$$S = a_i x_i = a_j x_j = a_m x_m = a_1 x_1 + a_2 x_2 + \cdots + a_{n_d} x_{n_d}.$$

This convention was introduced by Albert Einstein in the famous 1916 paper in which he outlined the principles of general relativity [Ein16]. It is therefore called *Einstein's summation convention* or just the *summation convention* for short.[13]

[12] This section follows the introduction to indicial notation in [LRK78].

[13] Although the summation convention is an extremely simple idea, it is also extremely useful and is therefore widely used and quoted. This amused Einstein who is reported to have joked with a friend that apparently "I have made a great discovery in mathematics; I have suppressed the summation sign every time that summation must be made over an index which occurs twice . . ." [Wei11].

Example 2.1 (The Einstein summation convention for $n_d = 3$) Several examples are:

1. $a_i x_i = a_1 x_1 + a_2 x_2 + a_3 x_3$.
2. $a_i a_i = a_1^2 + a_2^2 + a_3^2$.
3. $\sigma_{ii} = \sigma_{11} + \sigma_{22} + \sigma_{33}$.

It is important to point out that the summation convention only applies to indices that appear twice in a product of variables. A product containing more than two occurrences of a dummy index, such as $a_i b_i x_i$, is meaningless. If the objective here is to sum over index i, this would have to be written as $\sum_{i=1}^{n_d} a_i b_i x_i$. The summation convention does, however, generalize to the case where there are multiple dummy indices in a product. For example a double sum over dummy indices i and j is

$$\begin{aligned} A_{ij} x_i y_j = & A_{11} x_1 y_1 + A_{12} x_1 y_2 + A_{13} x_1 y_3 \\ & + A_{21} x_2 y_1 + A_{22} x_2 y_2 + A_{23} x_2 y_3 \\ & + A_{31} x_3 y_1 + A_{32} x_3 y_2 + A_{33} x_3 y_3. \end{aligned}$$

We see how the summation convention provides a very efficient shorthand notation for writing complex expressions. Finally, there may be situations where although an index appears twice in a product, we do *not* wish to sum over it. For example, say we wish to state that the diagonal components of a second-order tensor are zero: $A_{11} = A_{22} = A_{33} = 0$. In order to temporarily "deactivate" the summation convention we write:

$$A_{ii} = 0 \quad \text{(no sum)} \qquad \text{or} \qquad A_{\underline{i}\,\underline{i}} = 0.$$

2.2.3 Free indices

An index that appears only once in each product term of an equation is referred to as a *free index*. A free index takes on the values $1, 2, \ldots, n_d$, one at a time. For example,

$$A_{ij} x_j = b_i.$$

Here i is a free index and j is a dummy index. Since i can take on n_d separate values, the above expression represents the following system of n_d equations:

$$\begin{aligned} A_{11} x_1 + A_{12} x_2 + \cdots + A_{1 n_d} x_{n_d} &= b_1, \\ A_{21} x_1 + A_{22} x_2 + \cdots + A_{2 n_d} x_{n_d} &= b_2, \\ \vdots \qquad\qquad & \quad \vdots \\ A_{n_d 1} x_1 + A_{n_d 2} x_2 + \cdots + A_{n_d n_d} x_{n_d} &= b_{n_d}. \end{aligned}$$

Naturally, all terms in an expression must have the same free indices (or no indices at all). The expression $A_{ij} x_j = b_k$ is meaningless. However, $A_{ij} x_j = c$ (where c is a scalar) is fine. There can be as many free indices as necessary. For example, the expression $D_{ijk} x_k = A_{ij}$ contains the two free indices i and j and therefore represents n_d^2 equations.

2.2.4 Matrix notation

Indicial operations involving tensors of rank two or less can be represented as matrix operations. For example, the product $A_{ij}x_j$ can be expressed as a matrix multiplication. For $n_\mathrm{d} = 3$ we have

$$A_{ij}x_j = \mathsf{A}\mathsf{x} = \begin{bmatrix} A_{11} & A_{12} & A_{13} \\ A_{21} & A_{22} & A_{23} \\ A_{31} & A_{32} & A_{33} \end{bmatrix} \begin{bmatrix} x_1 \\ x_2 \\ x_3 \end{bmatrix}.$$

We use a sans serif font to denote matrices to distinguish them from tensors. Thus, A is a rectangular table of numbers. The entries of A are equal to the components of the tensor A, i.e. $\mathsf{A} = [A]$, so that $\mathsf{A}_{ij} = A_{ij}$. Column matrices are denoted by lower-case letters and rectangular matrices by upper-case letters.

The expression $A_{ji}x_j$ can be computed in a similar manner, but the entries of A must be *transposed* before performing the matrix multiplication, i.e. its rows and columns must be swapped. Thus, (for $n_\mathrm{d} = 3$)

$$A_{ji}x_j = \mathsf{A}^T\mathsf{x} = \begin{bmatrix} A_{11} & A_{21} & A_{31} \\ A_{12} & A_{22} & A_{32} \\ A_{13} & A_{23} & A_{33} \end{bmatrix} \begin{bmatrix} x_1 \\ x_2 \\ x_3 \end{bmatrix},$$

where the superscript T denotes the transpose operation. Similarly, the sum $a_i x_i$ can be written

$$a_i x_i = \mathsf{a}^T\mathsf{x} = \begin{bmatrix} a_1 & a_2 & a_3 \end{bmatrix} \begin{bmatrix} x_1 \\ x_2 \\ x_3 \end{bmatrix}.$$

The transpose operation has the important property that

$$(\mathsf{AB})^T = \mathsf{B}^T\mathsf{A}^T.$$

This implies that $(\mathsf{ABC})^T = \mathsf{C}^T\mathsf{B}^T\mathsf{A}^T$, and so on.

Another example of a matrix operation is the expression, $A_{ii} = A_{11} + A_{22} + \cdots + A_{n_\mathrm{d}n_\mathrm{d}}$, which is defined as the *trace* of the matrix A. In matrix notation this is denoted as $\mathrm{tr}\,\mathsf{A}$.

2.2.5 Kronecker delta

The Kronecker delta[14] is defined as follows:

$$\delta_{ij} = \begin{cases} 1 & \text{if } i = j, \\ 0 & \text{if } i \neq j. \end{cases} \tag{2.3}$$

[14] The Kronecker delta is named after the German mathematician and logician Leopold Kronecker (1823–1891). Kronecker believed all mathematics should be founded on whole numbers, saying "God made the integers, all else is the work of man" [Wik10].

In matrix form, δ_{ij} are the entries of the *identity matrix* $\mathbf{I}$ (for $n_\text{d} = 3$),

$$\mathbf{I} = \begin{bmatrix} 1 & 0 & 0 \\ 0 & 1 & 0 \\ 0 & 0 & 1 \end{bmatrix}. \tag{2.4}$$

Most often the Kronecker delta appears in expressions as a result of a differentiation of a tensor with respect to its components. For example, $\partial x_i / \partial x_j = \delta_{ij}$. This is correct as long as the components of the tensor are independent.

An important property of δ_{ij} is *index substitution*:

$$a_i \delta_{ij} = a_j.$$

Proof

$$a_i \delta_{ij} = a_1 \delta_{1j} + a_2 \delta_{2j} + a_3 \delta_{3j} = \begin{cases} a_1 & \text{if } j = 1 \\ a_2 & \text{if } j = 2 \\ a_3 & \text{if } j = 3 \end{cases} = a_j.$$

□

Example 2.2 (The Kronecker delta for $n_\text{d} = 3$) Several examples are:

1. $A_{ij}\delta_{ij} = A_{ii} = A_{jj} = A_{11} + A_{22} + A_{33}$.
2. $\delta_{ii} = \delta_{11} + \delta_{22} + \delta_{33} = 3$.
3. $A_{ij} - A_{ik}\delta_{jk} = A_{ij} - A_{ij} = 0$.

2.2.6 Permutation symbol

The permutation symbol[15] ϵ_{ijk} for $n_\text{d} = 3$ is defined as follows:[16]

$$\epsilon_{ijk} = \begin{cases} 1 & \text{if } i, j, k \text{ form an even permutation of } 1, 2, 3, \\ -1 & \text{if } i, j, k \text{ form an odd permutation of } 1, 2, 3, \\ 0 & \text{if } i, j, k \text{ do not form a permutation of } 1, 2, 3. \end{cases} \tag{2.5}$$

Thus, $\epsilon_{123} = \epsilon_{231} = \epsilon_{312} = 1$, $\epsilon_{321} = \epsilon_{213} = \epsilon_{132} = -1$, and $\epsilon_{111} = \epsilon_{112} = \epsilon_{113} = \cdots = \epsilon_{333} = 0$. (See Fig. 2.1 for a convenient way to remember the sign of the permutation symbol.) Some properties of the permutation symbol are given below:

1. Useful identities:

$$\epsilon_{ijk}\delta_{ij} = \epsilon_{iik} = 0, \qquad \epsilon_{ijk}\epsilon_{mjk} = 2\delta_{im}, \qquad \epsilon_{ijk}\epsilon_{ijk} = 6. \tag{2.6}$$

[15] The permutation symbol is also known as the Levi–Civita symbol or the alternating symbol.

[16] It is possible to generalize the definition of the permutation symbol to arbitrary dimensionality, but since we deal primarily with three-dimensional space we limit ourselves to this special case.

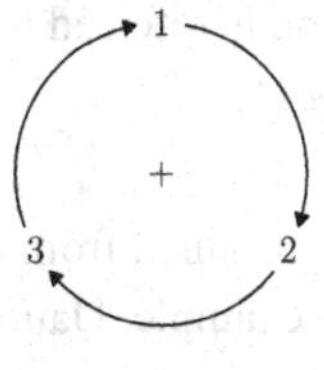

Fig. 2.1 A convenient mnemonic for the sign of the permutation symbol. A triplet of indices obtained by traversing the circle in a clockwise direction result in a positive permutation symbol. The reverse gives the negative.

2. The permutation symbol provides an expression for the determinant of a matrix:

$$\epsilon_{mnp} \det \mathbf{A} = \epsilon_{ijk} A_{im} A_{jn} A_{kp} = \epsilon_{ijk} A_{mi} A_{nj} A_{pk}. \tag{2.7}$$

These identities can be proven by substitution. Note that Eqn. (2.7) demonstrates the fact that $\det \mathbf{A} = \det \mathbf{A}^T$. A separate expression for $\det \mathbf{A}$ can be obtained by multiplying the last expression in Eqn. (2.7) by $\frac{1}{6}\epsilon_{mnp}$ and using Eqn. $(2.6)_3$:

$$\det \mathbf{A} = \frac{1}{6} \epsilon_{ijk} \epsilon_{mnp} A_{mi} A_{nj} A_{pk} = \epsilon_{ijk} A_{1i} A_{2j} A_{3k}, \tag{2.8}$$

where the last expression is obtained by expanding out the m, n, p indices and using the symmetries of the permutation tensor.

3. The derivative of the determinant of a matrix with respect to the matrix entries will be required later. To obtain this, start with the first equality in Eqn. (2.8). The derivative of this is

$$\begin{aligned}\frac{\partial(\det \mathbf{A})}{\partial A_{rs}} &= \frac{1}{6} \epsilon_{ijk} \epsilon_{mnp} \left[\delta_{rm} \delta_{is} A_{nj} A_{pk} + \delta_{rn} \delta_{js} A_{mi} A_{pk} + \delta_{rp} \delta_{ks} A_{mi} A_{nj}\right] \\ &= \frac{1}{2} \epsilon_{sjk} \epsilon_{rnp} A_{nj} A_{pk}. \end{aligned} \tag{2.9}$$

Passage from the first to second lines above is accomplished by noting through appropriate dummy index substitution that the three terms in the first line are equal. Equation (2.9) is concise, but it is component based. We continue the derivation to obtain a more general matrix expression. Replace ϵ_{rnp} in Eqn. (2.9) with $\epsilon_{qnp}\delta_{qr}$. Then assuming that $\det \mathbf{A} \neq 0$, there exists $\mathbf{A}^{-1}$ such that

$$\delta_{qr} = A_{qi} A^{-1}_{ir}.$$

This gives

$$\frac{\partial(\det \mathbf{A})}{\partial A_{rs}} = \left(\frac{1}{2} \epsilon_{sjk} A^{-1}_{ir}\right) \left(\epsilon_{qnp} A_{qi} A_{nj} A_{pk}\right).$$

Using Eqn. (2.7) followed by Eqn. $(2.6)_2$, we obtain the final expression[17]

$$\frac{\partial(\det \mathbf{A})}{\partial \mathbf{A}} = \mathbf{A}^{-T} \det \mathbf{A}. \tag{2.10}$$

[17] Although Eqn. (2.10) has been derived for the special case of $n_d = 3$, it is correct for any value of n_d.

4. The following relation is referred to as the ϵ–δ identity:

$$\epsilon_{ijk}\epsilon_{mnk} = \delta_{im}\delta_{jn} - \delta_{in}\delta_{jm}. \qquad (2.11)$$

This relation can be obtained from the determinant relation (Eqn. (2.7)) for the special case $\mathbf{A} = \mathbf{I}$ (See, for example, [Jau67]).

The permutation symbol plays an important role in vector cross products. We will see this in Section 2.3.

Now that we have explained the rules for tensor component interactions, we turn to the matter of the definition of a tensor.

2.3 What is a tensor?

The answer to the question "What is a tensor?" is not simple. Tensors are abstract entities that behave according to certain transformation rules. In fact, many books *define* tensors in terms of the transformation rules that they must obey in order to be invariant under coordinate system transformations. We prefer the linear algebra approach where tensors are defined independently of coordinate systems. The transformation rules are then an output of the definition rather than part of it.

So how do we define a tensor? Let us begin by considering the more familiar case of a vector, we can then generalize this definition to tensors of arbitrary rank. The typical high-school definition of a vector is "an entity with a magnitude and a direction," often stressed by the teacher by drawing an arrow on the board. This is clearly only a partial definition, since many things that are not vectors have a magnitude and a direction. This book, for example, has a magnitude (the number of pages in it) and a direction (front to back), yet it is not what we would normally consider a vector. It turns out that an indispensable part of the definition is the parallelogram law that defines how vectors are added together. This suggests that an *operational* approach must be taken to define vectors. However, if this is the case, then vectors can only be defined as a group and not individually. This leads to the idea of a *vector space*.

2.3.1 Vector spaces and the inner product and norm

A real vector space V is a set, defined over the field of real numbers $\mathbb{R}$, where the following two operations have been defined:

1. *vector addition* for any two vectors $\boldsymbol{a}, \boldsymbol{b} \in V$, we have $\boldsymbol{a} + \boldsymbol{b} = \boldsymbol{c} \in V$,
2. *scalar multiplication* for any scalar $\lambda \in \mathbb{R}$ and vector $\boldsymbol{a} \in V$, we have $\lambda \boldsymbol{a} = \boldsymbol{c} \in V$,

with the following properties[18] $\forall\, \boldsymbol{a}, \boldsymbol{b}, \boldsymbol{c} \in V$ and $\forall\, \lambda, \mu \in \mathbb{R}$:

[18] We use (but try not to overuse) the standard mathematical notation. $\forall$ should be read "for all" or "for every," $\in$ should be read "in," iff should be read "if and only if." The symbol "$\equiv$" means "equal by definition."

1. $\boldsymbol{a}+\boldsymbol{b}=\boldsymbol{b}+\boldsymbol{a}$	addition is commutative
2. $\boldsymbol{a}+(\boldsymbol{b}+\boldsymbol{c})=(\boldsymbol{a}+\boldsymbol{b})+\boldsymbol{c}$	addition is associative
3. $\boldsymbol{a}+\mathbf{0}=\boldsymbol{a}$	addition has an identity element $\mathbf{0}$
4. $\boldsymbol{a}+(-\boldsymbol{a})=\mathbf{0}$	addition has an additive inverse
5. $\lambda\boldsymbol{a}=\boldsymbol{a}\lambda$	multiplication is commutative
6. $\lambda(\mu\boldsymbol{a})=(\lambda\mu)\boldsymbol{a}$	multiplication is associative
7. $1\boldsymbol{a}=\boldsymbol{a}$	multiplication has an identity element 1
8. $(\lambda+\mu)(\boldsymbol{a}+\boldsymbol{b})=\lambda\boldsymbol{a}+\lambda\boldsymbol{b}+\mu\boldsymbol{a}+\mu\boldsymbol{b}$	distributive properties of addition and multiplication

At this point the definition is completely general and abstract. It is possible to invent many vector objects and definitions for addition and multiplication that satisfy these rules. An example that may help to show the abstract nature of a vector space is useful. Consider the set of all continuously differentiable functions with derivatives of all orders, $f(x)$, on the interval $\mathcal{X}=[0,1]$ such that $f(0)=f(1)=0$. It is easy to show that this set, called $\mathcal{C}^\infty(\mathcal{X})$, is in fact a vector space under the usual definitions of function addition and multiplication by a scalar.

The vectors that are familiar to us from the physical world have additional properties associated with the geometry of finite-dimensional space, such as distances and angles. The definition of the vector space must be extended to include these concepts. The result is the *Euclidean space* named after the Greek mathematician Euclid who laid down the foundations of "Euclidean geometry." We define these properties separately beginning with the concept of a finite-dimensional space.

Finite-dimensional spaces and basis vectors The dimensionality of a space is related to the concept of linear dependence. The m vectors $\boldsymbol{a}_1,\dots,\boldsymbol{a}_m\in V$ are *linearly dependent* if and only if there exist $\lambda_1,\dots,\lambda_m\in\mathbb{R}$ not all equal to zero, such that

$$\lambda_1\boldsymbol{a}_1+\cdots+\lambda_{\underline{m}}\boldsymbol{a}_{\underline{m}}=\mathbf{0}.$$

(Recall that the underline on the subscripts implies that the summation convention is not applied, see Section 2.2.2.) Otherwise, the vectors are *linearly independent*. The largest possible number of linearly-independent vectors is the dimensionality of the vector space. (For example, in a three-dimensional vector space there can be at most three linearly independent vectors.) This is denoted by dim V. We limit ourselves to vector spaces for which dim V is finite.

Consider an n_d-dimensional vector space V^{n_d}. Any set of n_d linearly independent vectors can be selected as a *basis* of V^{n_d}. A basis is useful because every vector in V can be written as a unique linear combination of the basis vectors. Basis vectors are commonly denoted by $\boldsymbol{e}_i$, $i=1,\dots,n_d$. Any other vector $\boldsymbol{a}\in V^{n_d}$ can be expressed as

$$\boldsymbol{a}=a_1\boldsymbol{e}_1+\cdots+a_{n_d}\boldsymbol{e}_{n_d}=a_i\boldsymbol{e}_i, \tag{2.12}$$

where a_i are called the *components* of vector $\boldsymbol{a}$ with respect to the basis $\boldsymbol{e}_i$. The basis vectors are said to *span* the vector space, since any other vector in the space can be represented as a linear combination of them. The proof for Eqn. (2.12) is straightforward:

Proof The basis vectors $(\boldsymbol{e}_1, \ldots, \boldsymbol{e}_{n_d})$ are linearly independent, therefore the set $(\boldsymbol{a}, \boldsymbol{e}_1, \ldots, \boldsymbol{e}_{n_d})$ must be linearly dependent. Hence, $\lambda_0 \boldsymbol{a} + \lambda_1 \boldsymbol{e}_1 + \cdots + \lambda_{n_d} \boldsymbol{e}_{n_d} = \boldsymbol{0}$. If $\lambda_0 = 0$, then the only solution is $\lambda_1 = \lambda_2 = \cdots = \lambda_{n_d} = 0$. This cannot be true since $\boldsymbol{a} \neq \boldsymbol{0}$. Thus, $\lambda_0 \neq 0$, and $a_i = -\lambda_i/\lambda_0$. □

The choice of basis vectors is not unique; however, the components of a vector in a particular basis are unique. This is easy to show by assuming the contrary and using the linear dependence of the basis vectors. Next we introduce the concept of *multilinear functions* that will be important for the definition of the inner product and later for the general definition of tensors.

Multilinear functions Let us begin by considering a scalar *linear function* of one variable. A real function $f(x)$ is linear in x if it is *additive*: $f(x + x') = f(x) + f(x')\ \forall\, x, x' \in \mathbb{R}$, and *homogeneous*: $f(\lambda x) = \lambda f(x)\ \forall\, x, \lambda \in \mathbb{R}$. These two conditions can be combined into the single requirement:

$$f[\lambda x + \mu x'] = \lambda f[x] + \mu f[x'], \quad \forall\, x, x', \lambda, \mu \in \mathbb{R},$$

where the square brackets are used to indicate that f is a linear function of its argument. Clearly, $f[x] = Cx$, where C is a constant, is a linear function, whereas $g(x) = Cx + D$ is not linear since $g(x + x') = C(x + x') + D \neq g(x) + g(x') = C(x + x') + 2D$.

The generalization of scalar linear functions of one variable to multilinear functions of n variables is straightforward. A *multilinear function* or *n-linear function* is linear with respect to each of its n independent variables. For example, a *bilinear* function must satisfy the linearity condition for both arguments:

$$f[\lambda x + \mu x', y] = \lambda f[x, y] + \mu f[x', y], \quad \forall\, x, x', y, \lambda, \mu \in \mathbb{R},$$
$$f[x, \lambda y + \mu y'] = \lambda f[x, y] + \mu f[x, y'], \quad \forall\, x, y, y', \lambda, \mu \in \mathbb{R}.$$

As before, $f[x, y] = Cxy$ is a bilinear function, while $g(x, y) = Cxy + D$ is not. In general, for an n-linear function we require $\forall\, x_i, x_i', \lambda, \mu \in \mathbb{R}$:

$$f[x_1, \ldots, \lambda x_i + \mu x_i', \ldots, x_n] = \lambda f[x_1, \ldots, x_i, \ldots, x_n] + \mu f[x_1, \ldots, x_i', \ldots, x_n].$$

The concept of a linear function also generalizes to functions of vector arguments. In this context the term *linear mapping* is often used. A real-valued linear mapping, $\boldsymbol{f} : V \to \mathbb{R}$, is a transformation that takes a vector $\boldsymbol{a}$ from V and returns a scalar in $\mathbb{R}$ that satisfies the conditions:

$$\boldsymbol{f}[\lambda \boldsymbol{a} + \mu \boldsymbol{a}'] = \lambda \boldsymbol{f}[\boldsymbol{a}] + \mu \boldsymbol{f}[\boldsymbol{a}'], \quad \forall\, \boldsymbol{a}, \boldsymbol{a}' \in V, \forall\, \lambda, \mu \in \mathbb{R}.$$

A bilinear mapping, $\boldsymbol{f} : V \times V \to \mathbb{R}$, is linear with respect to both arguments:

$$\boldsymbol{f}[\lambda \boldsymbol{a} + \mu \boldsymbol{a}', \boldsymbol{b}] = \lambda \boldsymbol{f}[\boldsymbol{a}, \boldsymbol{b}] + \mu \boldsymbol{f}[\boldsymbol{a}', \boldsymbol{b}], \quad \forall\, \boldsymbol{a}, \boldsymbol{a}', \boldsymbol{b} \in V, \forall\, \lambda, \mu \in \mathbb{R},$$
$$\boldsymbol{f}[\boldsymbol{a}, \lambda \boldsymbol{b} + \mu \boldsymbol{b}'] = \lambda \boldsymbol{f}[\boldsymbol{a}, \boldsymbol{b}] + \mu \boldsymbol{f}[\boldsymbol{a}, \boldsymbol{b}'], \quad \forall\, \boldsymbol{a}, \boldsymbol{b}, \boldsymbol{b}' \in V, \forall\, \lambda, \mu \in \mathbb{R}.$$

In the general case, a multilinear mapping of n arguments (also called an n-linear mapping), $f : \underbrace{V \times \cdots \times V}_{n \text{ times}} \to \mathbb{R}$, satisfies:

$$f[a_1, \ldots, \lambda a_i + \mu a_i', \ldots, a_n] = \lambda f[a_1, \ldots, a_i, \ldots, a_n] + \mu f[a_1, \ldots, a_i', \ldots, a_n],$$

$\forall\, a_i, a_i' \in V$ and $\forall\, \lambda, \mu \in \mathbb{R}$.

We now turn to the definition of the Euclidean space.

Euclidean space The real coordinate space $\mathbb{R}^{n_d}$ is an n_d-dimensional vector space defined over the field of real numbers. A vector in $\mathbb{R}^{n_d}$ is represented by a set of n_d real components relative to a given basis. Thus for $a \in \mathbb{R}^{n_d}$ we have $a = (a_1, \ldots, a_{n_d})$, where $a_i \in \mathbb{R}$. Addition and multiplication are defined for $\mathbb{R}^{n_d}$ in terms of the corresponding operations familiar to us from the algebra of real numbers:

1. *Addition:* $a + b = (a_1, \ldots, a_{n_d}) + (b_1, \ldots, b_{n_d}) = (a_1 + b_1, \ldots, a_{n_d} + b_{n_d})$.
2. *Multiplication:* $\lambda a = \lambda(a_1, \ldots, a_{n_d}) = (\lambda a_1, \ldots, \lambda a_{n_d})$.

These definitions clearly satisfy the requirements given above for the addition and multiplication operations for vector spaces.

In order for $\mathbb{R}^{n_d}$ to be a Euclidean space it must possess an *inner product*, which is related to angles between vectors, and it must possess a *norm*, which provides a measure for the length of a vector.[19] In this book we will be concerned primarily with three-dimensional Euclidean space for which $n_d = 3$.

Inner product and norm An inner product is a real-valued bilinear mapping. The inner product of two vectors a and b is denoted by $\langle a, b \rangle$. An inner product function must satisfy the following properties $\forall\, a, b, c \in V$ and $\forall\, \lambda, \mu \in \mathbb{R}$:

1.	$\langle \lambda a + \mu b, c \rangle = \lambda \langle a, c \rangle + \mu \langle b, c \rangle$	linearity with respect to first argument
2.	$\langle a, b \rangle = \langle b, a \rangle$	symmetry
3.	$\langle a, a \rangle \geq 0$ and $\langle a, a \rangle = 0$ iff $a = 0$	positivity

For $\mathbb{R}^{n_d}$ the standard choice for an inner product is the *dot product*:

$$\langle a, b \rangle = a \cdot b. \tag{2.13}$$

[19] Some authors use $\mathbb{E}^{n_d}$ to denote a Euclidean space to distinguish it from a real coordinate space without an inner product and norm. Since this distinction is not going to play a role in this book, we reduce notation and denote a Euclidean space by $\mathbb{R}^{n_d}$ with the existence of a norm and inner product implied.

The Euclidean norm is defined as[20]

$$\|\boldsymbol{a}\| = \sqrt{\boldsymbol{a}\cdot\boldsymbol{a}}. \tag{2.14}$$

This notation distinguishes the norm from the absolute value of a scalar, $|s| = \sqrt{s^2}$. A shorthand notation denoting $\boldsymbol{a}^2 \equiv \boldsymbol{a}\cdot\boldsymbol{a}$ is sometimes adopted. A vector $\boldsymbol{a}$ satisfying $\|\boldsymbol{a}\| = 1$ is called a *unit vector*.

A geometrical interpretation of the dot product is

$$\boldsymbol{a}\cdot\boldsymbol{b} = \|\boldsymbol{a}\|\ \|\boldsymbol{b}\| \cos\theta(\boldsymbol{a},\boldsymbol{b}), \tag{2.15}$$

where $\theta(\boldsymbol{a},\boldsymbol{b})$ is the angle between vectors $\boldsymbol{a}$ and $\boldsymbol{b}$ and the norm provides a measure for the length of a vector. Two vectors, $\boldsymbol{a}$ and $\boldsymbol{b}$, that are perpendicular to each other satisfy the condition $\boldsymbol{a}\cdot\boldsymbol{b} = 0$. An additional important property that can be proven using the three defining properties of an inner product given above is the *Schwarz inequality*:

$$|\boldsymbol{a}\cdot\boldsymbol{b}| \le \|\boldsymbol{a}\|\ \|\boldsymbol{b}\| \qquad \forall\, \boldsymbol{a},\boldsymbol{b} \in \mathbb{R}^{n_d}.$$

The property of scalar multiplication and the definition of the norm allow us to write a vector as a product of a magnitude and a direction:

$$\boldsymbol{v} = \|\boldsymbol{v}\| \frac{\boldsymbol{v}}{\|\boldsymbol{v}\|} = \|\boldsymbol{v}\|\, \boldsymbol{e}_v, \tag{2.16}$$

where $\boldsymbol{e}_v$ is the unit vector in the direction of $\boldsymbol{v}$. For example, if $\boldsymbol{v}$ is the velocity vector, $\|\boldsymbol{v}\|$ is the magnitude of the velocity (absolute speed) and $\boldsymbol{e}_v$ is the direction of motion.

2.3.2 Coordinate systems and their bases

In the definition of a frame of reference in Section 2.1, we stated that positions are measured relative to some specified physical object. However, the actual act of measurement requires the definition of a *coordinate system* – a standardized scheme that assigns a unique set of real numbers, the "coordinates," to each position. The idea of "positions" is in turn related to the concept of a "point space" as described next.

Euclidean point space Mathematically, the space associated with a frame of reference can be regarded as a set E of *points*, which are defined through their relation with a Euclidean vector space $\mathbb{R}^{n_d}$ (called the *translation space* of E). For every pair of points x, y in E, there exists a vector $\boldsymbol{v}(x,y)$ in $\mathbb{R}^{n_d}$ that satisfies the following conditions [Ogd84]:

$$\boldsymbol{v}(x,y) = \boldsymbol{v}(x,z) + \boldsymbol{v}(z,y) \qquad \forall x,y,z \in E, \tag{2.17}$$

$$\boldsymbol{v}(x,y) = \boldsymbol{v}(x,z) \qquad \text{if and only if } y = z. \tag{2.18}$$

20 An important theorem states that for a finite-dimensional space $\mathbb{R}^{n_d}$, all norms are equivalent in the sense that given two definitions for norms, 1 and 2, the results of one are bounded by the other, i.e. $m\,\|\boldsymbol{a}\|_1 \le \|\boldsymbol{a}\|_2 \le M\,\|\boldsymbol{a}\|_1, \forall \boldsymbol{a} \in \mathbb{R}^{n_d}$, where m and M are positive real numbers. This means that we can adopt the Euclidean norm without loss of generality.

A set satisfying these conditions is called a *Euclidean point space*. A *position vector* $\boldsymbol{x}$ for a point x is defined by singling out one of the points as the *origin* o and writing:

$$\boldsymbol{x} \equiv \boldsymbol{v}(x, o). \tag{2.19}$$

Equations (2.17) and (2.18) imply that every point x in E is uniquely associated with a vector $\boldsymbol{x}$ in $\mathbb{R}^{n_d}$. The vector connecting two points is given by

$$\boldsymbol{x} - \boldsymbol{y} = \boldsymbol{v}(x, o) - \boldsymbol{v}(y, o).$$

The distance between two points and the angles formed by three points can be computed using the norm and inner product of the corresponding translation space.

We now turn to the definition of coordinate systems.

Coordinate systems The most general type of coordinate systems we will consider are called *curvilinear coordinate systems*. These consist of an *origin* relative to which positions are measured (as described above), and a set of "coordinate curves" that correspond to paths through space along which all but one of the coordinates are constant. At each position in a three-dimensional space a set of three coordinate curves intersect. The tangent vectors to these coordinate curves do not all lie in a single plane and therefore form a *basis* (as defined in Section 2.3.1). The important point to understand is that for curvilinear coordinates, the basis vectors change from position to position. Examples of curvilinear coordinate systems include the polar cylindrical and spherical systems, both of which are discussed further in Section 2.6.3. A special type of a curvilinear coordinate system is a *rectilinear coordinate system* where the coordinate curves are straight lines.[21] The basis vectors of rectilinear coordinate systems point along the coordinate lines which are called *axes* in this case. In contrast to a general curvilinear coordinate system, the basis vectors of a rectilinear coordinate system are independent of position in space. An infinite number of rectilinear coordinate systems can be associated with a given frame of reference, differing by their origin and the orientation of their axes (or basis vectors). If the axes are orthogonal to each other, the term *Cartesian*[22] coordinate system is used (see Fig. 2.2).

Orthonormal basis and Cartesian coordinates The basis of a Cartesian coordinate system is *orthogonal*, i.e. all basis vectors are perpendicular to each other. If, in addition, the basis vectors have magnitude unity, the basis is called *orthonormal*. The requirements for an orthonormal basis are expressed mathematically by the condition

$$\boldsymbol{e}_i \cdot \boldsymbol{e}_j = \delta_{ij}, \tag{2.20}$$

where $\boldsymbol{e}_i$ are the basis vectors (see Fig. 2.2) and δ_{ij} is the Kronecker delta defined in Eqn. (2.3). By convention, we choose basis vectors that form a right-handed triad (this

21 Although we most often encounter the prefix "rect" in the word *rectangle* (where it means "right" as in a 90 degree angle), its occurrence in the word "rectilinear" does not refer to angles at all. In fact, in this case the prefix *recti* has the alternative meaning "straight," and thus, *rectilinear* means "characterized by straight lines."

22 "Cartesian" refers to the French mathematician René Descartes who among other things worked on developing an algebra for Euclidean geometry leading to the field of analytical geometry.

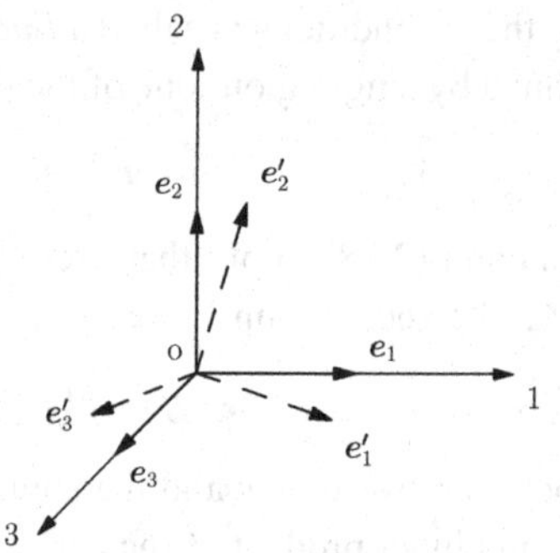

Fig. 2.2 The Cartesian coordinate system. The three axes and basis vectors e_i are shown along with an alternative rotated set of basis vectors e'_i. The origin of the coordinate system is o.

means that if we curl the fingers of the right hand, rotating them from e_1 towards e_2, the thumb will point in the positive direction of e_3).

In an orthonormal basis, the indicial expression for the dot product is

$$\boldsymbol{a} \cdot \boldsymbol{b} = (a_i \boldsymbol{e}_i) \cdot (b_j \boldsymbol{e}_j) = a_i b_j (\boldsymbol{e}_i \cdot \boldsymbol{e}_j) = a_i b_j \delta_{ij} = a_i b_i,$$

where we have used Eqn. (2.20) and the index substitution property of δ_{ij}. Therefore,

$$\boldsymbol{a} \cdot \boldsymbol{b} = a_i b_i. \tag{2.21}$$

The component of a vector along a basis vector direction is obtained by dotting the vector with the basis vector. Consider $\boldsymbol{a} = a_j \boldsymbol{e}_j$, and dot both sides with $\boldsymbol{e}_i$:

$$\boldsymbol{a} \cdot \boldsymbol{e}_i = a_j (\boldsymbol{e}_j \cdot \boldsymbol{e}_i) = a_j \delta_{ji} = a_i.$$

Thus, the standard method for obtaining vector components in an orthonormal basis is

$$a_i = \boldsymbol{a} \cdot \boldsymbol{e}_i. \tag{2.22}$$

Nonorthogonal bases and covariant and contravariant components The definitions given above for an orthonormal basis can be extended to the nonorthogonal case. In $\mathbb{R}^3$, any set of three noncollinear, nonplanar and nonzero vectors form a basis. There are no other constraints on the magnitude of the basis vectors or the angles between them. A general basis consisting of vectors that are not perpendicular to each other and may have magnitudes different from 1 is called a *nonorthogonal basis*. An example of such a basis is the set of lattice vectors that define the structure of a crystal (see Section 3.3 in [TM11]). To distinguish such a basis from an orthonormal basis, we denote its basis vectors with $\{\boldsymbol{g}_i\}$ instead of $\{\boldsymbol{e}_i\}$. Since the vectors $\boldsymbol{g}_i$ are not orthogonal, a *reciprocal*[23] basis $\{\boldsymbol{g}^i\}$ can be defined through

$$\boldsymbol{g}^i \cdot \boldsymbol{g}_j = \delta^i_j, \tag{2.23}$$

23 The reciprocal basis vectors of continuum mechanics are closely related to the reciprocal lattice vectors of solid state physics discussed in Section 3.7.1 of [TM11]. The only difference is a 2π factor introduced in the physics definition to simplify the form of plane wave expressions.

where δ^i_j has the same definition as the Kronecker delta defined in Eqn. (2.3). Note that the subscript and superscript placement of the indices is used to distinguish between a basis and its reciprocal partner. The existence of these two closely related bases leads to the existence of two sets of components for a given vector $\boldsymbol{a}$:

$$\boldsymbol{a} = a^i \boldsymbol{g}_i = a_j \boldsymbol{g}^j. \tag{2.24}$$

Here a^i are the *contravariant* components of $\boldsymbol{a}$, and a_i are the *covariant* components of $\boldsymbol{a}$. The connections between covariant and contravariant components are obtained by dotting Eqn. (2.24) with either $\boldsymbol{g}^k$ or $\boldsymbol{g}_k$, which gives

$$a^k = g^{jk} a_j \qquad \text{and} \qquad a_k = g_{ik} a^i, \tag{2.25}$$

where[24] $g_{ij} = \boldsymbol{g}_i \cdot \boldsymbol{g}_j$ and $g^{ij} = \boldsymbol{g}^i \cdot \boldsymbol{g}^j$. The processes in Eqn. (2.25) are called *raising* or *lowering* an index.

The existence of the parallel covariant and contravariant descriptions means that the dot product can be expressed in different ways. In contravariant components, we have

$$\boldsymbol{a} \cdot \boldsymbol{b} = (a^i \boldsymbol{g}_i) \cdot (b^j \boldsymbol{g}_j) = a^i b^j (\boldsymbol{g}_i \cdot \boldsymbol{g}_j) = a^i b^j g_{ij}. \tag{2.26}$$

Similarly, in covariant components

$$\boldsymbol{a} \cdot \boldsymbol{b} = a_i b_j g^{ij}. \tag{2.27}$$

Continuum mechanics can be phrased entirely in terms of nonorthogonal bases, and more generally in terms of curvilinear coordinate systems. However, the general derivation leads to notational complexity that can obscure the main physical concepts underlying the theory. We therefore mostly limit ourselves to Cartesian coordinate systems in this book except where necessary.

2.3.3 Cross product

We have already encountered the dot product that maps two vectors to a scalar. The *cross product* is a binary operation that maps two vectors to a new vector that is orthogonal to both with magnitude equal to the area of the parallelogram spanned by the original vectors. The cross product is denoted by the $\times$ symbol, so that $\boldsymbol{c} = \boldsymbol{a} \times \boldsymbol{b} = A(\boldsymbol{a}, \boldsymbol{b})\boldsymbol{n}$, where $A(\boldsymbol{a}, \boldsymbol{b}) = \|\boldsymbol{a}\| \, \|\boldsymbol{b}\| \sin\theta(\boldsymbol{a}, \boldsymbol{b})$ is the area spanned by $\boldsymbol{a}$ and $\boldsymbol{b}$ and $\boldsymbol{n}$ is a unit vector normal to the plane defined by them. This definition for the cross product is not complete since there are two possible opposite directions for the normal (see Fig. 2.3). The solution is to append to the definition the requirement that $(\boldsymbol{a}, \boldsymbol{b}, \boldsymbol{a} \times \boldsymbol{b})$ form a right-handed set.

The cross product has the following properties $\forall\, \boldsymbol{a}, \boldsymbol{b}, \boldsymbol{c} \in \mathbb{R}^3$ and $\forall\, \lambda, \mu \in \mathbb{R}$:

1.	$\boldsymbol{a} \times \boldsymbol{b} = -(\boldsymbol{b} \times \boldsymbol{a})$	anticommutative
2.	$(\lambda\boldsymbol{a} + \mu\boldsymbol{b}) \times \boldsymbol{c} = \lambda(\boldsymbol{a} \times \boldsymbol{c}) + \mu(\boldsymbol{b} \times \boldsymbol{c})$	bilinear mapping
	$\boldsymbol{a} \times (\lambda\boldsymbol{b} + \mu\boldsymbol{c}) = \lambda(\boldsymbol{a} \times \boldsymbol{b}) + \mu(\boldsymbol{a} \times \boldsymbol{c})$	
3.	$\boldsymbol{a} \cdot (\boldsymbol{a} \times \boldsymbol{b}) = 0$ and $\boldsymbol{b} \cdot (\boldsymbol{a} \times \boldsymbol{b}) = 0$	perpendicularity

[24] The quantities g_{ij} and g^{ij} are the components of the metric tensor $\boldsymbol{g}$ discussed further in Section 2.3.6.

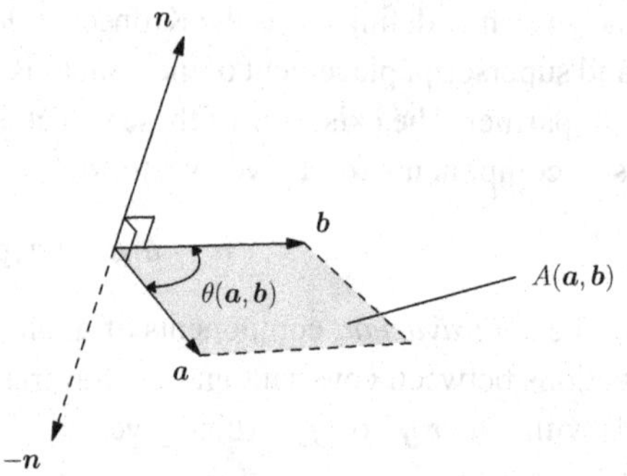

Fig. 2.3 The cross product between vectors $\boldsymbol{a}$ and $\boldsymbol{b}$. The direction of $\boldsymbol{a} \times \boldsymbol{b}$ resulting in a right-handed triad is $\boldsymbol{n}$. The magnitude of $\boldsymbol{a} \times \boldsymbol{b}$ is equal to the area of the shaded parallelogram.

Furthermore, if $\boldsymbol{a} \times \boldsymbol{b} = \boldsymbol{0}$ and neither $\boldsymbol{a}$ nor $\boldsymbol{b}$ is zero, then we must have $\boldsymbol{b} = \lambda \boldsymbol{a}$, $\lambda \in \mathbb{R}$, i.e. $\boldsymbol{a}$ is parallel to $\boldsymbol{b}$. To obtain the indicial expression for $\boldsymbol{a} \times \boldsymbol{b}$ in $\mathbb{R}^3$ we begin by noting that for a right-handed orthonormal basis

$$\begin{array}{lll} \boldsymbol{e}_1 \times \boldsymbol{e}_2 = \boldsymbol{e}_3, & \boldsymbol{e}_2 \times \boldsymbol{e}_3 = \boldsymbol{e}_1, & \boldsymbol{e}_3 \times \boldsymbol{e}_1 = \boldsymbol{e}_2, \\ \boldsymbol{e}_2 \times \boldsymbol{e}_1 = -\boldsymbol{e}_3, & \boldsymbol{e}_3 \times \boldsymbol{e}_2 = -\boldsymbol{e}_1, & \boldsymbol{e}_1 \times \boldsymbol{e}_3 = -\boldsymbol{e}_2, \\ \boldsymbol{e}_1 \times \boldsymbol{e}_1 = \boldsymbol{0}, & \boldsymbol{e}_2 \times \boldsymbol{e}_2 = \boldsymbol{0}, & \boldsymbol{e}_3 \times \boldsymbol{e}_3 = \boldsymbol{0}. \end{array}$$

This can be written in shorthand using the permutation symbol (Eqn. (2.5)):

$$\boldsymbol{e}_i \times \boldsymbol{e}_j = \epsilon_{ijk} \boldsymbol{e}_k. \tag{2.28}$$

Now consider $\boldsymbol{a} \times \boldsymbol{b} = (a_i \boldsymbol{e}_i) \times (b_j \boldsymbol{e}_j) = a_i b_j (\boldsymbol{e}_i \times \boldsymbol{e}_j)$. Using Eqn. (2.28), we have

$$\boldsymbol{a} \times \boldsymbol{b} = \epsilon_{ijk} a_i b_j \boldsymbol{e}_k, \tag{2.29}$$

which is the indicial form of the cross product. Equation (2.29) can also be written in a convenient form as a determinant of a matrix:

$$\boldsymbol{a} \times \boldsymbol{b} = \det \begin{bmatrix} \boldsymbol{e}_1 & \boldsymbol{e}_2 & \boldsymbol{e}_3 \\ a_1 & a_2 & a_3 \\ b_1 & b_2 & b_3 \end{bmatrix}.$$

Another useful operation is the *triple product* $(\boldsymbol{a} \times \boldsymbol{b}) \cdot \boldsymbol{c}$, which is equal to the volume of a parallelepiped spanned by the vectors $\boldsymbol{a}$, $\boldsymbol{b}$, $\boldsymbol{c}$ forming a right-handed triad. This can be readily shown using elementary geometry. In indicial notation we have $(\boldsymbol{a} \times \boldsymbol{b}) \cdot \boldsymbol{c} = (\epsilon_{ijk} a_i b_j \boldsymbol{e}_k) \cdot c_m \boldsymbol{e}_m = \epsilon_{ijk} a_i b_j c_m (\boldsymbol{e}_k \cdot \boldsymbol{e}_m)$. Using Eqn. (2.20) this becomes

$$(\boldsymbol{a} \times \boldsymbol{b}) \cdot \boldsymbol{c} = \epsilon_{ijk} a_i b_j c_k, \tag{2.30}$$

or in determinant form

$$(\boldsymbol{a} \times \boldsymbol{b}) \cdot \boldsymbol{c} = \det \begin{bmatrix} c_1 & c_2 & c_3 \\ a_1 & a_2 & a_3 \\ b_1 & b_2 & b_3 \end{bmatrix}.$$

2.3.4 Change of basis

We noted earlier that the choice of basis vectors $\boldsymbol{e}_i$ is not unique. There are, in fact, an infinite number of equivalent basis sets. Consider two orthonormal bases $\boldsymbol{e}_\alpha$ and $\boldsymbol{e}'_i$ as shown in Fig. 2.2. For the sake of clarity, we adopt Sokolnikoff notation where (with a wink to ancient history) Greek indices refer to the "original" basis and Roman indices refer to the "new" basis. We wish to find the relationship between $\boldsymbol{e}_\alpha$ and $\boldsymbol{e}'_i$. Since the vectors $\boldsymbol{e}_\alpha$ are linearly independent, it must be possible to write any other vector, including the vectors $\boldsymbol{e}'_i$, as a linear combination of them. Consequently, the two bases are related through the linear *transformation matrix* $\mathbf{Q}$:

$$\boldsymbol{e}'_i = \mathrm{Q}_{\alpha i}\boldsymbol{e}_\alpha \quad \Leftrightarrow \quad \begin{bmatrix} \boldsymbol{e}'_1 \\ \boldsymbol{e}'_2 \\ \boldsymbol{e}'_3 \end{bmatrix} = \begin{bmatrix} \mathrm{Q}_{11} & \mathrm{Q}_{12} & \mathrm{Q}_{13} \\ \mathrm{Q}_{21} & \mathrm{Q}_{22} & \mathrm{Q}_{23} \\ \mathrm{Q}_{31} & \mathrm{Q}_{32} & \mathrm{Q}_{33} \end{bmatrix}^T \begin{bmatrix} \boldsymbol{e}_1 \\ \boldsymbol{e}_2 \\ \boldsymbol{e}_3 \end{bmatrix}, \tag{2.31}$$

where $\mathrm{Q}_{\alpha i} = \boldsymbol{e}_\alpha \cdot \boldsymbol{e}'_i$. Note the transpose operation on the matrix $\mathbf{Q}$ in Eqn. (2.31).[25] Since the basis vectors are unit vectors, the entries of $\mathbf{Q}$ are directional cosines, $\mathrm{Q}_{\alpha i} = \cos\theta(\boldsymbol{e}'_i, \boldsymbol{e}_\alpha)$. The columns of $\mathbf{Q}$ are the components of the new basis $\boldsymbol{e}'_i$ with respect to the original basis $\boldsymbol{e}_\alpha$. Note that $\mathbf{Q}$ is not symmetric since the representation of $\boldsymbol{e}'_i$ in basis $\boldsymbol{e}_\alpha$ is not the same as the representation of $\boldsymbol{e}_\alpha$ in $\boldsymbol{e}'_i$.

As an example, consider a rotation by angle θ about the 3-axis. The new basis vectors are given by $\boldsymbol{e}'_1 = \cos\theta\boldsymbol{e}_1 + \cos(90-\theta)\boldsymbol{e}_2$, $\boldsymbol{e}'_2 = \cos(90+\theta)\boldsymbol{e}_1 + \cos\theta\boldsymbol{e}_2$, $\boldsymbol{e}'_3 = \boldsymbol{e}_3$. The corresponding transformation matrix is

$$\mathbf{Q} = \begin{bmatrix} \cos\theta & -\sin\theta & 0 \\ \sin\theta & \cos\theta & 0 \\ 0 & 0 & 1 \end{bmatrix},$$

where we have used some elementary trigonometry.

Properties of Q The transformation matrix has special properties due to the orthonormality of the basis vectors that it relates. Beginning from the orthonormality of $\boldsymbol{e}'_i$ and using the transformation in Eqn. (2.31), we have

$$\delta_{ij} = \boldsymbol{e}'_i \cdot \boldsymbol{e}'_j = (\mathrm{Q}_{\alpha i}\boldsymbol{e}_\alpha) \cdot (\mathrm{Q}_{\beta j}\boldsymbol{e}_\beta) = \mathrm{Q}_{\alpha i}\mathrm{Q}_{\beta j}(\boldsymbol{e}_\alpha \cdot \boldsymbol{e}_\beta) = \mathrm{Q}_{\alpha i}\mathrm{Q}_{\beta j}\delta_{\alpha\beta} = \mathrm{Q}_{\alpha i}\mathrm{Q}_{\alpha j}.$$

[25] Some authors define the transformation matrix as the transpose of our definition. We adopt this definition because it is consistent with the concept of tensor rotation discussed later in Section 2.5.1. Also, if nonorthonormal bases are used, then Q^{-1} must be substituted for Q^T in Eqn. (2.31).

We have shown that

$$Q_{\alpha i}Q_{\alpha j} = \delta_{ij} \quad \Leftrightarrow \quad \mathbf{Q}^T\mathbf{Q} = \mathbf{I}. \tag{2.32}$$

Similarly, we can show that $Q_{\alpha i}Q_{\beta i} = \delta_{\alpha\beta}$ (i.e. $\mathbf{Q}\mathbf{Q}^T = \mathbf{I}$). Consequently,

$$\mathbf{Q}^T = \mathbf{Q}^{-1}. \tag{2.33}$$

In addition, we can show that the determinant of $\mathbf{Q}$ equals only ± 1.

Proof

$$\det(\mathbf{Q}\mathbf{Q}^T) = \det \mathbf{I} \rightarrow \det \mathbf{Q} \det \mathbf{Q}^T = 1 \rightarrow (\det \mathbf{Q})^2 = 1 \rightarrow \det \mathbf{Q} = \pm 1.$$

□

Based on the sign of its determinant, $\mathbf{Q}$ can have two different physical significances. If $\det \mathbf{Q} = +1$, then the transformation defined by $\mathbf{Q}$ corresponds to a rotation, otherwise it corresponds to a rotation plus a reflection. Only a rotation satisfies the requirement that the handedness of the basis is retained following the transformation; transformation matrices are therefore normally limited to this case.

Matrices satisfying Eqn. (2.33) are called *orthogonal* matrices. Orthogonal matrices with a positive determinant (i.e. rotations) are called *proper orthogonal*. The set of all 3×3 orthogonal matrices $O(3)$ forms a group under matrix multiplication called the *orthogonal group*. Similarly, the set of 3×3 proper orthogonal matrices form a group under matrix multiplication called the *special orthogonal group*, which is denoted $SO(3)$.

We say that a set S constitutes a group G with respect to a particular binary operation $\star$, if it is closed with respect to that operation (i.e. $\forall\, \boldsymbol{a}, \boldsymbol{b} \in S,\ \boldsymbol{a} \star \boldsymbol{b} \in S$) and it satisfies the following three conditions $\forall\, \boldsymbol{a}, \boldsymbol{b}, \boldsymbol{c} \in S$:

1.	$(\boldsymbol{a} \star \boldsymbol{b}) \star \boldsymbol{c} = \boldsymbol{a} \star (\boldsymbol{b} \star \boldsymbol{c})$	associativity
2.	$\boldsymbol{a} \star \mathbf{1} = \boldsymbol{a}$	existence of a right identity element $\mathbf{1}$
3.	$\boldsymbol{a} \star \boldsymbol{a}^{-1} = \mathbf{1}, \quad \boldsymbol{a}^{-1} \in S$	existence of a right inverse element

It is straightforward to show from these properties that $\mathbf{1}$ is also the left identity element:

Proof Let $\boldsymbol{c}$ be the unique element in S associated with the product of $\mathbf{1}$ and $\boldsymbol{a}$, i.e. $\boldsymbol{c} = \mathbf{1} \star \boldsymbol{a}$. Multiplying both sides of this equation on the right by $\boldsymbol{a}^{-1}$ we find $\boldsymbol{c} \star \boldsymbol{a}^{-1} = (\mathbf{1} \star \boldsymbol{a}) \star \boldsymbol{a}^{-1}$. Using the associativity of the $\star$ operation, the existence of a right inverse element and finally the existence of a right identity element leads to

$$\boldsymbol{c} \star \boldsymbol{a}^{-1} = (\mathbf{1} \star \boldsymbol{a}) \star \boldsymbol{a}^{-1} = \mathbf{1} \star (\boldsymbol{a} \star \boldsymbol{a}^{-1}) = \mathbf{1} \star \mathbf{1} = \mathbf{1}.$$

The last equality ($\boldsymbol{c} \star \boldsymbol{a}^{-1} = \mathbf{1}$) shows that $\boldsymbol{c} = \boldsymbol{a}$ because $\boldsymbol{a}^{-1}$ is the unique right inverse of $\boldsymbol{a}$. Substituting this into our starting equation we find $\mathbf{1} \star \boldsymbol{a} = \boldsymbol{a}$, which proves that $\mathbf{1}$ is the left identity. □

The proof that a^{-1} is also the left inverse element of a follows a similar line of reasoning.

It is also straightforward to prove that $O(3)$ is a group:

Proof First, for $O(3)$ to be closed with respect to matrix multiplication, we need to show that $\forall\, \mathbf{A}, \mathbf{B} \in O(3)$ we have $\mathbf{AB} \in O(3)$: $(\mathbf{AB})(\mathbf{AB})^T = \mathbf{ABB}^T\mathbf{A}^T = \mathbf{AIA}^T = \mathbf{AA}^T = \mathbf{I}$, so $\mathbf{AB}$ is orthogonal. The remaining three properties are also satisfied. Associativity is satisfied because matrix multiplication is a linear operation. The identity element is $\mathbf{I}$. The inverse element is guaranteed to exist $\forall\, \mathbf{A} \in O(3)$ since $\det \mathbf{A} \neq 0$, and it belongs to $O(3)$, since $(\mathbf{A}^{-1})^T = (\mathbf{A}^T)^T = \mathbf{A} = (\mathbf{A}^{-1})^{-1}$. □

The proof that $SO(3)$ is a group is similar and left as an exercise for the reader. The fact that $O(3)$ and $SO(3)$ are groups is not critical for us at this juncture. However, it is useful to introduce the concept of groups, since groups will appear repeatedly in different settings in the remainder of the book. It is exactly this ubiquitousness of groups that makes them important. The general framework of *group theory* provides a powerful methodology for establishing useful properties of groups. See, for example, [Mil72, McW02, Ros08].

2.3.5 Vector component transformation

We are now in a position to derive the transformation rule for vector components under a change of basis. We require that a vector be *invariant* with respect to component transformation. Thus, for vector a we require $a = a_\alpha e_\alpha = a'_i e'_i$, where a_α are the components of a in basis e_α and a'_i are the components in e'_i. Making use of the transformation rule for basis vectors in Eqn. (2.31), we have

$$a = a_\alpha e_\alpha = a'_i e'_i = a'_i (Q_{\alpha i} e_\alpha),$$

which can be rewritten as

$$(a_\alpha - Q_{\alpha i} a'_i) e_\alpha = \mathbf{0}.$$

The basis vectors e_α are linearly independent, therefore the coefficients must be zero:

$$a_\alpha = Q_{\alpha i} a'_i \quad \Leftrightarrow \quad [a] = \mathbf{Q}\,[a]'. \tag{2.34}$$

The prime on $[a]'$ means that the components of a in the matrix representation are given in the basis $\{e'_i\}$. The inverse relation is obtained by applying $Q_{\alpha j}$ to both sides and making use of the orthogonality relation for $\mathbf{Q}$ in Eqn. (2.32):

$$a'_i = Q_{\alpha i} a_\alpha \quad \Leftrightarrow \quad [a]' = \mathbf{Q}^T\,[a]. \tag{2.35}$$

It is possible to use the transformation rules in Eqns. (2.34) and (2.35) as the *definition* of a vector, by stating that a 3-tuple whose components transform in this way is a vector. This

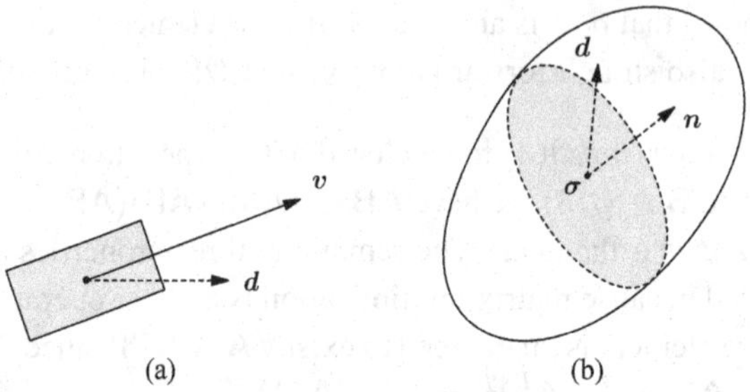

Fig. 2.4 The concept of a tensor. (a) The velocity $\boldsymbol{v}$ is a first-order tensor which returns the speed along any direction $\boldsymbol{d}$. Thus, if $\boldsymbol{v}$ is the velocity of a vehicle, then $\boldsymbol{v}$ evaluated at $\boldsymbol{d}$ is the speed with which the vehicle is moving in the $\boldsymbol{d}$-direction. (b) The stress $\boldsymbol{\sigma}$ is a second-order tensor which returns the force per unit area along direction $\boldsymbol{d}$ when bisecting a body by a plane with normal $\boldsymbol{n}$.

seems less transparent than the operational approach based on linear algebra that we have adopted here.[26]

2.3.6 Generalization to higher-order tensors

We now have a clear definition for vectors, which we would like to generalize to higher-order tensors. To do so requires us to consider vectors in a different manner.

Before going on to the technical definition which involves some subtle concepts in linear algebra, a loose "hand-wavy" explanation may be helpful. We have stressed the fact that a vector exists separately of a particular coordinate system. In this view, a vector is like the proverbial "arrow," oriented in space and projecting shadows of itself onto different coordinate system bases. An alternative view is to consider the vector more abstractly as an entity that carries with it all of the information related to the physical quantity that it represents. For example, the velocity vector tells us everything about the velocity of some object. In particular, it can tell us how fast an object is moving in any direction as illustrated in Fig. 2.4(a). Therefore, we can think of the vector as a velocity "function" that takes a direction and returns a speed. It turns out that these two views are distinct but intimately tied to each other. Thus, every "arrow" vector is uniquely associated with a "function" vector. The former is our standard vector. We call the latter a *first-order tensor*.

Now while some physical variables are only associated with a single direction, like velocity, others require more. Unlike the "arrow" definition, the "function" viewpoint of vectors readily generalizes to higher-order physical quantities; one simply adds more arguments. For example, obtaining the stress at a point involves a two-step process as illustrated in Fig. 2.4(b). First, an imaginary plane (defined by its normal) for bisecting the body is specified, and then a direction along which the stress is required. The stress tensor therefore takes two arguments: a normal to a plane and a direction in space. This is called a *second-order tensor*. A tensor of any order can be defined in exactly the same way.

[26] Broccoli analogy: Defining a vector based on the way in which its components transform is similar to defining what broccoli is according to its taste. This approach provides a definite test (if it tastes like broccoli, then it must be broccoli), but clearly this is not the most fundamental definition for this vegetable.

Thus, the conceptual procedure we follow is to: (1) provide an independent definition for vectors as members of a vector space; (2) define first-order tensors through their connection with vectors; and (3) extend the first-order tensor definition to tensors of any order. With these ideas in the back of our mind, let us now turn to the more technical presentation.[27]

We defined a vector as a member of a finite-dimensional Euclidean space and saw that it could be represented as a set of components on a given basis. For example, a velocity vector $\boldsymbol{v}$ is expressed as $v_i \boldsymbol{e}_i$, where the component v_i is the speed along direction $\boldsymbol{e}_i$. The speed s_d along any direction $\boldsymbol{d}$ (where $\|\boldsymbol{d}\| = 1$) is then obtained by projecting $\boldsymbol{v}$ along $\boldsymbol{d}$:

$$s_d = \boldsymbol{v} \cdot \boldsymbol{d}. \tag{2.36}$$

Interpreted in this way, a vector is like a machine that operates on a direction and returns the speed along it. Alternatively, we can view the projection operation in Eqn. (2.36) more abstractly as a linear mapping that takes a direction and returns a real number (speed):

$$s_d = \boldsymbol{v}^*[\boldsymbol{d}], \tag{2.37}$$

where $\boldsymbol{v}^* : \mathbb{R}^{n_d} \to \mathbb{R}$. We have replaced the vector $\boldsymbol{v}$ with a linear mapping $\boldsymbol{v}^*[\,]$ that provides the same "service." The set of linear mappings from $\mathbb{R}^{n_d}$ to $\mathbb{R}$ forms a new vector space $\mathbb{R}^{n_d *}$, called the *dual space*[28] of $\mathbb{R}^{n_d}$. The elements of $\mathbb{R}^{n_d *}$ are called *dual vectors* or *covectors*[29] to distinguish them from vectors belonging to the original vector space $\mathbb{R}^{n_d}$.

It can be shown that every vector $\boldsymbol{v} \in \mathbb{R}^{n_d}$ is uniquely associated with a covector $\boldsymbol{v}^* \in \mathbb{R}^{n_d *}$ and vice versa, so that $\mathbb{R}^{n_d *}$ is *isomorphic*[30] to $\mathbb{R}^{n_d}$. Hence vectors and covectors occupy two parallel universes. In one we have the standard definition of a vector and in the other, vectors are replaced by linear mappings. The connection between the two representations follows from the requirement that s_d in Eqns. (2.36) and (2.37) is the same:

$$\boldsymbol{v}^*[\boldsymbol{d}] = \boldsymbol{v} \cdot \boldsymbol{d}. \tag{2.38}$$

Thus, we can fully characterize the linear mapping $\boldsymbol{v}^*$ by the vector $\boldsymbol{v}$. What about the reverse direction? Given the linear mapping $\boldsymbol{v}^*$, how can we determine the associated vector $\boldsymbol{v}$ (assuming that it is not known)? To answer this question, we begin by focusing on the left-hand side of Eqn. (2.38) and use the linearity of $\boldsymbol{v}^*$ to obtain

$$\boldsymbol{v}^*[\boldsymbol{d}] = \boldsymbol{v}^*[d_i \boldsymbol{e}_i] = d_i \boldsymbol{v}^*[\boldsymbol{e}_i].$$

Using this in Eqn. (2.38) along with the component forms of $\boldsymbol{v}$ and $\boldsymbol{d}$ on the right gives

$$\boldsymbol{v}^*[\boldsymbol{e}_i] d_i = v_i d_j (\boldsymbol{e}_i \cdot \boldsymbol{e}_j).$$

For an *orthonormal* basis, $\boldsymbol{e}_i \cdot \boldsymbol{e}_j = \delta_{ij}$, so that $\boldsymbol{v}^*[\boldsymbol{e}_i] d_i = v_i d_i$. Since $\boldsymbol{d}$ is arbitrary, this implies that

$$\boldsymbol{v}^*[\boldsymbol{e}_i] = v_i. \tag{2.39}$$

[27] Since our objective is to convey to the reader the true concept of a tensor in the simplest possible manner, the presentation given below is limited to the special case of an orthonormal Cartesian coordinate system. For a more general discussion, applicable to arbitrary coordinate systems, see, for example, [Ogd84, Section 1.4.3].

[28] For a more thorough introduction to dual spaces, consult books on linear algebra ([LL09] has a succinct introduction and worked examples) or books on tensor theory ([BG68] is particularly clear).

[29] The term "1-form" is used for members of the dual space in differential geometry.

[30] Two vector spaces are said to be *isomorphic* if there exists a one-to-one and onto linear transformation mapping vectors from one space to the vectors in the other.

Thus, the components v_i of the vector $\boldsymbol{v}$ are obtained by evaluating the associated linear mapping $\boldsymbol{v}^*$ on the orthonormal Cartesian basis $\{\boldsymbol{e}_i\}$. This means that given $\boldsymbol{v}^*$, we can always revert to a vector representation, $\boldsymbol{v} = v_i \boldsymbol{e}_i$, where we define $v_i \equiv \boldsymbol{v}^*[\boldsymbol{e}_i]$.

Now we come to the point. We *define* the mapping $\boldsymbol{v}^*[\,]$ to be a *first-order tensor*. Thus, a first-order tensor is a linear mapping of a vector to a real number.[31] This definition may seem like a useless exercise given the fact that a first-order tensor and a vector are isomorphic, and in fact, have identical components in a Cartesian system. So what has been gained? The advantage is that the definition given above for a first-order tensor (unlike the definition of a vector) can be readily generalized to a tensor of any order:[32]

An n-th order *tensor* is a real-valued n-linear function of vectors.

In a more precise mathematical notation this says that an nth order tensor is a mapping

$$\boldsymbol{T} : \underbrace{\mathbb{R}^{n_d} \times \cdots \times \mathbb{R}^{n_d}}_{n \text{ times}} \to \mathbb{R}.$$

This constitutes a definition for tensors because vectors have been defined independently. Thus, through the isomorphism between vectors and real-valued linear mappings, a definition for tensors of any rank is obtained. Given this definition, a second-order tensor $\boldsymbol{T}$ is a bilinear function of two vector arguments, $\boldsymbol{T}[\boldsymbol{a}, \boldsymbol{b}]$. Just as for a first-order tensor, the components of a second-order tensor in a particular basis $\{\boldsymbol{e}_i\}$ are *defined* as

$$T_{ij} \equiv \boldsymbol{T}[\boldsymbol{e}_i, \boldsymbol{e}_j]. \tag{2.40}$$

Given two vectors, $\boldsymbol{a} = a_i \boldsymbol{e}_i$ and $\boldsymbol{b} = b_j \boldsymbol{e}_j$, the real number returned by the second-order tensor $\boldsymbol{T}$ is

$$\boldsymbol{T}[\boldsymbol{a}, \boldsymbol{b}] = \boldsymbol{T}[a_i \boldsymbol{e}_i, b_j \boldsymbol{e}_j] = a_i b_j \boldsymbol{T}[\boldsymbol{e}_i, \boldsymbol{e}_j] = a_i b_j T_{ij}. \tag{2.41}$$

Consider, for example, the stress tensor $\boldsymbol{\sigma}$ mentioned above. This can be written as $\boldsymbol{\sigma}[\boldsymbol{d}, \boldsymbol{n}]$, where $\boldsymbol{d}$ is a direction in space and $\boldsymbol{n}$ is the normal to a plane (see Fig. 2.4(b)). The scalar $\sigma_{ij} d_i n_j$ corresponds to the stress acting on the plane defined by $\boldsymbol{n}$ in the direction $\boldsymbol{d}$.

2.3.7 Tensor component transformation

We have stressed the fact that tensors are objects that are invariant with respect to the choice of coordinate system. However, at a practical level, when performing calculations with tensors it is necessary to select a particular coordinate system and to represent the tensor in terms of its components in the corresponding basis. The invariance of the tensor manifests itself in the fact that the components of the tensor with respect to different bases

[31] We will use the terms "vector" and "first-order tensor" interchangeably in the remainder of the book. However, it should be clear from this discussion that these terms are isomorphic to each other, but not identical.

[32] Actually, a tensor is still more general than this definition. The n-linear function can operate on covectors as well as vectors. Thus, the more general definition states that a tensor is a real-valued multilinear function of order (r, s), where r is the number of covector arguments and s is the number of vector arguments. See, for example, [Ogd84, Section 1.4.3] for a particularly clear explanation.

cannot be chosen arbitrarily, but must satisfy certain transformation relations. We have already obtained these relations for vectors in Eqns. (2.34) and (2.35). We will now derive them for tensors of arbitrary rank. The definition of a tensor as a linear function of vectors makes this a very simple derivation. For a first-order tensor starting from the component definition we have

$$a'_i \equiv \boldsymbol{a}[\boldsymbol{e}'_i] = \boldsymbol{a}[Q_{\alpha i}\boldsymbol{e}_\alpha] = Q_{\alpha i}\boldsymbol{a}[\boldsymbol{e}_\alpha] = Q_{\alpha i}a_\alpha,$$

where we have used Eqn. (2.31) and the linearity of $\boldsymbol{a}$. The form is identical to the vector transformation relation in Eqn. (2.35). For a second-order tensor the derivation is completely analogous:

$$A'_{ij} \equiv \boldsymbol{A}[\boldsymbol{e}'_i, \boldsymbol{e}'_j] = \boldsymbol{A}[Q_{\alpha i}\boldsymbol{e}_\alpha, Q_{\beta j}\boldsymbol{e}_\beta] = Q_{\alpha i}Q_{\beta j}\boldsymbol{A}[\boldsymbol{e}_\alpha, \boldsymbol{e}_\beta] = Q_{\alpha i}Q_{\beta j}A_{\alpha\beta}.$$

Thus

$$A'_{ij} = Q_{\alpha i}Q_{\beta j}A_{\alpha\beta} \quad \Leftrightarrow \quad [A]' = \mathbf{Q}^T [A]\, \mathbf{Q}. \tag{2.42}$$

Similarly for an nth-order tensor

$$B'_{i_1 i_2 \ldots i_n} = Q_{\alpha_1 i_1} Q_{\alpha_2 i_2} \cdots Q_{\alpha_n i_n} B_{\alpha_1 \alpha_2 \ldots \alpha_n}. \tag{2.43}$$

For the general case, there is no direct notation equivalent to the matrix multiplication form of the first- and second-order tensors.

In many texts, the component transformation laws are given as the definition of a tensor. We see that in our case the transformation relations emerge naturally from a more fundamental definition. However, the transformation relations provide a practical test for determining whether a given quantity is a tensor or not.

Proving a quantity is a tensor We will see in the next section that tensor operations always lead to the sums of products between tensor components as given by the indicial notation defined in Section 2.2. Any quantity written in this form is a tensor provided the arguments are tensors. For example, consider the product $c_\alpha = A_{\alpha\beta}b_\beta$, where $\boldsymbol{A}$ is a second-order tensor and $\boldsymbol{b}$ is a first-order tensor. To prove that $\boldsymbol{c}$ is a first-order tensor, we need to show that it transforms like one, i.e. that $c'_i = Q_{\alpha i}c_\alpha$.

Proof The definition of $\boldsymbol{c}$ holds for any basis, so we may write $c'_i = A'_{ij}b'_j$. Since $\boldsymbol{A}$ and $\boldsymbol{b}$ are tensors, they transform as tensors must. Substituting in the transformation relations for first- and second-order tensors in Eqns. (2.35) and (2.42), we have

$$\begin{aligned} c'_i = A'_{ij}b'_j &= (Q_{\alpha i}Q_{\beta j}A_{\alpha\beta})(Q_{\gamma j}b_\gamma) \\ &= Q_{\alpha i}A_{\alpha\beta}b_\gamma(Q_{\beta j}Q_{\gamma j}) = Q_{\alpha i}A_{\alpha\beta}b_\gamma\delta_{\beta\gamma} = Q_{\alpha i}A_{\alpha\beta}b_\beta = Q_{\alpha i}c_\alpha, \end{aligned}$$

where we have used the orthogonality of $\mathbf{Q}$. □

The proof shown above can be generalized to the product of any number of tensors. Free indices already transform appropriately since they belong to tensors, while the transformation matrices associated with dummy indices disappear due to the orthogonality condition. In the interest of brevity, we will not give the general proof, but we will show some additional examples when discussing specific tensor operations.

2.4 Tensor operations

We now turn to the description and classification of tensor operations.[33] Tensor operations can be divided into categories: (1) *addition* of two tensors; (2) *magnification* of a tensor; (3) *transposition* of a tensor; (4) taking the *product* of two or more tensors to form a higher-order tensor; and (5) *contraction* of a tensor to form a lower-order tensor. Together, tensor products and tensor contraction lead to the idea of a *tensor basis*.

2.4.1 Addition

Addition is defined for tensors of the same rank. For example, for second-order tensors we write

$$\boldsymbol{C}[\boldsymbol{x}, \boldsymbol{y}] = \boldsymbol{A}[\boldsymbol{x}, \boldsymbol{y}] + \boldsymbol{B}[\boldsymbol{x}, \boldsymbol{y}].$$

To obtain the indicial form, substitute $\boldsymbol{x} = x_i \boldsymbol{e}_i$ and $\boldsymbol{y} = y_j \boldsymbol{e}_j$ and use the bilinearity of the tensors. Moving all terms to one side, using Eqn. (2.41) and combining terms, we have

$$x_i y_j \left(C_{ij} - A_{ij} - B_{ij}\right) = 0.$$

This must be true for all $\boldsymbol{x}$ and $\boldsymbol{y}$, thus

$$C_{ij} = A_{ij} + B_{ij} \quad \Leftrightarrow \quad \boldsymbol{C} = \boldsymbol{A} + \boldsymbol{B}.$$

The expression on the right is the direct notation for the addition operation. Indices i and j are free indices using the terminology of Section 2.2. In that section we noted that each term in a sum of tensor terms must have the same free indices. We see that this is simply a different statement of the fact that addition is only defined for tensors of the same rank.

2.4.2 Magnification

Magnification corresponds to a rescaling of a tensor by scalar multiplication. For example, for a second-order tensor $\boldsymbol{A}$ and a scalar λ, a new second-order tensor $\boldsymbol{B}$ is defined by

$$\boldsymbol{B}[\boldsymbol{x}, \boldsymbol{y}] = \lambda \boldsymbol{A}[\boldsymbol{x}, \boldsymbol{y}].$$

The indicial form is obtained in the same manner as for addition:

$$B_{ij} = \lambda A_{ij} \quad \Leftrightarrow \quad \boldsymbol{B} = \lambda \boldsymbol{A}.$$

The direct notation appears on the right.

[33] The classification given here is partly based on the presentations in [Jau67] and [Sal01].

2.4.3 Transpose

The transpose operation exchanges the positions of arguments of a tensor. It is normally applied to second-order tensors:

$$\boldsymbol{B}[\boldsymbol{x}, \boldsymbol{y}] = \boldsymbol{A}[\boldsymbol{y}, \boldsymbol{x}].$$

The indicial form and direct notation are

$$B_{ij} = A_{ji} \quad \Leftrightarrow \quad \boldsymbol{B} = \boldsymbol{A}^T.$$

We see from the indicial form that $[\boldsymbol{B}] = [\boldsymbol{A}]^T$, where the superscript T denotes the matrix transpose operation. The direct notation is adopted in analogy to the matrix notation.

2.4.4 Tensor products

Tensor products refer to the formation of a higher-order tensor by combining two or more tensors. For example, below we combine a second-order tensor $\boldsymbol{A}$ with a vector $\boldsymbol{v}$:

$$\boldsymbol{D}[\boldsymbol{x}, \boldsymbol{y}, \boldsymbol{z}] = \boldsymbol{A}[\boldsymbol{x}, \boldsymbol{y}]\boldsymbol{v}[\boldsymbol{z}].$$

Substituting in $\boldsymbol{x} = x_i\boldsymbol{e}_i$, $\boldsymbol{y} = y_j\boldsymbol{e}_j$, and $\boldsymbol{z} = z_k\boldsymbol{e}_k$, and using linearity we have

$$\begin{aligned} \boldsymbol{D}[x_i\boldsymbol{e}_i, y_j\boldsymbol{e}_j, z_k\boldsymbol{e}_k] &= \boldsymbol{A}[x_i\boldsymbol{e}_i, y_j\boldsymbol{e}_j]\boldsymbol{v}[z_k\boldsymbol{e}_k] \\ x_i y_j z_k \boldsymbol{D}[\boldsymbol{e}_i, \boldsymbol{e}_j, \boldsymbol{e}_k] &= x_i y_j z_k \boldsymbol{A}[\boldsymbol{e}_i, \boldsymbol{e}_j]\boldsymbol{v}[\boldsymbol{e}_k] \\ x_i y_j z_k D_{ijk} &= x_i y_j z_k A_{ij} v_k. \end{aligned}$$

The last equation must be true for any $\boldsymbol{x}$, $\boldsymbol{y}$ and $\boldsymbol{z}$, so we have

$$D_{ijk} = A_{ij}v_k \quad \Leftrightarrow \quad \boldsymbol{D} = \boldsymbol{A} \otimes \boldsymbol{v}. \tag{2.44}$$

Products of the form $A_{ij}v_k$ are called *tensor products*. In direct notation, this operation is denoted $\boldsymbol{A} \otimes \boldsymbol{v}$, where $\otimes$ is the tensor product symbol. The rank of the resulting tensor is equal to the sum of the ranks of the combined tensors. In this case, a third-order tensor is formed by combining a first- and second-order tensor.

A particularly interesting case is the formation of a second-order tensor by a tensor product of two vectors:

$$A_{ij} = a_i b_j \quad \Leftrightarrow \quad \boldsymbol{A} = \boldsymbol{a} \otimes \boldsymbol{b}. \tag{2.45}$$

This is called the *dyad*[34] of the vectors $\boldsymbol{a}$ and $\boldsymbol{b}$. Note that the order of the vectors in a dyad is important, i.e. $\boldsymbol{a} \otimes \boldsymbol{b} \neq \boldsymbol{b} \otimes \boldsymbol{a}$. In matrix notation the dyad is written as

$$[\boldsymbol{a} \otimes \boldsymbol{b}] = \begin{bmatrix} a_1 b_1 & a_1 b_2 & a_1 b_3 \\ a_2 b_1 & a_2 b_2 & a_2 b_3 \\ a_3 b_1 & a_3 b_2 & a_3 b_3 \end{bmatrix}.$$

Let us prove that $\boldsymbol{A} = \boldsymbol{a} \otimes \boldsymbol{b}$ is a tensor:

Proof

$$A'_{ij} = a'_i b'_j = (Q_{\alpha i} a_\alpha)(Q_{\beta j} b_\beta) = Q_{\alpha i} Q_{\beta j} a_\alpha b_\beta = Q_{\alpha i} Q_{\beta j} A_{\alpha\beta}.$$

□

Dyads lead to the important concept of a *tensor basis*. We return to this in Section 2.4.6 after we discuss tensor contraction.

2.4.5 Contraction

Contraction corresponds to the formation of a lower-order tensor from a given tensor by summing two of its vector arguments. Given a tensor $\boldsymbol{T}[\boldsymbol{x}_1, \dots, \boldsymbol{x}_m]$ of rank m, we define the *contraction* operation on arguments i and j as[35]

$$\text{Cont}_{ij}\, \boldsymbol{T} = \boldsymbol{T}[\boldsymbol{x}_1, \dots, \boldsymbol{x}_{i-1}, \boldsymbol{e}_k, \boldsymbol{x}_{i+1}, \dots, \boldsymbol{x}_{j-1}, \boldsymbol{e}_k, \boldsymbol{x}_{j+1}, \dots, \boldsymbol{x}_m], \quad (2.46)$$

where $(\boldsymbol{e}_1, \boldsymbol{e}_2, \boldsymbol{e}_3)$ is an orthonormal basis and the summation convention is applied to the index k. The result of the contraction is a new tensor of rank $m - 2$. For example, for a third-order tensor $\boldsymbol{D}$ there are three possible contraction operations:

$$\begin{aligned} \boldsymbol{u}[\boldsymbol{x}] &= \text{Cont}_{23}\, \boldsymbol{D} = \boldsymbol{D}[\boldsymbol{x}, \boldsymbol{e}_j, \boldsymbol{e}_j], \\ \boldsymbol{v}[\boldsymbol{y}] &= \text{Cont}_{13}\, \boldsymbol{D} = \boldsymbol{D}[\boldsymbol{e}_i, \boldsymbol{y}, \boldsymbol{e}_i], \\ \boldsymbol{w}[\boldsymbol{z}] &= \text{Cont}_{12}\, \boldsymbol{D} = \boldsymbol{D}[\boldsymbol{e}_i, \boldsymbol{e}_i, \boldsymbol{z}], \end{aligned}$$

where $\boldsymbol{u}$, $\boldsymbol{v}$ and $\boldsymbol{w}$ are first-order tensors (vectors). The corresponding indicial expressions are obtained by substituting in the component form for each of the vector arguments, $\boldsymbol{x} = x_i \boldsymbol{e}_i$, $\boldsymbol{y} = y_j \boldsymbol{e}_j$, $\boldsymbol{z} = z_k \boldsymbol{e}_k$, and using linearity:

$$u_i = D_{ijj}, \qquad v_j = D_{iji}, \qquad w_k = D_{iik}.$$

We see that in indicial notation, contraction corresponds to a summation over dummy indices. Each contraction over a pair of dummy indices results in a reduction in the rank of

[34] Some authors use the shorthand notation $\boldsymbol{ab}$ for the dyad of $\boldsymbol{a}$ and $\boldsymbol{b}$, and more generally use this type of juxtaposition to indicate tensor products (i.e. the tensor product in Eqn. (2.44) would be written $\boldsymbol{D} = \boldsymbol{Av}$). Although this notation is self-consistent, it clashes with the standard notation from matrix algebra and abstract linear algebra. Therefore, we prefer to use the $\otimes$ symbol.

[35] More formally, the contraction operation is only defined for pairs of arguments where one is a vector and the other is a covector, i.e. a member of the dual space. When dealing with orthonormal bases as we do here, the distinction is obscured. See, for example, [Sal01] for the more general discussion.

the tensor by two orders. There is no standard direct notation for tensor contraction. The exception is contraction operations that lead to scalar invariants. These are discussed at the end of this section.

Contracted multiplication Contraction operations can be applied to tensor products, leading to familiar multiplication operations from matrix algebra. Consider the operation $\boldsymbol{u} = \text{Cont}_{23}(\boldsymbol{A} \otimes \boldsymbol{v})$, where $\boldsymbol{A}$ is a second-order tensor and $\boldsymbol{u}$ and $\boldsymbol{v}$ are vectors. Written explicitly, this is

$$\boldsymbol{u}[\boldsymbol{x}] = \text{Cont}_{23}\left(\boldsymbol{A}[\boldsymbol{x}, \boldsymbol{y}]\boldsymbol{v}[\boldsymbol{z}]\right) = \boldsymbol{A}[\boldsymbol{x}, \boldsymbol{e}_j]\boldsymbol{v}[\boldsymbol{e}_j],$$

where $\boldsymbol{x}$, $\boldsymbol{y}$, $\boldsymbol{z}$ are vectors. Substituting in the component form of the vector arguments and using linearity, we have

$$u_i = A_{ij}v_j \quad \Leftrightarrow \quad \boldsymbol{u} = \boldsymbol{A}\boldsymbol{v}. \tag{2.47}$$

The indicial expression can be written in matrix form as $[\boldsymbol{u}] = [\boldsymbol{A}]\,[\boldsymbol{v}]$. The direct notation appearing on the right of the above equation is adopted in analogy to the matrix operation. The matrix operation also lends to this operation its name of *contracted multiplication*. An important special case of Eqn. (2.47) follows when $\boldsymbol{A}$ is a dyad. In this case, the contracted multiplication satisfies the following relation:

$$(a_i b_j)v_j = a_i(b_j v_j) \quad \Leftrightarrow \quad (\boldsymbol{a} \otimes \boldsymbol{b})\boldsymbol{v} = \boldsymbol{a}(\boldsymbol{b} \cdot \boldsymbol{v}). \tag{2.48}$$

This identity can be viewed as a definition for the dyad as an operation that linearly transforms a vector $\boldsymbol{v}$ into a vector parallel to $\boldsymbol{a}$ with magnitude $\|\boldsymbol{a}\|\,|\boldsymbol{b} \cdot \boldsymbol{v}|$.

The contraction operation in Eqn. (2.47) also leads to an alternative definition for a second-order tensor as a linear mapping transforming one vector to another. We will adopt this viewpoint later when discussing the properties of second-order tensors in Section 2.5. We use Eqn. (2.47) to define the *identity tensor* $\boldsymbol{I}$ as the second-order tensor that leaves any vector $\boldsymbol{v}$ unchanged when operating on it:

$$\boldsymbol{I}\boldsymbol{v} = \boldsymbol{v}.$$

In component form this is $I_{ij}v_j = v_i$. Using $v_i = \delta_{ij}v_j$, this gives $(I_{ij} - \delta_{ij})v_j = 0$. This must be true for any v_j, therefore, $I_{ij} = \delta_{ij}$. Thus, the components of the identity tensor (with respect to an orthonormal basis) are equal to the entries of the identity matrix introduced in Eqn. (2.4):

$$[\boldsymbol{I}] = \mathbf{I}. \tag{2.49}$$

Next, consider the operation $\boldsymbol{C} = \text{Cont}_{23}(\boldsymbol{A} \otimes \boldsymbol{B})$, where $\boldsymbol{A}$, $\boldsymbol{B}$ and $\boldsymbol{C}$ are second-order tensors. Written explicitly this is

$$\boldsymbol{C}[\boldsymbol{x}, \boldsymbol{y}] = \text{Cont}_{23}\left(\boldsymbol{A}[\boldsymbol{x}, \boldsymbol{u}]\boldsymbol{B}[\boldsymbol{v}, \boldsymbol{y}]\right) = \boldsymbol{A}[\boldsymbol{x}, \boldsymbol{e}_k]\boldsymbol{B}[\boldsymbol{e}_k, \boldsymbol{y}],$$

where $\boldsymbol{u}$, $\boldsymbol{v}$, $\boldsymbol{x}$ and $\boldsymbol{y}$ are vectors. Substituting the component form of the vector arguments and using linearity, we have

$$C_{ij} = A_{ik}B_{kj} \quad \Leftrightarrow \quad \boldsymbol{C} = \boldsymbol{A}\boldsymbol{B}. \tag{2.50}$$

On the right is the direct notation, which is again borrowed from matrix algebra. A series of multiplications by the same tensor is denoted by an exponent:

$$\boldsymbol{A}^2 = \boldsymbol{A}\boldsymbol{A}, \qquad \boldsymbol{A}^3 = (\boldsymbol{A}^2)\boldsymbol{A} = \boldsymbol{A}\boldsymbol{A}\boldsymbol{A}, \quad \text{etc.}$$

It makes sense to think of the tensor $\boldsymbol{C}$ in Eqn. (2.50) as a *composition* of the tensors $\boldsymbol{A}$ and $\boldsymbol{B}$. The term "composition" is used here in the sense of a "function composition," where one function is applied to the results of the other. For example, the real function $h : x \to z$ is a composition of $f : y \to z$ and $g : x \to y$, if $h(x) = f(g(x))$. This is denoted $h = f \circ g$. For the tensor $\boldsymbol{C}$ this interpretation follows from the definition in Eqn. (2.47). Thus,

$$\boldsymbol{u} = \boldsymbol{C}\boldsymbol{v} = (\boldsymbol{A}\boldsymbol{B})\boldsymbol{v} = \boldsymbol{A}(\boldsymbol{B}\boldsymbol{v}).$$

We see that applying $\boldsymbol{C}$ to $\boldsymbol{v}$ is the same as first applying $\boldsymbol{B}$ and then applying $\boldsymbol{A}$ to the result $\boldsymbol{B}\boldsymbol{v}$. Thus, $\boldsymbol{C}$ is a composition of $\boldsymbol{A}$ and $\boldsymbol{B}$.

Many other contractions are possible. For example, following the procedure outlined above, the operation $\boldsymbol{C} = \text{Cont}_{24}(\boldsymbol{A} \otimes \boldsymbol{B})$ leads to

$$C_{ij} = A_{ik}B_{jk} \quad \Leftrightarrow \quad \boldsymbol{C} = \boldsymbol{A}\boldsymbol{B}^T, \tag{2.51}$$

where the superscript T corresponds to the transpose operation defined in Section 2.4.3. In similar fashion we also obtain

$$C_{ij} = A_{ki}B_{kj} \quad \Leftrightarrow \quad \boldsymbol{C} = \boldsymbol{A}^T\boldsymbol{B} \quad \text{and} \quad C_{ij} = A_{ki}B_{jk} \quad \Leftrightarrow \quad \boldsymbol{C} = \boldsymbol{A}^T\boldsymbol{B}^T. \tag{2.52}$$

The definition of tensor contraction allows us to define the inverse $\boldsymbol{A}^{-1}$ of a second-order tensor $\boldsymbol{A}$ through the relation

$$\boldsymbol{A}^{-1}\boldsymbol{A} = \boldsymbol{A}\boldsymbol{A}^{-1} = \boldsymbol{I}, \tag{2.53}$$

where $\boldsymbol{I}$ is the identity tensor defined above. In indicial form this is $A^{-1}_{ij} A_{jk} = A_{ij} A^{-1}_{jk} = \delta_{ik}$, and in matrix form it is $\left[\boldsymbol{A}^{-1}\right][\boldsymbol{A}] = [\boldsymbol{A}]\left[\boldsymbol{A}^{-1}\right] = [\boldsymbol{I}]$. Comparing the last expression with Eqn. (2.53), we see that $\left[\boldsymbol{A}^{-1}\right] = [\boldsymbol{A}]^{-1}$. Consistent with this, the determinant of a second-order tensor is defined as the determinant of its components matrix:

$$\det \boldsymbol{A} \equiv \det [\boldsymbol{A}].$$

We will see later that $\det \boldsymbol{A}$ is a scalar invariant and is therefore independent of the coordinate system basis.

Given the above definitions, the expression in Eqn. (2.10) for the derivative of the determinant of a square matrix can be rewritten for a tensor as

$$\frac{\partial(\det \boldsymbol{A})}{\partial \boldsymbol{A}} = \boldsymbol{A}^{-T} \det \boldsymbol{A}, \tag{2.54}$$

where $\boldsymbol{A}^{-T} = (\boldsymbol{A}^{-1})^T$.

Scalar contraction Of particular interest are contraction operations that result in the formation of a zeroth-order tensor (i.e. a scalar invariant). Any tensor of even order can be reduced to a scalar by repeated contraction. For a second-order tensor $\boldsymbol{A}$, one contraction operation leads to a scalar:

$$\text{Cont}_{12}\, \boldsymbol{A} = \boldsymbol{A}[\boldsymbol{e}_i, \boldsymbol{e}_i] = A_{ii}. \tag{2.55}$$

We see from the indicial expression that this contraction corresponds to the trace of the matrix of components of $\boldsymbol{A}$, since $A_{ii} = \text{tr}\,[\boldsymbol{A}]$. For this reason the direct notation for this operation is also denoted by the trace:

$$\text{tr}\, \boldsymbol{A} = \text{Cont}_{12}\, \boldsymbol{A} = \text{tr}\,[\boldsymbol{A}] = A_{ii}. \tag{2.56}$$

It is straightforward to show that $\text{tr}\, \boldsymbol{A}$ is a zeroth-order tensor.

Proof

$$A'_{ii} = Q_{\alpha i} Q_{\beta i} A_{\alpha\beta} = \delta_{\alpha\beta} A_{\alpha\beta} = A_{\alpha\alpha}.$$

□

We see that a scalar invariant is indeed invariant with respect to coordinate basis transformation. This is as it should be since a scalar invariant is a tensor. This brings up the interesting point that not every scalar is a zeroth-order tensor. For example, a single component of a tensor is a scalar but it is *not* a zeroth-order tensor, since it is not invariant with respect to coordinate system transformation.

Scalar contraction can also be applied to contracted multiplication. We have already seen an example of this in the dot product of two vectors, $\boldsymbol{a} \cdot \boldsymbol{b} = a_i b_i$. The dot product was defined in Section 2.3 as part of the definition of vector spaces. In terms of contraction, we can write the dot product as $\boldsymbol{a} \cdot \boldsymbol{b} = \mathrm{Cont}_{12}(\boldsymbol{a} \otimes \boldsymbol{b})$.

Other important examples of contractions leading to scalar invariants are the double contraction operations of two second-order tensors, $\boldsymbol{A}$ and $\boldsymbol{B}$, which can take two forms:

$$\boldsymbol{A} : \boldsymbol{B} = \mathrm{tr}[\boldsymbol{A}^T \boldsymbol{B}] = \mathrm{tr}[\boldsymbol{B}^T \boldsymbol{A}] = \mathrm{tr}[\boldsymbol{A}\boldsymbol{B}^T] = \mathrm{tr}[\boldsymbol{B}\boldsymbol{A}^T] = A_{ij} B_{ij}, \tag{2.57}$$

$$\boldsymbol{A} \cdot\cdot \boldsymbol{B} = \mathrm{tr}[\boldsymbol{A}\boldsymbol{B}] = \mathrm{tr}[\boldsymbol{B}^T \boldsymbol{A}^T] = \mathrm{tr}[\boldsymbol{B}\boldsymbol{A}] = \mathrm{tr}[\boldsymbol{A}^T \boldsymbol{B}^T] = A_{ij} B_{ji}. \tag{2.58}$$

The symbols $\cdot$, $:$ and $\cdot\cdot$ are the direct notation for the contraction operations.[36] It is worth pointing out that the double contraction $\boldsymbol{A} : \boldsymbol{B}$ is an inner product in the space of second-order tensors. The corresponding norm is $\|\boldsymbol{A}\| = (\boldsymbol{A} : \boldsymbol{A})^{1/2}$. (For this reason some books, like [Gur81], denote this contraction with the dot product, $\boldsymbol{A} \cdot \boldsymbol{B}$.) The definition of the double contraction operation is also extended to describe contraction of a fourth-order tensor $\boldsymbol{E}$ with a second-order tensor $\boldsymbol{A}$:

$$[\boldsymbol{E} : \boldsymbol{A}]_{ij} = E_{ijkl} A_{kl}, \qquad [\boldsymbol{E} \cdot\cdot \boldsymbol{A}]_{ij} = E_{ijkl} A_{lk}. \tag{2.59}$$

Finally, we note that when scalar contraction is applied to a contracted multiplication of the *same* vectors ($\boldsymbol{a} = \boldsymbol{b}$) or the *same* tensors ($\boldsymbol{A} = \boldsymbol{B}$) the results are scalar invariants of the tensors themselves. From the dot product we obtain the length squared of the vector $a_i a_i$ and from the tensor contractions, $A_{ij} A_{ij}$ and $A_{ij} A_{ji}$.

2.4.6 Tensor basis

We conclude the discussion of tensor operations by showing how tensor products of vectors, i.e. dyads, triads and so on, can be used to define a basis for tensors of rank two and above. Let us first consider the case of a second-order tensor. Since a dyad is a second-order tensor, an interesting question is whether *any* second-order tensor $\boldsymbol{A}$ can be written as a dyad. The answer is no, since dyads are not general tensors; they satisfy the identity

$$\det(\boldsymbol{a} \otimes \boldsymbol{b}) = 0,$$

e.g. in two dimensions

$$\det \begin{bmatrix} a_1 b_1 & a_1 b_2 \\ a_2 b_1 & a_2 b_2 \end{bmatrix} = a_1 b_1 a_2 b_2 - a_1 b_2 a_2 b_1 = 0.$$

[36] This convention is not universally adopted. Some authors reverse the meaning of $:$ and $\cdot\cdot$. Others do not use the double dot notation at all and use $\cdot$ to denote scalar contraction for both vectors and second-order tensors.

However, an arbitrary tensor *can* be written as a linear combination of dyads, which is called a *dyadic*. In two dimensions, two terms are required:

$$\boldsymbol{A} = \boldsymbol{a} \otimes \boldsymbol{b} + \boldsymbol{c} \otimes \boldsymbol{d},$$

where the pair of vectors $\boldsymbol{a}$ and $\boldsymbol{c}$ and the pair of vectors $\boldsymbol{b}$ and $\boldsymbol{d}$ are linearly independent (see Section 2.3). In three dimensions, three terms are required:

$$\boldsymbol{A} = \boldsymbol{a} \otimes \boldsymbol{b} + \boldsymbol{c} \otimes \boldsymbol{d} + \boldsymbol{e} \otimes \boldsymbol{f}, \tag{2.60}$$

where the triads $\boldsymbol{a},\boldsymbol{c},\boldsymbol{e}$ and $\boldsymbol{b},\boldsymbol{d},\boldsymbol{f}$ are linearly independent. Let us prove that a dyadic of two dyads is insufficient to represent an arbitrary second-order tensor on $\mathbb{R}^3$.

Proof Start with $\boldsymbol{A} = \boldsymbol{a} \otimes \boldsymbol{b} + \boldsymbol{c} \otimes \boldsymbol{d}$ and apply an arbitrary vector $\boldsymbol{v}$ to both sides:

$$\boldsymbol{A}\boldsymbol{v} = (\boldsymbol{a} \otimes \boldsymbol{b})\boldsymbol{v} + (\boldsymbol{c} \otimes \boldsymbol{d})\boldsymbol{v} = \boldsymbol{a}(\boldsymbol{b} \cdot \boldsymbol{v}) + \boldsymbol{c}(\boldsymbol{d} \cdot \boldsymbol{v}),$$

where we have used the identity in Eqn. (2.48). The above equation suggests that the vector formed by $\boldsymbol{A}$ operating on any vector $\boldsymbol{v}$ will always lie in the plane defined by $\boldsymbol{a}$ and $\boldsymbol{c}$. This is clearly not generally correct in three dimensions. □

The dyadic description does not provide a unique decomposition for $\boldsymbol{A}$ since there are more vector components than tensor components. However, it can be used to provide a basis description for tensors analogous to the $\boldsymbol{a} = a_i \boldsymbol{e}_i$ of vectors:

$$\boldsymbol{A} = \boldsymbol{a} \otimes \boldsymbol{b} = (a_i \boldsymbol{e}_i) \otimes (b_j \boldsymbol{e}_j) = a_i b_j (\boldsymbol{e}_i \otimes \boldsymbol{e}_j).$$

This expression was written for the special case of a single dyad; in the general case of a dyadic with n_{d} dyads, the components of the vectors combine to give the general form,

$$\boldsymbol{A} = A_{ij} (\boldsymbol{e}_i \otimes \boldsymbol{e}_j). \tag{2.61}$$

For instance, in the case of $n_{\mathrm{d}} = 3$, the components A_{ij} would be made up of combinations of the components of the vectors $\boldsymbol{a}$, $\boldsymbol{b}$, $\boldsymbol{c}$, $\boldsymbol{d}$, $\boldsymbol{e}$ and $\boldsymbol{f}$ from Eqn. (2.60). The dyads $\boldsymbol{e}_i \otimes \boldsymbol{e}_j$ can be thought of as the "basis tensors" relative to which the components of $\boldsymbol{A}$ are given. It is straightforward to show that $\boldsymbol{e}_i \otimes \boldsymbol{e}_j$ form a linearly independent basis.

The basis description can be used to obtain an expression for the components of $\boldsymbol{A}$. Replace the dummy indices in Eqn. (2.61) with m and n, apply $\boldsymbol{e}_j$ to both sides, and then use Eqn. (2.48) to obtain

$$\boldsymbol{A}\boldsymbol{e}_j = A_{mn} (\boldsymbol{e}_m \otimes \boldsymbol{e}_n) \boldsymbol{e}_j = A_{mn} (\boldsymbol{e}_n \cdot \boldsymbol{e}_j) \boldsymbol{e}_m = A_{mn} \delta_{nj} \boldsymbol{e}_m = A_{mj} \boldsymbol{e}_m.$$

Next, dot both sides with $\boldsymbol{e}_i$ to obtain the component relation

$$A_{ij} = \boldsymbol{e}_i \cdot \boldsymbol{A}\boldsymbol{e}_j. \tag{2.62}$$

The concept of a tensor basis naturally extends to higher-order tensors. For example, the basis descriptions for a third-order tensor $\boldsymbol{D}$ and a fourth-order tensor $\boldsymbol{E}$ are

$$\boldsymbol{D} = D_{ijk}(\boldsymbol{e}_i \otimes \boldsymbol{e}_j \otimes \boldsymbol{e}_k), \qquad \boldsymbol{E} = E_{ijkl}(\boldsymbol{e}_i \otimes \boldsymbol{e}_j \otimes \boldsymbol{e}_k \otimes \boldsymbol{e}_l).$$

We see that all tensors can be represented as a linear combination of tensor products of vectors. This provides an alternative approach to defining tensor operations which many books adopt. Rather than defining operations for general tensors of arbitrary rank as we have done, one defines operations for dyads, triads and so on, and from these builds up the more general tensor operations.

As an example, consider the trace operation introduced above for the scalar contraction of a second-order tensor. It is also possible to define the trace operator without reference to contraction by the following relation:

$$\operatorname{tr}[\boldsymbol{a} \otimes \boldsymbol{b}] = \boldsymbol{a} \cdot \boldsymbol{b} \qquad \forall \boldsymbol{a}, \boldsymbol{b} \in \mathbb{R}^{n_d}.$$

We can see that this definition is consistent with the contraction definition of the trace of a second-order tensor $\boldsymbol{A}$:

$$\operatorname{tr} \boldsymbol{A} = \operatorname{tr}\left[A_{ij}(\boldsymbol{e}_i \otimes \boldsymbol{e}_j)\right] = A_{ij} \operatorname{tr}\left[\boldsymbol{e}_i \otimes \boldsymbol{e}_j\right] = A_{ij}\left(\boldsymbol{e}_i \cdot \boldsymbol{e}_j\right) = A_{ij}\delta_{ij} = A_{ii} = \operatorname{tr}\left[\boldsymbol{A}\right].$$

In similar fashion, all contraction operations can be defined. See, for example, [Hol00].

2.5 Properties of tensors

Most of the tensors that we will be dealing with are second-order tensors. It is therefore worthwhile to review the properties of such tensors. Before we do so, we provide an alternative definition for a second-order tensor, which is less general than the definition given in Section 2.3, but which is helpful when discussing some of the properties of second-order tensors. The definition is:

> A second-order tensor $\boldsymbol{T}$ is a linear mapping transforming a vector $\boldsymbol{v}$ into a vector $\boldsymbol{w}$, defined by $\boldsymbol{w} = \boldsymbol{T}\boldsymbol{v}$.

In a more precise mathematical notation this says that a second-order tensor is a mapping

$$\boldsymbol{T} : \mathbb{R}^{n_d} \to \mathbb{R}^{n_d}.$$

We now turn to the properties of second-order tensors.

2.5.1 Orthogonal tensors

A second-order tensor $\boldsymbol{Q}$ is called *orthogonal* if for every pair of vectors $\boldsymbol{a}$ and $\boldsymbol{b}$, we have

$$(\boldsymbol{Q}\boldsymbol{a}) \cdot (\boldsymbol{Q}\boldsymbol{b}) = \boldsymbol{a} \cdot \boldsymbol{b}. \tag{2.63}$$

Geometrically, this means that $\boldsymbol{Q}$ preserves the magnitude of, and the angles between, the vectors on which it operates.[37] A necessary and sufficient condition for this is [Ogd84]

$$\boldsymbol{Q}^T\boldsymbol{Q} = \boldsymbol{Q}\boldsymbol{Q}^T = \boldsymbol{I}, \tag{2.64}$$

or equivalently

$$\boldsymbol{Q}^T = \boldsymbol{Q}^{-1}. \tag{2.65}$$

These conditions are completely analogous to the ones given for orthogonal matrices in Eqns. (2.32)–(2.33). As in that case, it can be shown that $\det \boldsymbol{Q} = \pm 1$. An orthogonal tensor $\boldsymbol{Q}$ is called *proper orthogonal* if $\det \boldsymbol{Q} = 1$, and *improper orthogonal* otherwise. When viewed as a linear mapping of vectors to vectors, $\boldsymbol{Q}$ is called an *orthogonal transformation*. A proper orthogonal transformation corresponds to a rotation. An improper orthogonal transformation involves a rotation and a reflection. The groups $O(3)$ and $SO(3)$ defined for orthogonal matrices in Section 2.3.4 also exist for orthogonal tensors.

Given the strong analogy between orthogonal matrices and orthogonal tensors, it is of interest to see how the proper orthogonal transformation matrix $\mathbf{Q}$ is related to a proper orthogonal tensor $\boldsymbol{Q}$ applying the associated rotation. Recall that the transformation matrix links two bases $\{\boldsymbol{e}_i\}$ and $\{\boldsymbol{e}'_i\}$ according to

$$\boldsymbol{e}'_j = \mathrm{Q}_{ij}\boldsymbol{e}_i. \tag{2.66}$$

This is an expression that decomposes the $\boldsymbol{e}'_j$ basis vectors into components with respect to the $\boldsymbol{e}_i$ basis vectors. A closely related, but distinct expression is a rigid-body rotation of the basis vectors $\boldsymbol{e}_j$ that maps them into the basis vectors $\boldsymbol{e}'_j$. This can be written as

$$\boldsymbol{e}'_j = \boldsymbol{R}\boldsymbol{e}_j, \tag{2.67}$$

where $\boldsymbol{R}$ is a proper orthogonal tensor. We wish to find the relation between the components of the rotation $\boldsymbol{R}$ in the original basis $\{\boldsymbol{e}_i\}$ and the components of the change of basis matrix Q_{ij}. Substituting $\boldsymbol{R} = R_{ik}\boldsymbol{e}_i \otimes \boldsymbol{e}_k$ into Eqn. (2.67) gives

$$\boldsymbol{e}'_j = R_{ik}(\boldsymbol{e}_i \otimes \boldsymbol{e}_k)\boldsymbol{e}_j = R_{ik}\boldsymbol{e}_i(\boldsymbol{e}_k \cdot \boldsymbol{e}_j) = R_{ik}\boldsymbol{e}_i\delta_{kj} = R_{ij}\boldsymbol{e}_i. \tag{2.68}$$

Comparing Eqns. (2.66) and (2.68), we see that $\mathrm{Q}_{ij} = R_{ij}$ or $\mathbf{Q} = [\boldsymbol{R}]$. In other words, given a transformation from basis $\{\boldsymbol{e}_i\}$ to basis $\{\boldsymbol{e}'_i\}$, the proper orthogonal tensor that rotates an individual basis vector has the same components in the original basis $\boldsymbol{e}_i$ as the transformation matrix that defines the transformation. Thus,

$$\begin{bmatrix} \boldsymbol{e}'_1 \\ \boldsymbol{e}'_2 \\ \boldsymbol{e}'_3 \end{bmatrix} = \mathbf{Q} \begin{bmatrix} \boldsymbol{e}_1 \\ \boldsymbol{e}_2 \\ \boldsymbol{e}_3 \end{bmatrix} \quad \text{and} \quad \boldsymbol{e}'_i = \boldsymbol{Q}\boldsymbol{e}_i, \tag{2.69}$$

[37] In fact, it is sufficient to require that $\boldsymbol{Q}$ preserves the magnitude of all vectors. From this property alone, it is possible to prove that $\boldsymbol{Q}$ also preserves the dot product, and thus the angles, between any two vectors.

where $[Q] = \mathbf{Q}$. It is important to understand that these two equations represent very different ideas. Equation $(2.69)_1$ is an example of writing a set of vectors as a linear combination of basis vectors, whereas Eqn. $(2.69)_2$ is an example of a rotation (which is a special type of linear mapping) taking a vector to a different vector.

2.5.2 Symmetric and antisymmetric tensors

A *symmetric* second-order tensor $\boldsymbol{S}$ satisfies the condition

$$S_{ij} = S_{ji} \quad \Leftrightarrow \quad \boldsymbol{S} = \boldsymbol{S}^T.$$

An *antisymmetric* tensor $\boldsymbol{A}$ (also called a *skew-symmetric* tensor) satisfies the condition

$$A_{ij} = -A_{ji} \quad \Leftrightarrow \quad \boldsymbol{A} = -\boldsymbol{A}^T.$$

From this definition it is clear that $A_{11} = A_{22} = A_{33} = 0$. Thus since the diagonal elements are zero and the off-diagonal elements are equal with a change of sign, an anti-symmetric tensor has only three independent components. It is therefore not surprising that there exists a unique one-to-one correspondence between an antisymmetric tensor $\boldsymbol{A}$ and a vector called the *axial vector* $\boldsymbol{w}$. The relation is defined by the condition:

$$\boldsymbol{A}\boldsymbol{a} = \boldsymbol{w} \times \boldsymbol{a} \qquad \forall \boldsymbol{a} \in \mathbb{R}^3. \tag{2.70}$$

This condition can be solved to obtain an explicit relation between $\boldsymbol{w}$ and $\boldsymbol{A}$ and vice versa:

$$w_k = -\frac{1}{2}\epsilon_{ijk}A_{ij} \quad \Leftrightarrow \quad A_{ij} = -\epsilon_{ijk}w_k. \tag{2.71}$$

The proof is left as an exercise for the reader (see Exercise 2.12). The axial vector is used in the definition of the differential curl operation in Section 2.6.

An important property related to the above definitions is that the contraction of any symmetric tensor $\boldsymbol{S}$ with an antisymmetric tensor $\boldsymbol{A}$ is zero, i.e. $\boldsymbol{S} : \boldsymbol{A} = S_{ij}A_{ij} = 0$.

Proof $S_{ij}A_{ij} = \frac{1}{2}S_{ij}(A_{ij} - A_{ji}) = \frac{1}{2}(S_{ij}A_{ij} - S_{ij}A_{ji})$. Now exchange the dummy indices i and j on the second term and use the fact that $S_{ij} = S_{ji}$. □

A tensor that is neither symmetric nor antisymmetric is called *anisotropic*. Any anisotropic tensor T_{ij} can be decomposed in a unique manner into a symmetric part $T_{(ij)}$ and a anti-symmetric part $T_{[ij]}$, so that $T_{ij} = T_{(ij)} + T_{[ij]}$, where

$$T_{(ij)} \equiv \frac{1}{2}(T_{ij} + T_{ji}), \qquad T_{[ij]} \equiv \frac{1}{2}(T_{ij} - T_{ji}).$$

2.5.3 Principal values and directions

A second-order tensor $\boldsymbol{G}$ maps a vector $\boldsymbol{v}$ to a new vector $\boldsymbol{w} = \boldsymbol{G}\boldsymbol{v}$. We now ask whether there are certain special directions, $\boldsymbol{v} = \boldsymbol{\Lambda}$, for which

$$\boldsymbol{w} = \boldsymbol{G}\boldsymbol{\Lambda} = \lambda\boldsymbol{\Lambda}, \quad \lambda \in \mathbb{R},$$

i.e. directions that are not changed (only magnified) by the operation of $\boldsymbol{G}$. Thus we seek solutions to the following equation:

$$G_{ij}\Lambda_j = \lambda\Lambda_i \quad \Leftrightarrow \quad \boldsymbol{G}\boldsymbol{\Lambda} = \lambda\boldsymbol{\Lambda}, \tag{2.72}$$

or equivalently

$$(G_{ij} - \lambda\delta_{ij})\Lambda_j = 0 \quad \Leftrightarrow \quad (\boldsymbol{G} - \lambda\boldsymbol{I})\boldsymbol{\Lambda} = \boldsymbol{0}. \tag{2.73}$$

A vector $\boldsymbol{\Lambda}^{\boldsymbol{G}}$ satisfying this requirement is called an *eigenvector* (principal direction) of $\boldsymbol{G}$ with $\lambda^{\boldsymbol{G}}$ being the corresponding *eigenvalue* (principal value).[38] The superscript "$\boldsymbol{G}$" denotes that these are the eigenvectors and eigenvalues specific to the tensor $\boldsymbol{G}$. Nontrivial solutions to Eqn. (2.73) require

$$\det(\boldsymbol{G} - \lambda\boldsymbol{I}) = 0.$$

For $n_{\mathrm{d}} = 3$, this is a cubic equation in λ that is called the *characteristic equation* of $\boldsymbol{G}$:

$$-\lambda^3 + I_1^{\boldsymbol{G}}\lambda^2 - I_2^{\boldsymbol{G}}\lambda + I_3^{\boldsymbol{G}} = 0, \tag{2.74}$$

where $I_1^{\boldsymbol{G}}, I_2^{\boldsymbol{G}}, I_3^{\boldsymbol{G}}$ are the *principal invariants* of $\boldsymbol{G}$:

$$I_1^{\boldsymbol{G}} = G_{ii} = \operatorname{tr}\boldsymbol{G}, \tag{2.75}$$

$$I_2^{\boldsymbol{G}} = \tfrac{1}{2}(G_{ii}G_{jj} - G_{ij}G_{ji}) = \frac{1}{2}\left[(\operatorname{tr}\boldsymbol{G})^2 - \operatorname{tr}\boldsymbol{G}^2\right] = \operatorname{tr}\boldsymbol{G}^{-1}\det\boldsymbol{G}, \tag{2.76}$$

$$I_3^{\boldsymbol{G}} = \epsilon_{ijk}G_{1i}G_{2j}G_{3k} = \det\boldsymbol{G}. \tag{2.77}$$

The characteristic equation (Eqn. (2.74)) has three solutions: $\lambda_\alpha^{\boldsymbol{G}}$ ($\alpha = 1, 2, 3$). Since the equation is cubic and has real coefficients, in general it has one real root and two complex conjugate roots. However, it can be proved that in the special case where $\boldsymbol{G}$ is symmetric ($\boldsymbol{G} = \boldsymbol{G}^T$), all three eigenvalues are real. Each eigenvalue $\lambda_\alpha^{\boldsymbol{G}}$ has an eigenvector $\boldsymbol{\Lambda}_\alpha^{\boldsymbol{G}}$ that is obtained by solving[39] Eqn. (2.73) after substituting in $\lambda = \lambda_\alpha^{\boldsymbol{G}}$ together with the normalization condition $\left\|\boldsymbol{\Lambda}_\alpha^{\boldsymbol{G}}\right\| = 1$.

An important theorem states that the eigenvectors corresponding to distinct eigenvalues of a symmetric tensor $\boldsymbol{S}$ are orthogonal. This together with the normalization condition means that

$$\boldsymbol{\Lambda}_\alpha^{\boldsymbol{S}} \cdot \boldsymbol{\Lambda}_\beta^{\boldsymbol{S}} = \delta_{\alpha\beta}. \tag{2.78}$$

38 It is also common to encounter eigenvalue equations on an infinite-dimensional vector space over the field of complex numbers (see Part II of [TM11]). For example, in quantum mechanics the tensor operator is not symmetric but *Hermitian*, which means that $\boldsymbol{H} = (\boldsymbol{H}^*)^T$, where $*$ represents the complex conjugate. Hermitian tensors are generalizations of symmetric tensors, and it can be shown that Hermitian tensors have real eigenvalues and orthogonal eigenvectors just like symmetric tensors.

39 Actually, for each distinct eigenvalue $\lambda_\alpha^{\boldsymbol{G}}$ there are two solutions to these equations. One is given by $\boldsymbol{\Lambda}_\alpha^{\boldsymbol{G}}$ and the other is given by its negative $-\boldsymbol{\Lambda}_\alpha^{\boldsymbol{G}}$. Both solutions are valid eigenvectors for the eigenvalue $\lambda_\alpha^{\boldsymbol{G}}$.

The proof that the eigenvectors are orthogonal is straightforward.

Proof Start with

$$(S_{ij} - \lambda_\alpha^{\boldsymbol{S}}\delta_{ij})\Lambda_{\alpha j}^{\boldsymbol{S}} = 0 \qquad \text{(no sum on } \alpha)$$

and multiply with $\Lambda_{\beta i}^{\boldsymbol{S}}$ to obtain

$$S_{ij}\Lambda_{\beta i}^{\boldsymbol{S}}\Lambda_{\alpha j}^{\boldsymbol{S}} - \lambda_\alpha^{\boldsymbol{S}}\Lambda_{\beta i}^{\boldsymbol{S}}\Lambda_{\alpha i}^{\boldsymbol{S}} = 0. \tag{2.79}$$

We adopt the convention of referring to the eigenvalue and eigenvector number with a Greek index and use Roman indices to refer to spatial directions. The summation convention does not apply to the Greek eigen indices. Now use the symmetry of $\boldsymbol{S}$ in the first term of Eqn. (2.79) to replace S_{ij} with S_{ji} and then swap the dummy indices i and j to obtain

$$S_{ij}\Lambda_{\alpha i}^{\boldsymbol{S}}\Lambda_{\beta j}^{\boldsymbol{S}} - \lambda_\alpha^{\boldsymbol{S}}\Lambda_{\beta i}^{\boldsymbol{S}}\Lambda_{\alpha i}^{\boldsymbol{S}} = 0.$$

The first term is equal to $\lambda_\beta^{\boldsymbol{S}}\Lambda_{\beta i}^{\boldsymbol{S}}\Lambda_{\alpha i}^{\boldsymbol{S}}$, where we have used Eqn. (2.79) with α and β swapped. We then have

$$(\lambda_\beta^{\boldsymbol{S}} - \lambda_\alpha^{\boldsymbol{S}})\Lambda_{\beta i}^{\boldsymbol{S}}\Lambda_{\alpha i}^{\boldsymbol{S}} = 0.$$

If $\alpha \neq \beta$ and the eigenvalues are distinct ($\lambda_\beta^{\boldsymbol{S}} \neq \lambda_\alpha^{\boldsymbol{S}}$), then the above equation is only satisfied if $\Lambda_{\beta i}^{\boldsymbol{S}}\Lambda_{\alpha i}^{\boldsymbol{S}} = 0$, i.e. the eigenvectors are orthogonal. □

In the situation where some eigenvalues are repeated the above proof does not hold. However, it is still possible to generate a set of three mutually orthogonal vectors, although the choice is not unique. If one root repeats ($\lambda_1^{\boldsymbol{S}} = \lambda_2^{\boldsymbol{S}} = \lambda \neq \lambda_3^{\boldsymbol{S}}$), then there exists a plane such that any vector $\boldsymbol{u}$ in the plane satisfies the eigen equation, $\boldsymbol{S}\boldsymbol{u} = \lambda\boldsymbol{u}$. If all roots are equal ($\lambda_1^{\boldsymbol{S}} = \lambda_2^{\boldsymbol{S}} = \lambda_3^{\boldsymbol{S}} = \lambda$), then the eigen equation is satisfied for any vector $\boldsymbol{v}$. A tensor satisfying this condition is called a *spherical tensor* or a second-order *isotropic tensor*. Isotropic tensors are discussed in Section 2.5.6.

The fact that it is always possible to construct a set of three mutually orthonormal eigenvectors for a symmetric second-order tensor $\boldsymbol{S}$ suggests using these eigenvectors as a basis for a Cartesian coordinate system.[40] This is referred to as the *principal coordinate system* of the tensor for which the eigenvectors form the *principal basis*. An important property of the eigenvectors that follows from this is the *completeness relation*:

$$\sum_{\alpha=1}^{3} \boldsymbol{\Lambda}_\alpha^{\boldsymbol{S}} \otimes \boldsymbol{\Lambda}_\alpha^{\boldsymbol{S}} = \boldsymbol{I}, \tag{2.80}$$

where $\boldsymbol{I}$ is the identity tensor. The proof is simple.

Proof Any vector $\boldsymbol{v} = v_i\boldsymbol{e}_i$ can be represented in the principal basis as

$$\boldsymbol{v} = \sum_{\alpha=1}^{3}(\boldsymbol{v} \cdot \boldsymbol{\Lambda}_\alpha^{\boldsymbol{S}})\boldsymbol{\Lambda}_\alpha^{\boldsymbol{S}}.$$

[40] The vectors should be suitably ordered so that a right-handed basis is obtained.

Dotting both sides of the equation with $\boldsymbol{e}_i$ gives

$$v_i = \sum_{\alpha=1}^{3} (v_j \Lambda^{\boldsymbol{S}}_{\alpha j}) \Lambda^{\boldsymbol{S}}_{\alpha i} = \left(\sum_{\alpha=1}^{3} \Lambda^{\boldsymbol{S}}_{\alpha i} \Lambda^{\boldsymbol{S}}_{\alpha j} \right) v_j.$$

Substituting in $v_j = \delta_{ij} v_i$ and rearranging gives

$$\left[\sum_{\alpha=1}^{3} \Lambda^{\boldsymbol{S}}_{\alpha i} \Lambda^{\boldsymbol{S}}_{\alpha j} - \delta_{ij} \right] v_j = 0.$$

This has to be true for all $\boldsymbol{v}$ and therefore Eqn. (2.80) is proved. □

The principal coordinate system is important because $\boldsymbol{S}$ has a particularly simple form in its principal basis. Using Eqn. (2.40), the components of $\boldsymbol{S}$ in the principal coordinate system are obtained as follows:

$$S_{\alpha\beta} = \boldsymbol{e}_\alpha \cdot (\boldsymbol{S}\boldsymbol{e}_\beta) = \boldsymbol{\Lambda}^{\boldsymbol{S}}_\alpha \cdot (\boldsymbol{S}\boldsymbol{\Lambda}^{\boldsymbol{S}}_\beta) = \boldsymbol{\Lambda}^{\boldsymbol{S}}_\alpha \cdot (\lambda^{\boldsymbol{S}}_\beta \boldsymbol{\Lambda}^{\boldsymbol{S}}_\beta) = \lambda^{\boldsymbol{S}}_\beta (\boldsymbol{\Lambda}^{\boldsymbol{S}}_\alpha \cdot \boldsymbol{\Lambda}^{\boldsymbol{S}}_\beta) = \lambda^{\boldsymbol{S}}_\beta \delta_{\alpha\beta} \qquad \text{(no sum)},$$

where we have used the eigen equation and the orthogonality of the eigenvectors. We have shown that in its principal coordinate system $\boldsymbol{S}$ is *diagonal* with components equal to its principal values:

$$[\boldsymbol{S}] = \begin{bmatrix} \lambda^{\boldsymbol{S}}_1 & 0 & 0 \\ 0 & \lambda^{\boldsymbol{S}}_2 & 0 \\ 0 & 0 & \lambda^{\boldsymbol{S}}_3 \end{bmatrix}.$$

This means that any symmetric tensor $\boldsymbol{S}$ may be represented as

$$\boldsymbol{S} = \sum_{\alpha=1}^{3} \lambda^{\boldsymbol{S}}_\alpha \boldsymbol{\Lambda}^{\boldsymbol{S}}_\alpha \otimes \boldsymbol{\Lambda}^{\boldsymbol{S}}_\alpha. \tag{2.81}$$

This is called the *spectral decomposition* of $\boldsymbol{S}$. The invariants of $\boldsymbol{S}$ given in Eqns. (2.75)–(2.77) take on a particularly simple form in the principal coordinate system:

$$I^{\boldsymbol{S}}_1 = \lambda^{\boldsymbol{S}}_1 + \lambda^{\boldsymbol{S}}_2 + \lambda^{\boldsymbol{S}}_3, \quad I^{\boldsymbol{S}}_2 = \lambda^{\boldsymbol{S}}_1\lambda^{\boldsymbol{S}}_2 + \lambda^{\boldsymbol{S}}_2\lambda^{\boldsymbol{S}}_3 + \lambda^{\boldsymbol{S}}_3\lambda^{\boldsymbol{S}}_1, \quad I^{\boldsymbol{S}}_3 = \lambda^{\boldsymbol{S}}_1\lambda^{\boldsymbol{S}}_2\lambda^{\boldsymbol{S}}_3. \tag{2.82}$$

2.5.4 Cayley–Hamilton theorem

The Cayley–Hamilton theorem states that any second-order tensor $\boldsymbol{T}$ on $\mathbb{R}^3$ satisfies its own characteristic equation:[41]

$$-\boldsymbol{T}^3 + I^{\boldsymbol{T}}_1 \boldsymbol{T}^2 - I^{\boldsymbol{T}}_2 \boldsymbol{T} + I^{\boldsymbol{T}}_3 \boldsymbol{I} = \boldsymbol{0}, \tag{2.83}$$

or in indicial form

$$-T_{im}T_{mn}T_{nj} + I^{\boldsymbol{T}}_1 T_{im}T_{mj} - I^{\boldsymbol{T}}_2 T_{ij} + I^{\boldsymbol{T}}_3 \delta_{ij} = 0.$$

[41] More generally the Cayley–Hamilton theorem holds for second-order tensors on $\mathbb{R}^{n_d}$ for any n_d.

A general proof of the Cayley–Hamilton theorem is quite lengthy. However, for the case of a symmetric tensor $\boldsymbol{S}$ one can easily obtain the following.

Proof Taking the spectral decomposition of $\boldsymbol{S}$, Eqn. (2.81) (where the $\boldsymbol{\Lambda}_\alpha^{\boldsymbol{S}}$ are chosen orthonormal), and substituting into Eqn. (2.83) we find

$$\sum_{\alpha=1}^{3}\left[-(\lambda_\alpha^{\boldsymbol{S}})^3 + I_1^{\boldsymbol{S}}(\lambda_\alpha^{\boldsymbol{S}})^2 - I_2^{\boldsymbol{S}}\lambda_\alpha^{\boldsymbol{S}} + I_3^{\boldsymbol{S}}\right]\boldsymbol{\Lambda}_\alpha^{\boldsymbol{S}}\otimes\boldsymbol{\Lambda}_\alpha^{\boldsymbol{S}} = \boldsymbol{0}. \tag{2.84}$$

The scalar term in square brackets is observed to be identically zero by the definition of the eigenvalues of $\boldsymbol{S}$ (see Eqn. (2.74)). □

The main consequence of the Cayley–Hamilton theorem is that a second-order tensor $\boldsymbol{T}$ raised to the power $n \geq 3$ can be expressed in terms of $\boldsymbol{I}$, $\boldsymbol{T}$, $\boldsymbol{T}^2$ with coefficients that depend only on $I_1^{\boldsymbol{T}}$, $I_2^{\boldsymbol{T}}$, $I_3^{\boldsymbol{T}}$. For example, $\boldsymbol{T}^3$ follows immediately from Eqn. (2.83):

$$\boldsymbol{T}^3 = I_1^{\boldsymbol{T}}\boldsymbol{T}^2 - I_2^{\boldsymbol{T}}\boldsymbol{T} + I_3^{\boldsymbol{T}}\boldsymbol{I}.$$

To get $\boldsymbol{T}^4$, multiply the above by $\boldsymbol{T}$ and then substitute $\boldsymbol{T}^3$ into the right-hand side:

$$\begin{aligned}\boldsymbol{T}^4 &= I_1^{\boldsymbol{T}}\boldsymbol{T}^3 - I_2^{\boldsymbol{T}}\boldsymbol{T}^2 + I_3^{\boldsymbol{T}}\boldsymbol{T},\\ &= ((I_1^{\boldsymbol{T}})^2 - I_2^{\boldsymbol{T}})\boldsymbol{T}^2 + (I_3^{\boldsymbol{T}} - I_1^{\boldsymbol{T}}I_2^{\boldsymbol{T}})\boldsymbol{T} + I_1^{\boldsymbol{T}}I_3^{\boldsymbol{T}}\boldsymbol{I}.\end{aligned}$$

An expression for any higher power of $\boldsymbol{T}$ can be obtained in the same manner.

2.5.5 The quadratic form of symmetric second-order tensors

A scalar functional form that often comes up with the application of tensors is the *quadratic form* $Q(\boldsymbol{x})$ associated with symmetric second-order tensors:

$$Q(\boldsymbol{x}) \equiv S_{ij}x_i x_j.$$

Special terminology is used to describe $\boldsymbol{S}$ if something definitive can be said about the sign of $Q(\boldsymbol{x})$, regardless of the choice of $\boldsymbol{x}$:

$$Q(\boldsymbol{x})\begin{cases} > 0 \quad \forall \boldsymbol{x}\in\mathbb{R}^{n_d}, \boldsymbol{x}\neq\boldsymbol{0} & \boldsymbol{S} \text{ is } \textit{positive definite},\\ \geq 0 \quad \forall \boldsymbol{x}\in\mathbb{R}^{n_d}, \boldsymbol{x}\neq\boldsymbol{0} & \boldsymbol{S} \text{ is } \textit{positive semi-definite},\\ < 0 \quad \forall \boldsymbol{x}\in\mathbb{R}^{n_d}, \boldsymbol{x}\neq\boldsymbol{0} & \boldsymbol{S} \text{ is } \textit{negative definite},\\ \leq 0 \quad \forall \boldsymbol{x}\in\mathbb{R}^{n_d}, \boldsymbol{x}\neq\boldsymbol{0} & \boldsymbol{S} \text{ is } \textit{negative semi-definite}.\end{cases}$$

Of these, positive definiteness will be the most important to us. A useful theorem states that $\boldsymbol{S}$ is positive definite if and only if all of its eigenvalues are positive (i.e. $\lambda_\alpha^{\boldsymbol{S}} > 0, \forall\alpha$).

Proof Write the quadratic form of $\boldsymbol{S}$ in its principal coordinate system:

$$Q(\boldsymbol{x}) = S_{\alpha\beta}x_\alpha x_\beta = \sum_{\gamma=1}^{n_d}\lambda_\gamma^{\boldsymbol{S}}(x_\gamma)^2.$$

This will be greater than zero for any $\boldsymbol{x}\neq\boldsymbol{0}$ provided that all $\lambda_\gamma^{\boldsymbol{S}} > 0$. □

The term "positive definite" is a generalization of the concept of positivity in scalars to second-order tensors. For example, just like a positive real number has a square root, so does a positive-definite tensor. Thus, if $\boldsymbol{S}$ is a symmetric positive-definite tensor we can always define a square root $\boldsymbol{R}$ of $\boldsymbol{S}$, such that $\boldsymbol{R}^2 = \boldsymbol{S}$. This is readily shown in the principal coordinate system of $\boldsymbol{S}$, where $\boldsymbol{R}$ can be expressed in terms of its spectral decomposition. For example, for $n_d = 3$,

$$\boldsymbol{R} \equiv \sum_{\alpha=1}^{3} \sqrt{\lambda_\alpha^{\boldsymbol{S}}} \left(\boldsymbol{\Lambda}_\alpha^{\boldsymbol{S}} \otimes \boldsymbol{\Lambda}_\alpha^{\boldsymbol{S}} \right).$$

We see from the definition of $\boldsymbol{R}$ that it has the same eigenvectors as $\boldsymbol{S}$, but its eigenvalues are the square roots of those of $\boldsymbol{S}$. This means that both $\boldsymbol{S}$ and $\boldsymbol{R}$ have the same principal coordinate system. In this system the components of $\boldsymbol{R}$ are:

$$[\boldsymbol{R}] = \begin{bmatrix} \sqrt{\lambda_1^{\boldsymbol{S}}} & 0 & 0 \\ 0 & \sqrt{\lambda_2^{\boldsymbol{S}}} & 0 \\ 0 & 0 & \sqrt{\lambda_3^{\boldsymbol{S}}} \end{bmatrix}.$$

From this it is obvious that $\boldsymbol{R}^2 = \boldsymbol{S}$. It is important to point out that the square root $\boldsymbol{R}$ is not unique, since each term $\sqrt{\lambda_i^{\boldsymbol{S}}}$ could be replaced with $-\sqrt{\lambda_i^{\boldsymbol{S}}}$ in the above definition. There are, in fact, 2^{n_d} possible expressions for $\boldsymbol{R}$, where n_d is the dimensionality of space. However, only one of these choices is positive definite (i.e. the one where all terms on the diagonal are greater than zero). We can therefore say that every positive-definite tensor has a unique positive-definite square root.

The quadratic form provides a geometrical interpretation for the eigenvalues and eigenvectors of a symmetric second-order tensor. To see this let us compute the extremal values of $Q(\boldsymbol{x}) = S_{ij}x_i x_j$, subject to the constraint $\|\boldsymbol{x}\| = 1$. To do so we introduce a modified quadratic form:

$$\tilde{Q}(\boldsymbol{x}) = S_{ij}x_i x_j - \mu(x_i x_i - 1),$$

where μ is a Lagrange multiplier. Extremal values are then associated with the solutions to the condition $\partial\tilde{Q}/\partial\boldsymbol{x} = \boldsymbol{0}$:

$$\frac{\partial \tilde{Q}}{\partial x_k} = S_{kj}x_j + S_{ik}x_i - 2\mu x_k = 0.$$

Making use of the symmetry of $\boldsymbol{S}$, this reduces to the eigen equation $\boldsymbol{S}\boldsymbol{x} = \mu\boldsymbol{x}$. We have shown that the extremal directions of $Q(\boldsymbol{x})$ are the eigenvectors of $\boldsymbol{S}$ and the corresponding Lagrange multipliers are its eigenvalues! The physical significance of the eigenvalues becomes apparent when we evaluate the quadratic form in the extremal directions:

$$Q(\boldsymbol{\Lambda}_\alpha^{\boldsymbol{S}}) = S_{ij}\,\Lambda_{\alpha i}^{\boldsymbol{S}}\Lambda_{\alpha j}^{\boldsymbol{S}} = \lambda_\alpha^{\boldsymbol{S}}\Lambda_{\alpha i}^{\boldsymbol{S}}\Lambda_{\alpha i}^{\boldsymbol{S}} = \lambda_\alpha^{\boldsymbol{S}} \qquad \text{(no sum on } \alpha\text{)},$$

where we have used the eigen equation and the fact that eigenvectors are normalized. We see that the eigenvalues are the extremal values associated with the extremal directions. Geometrically, we understand this by noting that $Q(\boldsymbol{x}) = S_{ij}x_i x_j$ represents an ellipsoid. The three eigenvectors point along the ellipsoid's primary axes and the three eigenvalues are the axes half-lengths.

2.5.6 Isotropic tensors

An *isotropic* tensor is a tensor whose components are unchanged by coordinate transformation.[42] For example, a second-order isotropic tensor must satisfy $T'_{ij} = T_{ij}$, where the primed and unprimed components refer to any two coordinate system bases. Substituting for T'_{ij} using Eqn. (2.42), we can write this requirement in mathematical form as

$$Q_{\alpha i} Q_{\beta j} T_{\alpha\beta} = T_{ij}, \quad \forall \mathbf{Q} \in SO(3).$$

This expression constitutes a constraint on the components of $\boldsymbol{T}$. Isotropy is important for constitutive relations where material symmetry implies that certain tensors must be isotropic (see Section 6.4). Let us explore the constraints imposed on the form of tensors of different rank by isotropy.

Zeroth-order tensors All zeroth-order tensors (scalar invariants) are isotropic.

Proof The proof is trivial since by definition for any scalar invariant s, $s = s'$. □

First-order tensors The only isotropic first-order tensor (vector) is the zero vector.

Proof We require,

$$v_i = Q_{\alpha i} v_\alpha, \quad \forall \mathbf{Q} \in SO(3). \tag{2.85}$$

This must be true for all $\mathbf{Q} \in SO(3)$, so in particular it has to be true for the following choice:

$$\mathbf{Q} = \begin{bmatrix} -1 & 0 & 0 \\ 0 & -1 & 0 \\ 0 & 0 & 1 \end{bmatrix}. \tag{2.86}$$

Substituting Eqn. (2.86) into Eqn. (2.85) gives $v_1 = -v_1$ and $v_2 = -v_2$, so we must have $v_1 = v_2 = 0$. We prove that $v_3 = 0$ by using either

$$\mathbf{Q} = \begin{bmatrix} -1 & 0 & 0 \\ 0 & 1 & 0 \\ 0 & 0 & -1 \end{bmatrix} \quad \text{or} \quad \mathbf{Q} = \begin{bmatrix} 1 & 0 & 0 \\ 0 & -1 & 0 \\ 0 & 0 & -1 \end{bmatrix}.$$

□

Second-order tensors All isotropic second-order tensors are proportional to the identity tensor $\boldsymbol{I}$.

Proof We require

$$T_{ij} = Q_{\alpha i} Q_{\beta j} T_{\alpha\beta}, \quad \forall \mathbf{Q} \in SO(3).$$

42 Technically for a tensor to be isotropic it must be invariant with respect to improper as well as proper orthogonal transformations. In other words, it must be unaffected by reflection as well as rotation. If a tensor is only invariant with respect to proper orthogonal transformations (rotations) it is called *hemitropic*. This distinction is only important for tensors of odd rank that can be hemitropic but not isotropic. Here we limit ourselves to proper orthogonal transformations that retain the handedness of the basis, but still use the terminology *isotropic*.

Using the following special choices for $\mathbf{Q}$,

$$\mathbf{Q} = \begin{bmatrix} 0 & 0 & -1 \\ -1 & 0 & 0 \\ 0 & 1 & 0 \end{bmatrix} \quad \text{and} \quad \mathbf{Q} = \begin{bmatrix} 0 & 0 & -1 \\ 1 & 0 & 0 \\ 0 & -1 & 0 \end{bmatrix},$$

we find that

$$\begin{bmatrix} T_{11} & T_{12} & T_{13} \\ T_{21} & T_{22} & T_{23} \\ T_{31} & T_{32} & T_{33} \end{bmatrix} = \begin{bmatrix} T_{22} & -T_{23} & T_{21} \\ -T_{32} & T_{33} & -T_{31} \\ T_{12} & -T_{13} & T_{11} \end{bmatrix} = \begin{bmatrix} T_{22} & -T_{23} & -T_{21} \\ -T_{32} & T_{33} & T_{31} \\ -T_{12} & T_{13} & T_{11} \end{bmatrix}.$$

Carefully examining these relations, we see that $T_{11} = T_{22} = T_{33}$ and that $T_{ij} = 0$, $\forall i \neq j$. In other words, we have proven that $T_{ij} = \alpha\delta_{ij}$, where α is any constant. No further restrictions on α are obtained by considering any of the remaining elements of $SO(3)$. □

Third-order tensors All isotropic third-order tensors are proportional to the permutation symbol:[43]

$$B_{ijk} = \beta\epsilon_{ijk}, \quad \beta \in \mathbb{R}.$$

In the interest of brevity we do not give the proof. For a proof, see, for example, [Jau67].

Fourth-order tensors All isotropic fourth-order tensors can be written in the following form:

$$C_{ijkl} = \alpha\delta_{ij}\delta_{kl} + \beta\delta_{ik}\delta_{jl} + \gamma\delta_{il}\delta_{jk},$$

where $\alpha, \beta, \gamma \in \mathbb{R}$ are constants. For a proof, see, for example, [Jau67]. The general theory for systematically obtaining such relations is known as *group representation theory* (see, for example, [JB67, Mil72, McW02]).

2.6 Tensor fields

The previous sections have discussed the definition and properties of tensors as discrete entities. In continuum mechanics, we most often encounter tensors as spatially and temporally varying fields over a given domain. For example, consider a (one-dimensional) rubber band that is tied to a rigid fixed wall at one end and pulled at a constant velocity v_{end} at the other. Clearly different points along the rubber band will experience different velocities ranging from zero at the support to v_{end} at the end whose position is changing with time. Consequently, the velocity in the rubber band is[44]

$$v(x,t) = \frac{x}{\ell(t)} v_{\text{end}}, \quad x \in [0, \ell(t)],$$

43 As noted earlier, the correct terminology for third-order tensors is *hemitropic*.

44 We assume that the rubber band is being stretched uniformly. In reality, the velocity distribution along the rubber band may not be linear.

where $\ell(t)$ is the length of the rubber band at time t. In this example, the rubber band is a one-dimensional structure and therefore the spatial dependence of the velocity is on the scalar x. For three-dimensional objects, a tensor field $\boldsymbol{T}$ defined over a domain Ω is a function[45] of the position vector $\boldsymbol{x} = x_i \boldsymbol{e}_i$ of points inside Ω:

$$\boldsymbol{T} = \boldsymbol{T}(\boldsymbol{x}, t) = \boldsymbol{T}(x_1, x_2, x_3, t), \quad \boldsymbol{x} \in \Omega(t).$$

Once we have accepted the concept of tensor fields, we can consider differentiation and integration of tensors. First, we focus our attention on the Cartesian coordinate system and introduce the differential operators in that context. In Section 2.6.3, we extend the discussion briefly into curvilinear coordinates, but only so far as to obtain the essential curvilinear results that we will need later in this book.

2.6.1 Partial differentiation of a tensor field

The partial differentiation of tensor fields with respect to their spatial arguments is readily expressed in component form:[46]

$$\frac{\partial s(\boldsymbol{x})}{\partial x_i}, \quad \frac{\partial v_i(\boldsymbol{x})}{\partial x_j}, \quad \frac{\partial T_{ij}(\boldsymbol{x})}{\partial x_k},$$

for a scalar s, vector $\boldsymbol{v}$ and second-order tensor $\boldsymbol{T}$. To simplify this notation and make it compatible with indicial notation, we introduce the *comma notation* for differentiation with respect to x_i:

$$(\cdot)_{,i} \equiv \frac{\partial(\cdot)}{\partial x_i}.$$

In this notation, the three expressions above are $s_{,i}$, $v_{i,j}$ and $T_{ij,k}$. Higher-order differentiation follows as expected: $\partial^2 s/(\partial x_i \partial x_j) = s_{,ij}$. The comma notation works in concert with the summation convention, e.g. $s_{,ii} = s_{,11} + s_{,22} + s_{,33}$ and $v_{i,i} = v_{1,1} + v_{2,2} + v_{3,3}$.

Example 2.3 (Using the comma notation for derivatives) Several examples are:

1. $x_{i,j} = \partial x_i / \partial x_j = \delta_{ij}$.
2. $(A_{ij} x_j)_{,i} = A_{ij} x_{j,i} = A_{ij} \delta_{ji} = A_{ii}$. (Here $\boldsymbol{A}$ is a constant.)
3. $(T_{ij}(\boldsymbol{x}) x_j)_{,i} = T_{ij,i} x_j + T_{ij} \delta_{ji} = T_{ij,i} x_j + T_{ii}$.

2.6.2 Differential operators in Cartesian coordinates

Four important differential operators are the *gradient*, *curl*, *divergence* and *Laplacian*. These operators involve derivatives of a tensor field with respect to its *vector* argument.

[45] Technically, when $\boldsymbol{T}$ is written as a function of components a different symbol should be used, e.g. $\boldsymbol{T} = \boldsymbol{T}(\boldsymbol{x}, t) = \breve{\boldsymbol{T}}(x_1, x_2, x_3, t)$, since the functional form is different. Here we use the same symbol for notational simplicity.

[46] Differentiation with respect to time is more subtle and will be discussed in Section 3.6.

This requires a generalization of the definition of a derivative. For a scalar function $s(r)$ of a scalar argument ($r \in \mathbb{R}$), we have

$$\frac{ds}{dr} \equiv \lim_{\epsilon \to 0} \frac{s(r+\epsilon) - s(r)}{\epsilon}.$$

For a scalar function $s(\boldsymbol{x})$ of a vector argument ($\boldsymbol{x} \in \mathbb{R}^3$), we define the derivative with respect to $\boldsymbol{x}$ through its role in computing the derivative in a given direction. The derivative of $s(\boldsymbol{x})$ in the direction of the vector $\boldsymbol{u}$ at point $\boldsymbol{x}_0$ is defined as

$$\langle \mathcal{D}_{\boldsymbol{x}} s(\boldsymbol{x}_0); \boldsymbol{u} \rangle \equiv \lim_{\eta \to 0} \frac{s(\boldsymbol{x}_0 + \eta \boldsymbol{u}) - s(\boldsymbol{x}_0)}{\eta} = \left. \frac{d}{d\eta} s(\boldsymbol{x}_0 + \eta \boldsymbol{u}) \right|_{\eta=0}, \tag{2.87}$$

where $\eta \in \mathbb{R}$. If $\boldsymbol{u}$ is a unit vector (i.e. $\|\boldsymbol{u}\| = 1$), then $\langle \mathcal{D}_{\boldsymbol{x}} s(\boldsymbol{x}_0); \boldsymbol{u} \rangle$ is called the *directional derivative* of s in direction $\boldsymbol{u}$. When this is not the case, we will use the term "nonnormalized directional derivative."[47]

Gradient To define the gradient, we introduce $\boldsymbol{x} = \boldsymbol{x}_0 + \eta \boldsymbol{u}$, and formally write

$$\langle \mathcal{D}_{\boldsymbol{x}} s(\boldsymbol{x}_0); \boldsymbol{u} \rangle = \left. \frac{d}{d\eta} s(\boldsymbol{x}(\eta)) \right|_{\eta=0} = \left. \frac{\partial s}{\partial \boldsymbol{x}} \cdot \frac{d\boldsymbol{x}}{d\eta} \right|_{\eta=0} = \frac{\partial s}{\partial \boldsymbol{x}} \cdot \boldsymbol{u},$$

where the chain rule was used. We call $\partial s / \partial \boldsymbol{x}$ the *gradient* of s and denote it by ∇s (or grad s). The gradient is thus defined by the relation

$$\langle \mathcal{D}_{\boldsymbol{x}} s(\boldsymbol{x}_0); \boldsymbol{u} \rangle = \nabla s \cdot \boldsymbol{u}. \tag{2.88}$$

Physically, the gradient provides the direction and magnitude of the maximum rate of increase of $s(\boldsymbol{x})$. The following example shows how the definition in Eqn. (2.88) can be used in practice to compute a gradient.

Example 2.4 (Computing a gradient) Let $s(\boldsymbol{x}) = \boldsymbol{A}\boldsymbol{x} \cdot \boldsymbol{x}$, where $\boldsymbol{A}$ is a constant second-order tensor. The nonnormalized directional derivative of s is

$$\begin{aligned}
\langle \mathcal{D}_{\boldsymbol{x}} s; \boldsymbol{u} \rangle &= \left. \frac{d}{d\eta} \left[\boldsymbol{A}(\boldsymbol{x} + \eta \boldsymbol{u}) \cdot (\boldsymbol{x} + \eta \boldsymbol{u}) \right] \right|_{\eta=0} \\
&= \left. \frac{d}{d\eta} \left[\boldsymbol{A}\boldsymbol{x} \cdot \boldsymbol{x} + \eta (\boldsymbol{A}\boldsymbol{x} \cdot \boldsymbol{u} + \boldsymbol{A}\boldsymbol{u} \cdot \boldsymbol{x}) + \eta^2 \boldsymbol{A}\boldsymbol{u} \cdot \boldsymbol{u} \right] \right|_{\eta=0} \\
&= \boldsymbol{A}\boldsymbol{x} \cdot \boldsymbol{u} + \boldsymbol{A}\boldsymbol{u} \cdot \boldsymbol{x} \\
&= (\boldsymbol{A}\boldsymbol{x} + \boldsymbol{A}^T \boldsymbol{x}) \cdot \boldsymbol{u}.
\end{aligned}$$

Comparing the above expression with Eqn. (2.88), we see that the gradient is

$$\nabla s = \boldsymbol{A}\boldsymbol{x} + \boldsymbol{A}^T \boldsymbol{x}.$$

47 The subscript $\boldsymbol{x}$ in $\langle \mathcal{D}_{\boldsymbol{x}} \cdot ; \cdot \rangle$ is included to explicitly indicate the independent quantity with respect to which the derivative is being taken. Here, the only choice is $\boldsymbol{x}$, but later (such as in Section 3.5) more options will be available.

The component form of ∇s relative to an orthonormal basis is obtained by rewriting $s(\boldsymbol{x})$ as a function of the components of $\boldsymbol{x}$, $s = s(x_1, x_2, x_3)$. Therefore

$$\langle \mathcal{D}_{\boldsymbol{x}} s(\boldsymbol{x}_0); \boldsymbol{u} \rangle = \left. \frac{ds}{d\eta} \right|_{\eta=0} = \left. \frac{\partial s}{\partial x_i} \frac{dx_i}{d\eta} \right|_{\eta=0} = \frac{\partial s}{\partial x_i} u_i,$$

where we have used $x_i = x_{0i} + \eta u_i$. Comparing this with Eqn. (2.88), we see that $[\nabla s]_i = \partial s / \partial x_i$, therefore

$$\nabla s = \frac{\partial s(\boldsymbol{x})}{\partial x_i} \boldsymbol{e}_i. \tag{2.89}$$

The gradient of a scalar field is a vector[48] (see Exercise 2.16). The definition of the gradient can be generalized to a tensor field $\boldsymbol{B}(\boldsymbol{x})$ of rank $m \geq 1$:

$$\nabla \boldsymbol{B} = \frac{\partial \boldsymbol{B}(\boldsymbol{x})}{\partial x_i} \otimes \boldsymbol{e}_i. \tag{2.90}$$

For example, for a vector $\boldsymbol{v}$ and second-order tensor $\boldsymbol{T}$:[49]

$$\nabla \boldsymbol{v} = \frac{\partial \boldsymbol{v}}{\partial x_j} \otimes \boldsymbol{e}_j = \frac{\partial (v_i \boldsymbol{e}_i)}{\partial x_j} \otimes \boldsymbol{e}_j = \frac{\partial v_i}{\partial x_j} (\boldsymbol{e}_i \otimes \boldsymbol{e}_j),$$

$$\nabla \boldsymbol{T} = \frac{\partial \boldsymbol{T}}{\partial x_k} \otimes \boldsymbol{e}_k = \frac{\partial [T_{ij} (\boldsymbol{e}_i \otimes \boldsymbol{e}_j)]}{\partial x_k} \otimes \boldsymbol{e}_k = \frac{\partial T_{ij}}{\partial x_k} (\boldsymbol{e}_i \otimes \boldsymbol{e}_j \otimes \boldsymbol{e}_k).$$

We see that the gradient operation increases the rank of the tensor by 1; $[\nabla \boldsymbol{v}]_{ij} = v_{i,j}$ are the components of a second-order tensor, and $[\nabla \boldsymbol{T}]_{ijk} = T_{ij,k}$ are the components of a third-order tensor.

Curl The *curl* of a tensor field $\boldsymbol{B}(\boldsymbol{x})$ of rank $m \geq 1$ is a tensor of the same rank denoted by curl $\boldsymbol{B}$. It is defined [Rub00]:

$$\operatorname{curl} \boldsymbol{B} \equiv -\frac{\partial \boldsymbol{B}(\boldsymbol{x})}{\partial x_i} \times \boldsymbol{e}_i. \tag{2.91}$$

[48] Actually, it is a vector field. We will often use the terms vector and vector field (and similarly tensor and tensor field) interchangeably, where the appropriate meaning is clear from the context.

[49] It is important to point out that a great deal of confusion exists in the continuum mechanics literature regarding the direct notation for differential operators. The notation we introduce here for the grad, curl and div operations is based on a linear algebraic view of tensor analysis. The same operations are often defined differently in other books. The confusion arises when the operations are applied to tensors of rank one and higher, where different definitions lead to different components being involved in the operation. For example, another popular notation for tensor calculus is based on the del differential operator, $\boldsymbol{\nabla} \equiv \boldsymbol{e}_i \partial / \partial x_i$. In this notation, the gradient, curl and divergence are denoted by $\boldsymbol{\nabla}\square$, $\boldsymbol{\nabla} \times \square$ and $\boldsymbol{\nabla} \cdot \square$. This notation is self-consistent; however, it is not equivalent to the notation used in this book. For example, according to this notation the gradient of a vector $\boldsymbol{v}$ is $\boldsymbol{\nabla v} = v_{j,i} \boldsymbol{e}_i \otimes \boldsymbol{e}_j$, which is the transpose of our definition. In our notation we retain an unbolded ∇ symbol for the gradient, but do not view it as a differential operator. Instead, we adopt the definition in the text which leads to the untransposed expression, $\nabla \boldsymbol{v} = v_{i,j} \boldsymbol{e}_i \otimes \boldsymbol{e}_j$. We will use the notation introduced here consistently throughout the book; however, the reader is warned to read the definitions carefully in other books or articles.

For example, for a vector $\boldsymbol{v}$

$$\operatorname{curl} \boldsymbol{v} = -\frac{\partial \boldsymbol{v}}{\partial x_j} \times \boldsymbol{e}_j = -\frac{\partial (v_i \boldsymbol{e}_i)}{\partial x_j} \times \boldsymbol{e}_j = -\frac{\partial v_i}{\partial x_j}(\boldsymbol{e}_i \times \boldsymbol{e}_j) = -\epsilon_{ijk}\frac{\partial v_i}{\partial x_j}\boldsymbol{e}_k = \epsilon_{kji}\frac{\partial v_i}{\partial x_j}\boldsymbol{e}_k,$$

where we have used Eqn. (2.28). The curl of a vector can alternatively be defined through the relation [Gur81]

$$(\nabla \boldsymbol{v} - \nabla \boldsymbol{v}^T)\boldsymbol{a} = (\operatorname{curl} \boldsymbol{v}) \times \boldsymbol{a}, \qquad \forall \boldsymbol{a} \in \mathbb{R}^3.$$

This definition implies that $\operatorname{curl} \boldsymbol{v}$ is the axial vector of the antisymmetric tensor $(\nabla \boldsymbol{v} - \nabla \boldsymbol{v}^T)$ (see Eqn. (2.70)). Therefore from Eqn. (2.71), we have

$$[\operatorname{curl} \boldsymbol{v}]_k = -\frac{1}{2}\epsilon_{ijk}(v_{i,j} - v_{j,i}) = -\frac{1}{2}\epsilon_{ijk}v_{i,j} + \frac{1}{2}\epsilon_{ijk}v_{j,i} = -\epsilon_{ijk}v_{i,j} = \epsilon_{kji}v_{i,j},$$

which is the same as the definition given above.

The curl of a vector field is related to the local rate of rotation of the field. It plays an important role in fluid dynamics where it characterizes the vorticity or spin of the flow (see Section 3.6). The definition of a curl can be extended to higher-order tensors; see, for example, [CG01].

Divergence The *divergence* of a tensor field $\boldsymbol{B}(\boldsymbol{x})$ of rank $m \geq 1$ is a tensor of rank $m - 1$ denoted by $\operatorname{div} \boldsymbol{B}$. The expressions for the divergence of a vector $\boldsymbol{v}$ and tensor $\boldsymbol{B}(\boldsymbol{x})$ of rank $m \geq 2$ are

$$\operatorname{div} \boldsymbol{v} \equiv \frac{\partial \boldsymbol{v}(\boldsymbol{x})}{\partial x_i} \cdot \boldsymbol{e}_i \quad \text{and} \quad \operatorname{div} \boldsymbol{B} \equiv \frac{\partial \boldsymbol{B}(\boldsymbol{x})}{\partial x_i}\boldsymbol{e}_i. \tag{2.92}$$

For example, for a vector $\boldsymbol{v}$ and second-order tensor $\boldsymbol{T}$

$$\operatorname{div} \boldsymbol{v} = \frac{\partial \boldsymbol{v}}{\partial x_j} \cdot \boldsymbol{e}_j = \frac{\partial (v_i \boldsymbol{e}_i)}{\partial x_j} \cdot \boldsymbol{e}_j = \frac{\partial v_i}{\partial x_j}(\boldsymbol{e}_i \cdot \boldsymbol{e}_j) = \frac{\partial v_i}{\partial x_j}\delta_{ij} = \frac{\partial v_i}{\partial x_i},$$

$$\operatorname{div} \boldsymbol{T} = \frac{\partial \boldsymbol{T}}{\partial x_k}\boldsymbol{e}_k = \frac{\partial [T_{ij}(\boldsymbol{e}_i \otimes \boldsymbol{e}_j)]}{\partial x_k}\boldsymbol{e}_k = \frac{\partial T_{ij}}{\partial x_k}(\boldsymbol{e}_i \otimes \boldsymbol{e}_j)\boldsymbol{e}_k = \frac{\partial T_{ij}}{\partial x_k}\boldsymbol{e}_i\delta_{jk} = \frac{\partial T_{ij}}{\partial x_j}\boldsymbol{e}_i,$$

where in the second expression we have used Eqn. (2.48). We see that the divergence of a vector is a scalar invariant, $\operatorname{div} \boldsymbol{v} = v_{i,i}$, and the divergence of a second-order tensor is a vector, $[\operatorname{div} \boldsymbol{T}]_i = T_{ij,j}$. In instances where the divergence is taken with respect to an argument other than $\boldsymbol{x}$ it will be denoted by a subscript. For example, the divergence with respect to $\boldsymbol{y}$ of a tensor $\boldsymbol{T}$ is denoted $\operatorname{div}_y \boldsymbol{T}$.

Two useful identities for the divergence of a vector and a second-order tensor that can also serve as definitions for these operations are [Gur81]

$$\operatorname{div} \boldsymbol{v} = \operatorname{tr} \nabla \boldsymbol{v}, \qquad \operatorname{div} \boldsymbol{T} \cdot \boldsymbol{a} = \operatorname{div}(\boldsymbol{T}^T \boldsymbol{a}),$$

where $\boldsymbol{a} \in \mathbb{R}^3$ is any constant vector.

The divergence of a tensor field is related to the net flow of the field per unit volume at a given point. This will be demonstrated in the next section where we discuss the divergence theorem.

Laplacian The *Laplacian* of a scalar field $s(\boldsymbol{x})$ is a scalar denoted by $\nabla^2 s$. The Laplacian is defined by the following relation:

$$\nabla^2 s \equiv \operatorname{div} \nabla s. \tag{2.93}$$

In component form, we have

$$\nabla^2 s = \frac{\partial}{\partial x_j}\left(\frac{\partial s}{\partial x_i}\boldsymbol{e}_i\right)\cdot \boldsymbol{e}_j = \frac{\partial^2 s}{\partial x_i \partial x_j}(\boldsymbol{e}_i \cdot \boldsymbol{e}_j) = \frac{\partial^2 s}{\partial x_i \partial x_j}\delta_{ij} = \frac{\partial^2 s}{\partial x_i \partial x_i}$$
$$= s_{,ii} = s_{,11} + s_{,22} + s_{,33},$$

where we have used Eqns. (2.89), $(2.92)_1$, (2.20) and the index substitution property of the Kronecker delta.

2.6.3 Differential operators in curvilinear coordinates

Often the geometry of a domain Ω makes it mathematically advantageous to use a set of curvilinear coordinates θ^i to describe the position of points within Ω. In such systems the basis vectors with respect to which the components of tensor fields are expressed depend on the position in space (see Section 2.3.2). This is in contrast to rectilinear (and in particular Cartesian) coordinate systems (with coordinates x_i), where the basis vectors are independent of position. Although it is straightforward to develop a general theory for tensor fields defined with respect to an arbitrary curvilinear coordinate system,[50] we will need only two specific results from this theory – the gradient of a vector and the divergence of a tensor:

$$\nabla \boldsymbol{v} = \frac{\partial \boldsymbol{v}}{\partial \theta^i} \otimes \boldsymbol{g}^i, \quad \operatorname{div} \boldsymbol{B} = \frac{\partial \boldsymbol{B}}{\partial \theta^i}\boldsymbol{g}^i, \tag{2.94}$$

with

$$\boldsymbol{g}^i \cdot \boldsymbol{g}_j = \delta^i_j, \quad \text{and} \quad \boldsymbol{g}_i \equiv \frac{\partial \boldsymbol{x}}{\partial \theta^i}. \tag{2.95}$$

The vectors $\boldsymbol{g}_i$ are called the "covariant basis vectors" and describe how the point in space changes as the coordinates change. The "contravariant basis vectors" $\boldsymbol{g}^i$ describe how the coordinates change as the point in space changes. The contravariant basis vectors are bi-orthogonal (reciprocal) to the covariant basis vectors, but are generally nonorthogonal (see Section 2.3.2). Further, it is important to note that, generally, $\boldsymbol{g}^i$ and $\boldsymbol{g}_i$ are functions of θ^i. The usual sums over i are implied in Eqn. (2.94) and the two quantities on the right-hand side of Eqn. $(2.94)_2$ are combined in a tensor contraction. So, if $\boldsymbol{B}$ is a tensor of order m, then $\operatorname{div} \boldsymbol{B}$ is a tensor of order $m-1$. It is easy to verify that if we take $\theta^i = x_i$ and

[50] See, for example, [TT60].

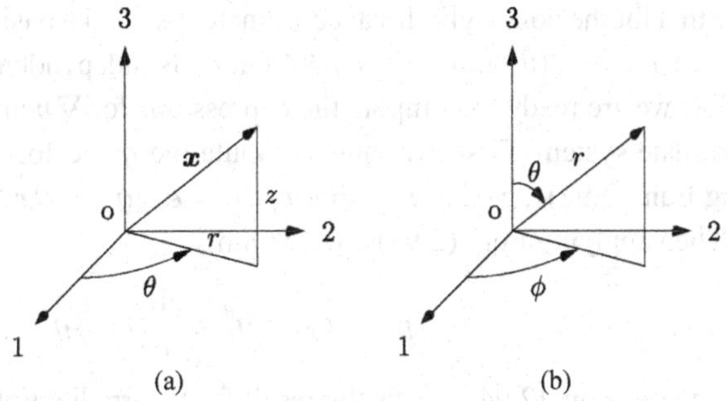

Fig. 2.5 Definitions of (a) the polar cylindrical and (b) the spherical coordinate systems.

$\boldsymbol{x} = x_i \boldsymbol{e}_i$, then Eqns. $(2.94)_1$ and $(2.94)_2$ reduce to Eqns. (2.90) and $(2.92)_2$, respectively. However, the expressions are not so simple in other coordinate systems.

Polar cylindrical coordinates This coordinate system specifies the position of points in space in terms of their distance r from a cylindrical axis, their angular orientation θ about that axis (measured relative to an arbitrary direction), and the distance z along the cylindrical axis from a chosen origin on the axis (see Fig. 2.5(a)). Thus, we have $(\theta^1, \theta^2, \theta^3) = (r, \theta, z)$. Consider a point $\boldsymbol{x}$, which in a Cartesian coordinate system has position components equal to its coordinates x_i, i.e. $\boldsymbol{x} = x_i \boldsymbol{e}_i$. In the polar cylindrical coordinate system this point will have components r, θ and z. The relationship between Cartesian and polar cylindrical coordinates is usually taken to be

$$x_1 = r\cos\theta, \quad x_2 = r\sin\theta, \quad x_3 = z. \tag{2.96}$$

The inverse relations are

$$r = (x_1^2 + x_2^2)^{1/2}, \quad \theta = \arctan(x_2/x_1), \quad z = x_3.$$

With these relations between the two coordinate systems, the point $\boldsymbol{x}$ can be written

$$\begin{aligned}\boldsymbol{x} = x_i \boldsymbol{e}_i &= x_1 \boldsymbol{e}_1 + x_2 \boldsymbol{e}_2 + x_3 \boldsymbol{e}_3 \\ &= (r\cos\theta)\boldsymbol{e}_1 + (r\sin\theta)\boldsymbol{e}_2 + z\boldsymbol{e}_3 \\ &= r(\cos\theta\boldsymbol{e}_1 + \sin\theta\boldsymbol{e}_2) + z\boldsymbol{e}_3 \\ &\equiv r\boldsymbol{e}_r + z\boldsymbol{e}_z,\end{aligned}$$

where the last line serves to define the radial and axial basis vectors, $\boldsymbol{e}_r$ and $\boldsymbol{e}_z$, respectively. The final basis vector, called the transverse basis vector, $\boldsymbol{e}_\theta$, can be obtained from the orthonormality condition and the condition that the ordered triplet $(\boldsymbol{e}_r, \boldsymbol{e}_\theta, \boldsymbol{e}_z)$ forms a right-handed system. Accordingly, we find $\boldsymbol{e}_\theta = -\sin\theta\boldsymbol{e}_1 + \cos\theta\boldsymbol{e}_2$. Thus,

$$\boldsymbol{e}_r = \cos\theta\boldsymbol{e}_1 + \sin\theta\boldsymbol{e}_2, \quad \boldsymbol{e}_\theta = -\sin\theta\boldsymbol{e}_1 + \cos\theta\boldsymbol{e}_2, \quad \boldsymbol{e}_z = \boldsymbol{e}_3. \tag{2.97}$$

Note that for the polar cylindrical coordinate system the basis vectors $\boldsymbol{e}_r$ and $\boldsymbol{e}_\theta$ are functions of θ, i.e. $\boldsymbol{e}_r = \boldsymbol{e}_r(\theta)$ and $\boldsymbol{e}_\theta = \boldsymbol{e}_\theta(\theta)$, but $\boldsymbol{e}_z$ is independent of position.

Now we are ready to compute the expressions for $\nabla \boldsymbol{v}$ and div $\boldsymbol{B}$ in the polar cylindrical coordinate system. First, we must compute the $\boldsymbol{g}^i$ vectors. Referring to Eqn. $(2.95)_2$ and using Eqn. $(2.97)_2$, we have $\boldsymbol{g}_r = \partial \boldsymbol{x}/\partial r = \boldsymbol{e}_r$, $\boldsymbol{g}_\theta = \partial \boldsymbol{x}/\partial \theta = r\boldsymbol{e}_\theta$, and $\boldsymbol{g}_z = \partial \boldsymbol{x}/\partial z = \boldsymbol{e}_z$. Then applying Eqn. $(2.95)_1$, we obtain

$$\boldsymbol{g}^r = \boldsymbol{e}_r, \quad \boldsymbol{g}^\theta = \frac{1}{r}\boldsymbol{e}_\theta, \quad \boldsymbol{g}^z = \boldsymbol{e}_z. \tag{2.98}$$

Writing out Eqn. $(2.94)_1$ gives the result for the gradient of a vector $\boldsymbol{v}$:

$$\begin{aligned}\nabla \boldsymbol{v} = {}& v_{r,r}\boldsymbol{e}_r \otimes \boldsymbol{e}_r + \frac{1}{r}(v_{r,\theta} - v_\theta)\boldsymbol{e}_r \otimes \boldsymbol{e}_\theta + v_{r,z}\boldsymbol{e}_r \otimes \boldsymbol{e}_z \\ & + v_{\theta,r}\boldsymbol{e}_\theta \otimes \boldsymbol{e}_r + \frac{1}{r}(v_{\theta,\theta} + v_r)\boldsymbol{e}_\theta \otimes \boldsymbol{e}_\theta + v_{\theta,z}\boldsymbol{e}_\theta \otimes \boldsymbol{e}_z \\ & + v_{z,r}\boldsymbol{e}_z \otimes \boldsymbol{e}_r + \frac{1}{r}v_{z,\theta}\boldsymbol{e}_z \otimes \boldsymbol{e}_\theta + v_{z,z}\boldsymbol{e}_z \otimes \boldsymbol{e}_z.\end{aligned} \tag{2.99}$$

For the divergence of a second-order tensor $\boldsymbol{T}$, we first write out Eqn. $(2.94)_2$ as

$$\operatorname{div} \boldsymbol{T} = \frac{\partial \boldsymbol{T}}{\partial r}\boldsymbol{e}_r + \frac{\partial \boldsymbol{T}}{\partial \theta}\frac{1}{r}\boldsymbol{e}_\theta + \frac{\partial \boldsymbol{T}}{\partial z}\boldsymbol{e}_z.$$

Substituting in the component expression for $\boldsymbol{T}$ (see Eqn. (2.61)) we have

$$\begin{aligned}\operatorname{div} \boldsymbol{T} = {}& \left(\frac{\partial T_{rr}}{\partial r}\boldsymbol{e}_r \otimes \boldsymbol{e}_r + \frac{\partial T_{r\theta}}{\partial r}\boldsymbol{e}_r \otimes \boldsymbol{e}_\theta + \cdots + \frac{\partial T_{zz}}{\partial r}\boldsymbol{e}_z \otimes \boldsymbol{e}_z\right. \\ & + T_{rr}\frac{\partial \boldsymbol{e}_r}{\partial r} \otimes \boldsymbol{e}_r + T_{r\theta}\frac{\partial \boldsymbol{e}_r}{\partial r} \otimes \boldsymbol{e}_\theta + \cdots + T_{zz}\frac{\partial \boldsymbol{e}_z}{\partial r} \otimes \boldsymbol{e}_z \\ & \left. + T_{rr}\boldsymbol{e}_r \otimes \frac{\partial \boldsymbol{e}_r}{\partial r} + T_{r\theta}\boldsymbol{e}_r \otimes \frac{\partial \boldsymbol{e}_\theta}{\partial r} + \cdots + T_{zz}\boldsymbol{e}_z \otimes \frac{\partial \boldsymbol{e}_z}{\partial r}\right)\boldsymbol{e}_r \\ & + \left(\frac{\partial T_{rr}}{\partial \theta}\boldsymbol{e}_r \otimes \boldsymbol{e}_r + \cdots + T_{zz}\boldsymbol{e}_z \otimes \frac{\partial \boldsymbol{e}_z}{\partial \theta}\right)\frac{1}{r}\boldsymbol{e}_\theta \\ & + \left(\frac{\partial T_{rr}}{\partial z}\boldsymbol{e}_r \otimes \boldsymbol{e}_r + \cdots + T_{zz}\boldsymbol{e}_z \otimes \frac{\partial \boldsymbol{e}_z}{\partial z}\right)\boldsymbol{e}_z.\end{aligned}$$

This equation has 81 terms. Performing the indicated differentiations and the various contractions results in the final form for the divergence in polar cylindrical coordinates:

$$\begin{aligned}\operatorname{div} \boldsymbol{T} = {}& \left(\frac{\partial T_{rr}}{\partial r} + \frac{1}{r}\frac{\partial T_{r\theta}}{\partial \theta} + \frac{T_{rr} - T_{\theta\theta}}{r} + \frac{\partial T_{rz}}{\partial z}\right)\boldsymbol{e}_r \\ & + \left(\frac{\partial T_{\theta r}}{\partial r} + \frac{1}{r}\frac{\partial T_{\theta\theta}}{\partial \theta} + \frac{T_{r\theta} + T_{\theta r}}{r} + \frac{\partial T_{\theta z}}{\partial z}\right)\boldsymbol{e}_\theta \\ & + \left(\frac{\partial T_{zr}}{\partial r} + \frac{1}{r}\frac{\partial T_{z\theta}}{\partial \theta} + \frac{T_{zr}}{r} + \frac{\partial T_{zz}}{\partial z}\right)\boldsymbol{e}_z.\end{aligned} \tag{2.100}$$

Spherical coordinates The spherical coordinate system identifies each point $\boldsymbol{x}$ in space by its distance r from a chosen origin and two angles: the inclination (or zenith) angle θ

between the position vector $\boldsymbol{x}$ and the $\boldsymbol{e}_3$ axis, and the polar (or azimuthal) angle ϕ between the $\boldsymbol{e}_1$ axis and the projection of $\boldsymbol{x}$ into the $\boldsymbol{e}_1$–$\boldsymbol{e}_2$ subspace (see Fig. 2.5(b)). Thus, we have[51] $(\theta^1, \theta^2, \theta^3) = (r, \theta, \phi)$. These are most easily understood through their relation to the Cartesian coordinates:

$$x_1 = r\sin\theta\cos\phi, \qquad x_2 = r\sin\theta\sin\phi, \qquad x_3 = r\cos\theta, \tag{2.101}$$

and the inverse relations

$$r = (x_1^2 + x_2^2 + x_3^2)^{1/2}, \quad \theta = \arccos(x_3/r), \quad \phi = \arctan(x_2/x_1).$$

The spherical coordinate basis vectors are given by

$$\begin{aligned}
\boldsymbol{e}_r &= \sin\theta\cos\phi\boldsymbol{e}_1 + \sin\theta\sin\phi\boldsymbol{e}_2 + \cos\theta\boldsymbol{e}_3,\\
\boldsymbol{e}_\theta &= \cos\theta\cos\phi\boldsymbol{e}_1 + \cos\theta\sin\phi\boldsymbol{e}_2 - \sin\theta\boldsymbol{e}_3,\\
\boldsymbol{e}_\phi &= -\sin\phi\boldsymbol{e}_1 + \cos\phi\boldsymbol{e}_2,
\end{aligned} \tag{2.102}$$

where the ordered triplet $(\boldsymbol{e}_r, \boldsymbol{e}_\theta, \boldsymbol{e}_\phi)$ forms a right-handed system. Thus, in the spherical coordinate system all three basis vectors are functions of ϕ and/or θ, and we have

$$\begin{aligned}
&\frac{\partial \boldsymbol{e}_r}{\partial\theta} = \boldsymbol{e}_\theta, && \frac{\partial \boldsymbol{e}_\theta}{\partial\theta} = -\boldsymbol{e}_r, && \frac{\partial \boldsymbol{e}_\phi}{\partial\theta} = 0,\\
&\frac{\partial \boldsymbol{e}_r}{\partial\phi} = \sin\theta\boldsymbol{e}_\phi, && \frac{\partial \boldsymbol{e}_\theta}{\partial\phi} = \cos\theta\boldsymbol{e}_\phi, && \frac{\partial \boldsymbol{e}_\phi}{\partial\phi} = -\sin\theta\boldsymbol{e}_r - \cos\theta\boldsymbol{e}_\theta.
\end{aligned}$$

From the position vector $\boldsymbol{x} = r\boldsymbol{e}_r$ and the above relations we find the vectors $\boldsymbol{g}^i$ to be

$$\boldsymbol{g}^r = \boldsymbol{e}_r, \quad \boldsymbol{g}^\theta = \frac{1}{r}\boldsymbol{e}_\theta, \quad \boldsymbol{g}^\phi = \frac{1}{r\sin\theta}\boldsymbol{e}_\phi. \tag{2.103}$$

Writing out Eqn. (2.94)$_1$ gives the gradient of a vector $\boldsymbol{v}$ in spherical coordinates:

$$\begin{aligned}
\nabla\boldsymbol{v} = {} & v_{r,r}\boldsymbol{e}_r\otimes\boldsymbol{e}_r + \frac{1}{r}(v_{r,\theta} - v_\theta)\boldsymbol{e}_r\otimes\boldsymbol{e}_\theta + \frac{1}{r\sin\theta}(v_{r,\phi} - \sin\theta v_\phi)\boldsymbol{e}_r\otimes\boldsymbol{e}_\phi\\
& + v_{\theta,r}\boldsymbol{e}_\theta\otimes\boldsymbol{e}_r + \frac{1}{r}(v_{\theta,\theta} + v_r)\boldsymbol{e}_\theta\otimes\boldsymbol{e}_\theta + \frac{1}{r\sin\theta}(v_{\theta,\phi} - \cos\theta v_\phi)\boldsymbol{e}_\theta\otimes\boldsymbol{e}_\phi\\
& + v_{\phi,r}\boldsymbol{e}_\phi\otimes\boldsymbol{e}_r + \frac{1}{r}v_{\phi,\theta}\boldsymbol{e}_\phi\otimes\boldsymbol{e}_\theta + \frac{1}{r\sin\theta}(v_{\phi,\phi} + \sin\theta v_r + \cos\theta v_\theta)\boldsymbol{e}_\phi\otimes\boldsymbol{e}_\phi.
\end{aligned} \tag{2.104}$$

[51] Unfortunately, there are many different conventions in use for the spherical coordinate system. Various names are often associated with the different conventions (see, for example, http://mathworld.wolfram.com/SphericalCoordinates.html), but it is not clear that these are always used consistently. It seems the best course of action is to be extremely careful when using reference materials and the spherical coordinate system. Always double check each author's definition of the coordinates.

For the divergence, substituting in the $\boldsymbol{g}^i$ vectors and the component expression for a second-order tensor $\boldsymbol{T}$ (see Eqn. (2.61)) into Eqn. $(2.94)_2$, we obtain

$$\begin{aligned}
\operatorname{div}\boldsymbol{T} = &\left(\frac{\partial T_{rr}}{\partial r}\boldsymbol{e}_r\otimes\boldsymbol{e}_r + \frac{\partial T_{r\theta}}{\partial r}\boldsymbol{e}_r\otimes\boldsymbol{e}_\theta + \cdots + \frac{\partial T_{\phi\phi}}{\partial r}\boldsymbol{e}_\phi\otimes\boldsymbol{e}_\phi\right.\\
&+ T_{rr}\frac{\partial \boldsymbol{e}_r}{\partial r}\otimes\boldsymbol{e}_r + T_{r\theta}\frac{\partial \boldsymbol{e}_r}{\partial r}\otimes\boldsymbol{e}_\theta + \cdots + T_{\phi\phi}\frac{\partial \boldsymbol{e}_\phi}{\partial r}\otimes\boldsymbol{e}_\phi\\
&\left.+ T_{rr}\boldsymbol{e}_r\otimes\frac{\partial \boldsymbol{e}_r}{\partial r} + T_{r\theta}\boldsymbol{e}_r\otimes\frac{\partial \boldsymbol{e}_\theta}{\partial r} + \cdots + T_{\phi\phi}\boldsymbol{e}_\phi\otimes\frac{\partial \boldsymbol{e}_\phi}{\partial r}\right)\boldsymbol{e}_r\\
&+\left(\frac{\partial T_{rr}}{\partial \theta}\boldsymbol{e}_r\otimes\boldsymbol{e}_r + \cdots + T_{\phi\phi}\boldsymbol{e}_\phi\otimes\frac{\partial \boldsymbol{e}_\phi}{\partial \theta}\right)\frac{1}{r}\boldsymbol{e}_\theta\\
&+\left(\frac{\partial T_{rr}}{\partial \phi}\boldsymbol{e}_r\otimes\boldsymbol{e}_r + \cdots + T_{\phi\phi}\boldsymbol{e}_\phi\otimes\frac{\partial \boldsymbol{e}_\phi}{\partial \phi}\right)\frac{1}{r\sin\theta}\boldsymbol{e}_\phi.
\end{aligned}$$

Again, this equation has 81 terms. Performing the indicated differentiations and the various contractions results in the final form for the divergence in spherical coordinates:

$$\begin{aligned}
\operatorname{div}\boldsymbol{T} = &\left(\frac{\partial T_{rr}}{\partial r} + \frac{1}{r}\left[\frac{\partial T_{r\theta}}{\partial \theta} + \csc\theta\frac{\partial T_{r\phi}}{\partial \phi} + 2T_{rr} - T_{\theta\theta} - T_{\phi\phi} + \cot\theta T_{r\theta}\right]\right)\boldsymbol{e}_r\\
&+\left(\frac{\partial T_{\theta r}}{\partial r} + \frac{1}{r}\left[\frac{\partial T_{\theta\theta}}{\partial \theta} + \csc\theta\frac{\partial T_{\theta\phi}}{\partial \phi} + \cot\theta(T_{\theta\theta} - T_{\phi\phi}) + T_{r\theta} + 2T_{\theta r}\right]\right)\boldsymbol{e}_\theta\\
&+\left(\frac{\partial T_{\phi r}}{\partial r} + \frac{1}{r}\left[\frac{\partial T_{\phi\theta}}{\partial \theta} + \csc\theta\frac{\partial T_{\phi\phi}}{\partial \phi} + T_{r\phi} + 2T_{\phi r} + \cot\theta(T_{\theta\phi} + T_{\phi\theta})\right]\right)\boldsymbol{e}_\phi.
\end{aligned} \tag{2.105}$$

2.6.4 Divergence theorem

In continuum mechanics, we often deal with integrals over the domain of the solid. There are a number of integral theorems that facilitate the evaluation of these integrals. These include Stokes' theorem relating line and surface integrals, and the divergence theorem relating surface and volume integrals. The latter is particularly important in continuum mechanics and is given in detail below.

Consider a closed volume Ω bounded by the surface $\partial\Omega$ with outward unit normal $\boldsymbol{n}(\boldsymbol{x})$ together with a smooth spatially varying vector field $\boldsymbol{w}(\boldsymbol{x})$ defined everywhere in Ω and on $\partial\Omega$. This is depicted schematically in Fig. 2.6(a), where the vector field is represented by arrows. The divergence theorem for the vector field $\boldsymbol{w}$ states

$$\int_{\partial\Omega} w_i n_i \, dA = \int_{\Omega} w_{i,i} \, dV \quad \Leftrightarrow \quad \int_{\partial\Omega} \boldsymbol{w}\cdot\boldsymbol{n}\, dA = \int_{\Omega} (\operatorname{div}\boldsymbol{w})\, dV, \tag{2.106}$$

where the integral over $\partial\Omega$ is a surface integral (dA is an infinitesimal surface element) and the integral over Ω is a volume integral (dV is an infinitesimal volume element). Physically, the surface term measures the flux of $\boldsymbol{w}$ out of Ω, while the volume term is a measure of sinks and sources of $\boldsymbol{w}$ inside Ω. The divergence theorem is therefore a conservation law

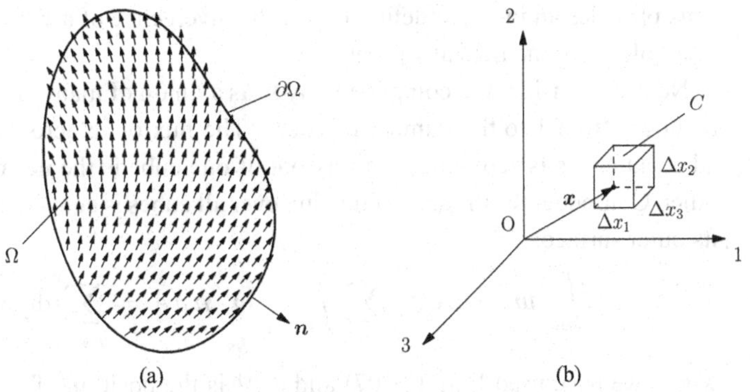

Fig. 2.6 (a) A domain Ω containing a spatially varying vector field $\boldsymbol{w}(\boldsymbol{x})$. (b) A small cube inside Ω.

for $\boldsymbol{w}$. This is easy to visualize for a fluid (where $\boldsymbol{w}$ is the fluid velocity), but it is true for any vector field.

There are different ways to prove the divergence theorem. A simple nonrigorous approach that provides some physical intuition is to demonstrate the theorem for an infinitesimal cube and to then construct the volume Ω as a union of such cubes. Consider a small cube C inside Ω with sides Δx_1, Δx_2, Δx_3 and one corner located at $\boldsymbol{x}$ (see Fig. 2.6(b)). The net flux across the faces of C is

$$\int_{\partial C} \boldsymbol{w} \cdot \boldsymbol{n}\, dA = \int_{\partial C_1} \boldsymbol{w} \cdot \boldsymbol{n}\, dA + \int_{\partial C_2} \boldsymbol{w} \cdot \boldsymbol{n}\, dA + \int_{\partial C_3} \boldsymbol{w} \cdot \boldsymbol{n}\, dA,$$

where ∂C_i are the faces perpendicular to $\boldsymbol{e}_i$. For example, for the ∂C_1 face[52]

$$\begin{aligned}\int_{\partial C_1} \boldsymbol{w} \cdot \boldsymbol{n}\, dA &= [\boldsymbol{w}(x_1, x_2, x_3) \cdot (-\boldsymbol{e}_1) + \boldsymbol{w}(x_1 + \Delta x_1, x_2, x_3) \cdot \boldsymbol{e}_1]\, \Delta x_2 \Delta x_3 \\ &= [w_1(x_1 + \Delta x_1, x_2, x_3) - w_1(x_1, x_2, x_3)]\, \Delta x_2 \Delta x_3.\end{aligned}$$

Similar expressions are obtained for the other two terms. Adding the terms and dividing by the volume of the cube, $\Delta V = \Delta x_1 \Delta x_2 \Delta x_3$, we have

$$\begin{aligned}\frac{1}{\Delta V} \int_{\partial C} \boldsymbol{w} \cdot \boldsymbol{n}\, dA = &\frac{w_1(x_1 + \Delta x_1, x_2, x_3) - w_1(x_1, x_2, x_3)}{\Delta x_1} \\ &+ \frac{w_2(x_1, x_2 + \Delta x_2, x_3) - w_2(x_1, x_2, x_3)}{\Delta x_2} \\ &+ \frac{w_3(x_1, x_2, x_3 + \Delta x_3) - w_3(x_1, x_2, x_3)}{\Delta x_3}.\end{aligned}$$

Now taking the limit $\Delta x_i \to 0$, the terms on the right become partial derivatives so that

$$\lim_{\Delta x_i \to 0} \frac{1}{\Delta V} \int_{\partial C} \boldsymbol{w} \cdot \boldsymbol{n}\, dA = \frac{\partial w_1}{\partial x_1} + \frac{\partial w_2}{\partial x_2} + \frac{\partial w_3}{\partial x_3} = \operatorname{div} \boldsymbol{w}. \tag{2.107}$$

52 Since we plan to take the limit $\Delta x_i \to 0$, we take $\boldsymbol{w}(\boldsymbol{x})$ to be constant on the cube faces. A more careful derivation would apply the mean-value theorem here (as is done, for example, in Section 4.2.3). However, this would clutter the notation, so we avoid it here.

This provides an intuitive definition for the divergence of a vector field as the net flow per unit volume of the field at a point.

Next we consider the complete body Ω as a union of many adjoining cubes $C^{(\alpha)}$, where α ranges from 1 to the number of cubes. The flux of $\boldsymbol{w}$ across an interface between two adjacent cubes is zero since the flux out of one cube is the negative of the flux out of the other. Consequently, the sum of the flux over all cubes is equal to the flux leaving Ω through its outer surface:

$$\int_{\partial\Omega} \boldsymbol{w} \cdot \boldsymbol{n}\, dA \approx \sum_{\alpha} \int_{\partial C^{(\alpha)}} \boldsymbol{w} \cdot \boldsymbol{n}\, dA = \sum_{\alpha} (\operatorname{div} \boldsymbol{w})(\boldsymbol{x}^{(\alpha)})\, \Delta V,$$

where we have used Eqn. (2.107) and $\boldsymbol{x}^{(\alpha)}$ is the position of a corner of cube $C^{(\alpha)}$. The equality is only approximate due to the discretization error associated with the finite size of the cubes. Taking $\Delta V \to 0$ we obtain the divergence theorem in Eqn. (2.106). As noted earlier this is not meant to be a rigorous proof (as any mathematician reading this will point out), however, it conveys the essence of the origin of the divergence theorem.

The divergence theorem can be generalized to a tensor field $\boldsymbol{B}$ of any rank:

$$\int_{\partial\Omega} \boldsymbol{B}\boldsymbol{n}\, dA = \int_{\Omega} \operatorname{div} \boldsymbol{B}\, dV. \tag{2.108}$$

In Cartesian component form, this can be written as

$$\int_{\partial\Omega} B_{ijk\ldots p} n_p\, dA = \int_{\Omega} B_{ijk\ldots p,p}\, dV. \tag{2.109}$$

For example, for a second-order tensor $\boldsymbol{T}$ this is

$$\int_{\partial\Omega} T_{ij} n_j\, dA = \int_{\Omega} T_{ij,j}\, dV.$$

Exercises

2.1 [SECTION 2.1] A rocket propels itself forward by "burning" fuel (mixing fuel with oxygen) and emitting the resulting hot gases at high velocity out of a nozzle at the rear of the rocket. As a result of the combustion process the mass of the rocket continuously decreases.

1. Show that the motion of the rocket is governed by the following equation:

$$m\frac{dv}{dt} + v^*_{\text{ex}}\frac{dm}{dt} = F(t),$$

where $v = v(t)$ is the velocity of the rocket, $m = m(t)$ is the mass of the rocket, v^*_{ex} is the velocity of the exhaust gas *relative* to that of the rocket, and $F(t)$ is the external force acting on the rocket. **Hint:** Compute the momentum of the rocket at time t and time $t + \Delta t$, i.e. $p(t)$ and $p(t + \Delta t) = p + \Delta p$. The mass of the rocket will be reduced by Δm during this interval. Account for the momentum of the exhaust gas. Obtain dp/dt through a limiting operation.

2. Compute the maximum velocity, $v_{\max}$, that the rocket can achieve under the following conditions. There is no external force acting on the rocket, $F(t) = 0$, the relative exhaust velocity, v^*_{ex}, and rate of change of mass, $\dot{m}$, are constant, the initial velocity is zero, $v(0) = 0$, the initial mass of the rocket is m_{init}, the final mass of the rocket (after the fuel is expended) is m_{fin}. Given your result, what is the best way for a rocket engineer to increase the maximum velocity?

2.2 [SECTION 2.2] Expand the following indicial expressions (all indices range from 1 to 3). Indicate the rank and the number of resulting expressions.

1. $a_i b_i$.
2. $a_i b_j$.
3. $\sigma_{ik} n_k$.
4. $A_{ij} x_i x_j$ ($\boldsymbol{A}$ is symmetric, i.e. $A_{ij} = A_{ji}$).

2.3 [SECTION 2.2] Simplify the following indicial expressions as much as possible (all indices range from 1 to 3).

1. $\delta_{mm} \delta_{nn}$.
2. $x_i \delta_{ik} \delta_{jk}$.
3. $B_{ij} \delta_{ij}$ ($\boldsymbol{B}$ is antisymmetric, i.e. $B_{ij} = -B_{ji}$).
4. $(A_{ij} B_{jk} - 2A_{im} B_{mk}) \delta_{ik}$.
5. Substitute $A_{ij} = B_{ik} C_{kj}$ into $\phi = A_{mk} C_{mk}$.
6. $\epsilon_{ijk} a_i a_j a_k$.

2.4 [SECTION 2.2] Write out the following expressions in indicial notation, if possible:

1. $A_{11} + A_{22} + A_{33}$.
2. $\mathbf{A}^T \mathbf{A}$, where $\mathbf{A}$ is a 3×3 matrix.
3. $A_{11}^2 + A_{22}^2 + A_{33}^2$.
4. $(u_1^2 + u_2^2 + u_3^2)(v_1^2 + v_2^2 + v_3^2)$.
5. $A_{11} = B_{11} C_{11} + B_{12} C_{21} \qquad A_{12} = B_{11} C_{12} + B_{12} C_{22}$
   $A_{21} = B_{21} C_{11} + B_{22} C_{21} \qquad A_{22} = B_{21} C_{12} + B_{22} C_{22}$.

2.5 [SECTION 2.2] Obtain an expression for $\partial \boldsymbol{A}^{-1} / \partial \boldsymbol{A}$, where $\boldsymbol{A}$ is a second-order tensor. This expression turns up in [TM11] when computing stress in statistical mechanics systems. **Hint:** Start with the identity $A^{-1}_{ik} A_{kj} = \delta_{ij}$. Use indicial notation in your derivation.

2.6 [SECTION 2.3] Show that, for two points with plane polar coordinates (r_1, θ_1) and (r_2, θ_2), the addition $(r, \theta) = (r_1 + r_2, \theta_1 + \theta_2)$ does not satisfy the vector parallelogram law and therefore (r, θ) are not the components of a vector.

2.7 [SECTION 2.3] A classical system of N particles is characterized by $n = 3N$ momentum coordinates, $p_1, \ldots, p_n$, and $n = 3N$ position coordinates, $q_1, \ldots, q_n$. The "Poisson bracket" between two functions, $f(\boldsymbol{q}, \boldsymbol{p})$ and $g(\boldsymbol{q}, \boldsymbol{p})$, is defined by

$$\{f, g\} = \frac{\partial f}{\partial q_i} \frac{\partial g}{\partial p_i} - \frac{\partial f}{\partial p_i} \frac{\partial g}{\partial q_i},$$

where the summation convention applies. The Poisson bracket is an important operator in statistical mechanics. Prove that $\{f, g\}$ is a bilinear operator (as defined in Section 2.3) with respect to its arguments.

2.8 [SECTION 2.3] Consider a coordinate transformation from x_α to x'_i. We have, $x_\alpha = \mathsf{Q}_{\alpha i} x'_i$, where $\mathsf{Q}_{\alpha i} = \boldsymbol{e}_\alpha \cdot \boldsymbol{e}'_i = \cos \theta(\boldsymbol{e}_\alpha, \boldsymbol{e}'_i)$. Here $\boldsymbol{e}_\alpha$ and $\boldsymbol{e}'_i$ are orthonormal basis vectors of the unprimed and primed coordinate systems, respectively, and $\theta(\boldsymbol{e}_\alpha, \boldsymbol{e}'_i)$ is the angle between $\boldsymbol{e}_\alpha$ and $\boldsymbol{e}'_i$ measured in the counterclockwise direction.

1. Calculate the coefficients $Q_{\alpha i}$ for the particular transformation given in the table below (the numbers are the angles between the basis vectors):

	e'_1	e'_2	e'_3
e_1	120°	120°	45°
e_2	45°	135°	90°
e_3	60°	60°	45°

2. Verify that $\mathbf{Q}$ is proper orthogonal.

2.9 [SECTION 2.3] Express the following expressions in terms of tensor components:
1. $\boldsymbol{v}[\mathbf{e}_1]$.
2. $\boldsymbol{v}[\mathbf{e}_3 + 2\mathbf{e}_2]$.
3. $\boldsymbol{v}[y\mathbf{e}_1 - x\mathbf{e}_2]$.
4. $\boldsymbol{T}[\mathbf{e}_2, \mathbf{e}_1]$.
5. $\boldsymbol{T}[\mathbf{e}_3, 5\mathbf{e}_3 + 4\mathbf{e}_1]$.
6. $\boldsymbol{T}[\mathbf{e}_1 + \mathbf{e}_2, \mathbf{e}_1 + \mathbf{e}_3]$.

2.10 [SECTION 2.3] Given that v_i, T_{ij} and M_{ijk} are the rectangular Cartesian components of rank one, two and three tensors, respectively, prove that the following are tensors:
1. $T_{ij}v_iv_j$.
2. $M_{ijj}T_{ik}$.
3. $M_{ijk}v_k$.

2.11 [SECTION 2.4] Scalar contractions of tensors were defined in Section 2.4. The simplest example is the dot product $\boldsymbol{a} \cdot \boldsymbol{b}$. How can this contraction be obtained from the definition of a tensor as a scalar-valued multilinear function of vectors where the vectors are written as $\boldsymbol{a}[\boldsymbol{x}]$ and $\boldsymbol{b}[\boldsymbol{x}]$?

2.12 [SECTION 2.5] Prove that any antisymmetric tensor $\boldsymbol{A}$ has a one-to-one relation to a unique axial vector $\boldsymbol{w}$ as shown in Eqn. (2.71). **Hint:** Start from the axial vector condition in Eqn. (2.70). Write it out in indicial notation and manipulate the expression until you obtain the left-hand side of Eqn. (2.71). The inverse relation is obtained by multiplying both sides by the permutation tensor and using the ϵ–δ identity (Eqn. (2.11)).

2.13 [SECTION 2.5] Consider the dyad $\boldsymbol{D} = \boldsymbol{a} \otimes \boldsymbol{a}$ constructed from the vector $\boldsymbol{a}$.
1. Write out the components of $\boldsymbol{D}$ in matrix form.
2. Compute the three principal invariants of $\boldsymbol{D}$: $I_1^{\boldsymbol{D}}, I_2^{\boldsymbol{D}}, I_3^{\boldsymbol{D}}$. Simplify your expressions as much as possible.
3. Compute the eigenvalues of $\boldsymbol{D}$.

2.14 [SECTION 2.5] Let tensor $\boldsymbol{A}$ be given by $\boldsymbol{A} = \alpha(\boldsymbol{I} - \mathbf{e}_1 \otimes \mathbf{e}_1) + \beta(\mathbf{e}_1 \otimes \mathbf{e}_2 + \mathbf{e}_2 \otimes \mathbf{e}_1)$, where α, β are scalars (not equal to zero) and $\mathbf{e}_1, \mathbf{e}_2$ are orthogonal unit vectors.
1. Show that the eigenvalues $\lambda_k^{\boldsymbol{A}}$ of $\boldsymbol{A}$ are

$$\lambda_1^{\boldsymbol{A}} = \alpha, \quad \lambda_{2,3}^{\boldsymbol{A}} = \alpha/2 \pm \left(\alpha^2/4 + \beta^2\right)^{1/2}.$$

2. Show that the associated normalized eigenvectors $\boldsymbol{\Lambda}_k^{\boldsymbol{A}}$ are

$$\boldsymbol{\Lambda}_1^{\boldsymbol{A}} = \begin{bmatrix} 0 \\ 0 \\ 1 \end{bmatrix}, \quad \boldsymbol{\Lambda}_2^{\boldsymbol{A}} = \frac{1}{\sqrt{1 + (\lambda_2^{\boldsymbol{A}}/\beta)^2}} \begin{bmatrix} 1 \\ \lambda_2^{\boldsymbol{A}}/\beta \\ 0 \end{bmatrix}, \quad \boldsymbol{\Lambda}_3^{\boldsymbol{A}} = \frac{1}{\sqrt{1 + (\lambda_3^{\boldsymbol{A}}/\beta)^2}} \begin{bmatrix} 1 \\ \lambda_3^{\boldsymbol{A}}/\beta \\ 0 \end{bmatrix}.$$

3. Under what conditions on α and β (if any) is $\boldsymbol{A}$ positive definite?

2.15 [Section 2.6] Solve the following problems related to indicial notation for tensor field derivatives. In all cases indices range from 1 to 3. All variables are tensors and functions of the variables that they are differentiated by unless explicitly noted. The comma notation refers to differentiation with respect to $\boldsymbol{x}$.

1. Write out explicit expressions (i.e. ones that only have numbers as indices) for the following indicial expressions. In each case, indicate the rank and the number of the resulting expressions.
   a. $\dfrac{\partial u_i}{\partial z_k}\dfrac{\partial z_k}{\partial x_j}$.
   b. $\sigma_{ij,j} + \rho b_i = \rho a_i$.
   c. $u_{k,j}\delta_{jk} - u_{i,i}$.
2. Expand out and then simplify the following indicial expressions as much as possible. Leave the expression in indicial form.
   a. $(T_{ij}x_j)_{,i} - T_{ii}$.
   b. $(x_m x_m x_i A_{ij})_{,k}$ ($\boldsymbol{A}$ is constant).
   c. $(S_{ij}T_{jk})_{,ik}$.
3. Write out the following expressions in indicial notation.
   a. $B_{i1}\dfrac{\partial c_1}{\partial x_j} + B_{i2}\dfrac{\partial c_2}{\partial x_j} + B_{i3}\dfrac{\partial c_3}{\partial x_j}$.
   b. $\operatorname{div}\boldsymbol{v}$, where $\boldsymbol{v}$ is a vector.
   c. $\dfrac{\partial^2 T_{11}}{\partial x_1^2} + \dfrac{\partial^2 T_{12}}{\partial x_1 \partial x_2} + \dfrac{\partial^2 T_{13}}{\partial x_1 \partial x_3} + \dfrac{\partial^2 T_{21}}{\partial x_2 \partial x_1} + \dfrac{\partial^2 T_{22}}{\partial x_2^2} + \dfrac{\partial^2 T_{23}}{\partial x_2 \partial x_3} + \dfrac{\partial^2 T_{31}}{\partial x_3 \partial x_1}$
   $+ \dfrac{\partial^2 T_{32}}{\partial x_3 \partial x_2} + \dfrac{\partial^2 T_{33}}{\partial x_3^2}$.

2.16 [Section 2.6] Let $f = f(x_1, x_2, x_3)$ be a scalar field, and define $h_\alpha \equiv \partial f/\partial x_\alpha = f_{,\alpha}$. Show that upon transformation from one set of rectangular Cartesian coordinates to another, the following equality is satisfied:

$$h'_i = Q_{\alpha i} h_\alpha.$$

This shows that h_α are the components of a vector: $\boldsymbol{h} = (\partial f/\partial x_\alpha)\boldsymbol{e}_\alpha = f_{,\alpha}\boldsymbol{e}_\alpha$. This vector is called the "gradient of $f(\boldsymbol{x})$" and it is denoted by ∇f. **Hint:** In the unprimed coordinate system, $(\cdot)_{,\alpha} = \partial(\cdot)/\partial x_\alpha$, and in the primed coordinate system, $(\cdot)_{,i} = \partial(\cdot)/\partial x'_i$. To switch from one to the other use the chain rule.

2.17 [Section 2.6] Prove the following identities, involving scalar fields ξ and η, vector fields $\boldsymbol{u}$ and $\boldsymbol{v}$, and tensor field $\boldsymbol{T}$, using indicial notation:

1. $\operatorname{curl}\nabla\eta = \boldsymbol{0}$.
2. $\operatorname{div}\operatorname{curl}\boldsymbol{u} = 0$.
3. $\nabla^2(\xi\eta) = \xi\nabla^2\eta + \eta\nabla^2\xi + 2\nabla\xi\cdot\nabla\eta$.
4. $\operatorname{div}(\boldsymbol{T}\boldsymbol{v}) = (\operatorname{div}\boldsymbol{T}^T)\cdot\boldsymbol{v} + \boldsymbol{T} : (\nabla\boldsymbol{v})^T$.

2.18 [Section 2.6] The divergence theorem for a region Ω bounded by a closed surface $\partial\Omega$ is given in Eqn. (2.109).

1. Apply Eqn. (2.109) to a vector field, $\boldsymbol{v} = (\xi\,\eta_{,i})\boldsymbol{e}_i$, where both ξ and η are scalar functions of $\boldsymbol{x}$ and obtain

$$\int_{\partial\Omega} \xi\,\eta_{,i} n_i \, dA = \int_{\Omega} (\xi_{,i}\eta_{,i} + \xi\,\eta_{,ii})\, dV, \qquad (*)$$

which is known as Green's first identity.

2. Interchange the roles of ξ and η in Eqn. ($*$) and subtract from the original version of Eqn. ($*$) to obtain
$$\int_{\partial\Omega} (\xi\,\eta_{,i} - \eta\,\xi_{,i})\,n_i\,dA = \int_{\Omega} (\xi\,\eta_{,ii} - \eta\,\xi_{,ii})\,dV, \qquad (**)$$
which is known as Green's second identity.
3. Write Eqns. ($*$) and ($**$) in coordinate-free (direct) notation, noting that $\nabla\xi\cdot\boldsymbol{n} = \langle \mathcal{D}_{\boldsymbol{x}}\xi;\boldsymbol{n}\rangle$ is a normal derivative.

3 Kinematics of deformation

Continuum mechanics deals with the change of shape (deformation) of bodies subjected to external mechanical and thermal loads. However, before we can discuss the physical laws governing deformation, we must develop measures that characterize and quantify it. This is the subject described by the *kinematics of deformation*. Kinematics does not deal with predicting the deformation resulting from a given loading, but rather with the machinery for describing all possible deformations a body can undergo.

3.1 The continuum particle

A material body B bounded by a surface ∂B is represented by a continuous distribution of an infinite number of *continuum particles*. On the macroscopic scale, each particle is a point of zero extent much like a point in a geometrical space. It should therefore not be thought of as a small piece of material. At the same time, it has to be realized that a continuum particle derives its properties from a finite-size region ℓ on the micro scale (see Fig. 3.1). One can think of the properties of the particle as an average over the atomic behavior within this domain. As one moves from one particle to its neighbor the microscopic domain moves over, largely overlapping the previous domain. In this way the smooth field-like behavior we expect in a continuum is obtained.[1] A fundamental assumption of continuum mechanics is that it is possible to define a length ℓ that is large relative to atomic length scales and at the same time much smaller than the length scale associated with variations in the continuum fields.[2] We revisit this issue and the limitations that it imposes on the validity of continuum theory in Section 6.6.

[1] This is the approach taken in Section 8.2 of [TM11], where statistical mechanics ideas are used to obtain microscopic expressions for the continuum fields. See also footnote 31 in that section.

[2] This microscopically-based view of continuum mechanics is not mandatory. Clifford Truesdell, one of the major figures in continuum mechanics who, together with Walter Noll, codified it and gave it its modern mathematical form, was a strong proponent of continuum mechanics as an independent theory eschewing perceived connections with other theories. For example, in his book with Richard Toupin, *The Classical Field Theories* [TT60], he states: "The corpuscular theories and field theories are mutually contradictory as direct models of nature. The field is indefinitely divisible; the corpuscle is not. To mingle the terms and concepts appropriate to these two distinct representations of nature, while unfortunately a common practice, leads to confusion if not to error. For example, to speak of an element of volume in a gas as 'a region large enough to contain many molecules but small enough to be used as a element of integration' is not only loose but also needless and bootless." This is certainly true as long as continuum mechanics is studied as an independent theory. However, when attempts are made to connect it with phenomena occurring on smaller scales, as in this book and to a larger extent in [TM11], it leads to a dead end. Truesdell even acknowledged this fact in the

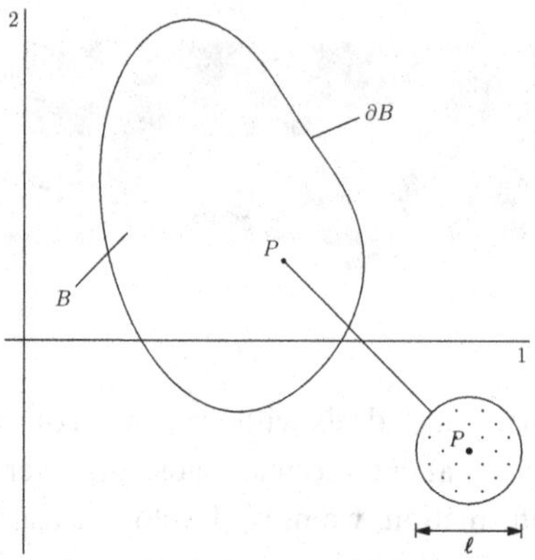

Fig. 3.1 A material body B with surface ∂B. A continuum particle P is shown together with a schematic representation of the atomic structure underlying the particle with length scale ℓ. The small dots in the atomic structure represent atoms.

3.2 The deformation mapping

A body B can take on many different shapes or *configurations* depending on the loading applied to it. We choose one of these configurations to be the *reference configuration* of the body and label it B_0. The reference configuration provides a convenient fixed state of the body to which other configurations can be compared to gauge their deformation. Any possible configuration of the body can be taken as its reference. Typically the choice is dictated by convenience to the analysis. Often, it corresponds to the state where no external loading is applied to the body.

We denote the position of a particle P in the reference configuration by $\boldsymbol{X} = \boldsymbol{X}(P)$. Since particles cannot be formed or destroyed, we can use the coordinates of a particle in the reference configuration as a label distinguishing this particle from all others. Once we have defined the reference configuration, the *deformed configuration* occupied by the body is described in terms of a *deformation mapping* function $\boldsymbol{\varphi}$ that maps the reference position of every particle $\boldsymbol{X} \in B_0$ to its deformed position $\boldsymbol{x}$:

$$x_i = \varphi_i(X_1, X_2, X_3) \quad \Leftrightarrow \quad \boldsymbol{x} = \boldsymbol{\varphi}(\boldsymbol{X}). \tag{3.1}$$

In the deformed configuration the body occupies a domain B, which is the union of all positions $\boldsymbol{x}$ (see Fig. 3.2). In the above, we have adopted the standard continuum mechanics

text immediately following the above quote where he discussed Noll's work on a microscopic definition of the stress tensor [Nol55]. Noll, following the work of Irving and Kirkwood [IK50], demonstrated that by defining continuum field variables as particular phase averages over the atomistic phase space, the continuum balance laws were exactly satisfied. Truesdell consequently (and perhaps grudgingly) concluded that "those who prefer to regard classical statistical mechanics as fundamental may nevertheless employ the field concept as exact in terms of *expected values*" [TT60]. Irving and Kirkwood and Noll's approach is discussed in Section 8.2 of [TM11].

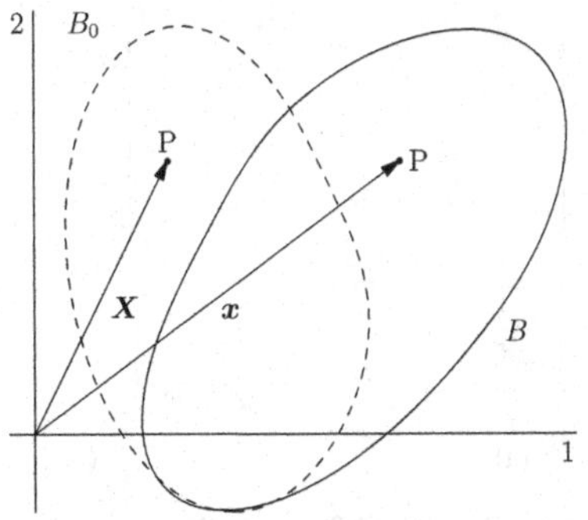

Fig. 3.2 The reference configuration B_0 of a body (dashed) and the deformed configuration B. The particle P located at position $\boldsymbol{X}$ in the reference configuration is mapped to a new point $\boldsymbol{x}$ in the deformed configuration.

convention of denoting all things associated with the reference configuration with upper-case letters (as in $\boldsymbol{X}$) or with a subscript 0 (as in B_0) and all things associated with the deformed configuration in lower-case (as in $\boldsymbol{x}$) or without a subscript (as in B).

In order to satisfy the condition that particles are not destroyed or created, φ must be a one-to-one mapping. This means that a single particle cannot be mapped to two positions and that two particles cannot be mapped to the same position. The fact that φ is one-to-one implies that it is invertible, i.e. it is always possible to define a unique inverse mapping, $\boldsymbol{X} = \varphi^{-1}(\boldsymbol{x})$, from B to B_0. This physically desirable property is not satisfied, in general, for an arbitrary function φ. The deformation mapping must satisfy the *inverse function theorem* as well as global invertibility conditions as described in Section 3.4.2.

Example 3.1 (Uniform stretching and simple shear) Two important examples of deformation mappings are shown in Fig. 3.3 and are detailed below:

1. Uniform stretching:

$$x_1 = \alpha_1 X_1, \quad x_2 = \alpha_2 X_2, \quad x_3 = \alpha_3 X_3,$$

where $\alpha_i > 0$ are the stretch parameters along the axes directions. When all three are equal ($\alpha_1 = \alpha_2 = \alpha_3 = \alpha$) the deformation is called a uniform *dilatation*, corresponding to a uniform contraction for $\alpha < 1$ and a uniform expansion for $\alpha > 1$.

2. Simple shear:

$$x_1 = X_1 + \gamma X_2, \quad x_2 = X_2, \quad x_3 = X_3,$$

where γ is the shear parameter measuring the amount of lateral motion per unit height. The shearing angle is given by $\tan^{-1}\gamma$ (see Fig. 3.3(c)). This deformation plays an important role in crystal plasticity where the passage of a dislocation can be described as a simple shear across an interatomic layer (see Section 6.5.5 of [TM11]). In general, for a simple shear in a direction $\boldsymbol{s}$ on a plane with normal $\boldsymbol{n}$, we have

$$\boldsymbol{x} = (\boldsymbol{I} + \gamma \boldsymbol{s} \otimes \boldsymbol{n})\boldsymbol{X}.$$

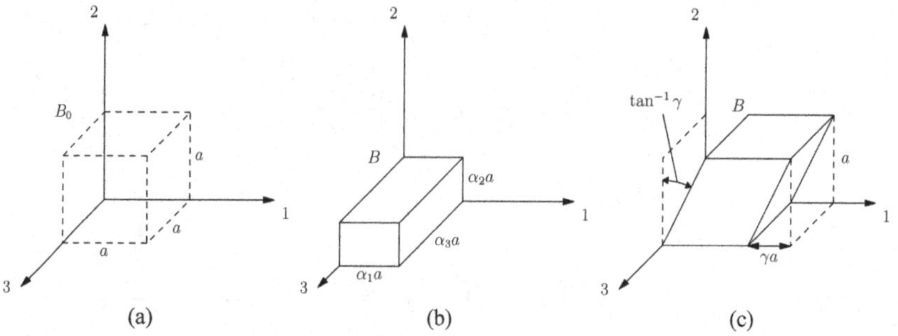

Fig. 3.3 Examples of deformation mappings. Frame (a) shows the reference configuration where the body is a cube (dashed). Frames (b) and (c) show the deformed configuration for uniform stretch and simple shear, respectively.

A time-dependent deformation mapping, $\varphi(\boldsymbol{X}, t)$, is called a *motion*. In this case the reference configuration is often associated with the motion at time $t = 0$, so that $\varphi(\boldsymbol{X}, 0) = \boldsymbol{X}$, and the deformed configuration is the motion at the "current" time t. For this reason the deformed configuration is often alternatively referred to as the *current configuration*.

3.3 Material and spatial field descriptions

Consider a scalar invariant field $\mathfrak{g}$ such as the temperature. We can write $\mathfrak{g}$ as a function over the deformed or the reference configuration:

$$\mathfrak{g} = g(\boldsymbol{x}, t) \quad \boldsymbol{x} \in \mathcal{B}, \qquad \text{or} \qquad \mathfrak{g} = \breve{g}(\boldsymbol{X}, t) \quad \boldsymbol{X} \in \mathcal{B}_0.$$

The two descriptions are linked by the deformation mapping $\breve{g}(\boldsymbol{X}, t) \equiv g(\varphi(\boldsymbol{X}, t), t)$. However, these are actually very different descriptions. In the first case, $\mathfrak{g} = g(\boldsymbol{x}, t)$ is written in terms of spatial positions. In other words, $g(\boldsymbol{x}, t)$ provides the temperature at a particular position in space regardless of which particle is occupying it at time t. This is referred to as a *spatial* or *Eulerian* description.

The second description is written in terms of *material particles* not spatial positions, i.e. $\mathfrak{g} = \breve{g}(\boldsymbol{X}, t)$ gives the temperature of particle $\boldsymbol{X}$ at time t regardless of where the particle is located in space. This is referred to as a *material* or *referential* description.[3] If the body occupies the reference state at $t = 0$, the term *Lagrangian* is used.[4]

For obvious reasons the coordinates of a particle in the reference configuration $\boldsymbol{X}$ are referred to as *material coordinates* and the coordinates of a spatial position $\boldsymbol{x}$ are referred

[3] There is actually a slight difference between the terms "material" and "referential." The former applies to the more abstract case where particles are identified by label (e.g. P), whereas the latter refers to the case where the positions of the particles in a reference configuration are used to identify them [TN65]. This subtle distinction is inconsequential for the discussion here.

[4] Rather unfortunately, the terms "Lagrangian" and "Eulerian" are historically inaccurate. According to Truesdell [Tru52, footnote 5 on p. 139], material descriptions were actually introduced by Euler, whereas the spatial description was introduced by d'Alembert.

to as *spatial coordinates*.[5] If the deformation mapping is available, then the link between the spatial and material coordinates is given by $\boldsymbol{x} = \boldsymbol{\varphi}(\boldsymbol{X})$.

A referential description is suitable for solids where a reference configuration can be readily defined and particles which are nearby in the reference configuration generally remain nearby in the deformed configuration. In contrast, a spatial description is advantageous for fluid flow where material particles can travel large relative distances, and thus, a reference configuration is all but meaningless.

3.3.1 Material and spatial tensor fields

As soon as one starts to consider higher-order tensor fields in the material and spatial descriptions an additional complication is encountered. As discussed in Section 2.3 an nth-order tensor is a real-valued n-linear function of vectors. Thus, in continuum mechanics there are three parts to every tensor field: (i) an n-linear function, (ii) the vector space(s) that serve as the domain(s) of the n-linear function and (iii) a point set (e.g. $\mathcal{B}_0$ or $\mathcal{B}$) over which the nth-order tensor field is defined.

To unambiguously define a tensor field, we must specify the vector spaces on which the tensor acts as well as the point set over which the field is defined (see page 26). In a general mathematical setting each point in space is associated with a distinct *tangent translation space*. To see this, first consider two material points with coordinates $\boldsymbol{X}'$ and $\boldsymbol{X}$ and form a *material vector* $\Delta\boldsymbol{X}$ by subtracting them:

$$\Delta X_I = X_I' - X_I \quad \Leftrightarrow \quad \Delta\boldsymbol{X} = \boldsymbol{X}' - \boldsymbol{X}.$$

We say that this is a vector in the tangent translation space at $\boldsymbol{X}$. Second, consider two spatial positions with coordinates $\boldsymbol{x}'$ and $\boldsymbol{x}$ and form the *spatial vector* $\Delta\boldsymbol{x}$:

$$\Delta x_i = x_i' - x_i \quad \Leftrightarrow \quad \Delta\boldsymbol{x} = \boldsymbol{x}' - \boldsymbol{x}.$$

We say this is a vector in the tangent translation space at $\boldsymbol{x}$.

When attention is restricted to Euclidean point spaces all tangent spaces become equivalent and we can simply speak of the translation space (as we did in Section 2.3.1). However, it is useful to retain the distinction between material vectors and spatial vectors, even when considering Euclidean point spaces. Thus, in the above equations we have extended, to the indices of coordinates and tensor components, the convention of using upper-case letters for all things associated with the reference configuration (material description) and lower-case letters for all things associated with the deformed configuration (spatial description). This component notation becomes especially important when curvilinear coordinate systems are used. In Section 8.2 we present an example using polar coordinates which illustrates many of the subtle aspects of working with both material and spatial quantities.

[5] In general, different coordinate systems may be used for the reference and deformed configurations. For instance, Cartesian coordinates would be best suited to describe the reference configuration when $\mathcal{B}_0$ is box-shaped. However, polar cylindrical coordinates would be best suited to describe the deformed configuration when $\mathcal{B}$ is a sector of a hollow circular cylinder. In cases such as these, one must be careful to keep track of the basis vectors and their dependence on the appropriate coordinates. See Section 8.2 for a number of examples where the use of different coordinate systems is mathematically convenient.

A vector field $\boldsymbol{A}[\Delta \boldsymbol{X}](\boldsymbol{X})$ – where we have explicitly indicated that the vector acts on a material vector $\Delta \boldsymbol{X}$ and is a field defined for points in the referential description – is called a *material vector* field. Similarly, a vector field $\boldsymbol{b}[\Delta \boldsymbol{x}](\boldsymbol{x})$ (acting on a spatial vector $\Delta \boldsymbol{x}$ at each spatial position $\boldsymbol{x}$) is called a *spatial vector* field. In component form, we write A_I and b_i, where the dependence on the field coordinates has been suppressed. Examples are the material and spatial surface normals $\boldsymbol{N}$ and $\boldsymbol{n}$ with components N_I and n_i, respectively, which will be discussed later. It is possible to convert a material vector to a spatial vector and a spatial vector to a material vector by processes which are referred to as *push-forward* and *pull-back* operations, respectively. These operations will be discussed further in Section 3.4 once the deformation gradient has been introduced. Finally, the introduction of upper-case and lower-case indices means that the summation convention introduced earlier now becomes case sensitive. Thus, $A_I A_I$ will be summed, but $b_i A_I$ will not.

The distinction between material and spatial vectors easily extends to higher-order tensors. For instance, suppose $\boldsymbol{A}[\boldsymbol{B}, \boldsymbol{C}]$ is a second-order tensor whose two vector arguments are both material vectors. Then we say that $\boldsymbol{A}$ is a *material tensor* or a tensor *in the reference configuration* and denote its components as A_{IJ}. Similarly, $\boldsymbol{a}[\boldsymbol{b}, \boldsymbol{c}]$, with components a_{ij}, is called a *spatial tensor* or a tensor *in the deformed configuration* because its arguments are both spatial vectors. However, for higher-order tensors, a third possibility exists where one of the tensor's arguments is a material vector and one is a spatial vector: $\boldsymbol{A}[\boldsymbol{B}, \boldsymbol{c}]$. Tensors of this type are called *mixed* or *two-point* tensors. The extension of the index notation to tensors of rank three and higher is straightforward. As used above, upper-case tensor symbols are typically used for two-point tensors to indicate that they have (at least) one material vector argument.

At the other extreme are scalar invariant (zeroth-order tensor) fields, which possess no indices to distinguish between material and spatial representations. The definition of tensors indicates that even zero-order tensors are associated with a vector space. For a scalar invariant field expressed in the spatial description, spatial vectors are the natural associated vector space and such an entity is referred to as a spatial scalar field. Similarly, a scalar invariant field expressed in the material description is called a material scalar field. With this definition the labeling of all tensor fields as spatial, material, or two-point is complete and justified. However, for scalar invariant fields the distinction is purely mathematical.

3.3.2 Differentiation with respect to position

The introduction of referential and spatial descriptions for tensor fields means that the indicial and direct notation introduced earlier for differentiation (see Section 2.6) must be suitably amended. When taking derivatives with respect to positions it is necessary to indicate whether the derivative is taken with respect to $\boldsymbol{X}$ or $\boldsymbol{x}$. In indicial notation, the comma notation refers to the index of the coordinate. Again, we find that the case convention for indices is necessary. Thus, differentiation with respect to the material and spatial coordinates can be unambiguously indicated using the comma notation already introduced as $\square_{,I}$ or $\square_{,i}$, where $\square$ represents the tensor field being differentiated. The direct notation for the gradient, curl and divergence operators with respect to the material

Table 3.1. The direct notation for the gradient, curl and divergence operators with respect to the material and spatial coordinates

Operator	Material coordinates	Spatial coordinates
gradient	$\nabla_0\Box$ or Grad $\Box$	$\nabla\Box$ or grad $\Box$
curl	Curl $\Box$	curl $\Box$
divergence	Div $\Box$	div $\Box$

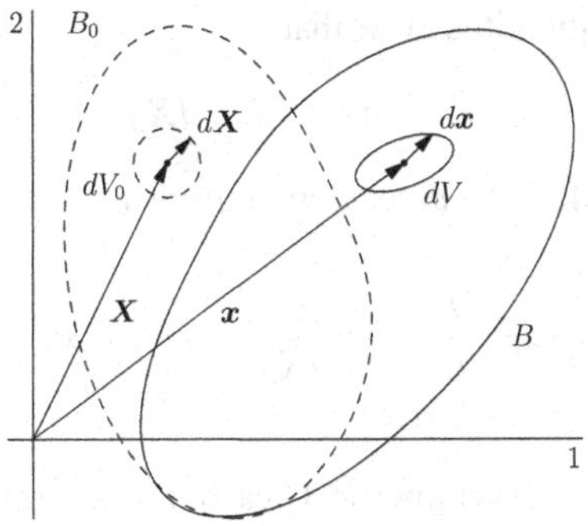

Fig. 3.4 Mapping of the local neighborhood of a material point $\boldsymbol{X}$ in the reference configuration to the deformed configuration. The infinitesimal material vector $d\boldsymbol{X}$ is mapped to the spatial vector $d\boldsymbol{x}$.

and spatial coordinates is given in Tab. 3.1. For example, $\nabla_0 \breve{g} = (\partial \breve{g}/\partial X_I)\boldsymbol{e}_I$ and $\nabla g = (\partial g/\partial x_i)\boldsymbol{e}_i$.

We defer the discussion of differentiation with respect to time until Section 3.6, where the time rate-of-change of kinematic variables is introduced.

3.4 Description of local deformation

The deformation mapping $\varphi(\boldsymbol{X})$ tells us how particles move, but it does not directly provide information on the change of shape of particles, i.e. strains in the material. This is important because materials resist changes to their shape and this information must be included in a physical model of deformation. To capture particle shape change, it is necessary to characterize the deformation in the infinitesimal neighborhood of a particle.

3.4.1 Deformation gradient

Figure 3.4 shows a body in the position it occupies in the reference and deformed configurations. A particle originally located at $\boldsymbol{X}$ is mapped to a deformed position $\boldsymbol{x}$. The infinitesimal environment or *neighborhood* of the particle in the reference configuration is the sphere of volume dV_0 mapped out by $\boldsymbol{X} + d\boldsymbol{X}$, where

$$d\boldsymbol{X} = \boldsymbol{M}\, dS, \tag{3.2}$$

is the differential of $\boldsymbol{X}$, $\boldsymbol{M}$ is a unit material vector allowed to point along all possible directions in the body and $dS = \|d\boldsymbol{X}\|$ is the magnitude of $d\boldsymbol{X}$ (or radius of dV_0). The neighborhood dV_0 is transformed by the deformation mapping to a distorted neighborhood dV in the deformed configuration. Expanding this mapping to first order we have

$$x_i + dx_i = \varphi_i(\boldsymbol{X} + d\boldsymbol{X}) = \varphi_i(\boldsymbol{X}) + \left.\frac{\partial \varphi_i}{\partial X_J}\right|_{\boldsymbol{X}} dX_J = x_i + F_{iJ}\, dX_J.$$

From this relation it is clear that

$$dx_i = F_{iJ}\, dX_J \quad \Leftrightarrow \quad d\boldsymbol{x} = \boldsymbol{F}\, d\boldsymbol{X}, \tag{3.3}$$

where $\boldsymbol{F}$ is called the *deformation gradient* and is given by

$$F_{iJ} = \frac{\partial \varphi_i}{\partial X_J} = \frac{\partial x_i}{\partial X_J} = x_{i,J} \quad \Leftrightarrow \quad \boldsymbol{F} = \frac{\partial \boldsymbol{\varphi}}{\partial \boldsymbol{X}} = \frac{\partial \boldsymbol{x}}{\partial \boldsymbol{X}} = \nabla_0 \boldsymbol{x}. \tag{3.4}$$

In general $\boldsymbol{F}$ is *not* symmetric. Clearly, the deformation gradient is a second-order two-point tensor. This requires that the material and spatial indices of $\boldsymbol{F}$ transform separately like vectors when separate coordinate transformations are performed for the material and spatial coordinate systems, respectively. For the special case where parallel Cartesian coordinate systems are used for the reference and deformed configurations, $\boldsymbol{F}$ satisfies the usual transformation relations for a second-order tensor.

Proof Start with $F'_{iJ} = \partial x'_i / \partial X'_J$ and substitute in

$$x'_i = \mathrm{Q}_{\alpha i} x_\alpha, \quad X'_J = \mathrm{Q}_{AJ} X_A,$$

giving

$$F'_{iJ} = \frac{\partial (\mathrm{Q}_{\alpha i} x_\alpha)}{\partial X_A} \frac{\partial X_A}{\partial X'_J} = \mathrm{Q}_{\alpha i} \frac{\partial x_\alpha}{\partial X_A} \mathrm{Q}_{AJ} = \mathrm{Q}_{\alpha i} \mathrm{Q}_{AJ} F_{\alpha A}.$$

□

The deformation gradient plays a key role in describing the *local* deformation in the vicinity of a particle. It fully characterizes the deformation of the neighborhood of $\boldsymbol{x}$ given by

$$d\boldsymbol{x} = \boldsymbol{m}\, ds, \tag{3.5}$$

where $\boldsymbol{m} = \boldsymbol{F}\boldsymbol{M} / \|\boldsymbol{F}\boldsymbol{M}\|$ is a unit spatial vector along the direction to which $\boldsymbol{M}$ is rotated by the local deformation and $ds = \|d\boldsymbol{x}\| = \|\boldsymbol{F}\boldsymbol{M}\|\, dS$ is the new infinitesimal magnitude. The ratio between ds and dS gives the *stretch* of the infinitesimal material line element originally oriented along $\boldsymbol{M}$:

$$\alpha = \frac{ds}{dS} = \|\boldsymbol{F}\boldsymbol{M}\|.$$

Deformation mappings for which the deformation gradient is constant in space are referred to as *homogeneous deformations* (also called *uniform deformations*).

Example 3.2 (Deformation gradients for uniform stretching and simple shear) The deformation gradients for the mappings given in Example 3.1 are given below.

(i) Uniform stretching:

$$[\boldsymbol{F}] = \begin{bmatrix} \alpha_1 & 0 & 0 \\ 0 & \alpha_2 & 0 \\ 0 & 0 & \alpha_3 \end{bmatrix}.$$

(ii) Simple shear:

$$[\boldsymbol{F}] = \begin{bmatrix} 1 & \gamma & 0 \\ 0 & 1 & 0 \\ 0 & 0 & 1 \end{bmatrix}.$$

We see that the deformation gradients are constant in space indicating that these are homogeneous deformations.

3.4.2 Volume changes

We can also compute the local change in volume due to the deformation. The volume of the spherical neighborhood in the reference configuration is $dV_0 = \frac{4}{3}\pi(dS)^3$. In the deformed configuration this sphere becomes an ellipsoid with volume $dV = \frac{4}{3}\pi abc$, where a, b, c are its half-lengths. To determine the half-lengths, consider the infinitesimal magnitude squared: $ds^2 = (\boldsymbol{F}\boldsymbol{M}) \cdot (\boldsymbol{F}\boldsymbol{M})(dS)^2 = \boldsymbol{M} \cdot (\boldsymbol{F}^T\boldsymbol{F})\boldsymbol{M}(dS)^2 = \boldsymbol{M} \cdot \boldsymbol{C}\boldsymbol{M}(dS)^2$, where $\boldsymbol{C}$ is a symmetric second-order material tensor called the *right Cauchy–Green deformation tensor*:

$$C_{IJ} = F_{kI}F_{kJ} \quad \Leftrightarrow \quad \boldsymbol{C} = \boldsymbol{F}^T\boldsymbol{F}. \tag{3.6}$$

The key role that this material tensor plays in describing local deformation will be discussed later. For now, we recall the discussion of quadratic forms (see Section 2.5.5) and note that consequently the eigenvalues of $\boldsymbol{C}$ correspond to the squares of the ellipsoid half-lengths (i.e. $a = \sqrt{\lambda_1^{\boldsymbol{C}}}, b = \sqrt{\lambda_2^{\boldsymbol{C}}}$ and $c = \sqrt{\lambda_3^{\boldsymbol{C}}}$). Substituting these values into dV and dividing by dV_0, we obtain the local ratio of deformed-to-reference volume:

$$\frac{dV}{dV_0} = \sqrt{\lambda_1^{\boldsymbol{C}}\lambda_2^{\boldsymbol{C}}\lambda_3^{\boldsymbol{C}}} = \sqrt{\det \boldsymbol{C}} = \sqrt{\det(\boldsymbol{F}^T\boldsymbol{F})} = \det \boldsymbol{F} = J, \tag{3.7}$$

where $J \equiv \det \boldsymbol{F}$ is the *Jacobian* of the deformation mapping. The Jacobian therefore gives the volume change of a particle. A volume preserving deformation satisfies $J = 1$ at all particles.

The definition of the Jacobian leads to a local condition for invertibility called the *inverse function theorem*. Assuming that φ is continuously differentiable and $J(\boldsymbol{X}) \neq 0$, then there exists a neighborhood of particle $\boldsymbol{X}$ where φ is a one-to-one mapping. Failure of this condition means that $dV/dV_0 \to 0$, which implies that the volume at the point shrinks

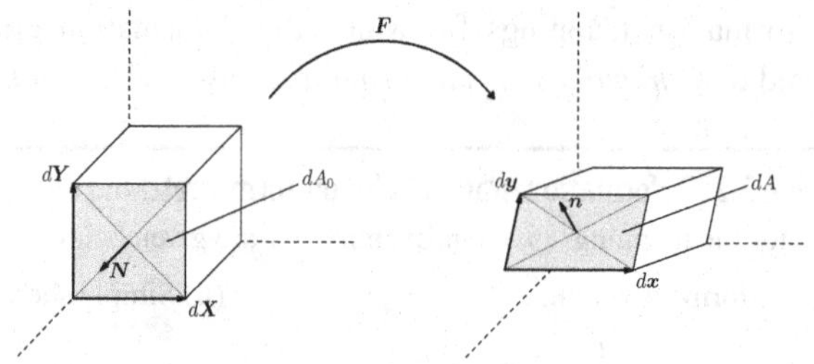

Fig. 3.5 The mapping of the infinitesimal area dA_0 in the reference configuration to dA in the deformed configuration.

to zero, a physically unacceptable situation. It is important to realize that even if the local invertibility condition is satisfied at all points in the body, this does not guarantee that φ is globally one-to-one. Consider, for example, a two-dimensional deformation taking a line segment into a pretzel shape. The mapping is locally invertible at all points, however, some distant points will be mapped to the same positions (the pretzel intersections) making the mapping globally not one-to-one. It is impossible to catch a violation like this using a local pointwise criterion.

Example 3.3 (Volume change for uniform stretching and simple shear) The Jacobians for the deformation gradients in Example 3.2 are given below.

(i) Uniform stretching:

$$J = \det \boldsymbol{F} = \alpha_1 \alpha_2 \alpha_3.$$

(ii) Simple shear:

$$J = \det \boldsymbol{F} = 1.$$

We see that uniform stretching is associated with a volume change, while simple shear is volume preserving. The latter is also true for an arbitrary simple shear along direction $\boldsymbol{s}$ on a plane with normal $\boldsymbol{n}$, for which the deformation gradient is $\boldsymbol{F} = \boldsymbol{I} + \gamma \boldsymbol{s} \otimes \boldsymbol{n}$. The proof is left as an exercise (see Exercise 3.4).

3.4.3 Area changes

We have seen that the determinant of the deformation gradient provides a measure for local volume change. We are also interested in local area changes that are important when discussing stress, which is defined as a force per unit area.

Consider two infinitesimal material vectors $d\boldsymbol{X}$ and $d\boldsymbol{Y}$ (see Fig. 3.5). The area dA_0 spanned by these vectors and the normal to the plane they define are, respectively,

$$dA_0 = \|d\boldsymbol{X} \times d\boldsymbol{Y}\|, \qquad \boldsymbol{N} = \frac{d\boldsymbol{X} \times d\boldsymbol{Y}}{\|d\boldsymbol{X} \times d\boldsymbol{Y}\|}. \tag{3.8}$$

Together these variables define an *element of oriented area* in the reference configuration: $d\boldsymbol{A}_0 = \boldsymbol{N}\, dA_0 = d\boldsymbol{X} \times d\boldsymbol{Y}$. As a result of the imposed deformation, characterized locally by the deformation gradient $\boldsymbol{F}$, the material vectors $d\boldsymbol{X}$ and $d\boldsymbol{Y}$ are mapped to the spatial vectors $d\boldsymbol{x}$ and $d\boldsymbol{y}$ of the deformed configuration. The corresponding element of oriented area in the deformed configuration is

$$\begin{aligned}[d\boldsymbol{A}]_i = n_i\, dA = [d\boldsymbol{x} \times d\boldsymbol{y}]_i &= \epsilon_{ijk} dx_j dy_k \\ &= \epsilon_{ijk}(F_{jJ} dX_J)(F_{kK} dY_K) \\ &= \epsilon_{ijk} F_{jJ} F_{kK}\, dX_J dY_K.\end{aligned}$$

Applying F_{iI} to both sides and using Eqn. (2.7) gives

$$n_i dA F_{iI} = (\det \boldsymbol{F}) \epsilon_{IJK} dX_J dY_K = J\, [d\boldsymbol{X} \times d\boldsymbol{Y}]_I = J N_I dA_0.$$

Finally, multiplying both sides by $\boldsymbol{F}^{-T}$, we obtain *Nanson's formula*[6] relating elements of oriented area in the reference and deformed configurations:

$$n_i\, dA = J F_{Ii}^{-1} N_I\, dA_0 \quad \Leftrightarrow \quad \boldsymbol{n}\, dA = J \boldsymbol{F}^{-T} \boldsymbol{N}\, dA_0. \tag{3.9}$$

This relation plays a key role in the derivation of material and mixed stress measures, which are tensors in the reference configuration and two-point tensors, respectively.

Example 3.4 (The effect of simple shear on oriented areas) Consider a simple shear deformation along the 1-direction (see Example 3.2). Simple shear is volume preserving, so $J = 1$. The changes in elements of oriented area oriented along the main directions of the axes in the reference configuration ($\boldsymbol{N}_1\, dA_0 = \boldsymbol{e}_1\, dA_0, \boldsymbol{N}_2\, dA_0 = \boldsymbol{e}_2\, dA_0$) are obtained from Nanson's formula:

$$\boldsymbol{n}_1\, dA_1 = J\boldsymbol{F}^{-T}\boldsymbol{N}_1\, dA_0 = \begin{bmatrix} 1 & 0 & 0 \\ -\gamma & 1 & 0 \\ 0 & 0 & 1 \end{bmatrix} \begin{bmatrix} 1 \\ 0 \\ 0 \end{bmatrix} dA_0 = \begin{bmatrix} 1 \\ -\gamma \\ 0 \end{bmatrix} dA_0 = (\boldsymbol{e}_1 - \gamma \boldsymbol{e}_2)\, dA_0,$$

$$\boldsymbol{n}_2\, dA_2 = J\boldsymbol{F}^{-T}\boldsymbol{N}_2\, dA_0 = \begin{bmatrix} 1 & 0 & 0 \\ -\gamma & 1 & 0 \\ 0 & 0 & 1 \end{bmatrix} \begin{bmatrix} 0 \\ 1 \\ 0 \end{bmatrix} dA_0 = \begin{bmatrix} 0 \\ 1 \\ 0 \end{bmatrix} dA_0 = \boldsymbol{e}_2\, dA_0.$$

This gives for the 1-direction $dA_1 = dA_0\sqrt{1+\gamma^2}$ and $\boldsymbol{n}_1 = (\boldsymbol{e}_1 - \gamma\boldsymbol{e}_2)/\sqrt{1+\gamma^2}$, and for the 2-direction $dA_2 = dA_0$ and $\boldsymbol{n}_2 = \boldsymbol{e}_2 = \boldsymbol{N}_2$. Thus an element of area oriented along the 1-direction in the reference configuration stretches and rotates with a simple shear applied in the 1-direction, while area oriented along the 2-direction is unaffected.

[6] This formula is named after Edward J. Nanson (1850–1936), a British mathematician educated at Trinity College Cambridge who immigrated to Australia and became a professor of mathematics at the University of Melbourne. Nanson derived the relation in the context of hydrodynamic theory [Nan74, Nan78]. Interestingly, Nanson became far better known for his reform of the Australian voting system. The voting system proposed by Nanson sometimes goes under the name of "Nanson's rule" [Nio87]. Thus, Nanson has left formulas and rules in very different disciplines.

3.4.4 Pull-back and push-forward operations

With the introduction of the deformation gradient we have two mixed tensors that map material vectors to spatial vectors: F_{iJ} and F^{-T}_{iJ}. Also, we can use F^T_{Ij} and F^{-1}_{Ij} to map spatial vectors to material vectors. Thus, the deformation gradient provides a natural mechanism for mapping between material components and spatial components of a tensor field. For example, consider the velocity field $\boldsymbol{v}(\boldsymbol{x},t)$, which is a spatial vector field with components $v_i(\boldsymbol{x},t)$. We may convert this to a material vector field by a so-called *pull-back operation*:

$$V_I(\boldsymbol{X},t) \equiv F^{-1}_{Ii}(\boldsymbol{X},t)v_i(\varphi(\boldsymbol{X},t),t),$$

where we have now used an upper-case V to emphasize that the pulled-back velocity field is associated with the particular reference configuration under consideration. The two-point tensor F^T_{Ii} can also be used to pull-back a spatial vector. However, it is important to note that this operation produces a different material tensor field than the one obtained by the pull-back operation using $\boldsymbol{F}^{-1}$. These material tensors have no particular physical significance; however, it is often convenient to work with such fields.[7]

In a similar manner we may identify two *push-forward operations* that convert material vector fields to spatial vector fields. For example, imagine a constant material vector field $\boldsymbol{G}(\boldsymbol{X})$ with unit length and which points in the 1-direction everywhere. That is, $\boldsymbol{G}(\boldsymbol{X}) = \boldsymbol{e}_1$, with components $G_I = \delta_{I1}$. The push-forward of $\boldsymbol{G}$ is

$$\breve{G}_i(\boldsymbol{x},t) \equiv F_{iI}(\varphi^{-1}(\boldsymbol{x},t),t)G_I(\varphi^{-1}(\boldsymbol{x},t)) = F_{i1}(\varphi^{-1}(\boldsymbol{x},t),t),$$

where we have used the notation $\breve{G}$, this one time, to indicate the change from the material form to the spatial form. Here we have retained the upper-case G in order to emphasize the fact that, even though the vector field is a spatial one, its value depends on the reference configuration. The push-forward operation has given us a spatial vector field $\breve{\boldsymbol{G}}$ which changes from position to position in B. At any particular position $\boldsymbol{x} \in B$, the magnitude of this vector is equal to the stretch ratio $\alpha = \left\|\boldsymbol{F}(\varphi^{-1}(\boldsymbol{x}))\boldsymbol{e}_1\right\|$ for the material particle currently located at $\boldsymbol{x}$. Further, we see that the direction of the spatial vector is no longer aligned with the 1-direction, but depends explicitly on the motion. This is an example of a vector field which is said to be *convected with the body*; it is sometimes also called an *embedded material field*. This is meant to indicate that changes in the value of the field are entirely due to changes in the deformation mapping.

The idea of pull-back and push-forward operations can easily be extended to higher-order tensor fields. For second-order spatial tensors four distinct types of pull-back operations may be defined corresponding to the two different mappings available for transforming each of the spatial vector arguments into material vector arguments. The situation is similar for push-forward operations. For even higher-order tensor fields the number of possible

[7] We will see an example of this in Section 4.4 where we derive the material form of the equations for the balance of linear and angular momentum. No additional physical content is gained by recasting these equations in the reference configuration, however, the resulting problem is often significantly easier to solve.

pull-back and push-forward operations becomes significant, but fortunately, most of these operations find little (or no) application within standard applications of the theory.

3.4.5 Polar decomposition theorem

The deformation gradient $\boldsymbol{F}$ represents an affine mapping[8] of the neighborhood of a material particle from the reference to deformed configuration. We state above that $\boldsymbol{F}$ provides a measure for the deformation of the neighborhood. This is a true statement but it is not precise. When we say "deformation" we are implicitly referring to changes in the shape of the neighborhood. This includes changes in lengths or *stretching* and changes in angles or *shearing* (see Example 3.1). However, the deformation gradient may also include a part that is simply a rotation of the neighborhood. Since rotation does not play a role in shape change, it would be useful to decompose $\boldsymbol{F}$ into its rotation and "shape-change" parts. It turns out that such a decomposition exists and is unique. This statement is called the *polar decomposition theorem.*

Polar decomposition theorem Any tensor $\boldsymbol{F}$ with positive determinant ($\det \boldsymbol{F} > 0$) can be uniquely expressed as

$$F_{iJ} = R_{iI}U_{IJ} = V_{ij}R_{jJ} \quad \Leftrightarrow \quad \boldsymbol{F} = \boldsymbol{R}\boldsymbol{U} = \boldsymbol{V}\boldsymbol{R}, \tag{3.10}$$

called the *right and left polar decompositions* of $\boldsymbol{F}$, where $\boldsymbol{R}$ is a proper orthogonal transformation (finite rotation) and $\boldsymbol{U}$ and $\boldsymbol{V}$ are symmetric positive-definite tensors called, respectively, the *right and left stretch tensors.*[9] This theorem is true for any second-order tensor (with positive determinant), but here it is applied to the two-point deformation gradient. Thus, we find that $\boldsymbol{R}$ is a two-point tensor and $\boldsymbol{U}$ and $\boldsymbol{V}$ are material and spatial second-order tensors, respectively. In accordance with our case convention, we have used upper-case U and V to indicate that these tensors are associated with the reference configuration.

Considering the right polar decomposition, it is natural to imagine a two-stage sequence where a material neighborhood first changes its shape and then is rotated into the deformed configuration, $\boldsymbol{F}d\boldsymbol{X} = \boldsymbol{R}(\boldsymbol{U}d\boldsymbol{X})$. For the left decomposition, the neighborhood is first rotated and then its shape is changed into the deformed configuration, $\boldsymbol{F}d\boldsymbol{X} = \boldsymbol{V}(\boldsymbol{R}d\boldsymbol{X})$. Although $\boldsymbol{U}$ is a material tensor and $\boldsymbol{V}$ is a spatial tensor, the two stretch tensors are equivalent in the sense that they both fully describe the deformation of the neighborhood of a particle. To see this, we begin by proving the polar decomposition theorem and, in the process of doing so, we introduce a number of important variables and relations.

[8] An affine mapping is a transformation that preserves collinearity, i.e. points that were originally on a straight line remain on a straight line. Strictly speaking, this includes rigid-body translation, however, the deformation gradient is insensitive to translation.

[9] The name "stretch tensor" is a bit unfortunate since $\boldsymbol{U}$ and $\boldsymbol{V}$ include information on both stretching and shear. We will see later that this terminology is related to the physical significance of the eigenvalues of these tensors.

Proof Start with Eqn. (3.3):

$$dx_i = F_{iJ}\, dX_J.$$

We require $\det \boldsymbol{F} > 0$, so $d\boldsymbol{x} = \boldsymbol{0}$ iff $d\boldsymbol{X} = \boldsymbol{0}$. Therefore, $d\boldsymbol{x} \cdot d\boldsymbol{x}$ is a positive-definite quadratic form:

$$dx_k dx_k = (F_{kI} dX_I)(F_{kJ} dX_J) = (F_{kI} F_{kJ}) dX_I dX_J = C_{IJ} dX_I dX_J > 0, \quad \forall d\boldsymbol{X} \neq \boldsymbol{0}.$$

$\boldsymbol{C}$ is symmetric and positive definite, which means that its square root exists. Let us make the enlightened "guess" that $\boldsymbol{U} = \sqrt{\boldsymbol{C}}$, or in other words that

$$\boldsymbol{C} = \boldsymbol{F}^T \boldsymbol{F} = \boldsymbol{U}^2 = \boldsymbol{U}^T \boldsymbol{U}, \tag{3.11}$$

where the last equality follows from the symmetry of $\boldsymbol{U}$. Later we will also require the determinants of $\boldsymbol{C}$ and $\boldsymbol{U}$:

$$\det \boldsymbol{C} = \det(\boldsymbol{F}^T \boldsymbol{F}) = \det(\boldsymbol{F}^T) \det \boldsymbol{F} = (\det \boldsymbol{F})^2, \tag{3.12}$$

$$\det \boldsymbol{U} = \det \sqrt{\boldsymbol{C}} = \det \boldsymbol{F}. \tag{3.13}$$

Now, if $\boldsymbol{F} = \boldsymbol{R}\boldsymbol{U}$, then we have

$$\boldsymbol{R} = \boldsymbol{F}\boldsymbol{U}^{-1}. \tag{3.14}$$

We need to prove that $\boldsymbol{R}$, defined in this manner, is proper orthogonal, i.e. that it satisfies the following two conditions:

1. $\boldsymbol{R}^T \boldsymbol{R} = \boldsymbol{I}$.

   Proof

$$\begin{aligned} \boldsymbol{R}^T \boldsymbol{R} &= (\boldsymbol{F}\boldsymbol{U}^{-1})^T \boldsymbol{F}\boldsymbol{U}^{-1} \\ &= \boldsymbol{U}^{-T} (\boldsymbol{F}^T \boldsymbol{F}) \boldsymbol{U}^{-1} \\ &= \boldsymbol{U}^{-T} (\boldsymbol{U}^T \boldsymbol{U}) \boldsymbol{U}^{-1} = (\boldsymbol{U}^{-T} \boldsymbol{U}^T)(\boldsymbol{U}\boldsymbol{U}^{-1}) = \boldsymbol{I}\boldsymbol{I} = \boldsymbol{I}, \end{aligned}$$

   where Eqn. (3.11) was used to go from the second to the third line. □

2. $\det \boldsymbol{R} = +1$.

   Proof

$$\det \boldsymbol{R} = \det(\boldsymbol{F}\boldsymbol{U}^{-1}) = \det \boldsymbol{F} \frac{1}{\det \boldsymbol{U}} = \det \boldsymbol{F} \frac{1}{\det \boldsymbol{F}} = 1,$$

   where Eqn. (3.13) was used. □

So far we have found one particular decomposition, $\boldsymbol{F} = \boldsymbol{R}\boldsymbol{U}$, with $\boldsymbol{U}$ defined in Eqn. (3.11), that satisfies the polar decomposition theorem. We must still prove that this is the only possible choice, i.e. that the decomposition is unique. Let us assume that there exists another decomposition $\bar{\boldsymbol{R}}\bar{\boldsymbol{U}}$ such that $\boldsymbol{F} = \boldsymbol{R}\boldsymbol{U} = \bar{\boldsymbol{R}}\bar{\boldsymbol{U}}$. Then, $\boldsymbol{F}^T \boldsymbol{F} = \boldsymbol{U}^2 = \bar{\boldsymbol{U}}^2$, so $\boldsymbol{U} = \bar{\boldsymbol{U}}$. The last step is correct since any positive-definite tensor has a unique positive-definite square root. The uniqueness of $\boldsymbol{R}$ then follows from $(\boldsymbol{R} - \bar{\boldsymbol{R}})\boldsymbol{U} = \boldsymbol{0}$. We have proven the right polar decomposition theorem.

The proof for the left polar decomposition is completely analogous and leads to the following definitions. The *left Cauchy–Green deformation tensor* $\boldsymbol{B}$ is

$$B_{ij} = F_{iK} F_{jK} \quad \Leftrightarrow \quad \boldsymbol{B} = \boldsymbol{F}\boldsymbol{F}^T. \tag{3.15}$$

$\boldsymbol{B}$ is symmetric and positive definite. The left stretch tensor $\boldsymbol{V}$ is defined through

$$\boldsymbol{B} = \boldsymbol{F}\boldsymbol{F}^T = \boldsymbol{V}^2, \tag{3.16}$$

so that $\boldsymbol{V} = \sqrt{\boldsymbol{B}}$. The determinants of $\boldsymbol{B}$ and $\boldsymbol{V}$ are

$$\det \boldsymbol{B} = (\det \boldsymbol{F})^2, \quad \det \boldsymbol{V} = \det \boldsymbol{F}. \tag{3.17}$$

Finally, we must prove that $\boldsymbol{R}$ is the same in both the right and left decompositions. We can prove this by contradiction. Assume that the rotations in the right and left decompositions are different:

$$\boldsymbol{F} = \boldsymbol{R}\boldsymbol{U} = \boldsymbol{V}\widetilde{\boldsymbol{R}}.$$

Now consider[10]

$$\boldsymbol{F} = \boldsymbol{R}\boldsymbol{U} = \boldsymbol{R}\boldsymbol{U}(\boldsymbol{R}^T\boldsymbol{R}) = (\boldsymbol{R}\boldsymbol{U}\boldsymbol{R}^T)\boldsymbol{R}.$$

The final expression has the same form as the left polar decomposition. By the uniqueness of the left polar decomposition we then have

$$\boldsymbol{V} = \boldsymbol{R}\boldsymbol{U}\boldsymbol{R}^T, \tag{3.18}$$

which is called the *congruence relation* and $\widetilde{\boldsymbol{R}} = \boldsymbol{R}$, which completes the proof of the polar decomposition theorem. □

In a practical calculation of the polar decomposition, it is necessary to compute $\boldsymbol{U}$ or $\boldsymbol{V}$, which are defined as the square roots of $\boldsymbol{C}$ and $\boldsymbol{B}$. A convenient approach is to use the spectral decomposition representation of the Cauchy–Green tensors. For example, to compute $\boldsymbol{U}$, we first write the spectral decomposition of $\boldsymbol{C}$ (see Eqn. (2.81)):

$$\boldsymbol{C} = \sum_{\alpha=1}^{3} \lambda_\alpha^{\boldsymbol{C}} \boldsymbol{\Lambda}_\alpha^{\boldsymbol{C}} \otimes \boldsymbol{\Lambda}_\alpha^{\boldsymbol{C}}, \tag{3.19}$$

where $\lambda_\alpha^{\boldsymbol{C}}$ and $\boldsymbol{\Lambda}_\alpha^{\boldsymbol{C}}$ are the eigenvalues and eigenvectors of $\boldsymbol{C}$. Then $\boldsymbol{U}$ follows as

$$\boldsymbol{U} = \sum_{\alpha=1}^{3} \sqrt{\lambda_\alpha^{\boldsymbol{C}}} \boldsymbol{\Lambda}_\alpha^{\boldsymbol{C}} \otimes \boldsymbol{\Lambda}_\alpha^{\boldsymbol{C}}. \tag{3.20}$$

[10] See Exercises 3.5 and 3.6.

Similarly, for $\boldsymbol{V}$ we have

$$\boldsymbol{V} = \sum_{\alpha=1}^{3} \sqrt{\lambda_\alpha^{\boldsymbol{B}}} \boldsymbol{\Lambda}_\alpha^{\boldsymbol{B}} \otimes \boldsymbol{\Lambda}_\alpha^{\boldsymbol{B}}, \tag{3.21}$$

where $\lambda_\alpha^{\boldsymbol{B}}$ and $\boldsymbol{\Lambda}_\alpha^{\boldsymbol{B}}$ are the eigenvalues and eigenvectors of $\boldsymbol{B}$. Now, using the congruence relation (Eqn. (3.18)) we have

$$\boldsymbol{V} = \boldsymbol{R}\boldsymbol{U}\boldsymbol{R}^T$$

$$\sum_{\alpha=1}^{3} \sqrt{\lambda_\alpha^{\boldsymbol{B}}} \Lambda_{\alpha i}^{\boldsymbol{B}} \Lambda_{\alpha j}^{\boldsymbol{B}} = R_{iI} \left(\sum_{\alpha=1}^{3} \sqrt{\lambda_\alpha^{\boldsymbol{C}}} \Lambda_{\alpha I}^{\boldsymbol{C}} \Lambda_{\alpha J}^{\boldsymbol{C}} \right) R_{jJ}$$

$$= \sum_{\alpha=1}^{3} \sqrt{\lambda_\alpha^{\boldsymbol{C}}} (R_{iI} \Lambda_{\alpha I}^{\boldsymbol{C}})(R_{jJ} \Lambda_{\alpha J}^{\boldsymbol{C}}).$$

Due to the uniqueness of the polar decomposition we have[11]

$$\lambda_\alpha^{\boldsymbol{B}} = \lambda_\alpha^{\boldsymbol{C}}, \qquad \boldsymbol{\Lambda}_\alpha^{\boldsymbol{B}} = \boldsymbol{R}\boldsymbol{\Lambda}_\alpha^{\boldsymbol{C}}. \tag{3.22}$$

Thus, the eigenvalues of $\boldsymbol{C}$ and $\boldsymbol{B}$ (as well as $\boldsymbol{U}$ and $\boldsymbol{V}$) are the same, and the eigenvectors are related through the rotational part of the deformation gradient.

Example 3.5 (Polar decomposition for uniform stretching and simple shear) Consider the deformation mappings given in Example 3.1. The deformation gradients associated with these mappings are given in Example 3.2. The right Cauchy–Green deformation tensors for these mappings are:

(i) uniform stretching:

$$[\boldsymbol{C}] = \begin{bmatrix} \alpha_1^2 & 0 & 0 \\ 0 & \alpha_2^2 & 0 \\ 0 & 0 & \alpha_3^2 \end{bmatrix};$$

(ii) simple shear:

$$[\boldsymbol{C}] = \begin{bmatrix} 1 & \gamma & 0 \\ \gamma & 1+\gamma^2 & 0 \\ 0 & 0 & 1 \end{bmatrix}.$$

Let us explore the right polar decomposition for these cases.

1. For uniform stretching, the eigenvalues and eigenvectors of $\boldsymbol{C}$ are

$$\lambda_1^{\boldsymbol{C}} = \alpha_1^2, \quad \lambda_2^{\boldsymbol{C}} = \alpha_2^2, \quad \lambda_3^{\boldsymbol{C}} = \alpha_3^2,$$

$$\left[\boldsymbol{\Lambda}_1^{\boldsymbol{C}}\right] = [1,0,0]^T, \quad \left[\boldsymbol{\Lambda}_2^{\boldsymbol{C}}\right] = [0,1,0]^T, \quad \left[\boldsymbol{\Lambda}_3^{\boldsymbol{C}}\right] = [0,0,1]^T.$$

The right stretch tensor follows from Eqn. (3.20) as

$$[\boldsymbol{U}] = \begin{bmatrix} \alpha_1 & 0 & 0 \\ 0 & \alpha_2 & 0 \\ 0 & 0 & \alpha_3 \end{bmatrix},$$

and then from Eqn. (3.14) $\boldsymbol{R} = \boldsymbol{I}$. In this simple case, the deformation gradient corresponds to pure stretching without rotation.

[11] More precisely, the most we can say is that $\boldsymbol{\Lambda}_\alpha^{\boldsymbol{B}} = \pm\boldsymbol{R}\boldsymbol{\Lambda}_\alpha^{\boldsymbol{C}}$. However, if we require that the eigenvectors of $\boldsymbol{B}$ and $\boldsymbol{C}$ individually both form right-handed systems, then choosing one of the eigenvectors, say the $\alpha = 1$ case, such that $\boldsymbol{\Lambda}_1^{\boldsymbol{B}} = \boldsymbol{R}\boldsymbol{\Lambda}_1^{\boldsymbol{C}}$ ensures that Eqn. (3.22) is satisfied for each $\alpha = 1, 2, 3$.

2. For simple shear, the eigenvalues and eigenvectors of $\boldsymbol{C}$ are

$$\lambda_1^C = 1 - \gamma\beta^-, \quad \lambda_2^C = 1 + \gamma\beta^+, \quad \lambda_3^C = 1,$$

$$\left[\boldsymbol{\Lambda}_1^C\right] = \frac{\left[-\beta^+, 1, 0\right]^T}{\sqrt{1+(\beta^+)^2}}, \quad \left[\boldsymbol{\Lambda}_2^C\right] = \frac{\left[\beta^-, 1, 0\right]^T}{\sqrt{1+(\beta^-)^2}}, \quad \left[\boldsymbol{\Lambda}_3^C\right] = [0,0,1]^T,$$

where $\beta^{\pm} = \frac{1}{2}(\sqrt{4+\gamma^2} \pm \gamma) \geq 1$. The right stretch tensor follows from Eqn. (3.20):

$$[U] = \frac{\sqrt{1-\gamma\beta^-}}{1+(\beta^+)^2}\begin{bmatrix} (\beta^+)^2 & -\beta^+ & 0 \\ -\beta^+ & 1 & 0 \\ 0 & 0 & 0 \end{bmatrix} + \frac{\sqrt{1+\gamma\beta^+}}{1+(\beta^-)^2}\begin{bmatrix} (\beta^-)^2 & \beta^- & 0 \\ \beta^- & 1 & 0 \\ 0 & 0 & 0 \end{bmatrix} + \begin{bmatrix} 0 & 0 & 0 \\ 0 & 0 & 0 \\ 0 & 0 & 1 \end{bmatrix}.$$

The rotation can be computed from Eqn. (3.14), but the analytical form is complex and we do not give it here.

We have shown that the deformation gradient can be uniquely decomposed into a finite rotation and stretch. But what is the physical significance of the stretch tensors and the related Cauchy–Green deformation tensors? This is discussed next.

3.4.6 Deformation measures and their physical significance

The right and left stretch tensors $\boldsymbol{U}$ and $\boldsymbol{V}$ characterize the shape change of a particle neighborhood, but they are inconvenient to work with because their components are irrational functions of $\boldsymbol{F}$ that are difficult to obtain. This is clearly demonstrated for the simple shear problem in Example 3.5. Instead, the right and left Cauchy–Green deformation tensors $\boldsymbol{C}$ and $\boldsymbol{B}$, which are uniquely related to the stretch tensors, are usually preferred. For solids,[12] the most convenient variable is $\boldsymbol{C}$. Next, we discuss the physical significance of the components of this tensor.

Let us start by considering changes in length of material vectors and see how this is related to the components of the material tensor $\boldsymbol{C}$. In Fig. 3.4, we show the mapping of the infinitesimal material vector $d\boldsymbol{X}$ in the reference configuration to the spatial vector $d\boldsymbol{x}$. The lengths squared of these two vectors are

$$dS^2 = dX_I dX_I,$$
$$ds^2 = dx_i dx_i = (F_{iI}dX_I)(F_{iJ}dX_J) = (F_{iI}F_{iJ})dX_I dX_J = C_{IJ}dX_I dX_J,$$

where we have used Eqns. (3.3) and (3.6). The change in squared length follows as

$$ds^2 - dS^2 = (C_{IJ} - \delta_{IJ})dX_I dX_J.$$

Next, we define the *Lagrangian strain tensor* $\boldsymbol{E}$ as

$$E_{IJ} = \frac{1}{2}(C_{IJ} - \delta_{IJ}) = \frac{1}{2}(F_{iI}F_{iJ} - \delta_{IJ}) \quad \Leftrightarrow \quad \boldsymbol{E} = \frac{1}{2}(\boldsymbol{C} - \boldsymbol{I}) = \frac{1}{2}(\boldsymbol{F}^T\boldsymbol{F} - \boldsymbol{I}). \tag{3.23}$$

[12] For fluids, measures of deformation are less important than rates of deformation that are discussed later.

The change in squared length is then

$$ds^2 - dS^2 = 2E_{IJ}dX_I dX_J.$$

The $\frac{1}{2}$ factor in the definition of the Lagrangian strain (which leads to the factor of 2 above) is introduced to agree with the infinitesimal definition of strain familiar from elasticity theory. We will see this later when we discuss linearization in Section 3.5.

The physical significance of the diagonal elements of $\boldsymbol{C}$ becomes apparent when considering the change in length of an infinitesimal material vector oriented along an axis direction. For example, consider $[d\boldsymbol{X}] = [dX_1, 0, 0]^T$ oriented along the 1-direction. The length of this vector in the reference configuration and that of its image in the deformed configuration are, respectively,

$$dS^2 = dX_I dX_I = (dX_1)^2,$$
$$ds^2 = C_{IJ}dX_I dX_J = C_{11}(dX_1)^2.$$

The stretch along the 1-direction is then

$$\alpha^{(1)} = \frac{ds}{dS} = \sqrt{C_{11}};$$

similarly for the 2- and 3-directions,

$$\alpha^{(2)} = \sqrt{C_{22}}, \qquad \alpha^{(3)} = \sqrt{C_{33}}.$$

We see that *the diagonal components of $\boldsymbol{C}$ are related to stretching of material elements oriented along the axis directions in the reference configuration.*

The physical significance of the off-diagonal elements of $\boldsymbol{C}$ can be explored by considering two material vectors $[d\boldsymbol{X}] = [dX_1, 0, 0]^T$ and $[d\boldsymbol{Y}] = [0, dX_2, 0]^T$, oriented along the 1- and 2-directions. The vectors are mapped to the spatial vectors $d\boldsymbol{x}$ and $d\boldsymbol{y}$. The vectors $d\boldsymbol{X}$ and $d\boldsymbol{Y}$ are orthogonal in the reference configuration. In the deformed configuration, the angle θ_{12} between $d\boldsymbol{x}$ and $d\boldsymbol{y}$ is given by

$$\begin{aligned}\cos\theta_{12} &= \frac{d\boldsymbol{x}\cdot d\boldsymbol{y}}{\|d\boldsymbol{x}\|\,\|d\boldsymbol{y}\|} = \frac{C_{IJ}dX_I dY_J}{[C_{KL}dX_K dX_L]^{1/2}[C_{MN}dY_M dY_N]^{1/2}} \\ &= \frac{C_{12}dX_1 dX_2}{(\sqrt{C_{11}}dX_1)(\sqrt{C_{22}}dX_2)} = \frac{C_{12}}{\sqrt{C_{11}}\sqrt{C_{22}}}.\end{aligned}$$

Similarly.

$$\cos\theta_{13} = \frac{C_{13}}{\sqrt{C_{11}}\sqrt{C_{33}}}, \qquad \cos\theta_{23} = \frac{C_{23}}{\sqrt{C_{22}}\sqrt{C_{33}}}.$$

We see that *the off-diagonal components of $\boldsymbol{C}$ are related to angle changes between pairs of elements oriented along the axis directions in the reference configuration.*

In its principal coordinate system, $\boldsymbol{C}$ is diagonal:

$$[\boldsymbol{C}] = \begin{bmatrix} \lambda_1^C & 0 & 0 \\ 0 & \lambda_2^C & 0 \\ 0 & 0 & \lambda_3^C \end{bmatrix},$$

where λ_α^C are the eigenvalues of C. Given the physical significance of the components of C, we see that λ_α^C are the squares of the stretches in the principal coordinate system, i.e. the squares of the *principal stretches*. In the principal coordinate system, the stretch tensor corresponds to uniform stretching along the principal directions.[13] Recall that the eigenvalues of the right stretch tensor are the square roots of the eigenvalues of C (see Eqn. (3.20)). Therefore, the eigenvalues of U are the principal stretches. This is the reason for the term "stretch tensor."

Example 3.6 (Lagrangian strain for uniform stretching and simple shear) Consider the deformation mappings given in Example 3.1. The right Cauchy–Green deformation tensors associated with these mappings are given in Example 3.5. The corresponding Lagrangian strain tensors are:

(i) uniform stretching:

$$[E] = \frac{1}{2}\begin{bmatrix} \alpha_1^2 - 1 & 0 & 0 \\ 0 & \alpha_2^2 - 1 & 0 \\ 0 & 0 & \alpha_3^2 - 1 \end{bmatrix};$$

(ii) simple shear:

$$[E] = \frac{1}{2}\begin{bmatrix} 0 & \gamma & 0 \\ \gamma & \gamma^2 & 0 \\ 0 & 0 & 0 \end{bmatrix}.$$

Let us explore the stretching and angle changes for these deformations.

1. For uniform stretching, the stretches of elements originally oriented along the axes are

$$\alpha^{(k)} = \sqrt{C_{\underline{k}\underline{k}}} = \sqrt{\alpha_k^2} = \alpha_k, \qquad k = 1, 2, 3.$$

Since C is diagonal, it is already expressed in its principal coordinate system and α_k are the principal stretches. The changes in angle between pairs of elements originally aligned with the axes are

$$\cos\theta_{k\ell} = \frac{C_{k\ell}}{\sqrt{C_{\underline{k}\underline{k}}}\sqrt{C_{\underline{\ell}\underline{\ell}}}} = \frac{0}{\alpha_k \alpha_\ell} = 0 \quad \Rightarrow \quad \theta_{k\ell} = 90^\circ \qquad k, \ell = 1, 2, 3,\ k \neq \ell.$$

As expected, elements originally aligned with the axes remain orthogonal under uniform stretching.

2. For simple shear, the stretches for elements originally oriented along the axes are

$$\alpha^{(1)} = 1, \quad \alpha^{(2)} = \sqrt{1+\gamma^2}, \quad \alpha^{(3)} = 1.$$

It is clear from Fig. 3.3(c) that there is no change in length in directions 1 and 3, while an application of Pythagoras' theorem gives $\alpha^{(2)}$. The changes in angle between elements originally aligned with the axes are

$$\cos\theta_{12} = \frac{\gamma}{\sqrt{1+\gamma^2}}, \qquad \cos\theta_{13} = \cos\theta_{23} = 0.$$

Again, these results are readily verified by considering the geometry of Fig. 3.3(c). The principal stretches and directions can be obtained from the eigenvalues and eigenvectors of C given in Example 3.5.

[13] An important point to keep in mind is that all symmetric tensors C have a principal orientation (as shown in Section 2.5.3). This means that *every* shape-changing deformation, including shear, is equivalent to three direct stretches along some set of orthogonal directions.

3.4.7 Spatial strain tensor

Consider the deformation from the perspective of the spatial description: $\boldsymbol{X} = \boldsymbol{\varphi}^{-1}(\boldsymbol{x})$. The local deformation in an infinitesimal neighborhood of a continuum particle is[14]

$$dX_I = \frac{\partial \varphi_I^{-1}}{\partial x_j} dx_j = F_{Ij}^{-1} dx_j. \tag{3.24}$$

The lengths squared of $d\boldsymbol{X}$ and $d\boldsymbol{x}$ are, respectively,

$$\begin{aligned} dS^2 &= dX_I dX_I = F_{Ii}^{-1} F_{Ij}^{-1} dx_i dx_j = (F_{iI} F_{jI})^{-1} dx_i dx_j = B_{ij}^{-1} dx_i dx_j, \\ ds^2 &= dx_i dx_i = \delta_{ij} dx_i dx_j. \end{aligned}$$

The change in length squared follows as

$$ds^2 - dS^2 = (\delta_{ij} - B_{ij}^{-1}) dx_i dx_j = 2 e_{ij} dx_i dx_j,$$

where we have defined the spatial *Euler–Almansi strain tensor*:

$$\begin{aligned} e_{ij} &= \frac{1}{2}(\delta_{ij} - B_{ij}^{-1}) & & & \boldsymbol{e} &= \frac{1}{2}(\boldsymbol{I} - \boldsymbol{B}^{-1}) \\ &= \frac{1}{2}(\delta_{ij} - F_{Ii}^{-1} F_{Ij}^{-1}) & &\Leftrightarrow & &= \frac{1}{2}(\boldsymbol{I} - \boldsymbol{F}^{-T} \boldsymbol{F}^{-1}). \end{aligned} \tag{3.25}$$

Although the Euler–Almansi strain tensor is associated with the reference configuration, and should therefore be represented with an upper-case letter, the use of a lower-case e to distinguish it from the Lagrangian strain tensor is conventional.

Example 3.7 (The spatial strain for uniform stretching and simple shear) Consider the deformation mappings given in Example 3.1. The deformation gradients associated with these mappings are given in Example 3.2. The left Cauchy–Green deformation tensors and the Euler–Almansi strain tensors for these mappings are

(i) uniform stretching:

$$[\boldsymbol{B}] = \begin{bmatrix} \alpha_1^2 & 0 & 0 \\ 0 & \alpha_2^2 & 0 \\ 0 & 0 & \alpha_3^2 \end{bmatrix},$$

$$[\boldsymbol{e}] = \frac{1}{2}\begin{bmatrix} 1-\alpha_1^{-2} & 0 & 0 \\ 0 & 1-\alpha_2^{-2} & 0 \\ 0 & 0 & 1-\alpha_3^{-2} \end{bmatrix};$$

(ii) simple shear:

$$[\boldsymbol{B}] = \begin{bmatrix} 1+\gamma^2 & \gamma & 0 \\ \gamma & 1 & 0 \\ 0 & 0 & 1 \end{bmatrix},$$

$$[\boldsymbol{e}] = \frac{1}{2}\begin{bmatrix} 0 & \gamma & 0 \\ \gamma & -\gamma^2 & 0 \\ 0 & 0 & 0 \end{bmatrix}.$$

Compare these with the material strain measures in Example 3.6.

[14] In Eqn. (3.24) we identify $\partial \boldsymbol{\varphi}^{-1}/\partial \boldsymbol{x}$ with $\boldsymbol{F}^{-1}$. It is easy to see that this is indeed the case. From Eqn. (3.3), we have that $dx_i = F_{iJ} dX_J$, where $F_{iJ} = \partial \varphi_i / \partial X_J$. We denote the inverse mapping from $\boldsymbol{x}$ to $\boldsymbol{X}$ as $\boldsymbol{X} = \boldsymbol{\varphi}^{-1}(\boldsymbol{x})$. Let us denote the gradient of $\boldsymbol{\varphi}^{-1}$ as $G_{Ji} = \partial \varphi_J^{-1}/\partial x_i$, such that $dX_J = G_{Jj} dx_j$. Substituting this into the expression for dx_i above, we have that $dx_i = F_{iJ} G_{Jj} dx_j$. This implies that $F_{iJ} G_{Jj} = \delta_{ij}$, which means that $\boldsymbol{G} = \boldsymbol{F}^{-1}$ as we have stated above.

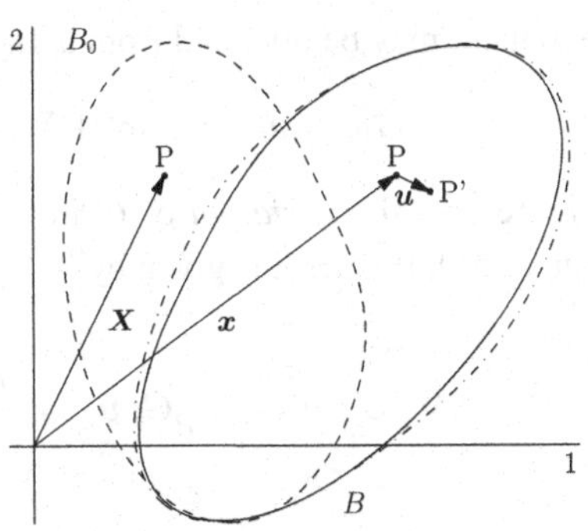

Fig. 3.6 The reference configuration B_0 of a body (dashed), deformed configuration B (solid), and a configuration obtained from B by an additional increment of deformation (dash-dotted). The particle P located at position $\boldsymbol{X}$ in the reference configuration is mapped to a position $\boldsymbol{x}$ in the deformed configuration and then to a new position $\boldsymbol{x} + \boldsymbol{u}$ by the deformation increment (or displacement) $\boldsymbol{u}$.

3.5 Linearized kinematics

The discussion so far has focused on a description of the deformation represented by a given mapping $\varphi(\boldsymbol{X})$. This mapping could, for instance, represent the deformation of the body that brings it into equilibrium with a set of forces or displacements that are prescribed on the boundary of the body. That is, $\varphi(\boldsymbol{X})$ often corresponds to the solution of an equilibrium *boundary-value problem* (as described later in Section 7.1). The nonlinear nature of continuum mechanics problems necessitates (in almost all cases) that these solutions be obtained numerically. Further, numerical solutions are usually obtained by an *incremental process*. In such a process the prescribed boundary values are applied in small parts (or increments) and an equilibrium solution is determined for each value of the boundary conditions until, finally, the desired solution is obtained. In order to successfully implement this solution procedure, it is important to know how to calculate the increments of all quantities that ultimately make up the mathematical equations to be solved during each step of the process. In particular, we will require the linearized or incremental expressions for the kinematic quantities already discussed. These include the deformation gradient, the Cauchy–Green stretch tensors, the Lagrangian strain tensor and the Jacobian.

If the resulting expressions are evaluated at the reference configuration, a linear theory for material deformation called *small-strain elasticity theory* is obtained. This theory is important because, when it is coupled with linear constitutive relations (see Section 6.5), its boundary-value problems can be solved analytically for many problems of interest.

The situation to be investigated is illustrated in Fig. 3.6, which presents a body in the deformed configuration B along with an additional small increment of deformation characterized by the *displacement* field $\boldsymbol{u}(\boldsymbol{X})$, thus $\varphi(\boldsymbol{X}) \rightarrow \varphi(\boldsymbol{X}) + \boldsymbol{u}(\boldsymbol{X})$. Let $G[\varphi]$ be a kinematic field which is a *functional*[15] of the deformation mapping. An approximation to

[15] A "functional," as opposed to a function, is a mapping that takes as its argument a function rather than a variable. To distinguish a functional from a function square brackets are used to enclose its arguments. (This is not to

G after the increment may be obtained from a Taylor expansion,

$$G[\boldsymbol{\varphi} + \boldsymbol{u}] \approx G[\boldsymbol{\varphi}] + \nabla_{\varphi} G[\boldsymbol{\varphi}] \cdot \boldsymbol{u} + \cdots,$$

where $\nabla_{\varphi} G$ represents the *variation* of G in the direction of the vector *field* $\boldsymbol{u}(\boldsymbol{X})$. In analogy to Eqn. (2.88), this term is given by[16]

$$\nabla_{\varphi} G \cdot \boldsymbol{u} = \langle \mathcal{D}_{\varphi} G; \boldsymbol{u} \rangle = \left. \frac{d}{d\eta} G[\boldsymbol{\varphi} + \eta \boldsymbol{u}] \right|_{\eta=0}. \tag{3.26}$$

The linear parts of some important kinematic fields are computed in the following example.

Example 3.8 (Linear parts of kinematic fields) Application of the nonnormalized directional derivative to important kinematic fields yields:

1. Deformation gradient $\boldsymbol{F}$:

$$\langle \mathcal{D}_{\varphi} \boldsymbol{F}; \boldsymbol{u} \rangle = \frac{d}{d\eta} \left[\frac{\partial}{\partial X_J} (\varphi_i + \eta u_i) \right] \bigg|_{\eta=0} = \frac{\partial u_i}{\partial X_J} = \nabla_0 \boldsymbol{u}. \tag{3.27}$$

2. Right Cauchy–Green deformation tensor $\boldsymbol{C}$:

$$\langle \mathcal{D}_{\varphi} \boldsymbol{C}; \boldsymbol{u} \rangle = \frac{d}{d\eta} \left[\frac{\partial}{\partial X_I} (\varphi_i + \eta u_i) \frac{\partial}{\partial X_J} (\varphi_i + \eta u_i) \right] \bigg|_{\eta=0} = \frac{\partial u_i}{\partial X_I} F_{iJ} + \frac{\partial u_i}{\partial X_J} F_{iI},$$

which in direct notation is

$$\langle \mathcal{D}_{\varphi} \boldsymbol{C}; \boldsymbol{u} \rangle = \boldsymbol{F}^T \nabla_0 \boldsymbol{u} + (\boldsymbol{F}^T \nabla_0 \boldsymbol{u})^T.$$

3. Left Cauchy–Green deformation tensor $\boldsymbol{B}$:

$$\langle \mathcal{D}_{\varphi} \boldsymbol{B}; \boldsymbol{u} \rangle = \frac{d}{d\eta} \left[\frac{\partial}{\partial X_I} (\varphi_i + \eta u_i) \frac{\partial}{\partial X_I} (\varphi_j + \eta u_j) \right] \bigg|_{\eta=0} = \frac{\partial u_i}{\partial X_I} F_{jI} + \frac{\partial u_j}{\partial X_I} F_{iI},$$

which in direct notation is

$$\langle \mathcal{D}_{\varphi} \boldsymbol{B}; \boldsymbol{u} \rangle = \nabla_0 \boldsymbol{u} \, \boldsymbol{F}^T + (\nabla_0 \boldsymbol{u} \, \boldsymbol{F}^T)^T.$$

4. Lagrangian strain tensor $\boldsymbol{E}$:

$$\langle \mathcal{D}_{\varphi} \boldsymbol{E}; \boldsymbol{u} \rangle = \langle \mathcal{D}_{\varphi} \frac{1}{2} (\boldsymbol{C} - \boldsymbol{I}); \boldsymbol{u} \rangle = \frac{1}{2} \langle \mathcal{D}_{\varphi} \boldsymbol{C}; \boldsymbol{u} \rangle = \frac{1}{2} \left[\boldsymbol{F}^T \nabla_0 \boldsymbol{u} + (\boldsymbol{F}^T \nabla_0 \boldsymbol{u})^T \right].$$

be confused with the notation for linear functions used in Section 2.3.1.) For example, $I[f] = \int_0^1 f(x)\, dx$ is a functional which given a function $f(x)$ returns its integral over the domain $[0, 1]$.

16 It is important to distinguish between the operators $\langle \mathcal{D}_{\varphi} G; \cdot \rangle$ and $\langle \mathcal{D}_{\boldsymbol{X}} G; \cdot \rangle$. The first deals with the variation of G as one changes the *function* $\boldsymbol{\varphi}(\boldsymbol{X})$ by adding the vector field $\boldsymbol{u}(\boldsymbol{X})$. The second deals with the derivative of G as one changes the *particle* $\boldsymbol{X}$ in the reference configuration. For the latter case, we treat G as a function (rather than a functional) of material coordinates, $G = G(\boldsymbol{X})$, and consider the derivative as $\boldsymbol{X} \rightarrow \boldsymbol{X} + \boldsymbol{u}$, where $\boldsymbol{u}$ is a vector not a vector field. Then $\langle \mathcal{D}_{\boldsymbol{X}} G; \boldsymbol{u} \rangle = \nabla_0 G \cdot \boldsymbol{u}$.

If the linearization is evaluated at the undeformed reference configuration ($\varphi = \mathbf{I}$), then the deformation gradient becomes the identity, $F_{iJ} = \delta_{iJ}$, and the distinction between the reference and deformed coordinates disappears. Thus, $\nabla_0 = \nabla$ and the case of tensor indices is immaterial. In this scenario $\langle \mathcal{D}_\varphi \mathbf{E}; \mathbf{u} \rangle$ is equal to the *small-strain tensor* $\boldsymbol{\epsilon}$ familiar from elasticity theory:

$$\epsilon_{ij} = \frac{1}{2}(u_{i,j} + u_{j,i}) \quad \Leftrightarrow \quad \boldsymbol{\epsilon} = \frac{1}{2}\left[\nabla \mathbf{u} + (\nabla \mathbf{u})^T\right]. \tag{3.28}$$

We can also see this by setting $\varphi = \mathbf{X} + \mathbf{u}$. The deformation gradient is then

$$\mathbf{F} = \frac{\partial \varphi}{\partial \mathbf{X}} = \mathbf{I} + \nabla_0 \mathbf{u}, \tag{3.29}$$

and the Lagrangian strain is

$$\mathbf{E} = \frac{1}{2}(\mathbf{F}^T\mathbf{F} - \mathbf{I}) = \frac{1}{2}\left[\nabla_0 \mathbf{u} + (\nabla_0 \mathbf{u})^T\right] + \frac{1}{2}(\nabla_0 \mathbf{u})^T \nabla_0 \mathbf{u}.$$

The nonlinear part $\frac{1}{2}(\nabla_0 \mathbf{u})^T \nabla_0 \mathbf{u}$ is neglected in the small-strain tensor. (Note that for $\mathbf{F} = \mathbf{I}$, we have $\nabla_0 = \nabla$.) In contrast to the Lagrangian strain tensor, the small-strain tensor is *not* invariant with respect to finite rotations. See Exercises 3.12 and 3.13 for a discussion of this point.

5. Jacobian J:

$$\begin{aligned}\langle \mathcal{D}_\varphi J; \mathbf{u} \rangle &= \langle \mathcal{D}_\varphi \det \mathbf{F}; \mathbf{u} \rangle \\ &= \frac{\partial \det \mathbf{F}}{\partial F_{iJ}} \langle \mathcal{D}_\varphi F_{iJ}; \mathbf{u} \rangle = (\det \mathbf{F}) F^{-1}_{Ji} u_{i,J} = J \operatorname{tr}((\nabla_0 \mathbf{u})\mathbf{F}^{-1}),\end{aligned}$$

where Eqns. (2.54) and (3.27) were used. If the linearization is about the undeformed reference configuration ($J = 1$),

$$\langle \mathcal{D}_\varphi J; \mathbf{u} \rangle = \operatorname{tr} \nabla \mathbf{u} = \operatorname{tr} \boldsymbol{\epsilon}.$$

This is called the *dilatation*. It is a small-strain measure for the local change in volume.

3.6 Kinematic rates

In order to study the dynamical behavior of materials, it is necessary to establish the time rate of change of the kinematic fields introduced so far in this chapter. To do so, we must first discuss time differentiation in the context of the referential and spatial descriptions.

3.6.1 Material time derivative

The difference between the referential and spatial descriptions of a continuous medium becomes particularly apparent when considering the time derivative of tensor fields. Consider the field $\mathfrak{g}$, which can be written within the referential or spatial descriptions (see Section 3.3),

$$\mathfrak{g} = g(\mathbf{x}, t) = \breve{g}(\mathbf{X}(\mathbf{x}, t), t).$$

Here $\mathfrak{g}$ represents the value of the field variable, while g and $\breve{g}$ represent the functional dependence of $\mathfrak{g}$ on specific arguments. There are two possibilities for taking a time derivative:

$$\left.\frac{\partial \breve{g}(\boldsymbol{X},t)}{\partial t}\right|_{\boldsymbol{X}} \quad \text{or} \quad \left.\frac{\partial g(\boldsymbol{x},t)}{\partial t}\right|_{\boldsymbol{x}},$$

where the notation $\square|_{\boldsymbol{X}}$ and $\square|_{\boldsymbol{x}}$ is used (this one time) to place special emphasis on the fact that $\boldsymbol{X}$ and $\boldsymbol{x}$, respectively, are held fixed during the partial differentiation. The first is called the *material time derivative* of $\mathfrak{g}$, since it corresponds to the rate of change of $\mathfrak{g}$ while following a particular material particle $\boldsymbol{X}$. The second derivative is called the *local rate of change* of $\mathfrak{g}$. This is the rate of change of $\mathfrak{g}$ at a fixed spatial position $\boldsymbol{x}$. The material time derivative is the appropriate derivative to use whenever considering the time rate of change of properties tied to the material itself, such as the rate of change of strain at a material particle. It is denoted by a superposed dot, $\dot{\square}$, or by $D\square/Dt$:

$$\dot{\mathfrak{g}} = \frac{D\mathfrak{g}}{Dt} = \frac{\partial \breve{g}(\boldsymbol{X},t)}{\partial t}. \tag{3.30}$$

For example consider the case where $\mathfrak{g}$ is the motion $\boldsymbol{x} = \boldsymbol{\varphi}(\boldsymbol{X},t)$. The first and second material time derivatives of $\boldsymbol{x}$ are the velocity and acceleration of a continuum particle $\boldsymbol{X}$:

$$\breve{v}_i(\boldsymbol{X},t) = \dot{\boldsymbol{x}} = \frac{\partial \varphi_i(\boldsymbol{X},t)}{\partial t}, \qquad \breve{a}_i(\boldsymbol{X},t) = \ddot{\boldsymbol{x}} = \frac{\partial^2 \varphi_i(\boldsymbol{X},t)}{\partial t^2}. \tag{3.31}$$

Although these fields are given as functions over the reference body $\mathcal{B}_0$, they are spatial vector fields and therefore lower-case symbols are appropriate. Expressed in the spatial description, these fields are

$$v_i(\boldsymbol{x},t) \equiv \breve{v}_i(\boldsymbol{X}(\boldsymbol{x},t),t), \qquad a_i(\boldsymbol{x},t) \equiv \breve{a}_i(\boldsymbol{X}(\boldsymbol{x},t),t). \tag{3.32}$$

In some cases, it may be necessary to compute the material time derivative within a spatial description. This can be readily done by using the chain rule,

$$\begin{aligned} \dot{\mathfrak{g}} = \frac{Dg(\boldsymbol{x},t)}{Dt} &= \frac{\partial g(\boldsymbol{x},t)}{\partial t} + \frac{\partial g(\boldsymbol{x},t)}{\partial x_j}\frac{\partial x_j(\boldsymbol{X},t)}{\partial t} \\ &= \frac{\partial g(\boldsymbol{x},t)}{\partial t} + \frac{\partial g(\boldsymbol{x},t)}{\partial x_j} v_j(\boldsymbol{x},t), \end{aligned} \tag{3.33}$$

where we have used Eqns. $(3.31)_1$ and $(3.32)_1$. To clarify how this expression is used in practice, let us take a specific example. Consider using a velocimeter (an instrument for measuring the velocity of a fluid) to measure the velocity, at a position $\boldsymbol{x}$, of a fluid flowing through a channel. The velocimeter's measurement $\boldsymbol{v}(\boldsymbol{x},t)$ provides the velocity at point $\boldsymbol{x}$ as a function of time. At each instant of time a different particle will be passing through the instrument. We wish to calculate the acceleration of the particle going through $\boldsymbol{x}$ at time t. This is *not* the local rate of change of $\boldsymbol{v}$,

$$\frac{\partial \boldsymbol{v}(\boldsymbol{x},t)}{\partial t},$$

which is the rate of change of the reading of the velocimeter at $\boldsymbol{x}$. To obtain the acceleration of the particle, we apply the material time derivative in Eqn. (3.33), which gives

$$a_i = \frac{\partial v_i}{\partial t} + l_{ij} v_j \quad \Leftrightarrow \quad \boldsymbol{a} = \frac{\partial \boldsymbol{v}}{\partial t} + \boldsymbol{l}\boldsymbol{v}, \tag{3.34}$$

where we have defined $\boldsymbol{l}$ as the spatial gradient of the velocity field,

$$l_{ij} = v_{i,j} \quad \Leftrightarrow \quad \boldsymbol{l} = \nabla \boldsymbol{v}. \tag{3.35}$$

This result shows that we are able to compute material time derivatives entirely from information available in the spatial description! Examples of material time differentiation in the referential and spatial descriptions are given below.

Example 3.9 (The material time derivative) Consider the motion:

$$x_1 = (1+t)X_1, \quad x_2 = (1+t)^2 X_2, \quad x_3 = (1+t^2)X_3.$$

In the referential description, velocity and acceleration functions can be readily computed:

$$[\breve{v}] = \left[\frac{\partial \boldsymbol{x}}{\partial t}\right] = \begin{bmatrix} X_1 \\ 2(1+t)X_2 \\ 2tX_3 \end{bmatrix}, \quad [\breve{a}] = \left[\frac{\partial \breve{\boldsymbol{v}}}{\partial t}\right] = \begin{bmatrix} 0 \\ 2X_2 \\ 2X_3 \end{bmatrix}.$$

Next let us consider the spatial description. What would be the velocity function measured at a point $\boldsymbol{x}$ in the spatial description? First, we invert the motion to determine which particles pass through $\boldsymbol{x}$ at time t. This gives

$$X_1 = x_1/(1+t), \quad X_2 = x_2/(1+t)^2, \quad X_3 = x_3/(1+t^2).$$

Next we substitute this into the velocity computed above (as in Eqn. (3.32)) to get

$$v_1 = x_1/(1+t), \quad v_2 = 2x_2/(1+t), \quad v_3 = 2tx_3/(1+t^2).$$

This is what a velocimeter located at $\boldsymbol{x}$ would measure. Now imagine that the velocimeter's measurement is the only information available[17] and we wish to compute the acceleration of a particle passing through it at time t. This is obtained from the material time derivative given in Eqn. (3.34):

$$a_1 = -\frac{x_1}{(1+t)^2} + \frac{1}{(1+t)}\frac{x_1}{(1+t)} = 0,$$

$$a_2 = -\frac{2x_2}{(1+t)^2} + \frac{2}{(1+t)}\frac{2x_2}{(1+t)} = \frac{2x_2}{(1+t)^2},$$

$$a_3 = 2x_3\frac{1+t^2-2t^2}{(1+t^2)^2} + \frac{2t}{(1+t^2)}\frac{2tx_3}{(1+t^2)} = \frac{2x_3}{1+t^2}.$$

Substituting in the motion $x_i(\boldsymbol{X}, t)$, we find as expected that this is exactly the same as the acceleration $\breve{\boldsymbol{a}}$ computed from the material description.

[17] Realistically, in order to ascertain the material accelerations, one would need to use the velocimeter to somehow estimate the velocity gradient, perhaps by taking velocity measurements at nearby points as well. In this example, however, we have the benefit of knowing the full mathematical form of the motion.

We can now turn to a calculation of the material rate of change of various kinematic measures and relations.

3.6.2 Rate of change of local deformation measures

Recall Eqn. (3.3) for the local deformation of an infinitesimal material neighborhood. The material time derivative of this relation is

$$\dot{\overline{dx_i}} = \dot{\overline{F_{iJ}\,dX_J}} = \dot{F}_{iJ}\,dX_J,$$

where the notation is meant to clarify that the dot is applied to the entire term beneath the overbar. Now,

$$\dot{F}_{iJ} = \dot{\overline{\left(\frac{\partial x_i}{\partial X_J}\right)}} = \frac{\partial \breve{v}_i}{\partial X_J} = \frac{\partial v_i}{\partial x_j}\frac{\partial x_j}{\partial X_J} = l_{ij}F_{jJ},$$

thus the rate of change of the deformation gradient is

$$\dot{F}_{iJ} = l_{ij}F_{jJ} \quad \Leftrightarrow \quad \dot{\boldsymbol{F}} = \boldsymbol{l}\boldsymbol{F}. \tag{3.36}$$

The rate of change of local deformation follows as

$$\dot{\overline{dx_i}} = l_{ij}F_{jJ}\,dX_J = l_{ij}\,dx_j. \tag{3.37}$$

We see that in a dynamical spatial setting, the velocity gradient plays a role similar to $\boldsymbol{F}$.

Rate of deformation and spin tensors Let us consider the material time derivative of the squared length of a spatial differential vector, $ds^2 = dx_i dx_i$. This is

$$\dot{\overline{ds^2}} = \dot{\overline{(dx_i dx_i)}} = 2dx_i\dot{\overline{dx_i}} = 2l_{ij}\,dx_i dx_j,$$

where Eqn. (3.37) has been used. The product $dx_i dx_j$ is symmetric and therefore only the symmetric part of $\boldsymbol{l}$ contributes to the above contraction since the contraction with the antisymmetric part is zero (see Section 2.5.2). The symmetric part of $\boldsymbol{l}$ is called the *rate of deformation tensor* and is denoted by $\boldsymbol{d}$:

$$d_{ij} \equiv \frac{1}{2}(l_{ij} + l_{ji}) = \frac{1}{2}(v_{i,j} + v_{j,i}). \tag{3.38}$$

The rate of change of the squared length is then

$$\dot{\overline{ds^2}} = 2d_{ij}\,dx_i dx_j. \tag{3.39}$$

The antisymmetric part of $\boldsymbol{l}$ plays an important role in fluid mechanics. It is called the *spin tensor* and it is denoted by $\boldsymbol{w}$:

$$w_{ij} \equiv \frac{1}{2}(l_{ij} - l_{ji}) = \frac{1}{2}(v_{i,j} - v_{j,i}). \tag{3.40}$$

Rate of change of stretch Recall from Section 3.4 that the stretch α along an infinitesimal line element is given by $\alpha = ds/dS$, where $ds = \sqrt{dx_i dx_i}$ and $dS = \sqrt{dX_I dX_I}$. We wish to compute the rate of change of α. We begin with the material time derivative of ds:

$$\dot{\overline{ds}} = \dot{\overline{\sqrt{dx_i dx_i}}} = \frac{d_{ij} dx_i dx_j}{ds},$$

where Eqn. (3.37) was used and the antisymmetric part of $\boldsymbol{l}$ was discarded. Substituting in $dx_i = m_i ds$, where $\boldsymbol{m}$ is a unit vector pointing along $d\boldsymbol{x}$, and rearranging we have

$$\frac{1}{ds}\dot{\overline{ds}} = d_{ij} m_i m_j. \tag{3.41}$$

Now, note that the material time derivative of α is $\dot{\alpha} = \dot{\overline{ds}}/dS$. Dividing through by α and using Eqn. (3.41) we have

$$\dot{\overline{\ln \alpha}} = d_{ij} m_i m_j, \tag{3.42}$$

where we have also used the identity

$$\dot{\overline{\ln \alpha}} = \dot{\alpha}/\alpha. \tag{3.43}$$

This is the *logarithmic rate of stretch* along direction $\boldsymbol{m}$. Another useful relation can be obtained between the rate of change of stretch and the velocity gradient. Start with $dx_i = F_{iJ} dX_J$ and substitute in $dx_i = m_i ds$ and $dX_I = M_I dS$. Dividing through by dS this is $\alpha m_i = F_{iJ} M_J$. Taking the material time derivative of this relation we have

$$\dot{\alpha} m_i + \alpha \dot{m}_i = \dot{F}_{iJ} M_J = l_{ij} F_{jJ} M_J = l_{ij} F_{jJ} \frac{dX_J}{dS} = l_{ij} \frac{dx_j}{dS} = l_{ij} \alpha m_j.$$

Dividing through by α we have

$$\frac{\dot{\alpha}}{\alpha} m_i + \dot{m}_i = l_{ij} m_j. \tag{3.44}$$

This relation is used next to clarify the physical significance of the eigenvalues and eigenvectors of the rate of deformation tensor.

Eigenvalues and eigenvectors of the rate of deformation tensor $\boldsymbol{d}$ We found earlier that the eigenvalues and eigenvectors of $\boldsymbol{C}$ (and $\boldsymbol{E}$) correspond to the principal stretches and directions of the material. It is of interest to similarly explore the significance of the eigenvalues and eigenvectors of $\boldsymbol{d}$. Consider Eqn. (3.44) for the special case where $\boldsymbol{m} = \boldsymbol{\Lambda}^d$ is an eigenvector of $\boldsymbol{d}$,

$$\frac{\dot{\alpha}}{\alpha}\Lambda_i^{\boldsymbol{d}} + \dot{\overline{\Lambda_i^{\boldsymbol{d}}}} = (d_{ij} + w_{ij})\Lambda_j^{\boldsymbol{d}}$$
$$= \lambda^{\boldsymbol{d}}\Lambda_i^{\boldsymbol{d}} + w_{ij}\Lambda_j^{\boldsymbol{d}}, \tag{3.45}$$

where we have used $d_{ij}\Lambda_j^{\boldsymbol{d}} = \lambda^{\boldsymbol{d}}\Lambda_i^{\boldsymbol{d}}$. Apply $\Lambda_i^{\boldsymbol{d}}$ to both sides of the equation:

$$\frac{\dot{\alpha}}{\alpha}\Lambda_i^{\boldsymbol{d}}\Lambda_i^{\boldsymbol{d}} + \dot{\overline{\Lambda_i^{\boldsymbol{d}}}}\Lambda_i^{\boldsymbol{d}} = \lambda^{\boldsymbol{d}}\Lambda_i^{\boldsymbol{d}}\Lambda_i^{\boldsymbol{d}} + w_{ij}\Lambda_i^{\boldsymbol{d}}\Lambda_j^{\boldsymbol{d}}. \tag{3.46}$$

The above relation can be simplified by making use of the normalization condition of the eigenvectors, $\Lambda_i^{\boldsymbol{d}}\Lambda_i^{\boldsymbol{d}} = 1$, and its material time derivative, $2\dot{\overline{\Lambda_i^{\boldsymbol{d}}}}\Lambda_i^{\boldsymbol{d}} = 0$. Also, $w_{ij}\Lambda_i^{\boldsymbol{d}}\Lambda_j^{\boldsymbol{d}} = 0$, since $\boldsymbol{w}$ is antisymmetric and $\Lambda_i^{\boldsymbol{d}}\Lambda_j^{\boldsymbol{d}}$ is symmetric. Using all of the above, Eqn. (3.46) simplifies to

$$\lambda^{\boldsymbol{d}} = \frac{\dot{\alpha}}{\alpha}, \tag{3.47}$$

which we recognize as the logarithmic rates of stretch from Eqn. (3.43). We have shown that the eigenvalues of $\boldsymbol{d}$ are the logarithmic rates of stretch for the directions that undergo pure instantaneous stretch. We can continue this analysis to gain insight into the physical significance of the spin tensor. Substituting Eqn. (3.47) into Eqn. (3.45) and simplifying gives

$$\dot{\overline{\Lambda_i^{\boldsymbol{d}}}} = w_{ij}\Lambda_j^{\boldsymbol{d}}. \tag{3.48}$$

The spin tensor is antisymmetric and is therefore associated with an axial vector $\boldsymbol{\psi}$ (see Eqn. (2.71)),

$$\psi_k = -\frac{1}{2}\epsilon_{ijk}w_{ij} = -\frac{1}{2}\epsilon_{ijk}v_{i,j}, \tag{3.49}$$

where we have used Eqn. (3.40). In direct notation this is $\boldsymbol{\psi} = \frac{1}{2}\mathrm{curl}\,\boldsymbol{v}$. The inverse of Eqn. (3.49) is $w_{ij} = -\epsilon_{ijk}\psi_k$. Substituting this into Eqn. (3.48) gives

$$\dot{\overline{\boldsymbol{\Lambda}^{\boldsymbol{d}}}} = \boldsymbol{\psi} \times \boldsymbol{\Lambda}^{\boldsymbol{d}}. \tag{3.50}$$

Thus, $\boldsymbol{\psi}$ (and $\boldsymbol{w}$) correspond to the instantaneous rotation experienced by the eigenvectors of $\boldsymbol{d}$. Motions for which $\boldsymbol{\psi} = \boldsymbol{0}$ are called *irrotational*. This is a particularly important concept in fluid mechanics. In an inviscid (nonviscous) fluid, flow remains irrotational if it starts out that way. Viscous fluid flow can only be irrotational if it is uniform and there are no boundaries.

Rate of change of strain The material time derivative of the Lagrangian strain tensor (Eqn. (3.23)) is

$$\dot{\boldsymbol{E}} = \frac{1}{2}(\dot{\boldsymbol{F}}^T\boldsymbol{F} + \boldsymbol{F}^T\dot{\boldsymbol{F}}).$$

Substituting in Eqn. (3.36) and using Eqn. (3.38), we have

$$\dot{E}_{IJ} = F_{iI} d_{ij} F_{jJ} \quad \Leftrightarrow \quad \dot{\boldsymbol{E}} = \boldsymbol{F}^T \boldsymbol{d} \boldsymbol{F}. \tag{3.51}$$

The material time derivative of the spatial Euler–Almansi strain tensor (Eqn. (3.25)) is a bit more tricky because it depends on the inverse of the deformation gradient,

$$\dot{e}_{ij} = -\frac{1}{2}\left(\dot{\overline{F_{Ii}^{-1}}} F_{Ij}^{-1} + F_{Ii}^{-1} \dot{\overline{F_{Ij}^{-1}}}\right). \tag{3.52}$$

We need to find $\dot{\overline{\boldsymbol{F}^{-1}}}$. Start with the identity $F_{iJ} F_{Jj}^{-1} = \delta_{ij}$ and take its material time derivative. This gives

$$\dot{F}_{iJ} F_{Jj}^{-1} + F_{iJ} \dot{\overline{F_{Jj}^{-1}}} = 0.$$

Apply F_{Ii}^{-1} to the above and use Eqn. (3.36) to obtain

$$\dot{\overline{F_{Ij}^{-1}}} = -F_{Ii}^{-1} l_{ij}. \tag{3.53}$$

Substitute Eqn. (3.53) into Eqn. (3.52) and use Eqn. (3.15) to obtain

$$\dot{e}_{ij} = \frac{1}{2}(l_{ki} B_{kj}^{-1} + B_{ik}^{-1} l_{kj}) \quad \Leftrightarrow \quad \dot{\boldsymbol{e}} = \frac{1}{2}(\boldsymbol{l}^T \boldsymbol{B}^{-1} + \boldsymbol{B}^{-1}\boldsymbol{l}). \tag{3.54}$$

An alternative relation is obtained by noting that $\boldsymbol{B}^{-1} = \boldsymbol{I} - 2e$. Substituting this into Eqn. (3.54) gives

$$\dot{\boldsymbol{e}} = \boldsymbol{d} - \boldsymbol{l}^T \boldsymbol{e} - \boldsymbol{e}\boldsymbol{l}. \tag{3.55}$$

Rate of change of volume The Jacobian provides a local measure for volume change. The material time derivative of the Jacobian is

$$\dot{J} = \dot{\overline{\det \boldsymbol{F}}} = \frac{\partial(\det \boldsymbol{F})}{\partial \boldsymbol{F}} : \dot{\boldsymbol{F}} = J\boldsymbol{F}^{-T} : \dot{\boldsymbol{F}}, \tag{3.56}$$

where we have used Eqn. (2.54) and the definition $J \equiv \det \boldsymbol{F}$. Substituting in Eqn. (3.36) and using the identity $\boldsymbol{A} : (\boldsymbol{B}\boldsymbol{C}) = (\boldsymbol{B}^T \boldsymbol{A}) : \boldsymbol{C}$, this simplifies to

$$\dot{J} = J\boldsymbol{I} : \boldsymbol{l}.$$

This relation leads to two alternative forms. In one case, we note that $\boldsymbol{I} : \boldsymbol{l} = \boldsymbol{I} : \nabla \boldsymbol{v} = \operatorname{div} \boldsymbol{v}$. In the other case, we note that $\boldsymbol{I} : \boldsymbol{l} = \operatorname{tr} \boldsymbol{l} = \operatorname{tr}(\boldsymbol{d} + \boldsymbol{w}) = \operatorname{tr} \boldsymbol{d}$ (since $\operatorname{tr} \boldsymbol{w} = 0$ because $\boldsymbol{w}$ is antisymmetric). Thus,

$$\dot{J} = J \operatorname{div} \boldsymbol{v} = J \operatorname{tr} \boldsymbol{l} = J \operatorname{tr} \boldsymbol{d}. \tag{3.57}$$

A motion that preserves volume, i.e. $\dot{J} = 0$, is called an *isochoric motion*. Thus the conditions for an isochoric motion are

$$\operatorname{div} \boldsymbol{v} = v_{k,k} = d_{kk} = 0. \tag{3.58}$$

These are key equations for incompressible fluid flow. For incompressible solids, the following requirement obtained from Eqn. (3.56) is more convenient:

$$\boldsymbol{F}^{-T} : \dot{\boldsymbol{F}} = 0. \tag{3.59}$$

Rate of change of oriented area The material time derivative of an element of oriented area (Eqn. (3.9)) is

$$\dot{\overline{d\boldsymbol{A}}} = \left[\dot{J}\boldsymbol{F}^{-T} + J\dot{\overline{\boldsymbol{F}^{-T}}}\right] d\boldsymbol{A}_0.$$

Substituting in Eqns. (3.57) and (3.53) and using Eqn. (3.9) gives

$$\dot{\overline{dA_i}} = [(v_{k,k})\delta_{ij} - l_{j,i}]\, dA_j \quad \Leftrightarrow \quad \dot{\overline{d\boldsymbol{A}}} = \left[(\operatorname{div} \boldsymbol{v})\boldsymbol{I} - \boldsymbol{l}^T\right] d\boldsymbol{A}. \tag{3.60}$$

3.6.3 Reynolds transport theorem

So far we have discussed the time rate of change of continuum fields. Now, we consider the rate of change of integral quantities. Consider an integral of the field $\mathfrak{g} = g(\boldsymbol{x}, t)$ over a subbody E of the body B:

$$I = \int_E g(\boldsymbol{x}, t)\, dV,$$

where $dV = dx_1 dx_2 dx_3$. The material time derivative of I is

$$\dot{I} = \frac{D}{Dt}\int_E g(\boldsymbol{x}, t)\, dV = \frac{D}{Dt}\int_{E_0} \breve{g}(\boldsymbol{X}, t) J\, dV_0,$$

where we have changed the integration variables from $\boldsymbol{x}$ to $\boldsymbol{X}$, $dV_0 = dX_1 dX_2 dX_3$ and E_0 is the domain occupied by E in the reference configuration. Since E_0 is constant in time the differentiation can be brought inside the integral:

$$\dot{I} = \int_{E_0} \dot{\overline{\breve{g}J}}\, dV_0 = \int_{E_0} [\dot{\breve{g}}J + \breve{g}\dot{J}]\, dV_0 = \int_{E_0} [\dot{\breve{g}} + \breve{g}(\operatorname{div} \breve{\boldsymbol{v}})] J\, dV_0,$$

where we have used Eqn. (3.57). We now change variables back to the spatial description:

$$\dot{I} = \frac{D}{Dt}\int_E g(\boldsymbol{x}, t)\, dV = \int_E [\dot{g} + g(\operatorname{div} \boldsymbol{v})]\, dV. \tag{3.61}$$

This equation is called the *Reynolds transport theorem*. This relation can be recast in a different form that sheds more light on its physical significance. Substituting Eqn. (3.33) into Eqn. (3.61) and simplifying gives

$$\dot{I} = \int_E \left[\frac{\partial g}{\partial t} + \operatorname{div}(g\boldsymbol{v}) \right] dV.$$

Next, we apply the divergence theorem (Eqn. (2.108)) to the second term to obtain

$$\dot{I} = \int_E \frac{\partial g}{\partial t}\, dV + \int_{\partial E} g\boldsymbol{v} \cdot \boldsymbol{n}\, dA. \qquad (3.62)$$

This alternative form for the Reynolds transport theorem states that the rate of change of I is equal to the production of g inside E plus the net transport of g across its boundary ∂E.

A useful corollary to the Reynolds transport theorem for extensive properties, i.e. properties that are proportional to mass, is given in Section 4.1.

Exercises

3.1 [SECTION 3.4] The most general two-dimensional homogeneous finite strain distribution is defined by giving the spatial coordinates as linear homogeneous functions:

$$x_1 = X_1 + aX_1 + bX_2, \quad x_2 = X_2 + cX_1 + dX_2.$$

1. Express the components of the right Cauchy–Green deformation tensor $\boldsymbol{C}$ and Lagrangian strain $\boldsymbol{E}$ in terms of the given constants a, b, c, d. Display your answers in two matrices.
2. Calculate ds^2 and $ds^2 - dS^2$ for $d\boldsymbol{X}$ with components (dL, dL).

3.2 [SECTION 3.4] The deformation of a plate in circular bending is given by

$$x_1 = (X_2 + R)\sin\frac{X_1}{R}, \qquad x_2 = X_2 - (X_2 + R)\left[1 - \cos\frac{X_1}{R}\right], \qquad x_3 = X_3,$$

where L is the length of the plate and R is the radius of curvature.

1. Given a rectangular plate in the reference configuration with length L in the 1-direction and height h in 2-direction, draw the shape of the plate in the deformed configuration for some radius of curvature R.
2. Determine the deformation gradient $\boldsymbol{F}(\boldsymbol{X})$ at any point in the plate.
3. Determine the Jacobian of the deformation $J(\boldsymbol{X})$ at any point in the plate.
4. Use the result for the Jacobian to show that the plate experiences expansion above the centerline and contraction below it.
5. Determine the element of oriented area at the end of the plate in the deformed configuration. In what direction is the end pointing and what is the change in its cross-sectional area?
6. Use the result for the oriented area to show that planes in the reference configuration remain plane in the deformed configuration.

3.3 [SECTION 3.4] In a two-dimensional finite strain experiment, a strain gauge gave stretch ratios α of 0.8 and 0.6 in the X_1- and X_2-directions, respectively, and 0.5 in the direction bisecting the angle between X_1 and X_2.

1. Show in general that the stretch of an element oriented along the unit vector $\boldsymbol{N}$ in the reference configuration is

$$\alpha(\boldsymbol{N}) = \sqrt{\boldsymbol{N} \cdot (\boldsymbol{C}\boldsymbol{N})}.$$

2. Determine the components of $\boldsymbol{C}$ and $\boldsymbol{E}$ at the position of the strain gauge.
3. Determine the new angle between elements initially parallel to the axes.
4. Determine the Jacobian of the deformation.

3.4 [SECTION 3.4] Prove that an arbitrary simple shear described by $\boldsymbol{F} = \boldsymbol{I} + \gamma \boldsymbol{s} \otimes \boldsymbol{N}$ is volume-preserving. Here γ is the shear parameter, $\boldsymbol{s}$ is the spatial shear direction, $\boldsymbol{N}$ is the material shear plane normal ($\boldsymbol{s} \cdot \boldsymbol{N} = 0$).

3.5 [SECTION 3.4] The identity $\boldsymbol{R}\boldsymbol{U}(\boldsymbol{R}^T\boldsymbol{R}) = (\boldsymbol{R}\boldsymbol{U}\boldsymbol{R}^T)\boldsymbol{R}$ was used to prove that the rotation $\boldsymbol{R}$ appearing in the right polar decomposition is the same as that appearing in the left polar decomposition. Verify this identity using indicial notation. (Compare with Exercise 3.6.)

3.6 [SECTION 3.4] Use indicial notation to show that the expression $\boldsymbol{R}\boldsymbol{U}(\boldsymbol{R}\boldsymbol{R}^T)$ is nonsensical. What is $\boldsymbol{R}\boldsymbol{U}(\boldsymbol{R}\boldsymbol{R}^T)$ equal to? (This exercise demonstrates that the congruence relation is unique.)

3.7 [SECTION 3.4] Consider the deformation defined by

$$x_1 = X_3 - X_1 - 2X_2, \quad x_2 = \sqrt{2}(X_1 - 2X_2), \quad x_3 = X_3 + X_1 + 2X_2.$$

1. Calculate the deformation gradient, $\boldsymbol{F}$.
2. Determine the polar decomposition of $\boldsymbol{F} = \boldsymbol{R}\boldsymbol{U}$.
3. Consider a line element $d\boldsymbol{X}$ lying along the X_1 axis with length dS. Under this deformation the line element is stretched and rotated into the line element $d\boldsymbol{x}$.
   a. Calculate the length, $ds = \|d\boldsymbol{x}\|$.
   b. Calculate the vector, $d\boldsymbol{y} = \boldsymbol{U}d\boldsymbol{X}$, and show $\|d\boldsymbol{y}\| = ds$. This shows that all the stretching is represented by $\boldsymbol{U}$.
   c. Explain why $d\boldsymbol{y}$ is parallel to $d\boldsymbol{X}$. Is this true in general?
   d. Calculate the vector $d\boldsymbol{z} = \boldsymbol{R}d\boldsymbol{X}$ and show that $\|d\boldsymbol{z}\| = dS$, i.e. $\boldsymbol{R}$ is a pure rotation with no change in length.

3.8 [SECTION 3.4] The following mapping is an example of a "pure stretch" deformation:

$$x_1 = (1+p)X_1 + qX_2, \qquad x_2 = qX_1 + (1+p)X_2, \qquad x_3 = X_3,$$

where $p > 0$ and $q > 0$ are constants.

1. The above deformation mapping is applied homogeneously to a body which in the reference configuration is a cube with sides a_0. Draw the shape of the cube in the deformed configuration. Provide the dimensions necessary to define the deformed shape.
2. Compute the components of the deformation gradient $\boldsymbol{F}$ and the Jacobian J of the deformation. What conditions do the parameters p and q need to satisfy in order for the following conditions to be met (each separately):
   a. The deformation is invertible. Give an example of a situation where this condition is not satisfied. Draw the result in the deformed configuration and describe the problem that occurs.
   b. The deformation is incompressible.
3. Compute the components of the right Cauchy–Green deformation tensor $\boldsymbol{C}$.
4. Compute the components of the right stretch tensor $\boldsymbol{U}$. **Hint:** First compute the eigenvalues and eigenvectors of $\boldsymbol{C}$ and then use the spectral decomposition of $\boldsymbol{U}$ to obtain the components of $\boldsymbol{U}$.

5. Compute the components of the rotation part $\boldsymbol{R}$ of the polar decomposition of $\boldsymbol{F}$. Given this result, why do you think the deformation is referred to as "pure stretch"? **Hint:** If all is well, this part should require no additional work.

3.9 [SECTION 3.4] The deformation gradient of a homogeneous deformation is given by

$$[\boldsymbol{F}] = \begin{bmatrix} \sqrt{3} & 1 & 0 \\ 0 & 2 & 0 \\ 0 & 0 & 1 \end{bmatrix}.$$

1. Write out the deformation mapping corresponding to this deformation gradient.
2. Compute the components of the right Cauchy–Green deformation tensor $\boldsymbol{C}$. Display your results in matrix form.
3. Compute the principal values (eigenvalues) and principal directions (eigenvectors) of $\boldsymbol{C}$.
4. Determine the polar decomposition, $\boldsymbol{F} = \boldsymbol{R}\boldsymbol{U}$. Write out in matrix form the components of $\boldsymbol{R}$ and $\boldsymbol{U}$ relative to the Cartesian coordinate system.
5. To interpret $\boldsymbol{F} = \boldsymbol{R}\boldsymbol{U}$, we write $\boldsymbol{x} = \boldsymbol{F}\boldsymbol{X} = \boldsymbol{R}\boldsymbol{y}$, where $\boldsymbol{y} = \boldsymbol{U}\boldsymbol{X}$. Now consider a unit circle in the reference configuration. To what does $\boldsymbol{U}$ map this circle in the intermediate configuration $\boldsymbol{y}$? Plot your result pointing out important directions. Next, $\boldsymbol{R}\boldsymbol{y}$ rotates the intermediate configuration to the deformed configuration $\boldsymbol{x}$. By what angle is the intermediate configuration rotated? Plot the change from the intermediate to the deformed configuration pointing out the angle of rotation.
6. Apply the congruence relation to obtain the left stretch tensor $\boldsymbol{V}$. Verify that $\boldsymbol{F} = \boldsymbol{R}\boldsymbol{U} = \boldsymbol{V}\boldsymbol{R}$.

3.10 [SECTION 3.5] Compute the small-strain tensors $\boldsymbol{\epsilon}$ corresponding to the Lagrangian strains for uniform stretch and simple shear in Example 3.6.

3.11 [SECTION 3.5] Prove that the material Lagrangian strain tensor and the spatial Euler–Almansi strain tensor for the uniform stretch and simple shear cases given in Example 3.6 and Example 3.7 are the same to first order when $|\alpha_i - 1| \ll 1$ $(i = 1, 2, 3)$ and $\gamma \ll 1$.

3.12 [SECTION 3.5] Consider a pure two-dimensional rotation by angle θ about the 3-axis. The deformation gradient for this case is

$$[\boldsymbol{F}] = \begin{bmatrix} \cos\theta & -\sin\theta & 0 \\ \sin\theta & \cos\theta & 0 \\ 0 & 0 & 1 \end{bmatrix}.$$

1. Show that the Lagrangian strain tensor $\boldsymbol{E}$ is zero for this case.
2. Compute the small-strain tensor and show that it is *not* zero for $\theta > 0$.
3. As an example, consider the case where $\theta = 30°$. Compute the small-strain tensor for this case. Discuss the applicability of the small-strain approximation.

3.13 [SECTION 3.5] Consider a pure rotation deformation, $\boldsymbol{\varphi}(\boldsymbol{X}) = \boldsymbol{R}\boldsymbol{X}$ with deformation gradient $\boldsymbol{F} = \boldsymbol{R}$. Superposed on this is a small increment of displacement $\boldsymbol{u}$: $\boldsymbol{\varphi} \to \boldsymbol{\varphi} + \boldsymbol{u}$. What is the condition on $\boldsymbol{u}$ to ensure that the perturbation is also a rotation? What does this imply for the small-strain tensor $\boldsymbol{\epsilon}$?

3.14 [SECTION 3.5] Consider the three-dimensional deformation mapping defined by

$$x_1 = aX_1, \qquad x_2 = bX_2 - cX_3, \qquad x_3 = cX_2 + bX_3,$$

where a, b and c are real-valued constants. The deformation is applied to a solid which in the reference configuration is a cube of edge length 1 and aligned with the coordinate directions.

1. Make a schematic drawing of the cube in the reference and deformed configurations, shown as a projection in the 2–3 plane. Calculate the positions of the corners of the cube and indicate the dimensions on the diagram.
2. Compute the deformation gradient $\boldsymbol{F}$. Under what conditions is the deformation invertible?
3. Compute the Lagrangian strain tensor $\boldsymbol{E}$. What happens to the Lagrangian strain tensor when $a = 1, b = \cos\theta, c = \sin\theta$? What does this correspond to physically?
4. Compute the small-strain tensor $\boldsymbol{\epsilon}$ relative to the reference configuration. What happens to the small-strain tensor when $a = 1, b = 1$ and $c \neq 0$? Explain your result.

3.15 [SECTION 3.6] Consider the pure stretch deformation given in Exercise 3.8. Assume that $p = p(t)$ and $q = q(t)$ are functions of time, so that

$$x_1 = (1 + p(t))X_1 + q(t)X_2, \qquad x_2 = q(t)X_1 + (1 + p(t))X_2, \qquad x_3 = X_3.$$

1. Compute the time-dependent deformation gradient $\boldsymbol{F}(t)$.
2. Compute the components of the rate of change of the deformation gradient $\dot{\boldsymbol{F}}$.
3. Compute the inverse deformation mapping, $\boldsymbol{X} = \boldsymbol{\varphi}^{-1}(\boldsymbol{x}, t)$.
4. Verify that $\dot{F}_{iJ} = l_{ij}F_{jJ}$. **Hint:** You will need to compute $\boldsymbol{l}$ for this deformation and show that the result obtained from $l_{ij}F_{jJ}$ is equal to the result obtained above.

3.16 [SECTION 3.6] Consider the motion $\boldsymbol{\varphi}$ of a body given by

$$x_1 = \frac{X_1{}^2 + X_2{}^2}{2B(1+t)}, \qquad x_2 = C\tan^{-1}\frac{X_2}{X_1}, \qquad x_3 = \frac{B}{C}(1+t)X_3,$$

for times $t \geq 0$ and where B and C are constants with dimensions of velocity and length, respectively.

1. What constraints does the requirement of *local invertibility* place on constants B and C?
2. Are the constraints you obtained above sufficient to ensure that the motion $\boldsymbol{\varphi}$ is a 1–1 mapping? Explain.

For the remainder of the problem assume that the reference domain is limited to $X_1 \in [0, W]$, $X_2 \in [0, H]$, $X_3 \in [0, D]$, where $W > 0$, $H > 0$ and $D > 0$.

3. Let us visualize the deformation for the special case $B = C = W = H = D = 1$. Consider a regular square grid with 0.1 spacing in the reference domain. Use a computer to plot the shape of the deformed grid in the deformed configuration in the plane $X_3 = 0$ at times $t = 0, 1, 2$.
4. Find the inverse motion, $\boldsymbol{X} = \boldsymbol{\varphi}^{-1}(\boldsymbol{x}, t)$.
5. Find the velocity field in both the referential and spatial descriptions.
6. Consider a scalar invariant field, $\mathfrak{g}$, given in the referential description by $\mathfrak{g} = G(\boldsymbol{X}, t) = AX_1X_2$, where A is a constant. Find the spatial description, $\mathfrak{g} = g(\boldsymbol{x}, t) = \breve{G}(\boldsymbol{\varphi}(\boldsymbol{X}, t), t)$.
7. Find the material time derivative of $\mathfrak{g}$ using both its Lagrangian and Eulerian representations.

3.17 [SECTION 3.6] Equation (3.44) provides a relation between the rate of change of stretch α, the velocity gradient $\boldsymbol{l}$ and a unit vector $\boldsymbol{m}$ defining a direction in the deformed configuration. Using this relation, show that the following identities are satisfied:

1. $\ddot{\alpha} + \alpha\ddot{m}_i m_i = \alpha a_{i,j} m_i m_j$,
2. $\ddot{\alpha}/\alpha = \dot{m}_i \dot{m}_i + a_{i,j} m_i m_j$,

where $a_i = \dot{v}_i$ and $a_{i,j} = \partial a_i / \partial x_j$ are the components of the acceleration gradient. **Hint:** Note that $\dot{\overline{v_{i,j}}} \neq a_{i,j}$. You will need to find the correct expression for $\dot{\overline{v_{i,j}}}$ as part of your derivation.

3.18 [SECTION 3.6] Given the velocity field

$$v_1 = \exp(x_3 - ct)\cos\omega t, \quad v_2 = \exp(x_3 - ct)\sin\omega t, \quad v_3 = c = \text{const}:$$

1. Show that the speed (magnitude of the velocity) of every particle is constant.
2. Calculate the acceleration components a_i. (Note that you can check whether your answer is reasonable by recalling that the previous part implies $a_i v_i = 0$.)
3. Find the logarithmic rate of stretching, $\dot{\alpha}/\alpha$, for a line element that is in the direction of $(1/\sqrt{2}, 0, 1/\sqrt{2})$ in the deformed configuration at $\boldsymbol{x} = 0$.
4. Integrate the velocity equations to find the motion $\boldsymbol{x} = \boldsymbol{\varphi}(\boldsymbol{X}, t)$ using the initial conditions that at $t = 0$, $\boldsymbol{x} = \boldsymbol{X}$. **Hint:** Integrate the v_3 equation first.

3.19 [Section 3.6] A spherical cavity of radius A at time $t = 0$ in an infinite body is centered at the origin. An explosion inside the cavity at $t = 0$ produces the motion

$$\boldsymbol{x} = \frac{f(R,t)}{R}\boldsymbol{X}, \qquad (*)$$

where $R = \|\boldsymbol{X}\| = \sqrt{X_I X_I}$ is the magnitude of the position vector in the reference configuration. The cavity wall has a radial motion given in Eqn. $(*)$ such that at time t the cavity is spherical with radius $a(t)$.

1. Find the deformation gradient, $\boldsymbol{F}$, and the Jacobian of the transformation, J.
2. Find the velocity and acceleration fields.
3. Show that if the motion is restricted to be isochoric, then $f(R,t) = (R^3 + a^3 - A^3)^{1/3}$.

4 Mechanical conservation and balance laws

In the previous chapter, we derived kinematic fields to describe the possible deformed configurations of a continuous medium. These fields on their own cannot predict the configuration a body will adopt as a result of a given applied loading. To do so requires a generalization of the laws of mechanics (originally developed for collections of particles) to a continuous medium, together with an application of the laws of thermodynamics. The result is a set of *universal* conservation and balance laws that apply to all bodies:

1. conservation of mass;
2. balance of linear and angular momentum;[1]
3. thermal equilibrium (zeroth law of thermodynamics);
4. conservation of energy (first law of thermodynamics);
5. second law of thermodynamics.

These equations introduce four new important quantities to continuum mechanics. The concept of *stress* makes its appearance in the derivation of the momentum balance equations. *Temperature*, *internal energy* and *entropy* star in the zeroth, first and second laws, respectively. In this chapter we focus on the mechanical conservation laws (mass and momentum) leaving the thermodynamic laws to the next chapter.

4.1 Conservation of mass

A basic principle of classical mechanics is that mass is a fixed quantity that cannot be formed or destroyed, but only deformed by applied loads. Thus, the total amount of mass in a closed system is conserved. For a system of particles this is a trivial statement that requires no further clarification. However, for a continuous medium it must be recast in terms of the mass density ρ, which is a measure of the distribution of mass in space.

[1] The balance of angular momentum (or moment of momentum) is taken to be a basic principle in continuum mechanics. This is at odds with some physics textbooks that view the balance of angular momentum as a property of systems of particles in which the internal forces are central. Truesdell discussed this in his article "Whence the Law of Moment and Momentum?" in [Tru68, p. 239]. He stated: "Few if any specialists in mechanics think of their subject in this way. By them, classical mechanics is based on three fundamental laws, asserting the conservation or balance of *force*, *torque*, and *work*, or in other terms, of *linear momentum*, *moment of momentum*, and *energy*." Interestingly, it is possible to show that the two mechanical balance laws can be derived from the balance of energy subject to certain invariance requirements. This was shown separately by Noll [Nol63] and Green and Rivlin [GR64]. The equivalence of the two approaches is discussed by Beatty in [Bea67].

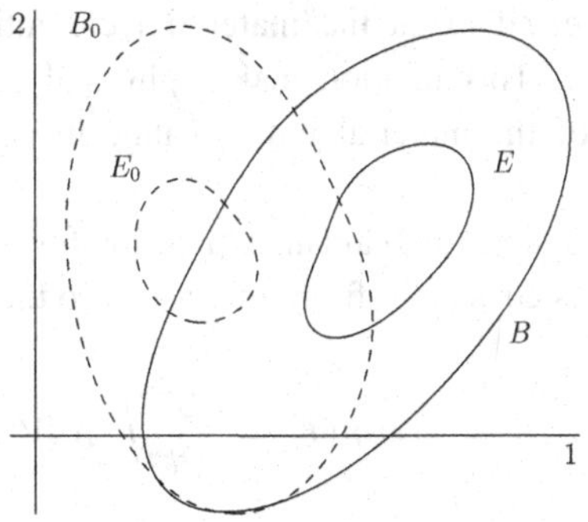

Fig. 4.1 A reference body B_0 and arbitrary subbody E_0 are mapped to B and E in the deformed configuration.

A continuum body occupies domains B_0 and B in the reference and deformed configurations, respectively (see Fig. 4.1). In the absence of diffusion, the principle of conservation of mass requires that the mass of any subbody E_0 remains unchanged by the deformation:

$$m_0(E_0) = m(E) \quad \forall E_0 \subset B_0.$$

Here $m_0(\cdot)$ and $m(\cdot)$ are the mass of a domain in the reference and deformed configurations, respectively. Let $\rho_0 \equiv dm_0/dV_0$ be the *reference* mass density, so that

$$m_0(E_0) = \int_{E_0} \rho_0 \, dV_0.$$

Note that $\rho_0 = \rho_0(\boldsymbol{X})$ is a material scalar invariant. Similarly $\rho \equiv dm/dV$ is the mass density in the deformed configuration, so that

$$m(E) = \int_E \rho \, dV,$$

where $\rho = \rho(\boldsymbol{x})$ is a spatial scalar invariant. With the above definitions, conservation of mass takes the following form:

$$\int_{E_0} \rho_0 \, dV_0 = \int_E \rho \, dV, \quad \forall E_0 \subset B_0.$$

Changing variables on the right from dV to dV_0 ($dV = J dV_0$) and rearranging gives

$$\int_{E_0} (J\breve{\rho} - \rho_0) \, dV_0 = 0 \quad \forall E_0 \subset B_0,$$

where $\breve{\rho}(\boldsymbol{X}) = \rho(\boldsymbol{\varphi}(\boldsymbol{X}))$ is the material description of the mass density. When the description (material or spatial) is clear from the context we will sometimes suppress the $\breve{\square}$ notation. Thus, in the above equation $\breve{\rho}$ becomes ρ. In order for this equation to be satisfied for all E_0 it must be satisfied pointwise, therefore

$$J\rho = \rho_0, \tag{4.1}$$

which is referred to as the material (referential) form[2] of the conservation of mass field equation. This relation makes physical sense. Since the total mass is conserved, the density of the material must change in correspondence with the local changes in volume.

It is also possible to obtain an expression for conservation of mass in the spatial description. If mass is conserved from one instant to the next, then

$$\dot{m}(E) = \frac{D}{Dt}\int_E \rho \, dV = 0 \quad \forall E \subset B.$$

Applying the Reynolds transport theorem (Eqn. (3.61)) gives

$$\int_E [\dot{\rho} + \rho(\operatorname{div} \boldsymbol{v})] \, dV = 0 \quad \forall E \subset B.$$

To be true for any subbody this must be satisfied pointwise:

$$\dot{\rho} + \rho v_{k,k} = 0 \quad \Leftrightarrow \quad \dot{\rho} + \rho(\operatorname{div} \boldsymbol{v}) = 0. \tag{4.2}$$

This is the spatial form of conservation of mass in terms of the material time derivative of the density field. This relation can also be obtained directly from the material description in Eqn. (4.1), by taking its material time derivative and using Eqn. (3.57).

An equivalent expression for Eqn. (4.2) is obtained by substituting in Eqn. (3.33) for the material time derivative:

$$\frac{\partial \rho}{\partial t} + (\rho v_k)_{,k} = 0 \quad \Leftrightarrow \quad \frac{\partial \rho}{\partial t} + \operatorname{div}(\rho \boldsymbol{v}) = 0. \tag{4.3}$$

This is the common form of the *continuity equation*. However, Eqn. (4.2) is also referred to by that name.

The continuity equation can be combined with the expression for material acceleration to form a new relation, which is used in Section 4.2. Starting with the material acceleration expression in Eqn. (3.34),

$$\rho a_i = \rho\left(\frac{\partial v_i}{\partial t} + \frac{\partial v_i}{\partial x_j} v_j\right),$$

[2] The term *material form* indicates that the corresponding partial differential equation is defined with respect to material coordinates. For example, here we have $J(\boldsymbol{X})\rho(\boldsymbol{X}) = \rho_0(\boldsymbol{X})$ for $\boldsymbol{X} \in B_0$.

we add to this the continuity equation (Eqn. (4.3), which is identically zero) multiplied by the velocity vector and then expand and recombine terms to obtain

$$\begin{aligned}
\rho a_i &= \rho\left(\frac{\partial v_i}{\partial t} + \frac{\partial v_i}{\partial x_j}v_j\right) + v_i\left(\frac{\partial \rho}{\partial t} + (\rho v_j)_{,j}\right) \\
&= \rho\left(\frac{\partial v_i}{\partial t} + \frac{\partial v_i}{\partial x_j}v_j\right) + v_i\left(\frac{\partial \rho}{\partial t} + \frac{\partial \rho}{\partial x_j}v_j + \rho\frac{\partial v_j}{\partial x_j}\right) \\
&= \left(\frac{\partial \rho}{\partial t}v_i + \rho\frac{\partial v_i}{\partial t}\right) + \left(\frac{\partial \rho}{\partial x_j}v_i v_j + \rho\frac{\partial v_i}{\partial x_j}v_j + \rho v_i\frac{\partial v_j}{\partial x_j}\right) \\
&= \frac{\partial}{\partial t}(\rho v_i) + \frac{\partial}{\partial x_j}(\rho v_i v_j).
\end{aligned}$$

Thus as long as the continuity equation holds the following identity is satisfied:

$$\rho a_i = \frac{\partial}{\partial t}(\rho v_i) + (\rho v_i v_j)_{,j} \quad \Leftrightarrow \quad \rho \boldsymbol{a} = \frac{\partial}{\partial t}(\rho \boldsymbol{v}) + \operatorname{div}(\rho \boldsymbol{v} \otimes \boldsymbol{v}). \tag{4.4}$$

This relation plays an important role in the definition of the microscopic stress tensor in Section 8.2 of [TM11].

4.1.1 Reynolds transport theorem for extensive properties

The conservation of mass can be used to obtain a useful corollary to the Reynolds transport theorem in Eqn. (3.61), which is reproduced here for convenience:

$$\frac{D}{Dt}\int_E g(\boldsymbol{x},t)\,dV = \int_E [\dot{g} + g(\operatorname{div}\boldsymbol{v})]\,dV,$$

for the special case where g is an *extensive property*, i.e. a property that is proportional to mass.[3] This means that $g = \rho\psi$, where ψ is a density field (g per unit mass). Substituting this into Eqn. (3.61) gives

$$\begin{aligned}
\frac{D}{Dt}\int_E \rho\psi\,dV &= \int_E \left[\dot{\overline{\rho\psi}} + \rho\psi(\operatorname{div}\boldsymbol{v})\right]\,dV \\
&= \int_E \left[\rho\dot{\psi} + \dot{\rho}\psi + \rho\psi(\operatorname{div}\boldsymbol{v})\right]\,dV \\
&= \int_E \left[\rho\dot{\psi} + \psi\left\{\dot{\rho} + \rho(\operatorname{div}\boldsymbol{v})\right\}\right]\,dV.
\end{aligned}$$

The expression in the curly brackets is zero due to conservation of mass (Eqn. (4.2)) and therefore

$$\frac{D}{Dt}\int_E \rho\psi\,dV = \int_E \rho\dot{\psi}\,dV. \tag{4.5}$$

This is the Reynolds transport theorem for extensive properties.

[3] See Section 5.1.3 for more on extensive properties.

4.2 Balance of linear momentum

4.2.1 Newton's second law for a system of particles

Anyone who has taken an undergraduate course in physics is familiar with the dynamics of runaway sand carts with the sand streaming off as the cart speeds away, or rockets whose solid core propellant burns away during the flight of the rocket. As explained in Section 2.1, such problems are described in classical mechanics by Newton's second law, also called the *balance of linear momentum*:

$$\frac{D}{Dt}\boldsymbol{L} = \boldsymbol{F}^{\text{ext}}, \tag{4.6}$$

where $\boldsymbol{L}$ is the linear momentum of the system and $\boldsymbol{F}^{\text{ext}}$ is the total external force acting on the system. Note the use of the *material* time derivative here. For a single particle with mass m,

$$\boldsymbol{L} = m\dot{\boldsymbol{r}}, \quad \boldsymbol{F}^{\text{ext}} = \boldsymbol{f},$$

where $\dot{\boldsymbol{r}}$ is the velocity of the particle and $\boldsymbol{f}$ is the force acting on it. If m is constant, then Eqn. (4.6) reduces to the more familiar form of Newton's second law

$$m\ddot{\boldsymbol{r}} = \boldsymbol{f}.$$

For a system of N particles with positions $\boldsymbol{r}^\alpha$ and velocities $\dot{\boldsymbol{r}}^\alpha$ ($\alpha = 1, 2, \dots, N$), Newton's second law holds individually for each particle,

$$\frac{d}{dt}(m^\alpha \dot{\boldsymbol{r}}^\alpha) = \boldsymbol{f}^\alpha,$$

where $\boldsymbol{f}^\alpha$ is the force on particle α and m^α is its mass. It also holds for the entire system of particles with

$$\boldsymbol{L} = \sum_{\alpha=1}^{N} m^\alpha \dot{\boldsymbol{r}}^\alpha, \quad \boldsymbol{F}^{\text{ext}} = \sum_{\alpha=1}^{N} \boldsymbol{f}^\alpha, \tag{4.7}$$

together with Eqn. (4.6). Examples where this formulation applies are celestial mechanics and a system of interacting atoms. The latter case is considered extensively in [TM11]. In particular, Section 4.3 of [TM11] describes the application of the Newtonian formulation to a system of particles and the more general Lagrangian and Hamiltonian formulations that include it.

The next step, which requires the extension of Newton's laws of motion from a system of particles to the differential equations for a continuous medium, involved the work of many researchers over a 100 year period following the publication of Newton's *Principia* in 1687 and culminating in Lagrange's masterpiece *Méchanique Analitique* published in 1788 [Tru68, Chapter II]. The main figures included the Bernoullis (John, James and Daniel), Leibniz, Euler, d'Alembert, Coulomb and Lagrange. The baton was then passed to Cauchy who developed the concept of stress in its current form. For a discussion of the history of

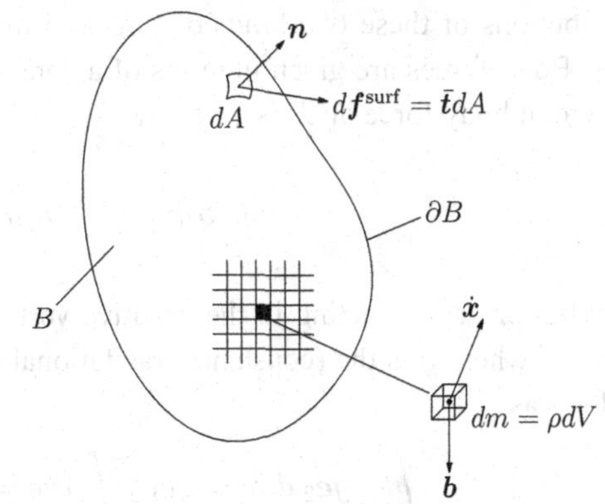

Fig. 4.2 A continuous body B with surface ∂B is divided into an infinite number of infinitesimal volume elements with mass dm and velocity $\dot{\boldsymbol{x}}$. Each volume element experiences a body force $\boldsymbol{b}$ per unit mass. In addition surface elements dA on ∂B with normal $\boldsymbol{n}$ experience forces $d\boldsymbol{f}^{\text{surf}}$ as a result of the interaction of B with its surroundings.

continuum mechanics see, for example, [SL09] and references therein. The theory resulting from these efforts is described in the next section.

4.2.2 Balance of linear momentum for a continuum system

Consider a continuous distribution of matter divided into infinitesimal volume elements as shown schematically in Fig. 4.2. The linear momentum of a single volume element is

$$d\boldsymbol{L} = \dot{\boldsymbol{x}}\, dm,$$

where dm is the mass of the element. Integrating this over the body gives the total linear momentum of B:

$$\boldsymbol{L}(B) = \int_B d\boldsymbol{L} = \int_B \dot{\boldsymbol{x}}\, dm = \int_B \dot{\boldsymbol{x}} \rho\, dV.$$

The balance of linear momentum follows from Eqn. (4.6) as

$$\frac{D}{Dt} \int_B \dot{\boldsymbol{x}} \rho\, dV = \boldsymbol{F}^{\text{ext}}(B), \tag{4.8}$$

where $\boldsymbol{F}^{\text{ext}}(B)$ is the total external force acting on B. As shown in Fig. 4.2, the forces on a continuous medium can be divided into two kinds:[4]

1. *body forces* – forces that act at a distance, such as gravity and electromagnetic fields;
2. *surface forces* – short-range interaction forces across ∂B resulting from the interaction of B with its surroundings.

[4] In reality, surface forces are also forces at a distance resulting from the interaction of atoms from the bodies coming into "contact." However, since the range of interactions is vastly smaller than typical macroscopic length scales, it is more convenient to treat these separately as surface forces rather than as short-range body forces.

The contributions of these two kinds of forces to the total linear momentum are treated separately. Body forces are given in terms of a density field, $\boldsymbol{b}(\boldsymbol{x})$, of body force per unit mass. The total body force on B is given by

$$\int_B \boldsymbol{b}\, dm = \int_B \boldsymbol{b}\rho\, dV. \tag{4.9}$$

For example, for gravity acting in the negative vertical direction, the body force density is $\boldsymbol{b} = -g\boldsymbol{e}_2$, where g is the (constant) gravitational acceleration. The gravitational body force follows as

$$\int_B -g\boldsymbol{e}_2\, dm = -g\boldsymbol{e}_2 \int_B dm = -m(B)g\boldsymbol{e}_2,$$

where $m(B)$ is the total mass of B.

Surface forces (also called *contact forces*) are defined in terms of a surface density field of force per unit area called the *traction* field. Consider an element of area in the deformed configuration ΔA on the surface of a deformed body. The resultant of the external interaction forces across this surface is[5] $\Delta\boldsymbol{f}^{\mathrm{surf}}$. The limit of this force per unit *spatial* area is defined as the external *traction* or *stress vector* $\bar{\boldsymbol{t}}$ (see Fig. 4.2):

$$\bar{\boldsymbol{t}} \equiv \lim_{\Delta A \to 0} \frac{\Delta\boldsymbol{f}^{\mathrm{surf}}}{\Delta A} = \frac{d\boldsymbol{f}^{\mathrm{surf}}}{dA}. \tag{4.10}$$

It is a fundamental assumption of continuum mechanics that this limit exists, is finite and is independent of how the surface area is brought to zero. The total surface force on B is

$$\int_{\partial B} d\boldsymbol{f}^{\mathrm{surf}} = \int_{\partial B} \bar{\boldsymbol{t}}\, dA,$$

and consequently the total force on the body B can now be written as a sum of the body force and surface force contributions:

$$\boldsymbol{F}^{\mathrm{ext}}(B) = \int_B \rho\boldsymbol{b}\, dV + \int_{\partial B} \bar{\boldsymbol{t}}\, dA. \tag{4.11}$$

Substituting this into Eqn. (4.8) gives

$$\frac{D}{Dt}\int_B \rho\dot{\boldsymbol{x}}\, dV = \int_B \rho\boldsymbol{b}\, dV + \int_{\partial B} \bar{\boldsymbol{t}}\, dA.$$

[5] From a microscopic perspective, the force $\Delta\boldsymbol{f}^{\mathrm{surf}}$ is taken to be the force resultant of all atomic interactions across ΔA. Notice that a term $\Delta\boldsymbol{m}^{\mathrm{surf}}$ accounting for the moment resultant of this microscopic distribution has not been included. This is correct as long as electrical and magnetic effects are neglected (we see this in Section 8.2 of [TM11] where we derive the microscopic stress tensor for a system of atoms interacting classically). If $\Delta\boldsymbol{m}^{\mathrm{surf}}$ is included in the formulation it leads to the presence of *couple stresses*, i.e. a field of distributed moments per unit area across surfaces. Theories that include this effect are called *multipolar*. Couple stresses can be important for magnetic materials in a magnetic field and polarized materials in an electric field. See, for example, [Jau67] or [Mal69] for more information on multipolar theories.

Applying the Reynolds transport theorem (Eqn. (4.5)) gives the spatial form of the global balance of linear momentum for B:

$$\int_B \rho \ddot{\boldsymbol{x}}\, dV = \int_B \rho \boldsymbol{b}\, dV + \int_{\partial B} \bar{\boldsymbol{t}}\, dA. \tag{4.12}$$

4.2.3 Cauchy's stress principle

In order to obtain a local expression for the balance of linear momentum it is first necessary to obtain an expression like that in Eqn. (4.12) for an arbitrary *internal* subbody E. This is not a problem for the body force term, but the external traction $\bar{\boldsymbol{t}}$ is defined explicitly in terms of the external forces acting on B across its outer surfaces. This dilemma was addressed by Cauchy in 1822 through his famous *stress principle* that lies at the heart of continuum mechanics. Cauchy's realization was that there is no inherent difference between external forces acting on the physical surfaces of a body and internal forces acting across virtual surfaces within the body. In both cases these can be described in terms of traction distributions. This makes sense since in the end external tractions characterize the interaction of a body with its surroundings (other material bodies) just like internal tractions characterize the interactions of two parts of a material body across an internal surface. A concise definition for Cauchy's stress principle is

Cauchy's stress principle Material interactions across an internal surface in a body can be described as a distribution of tractions in the same way that the effect of external forces on physical surfaces of the body are described.

This may appear to be a very simple, almost trivial, observation. However, it cleared up the confusion resulting from nearly 100 years of failed and partly failed attempts to understand internal forces that preceded Cauchy. Cauchy's principle paved the way for the continuum theory of solids and fluids.

To proceed, we consider a small pillbox-shaped[6] body P inside B, as shown in Fig. 4.3, and write the balance of linear momentum for it:

$$\int_P \rho \ddot{\boldsymbol{x}}\, dV = \int_P \rho \boldsymbol{b}\, dV + \int_{\partial P} \boldsymbol{t}\, dA.$$

Note the absence of the bar over the traction; $\boldsymbol{t}$ is now the *internal* traction evaluated on the surfaces of P regarded as a subbody of B. Rearranging this expression and dividing the boundaries of P into the top and bottom faces and cylindrical circumference (as shown in Fig. 4.3), we have

$$\int_P \rho(\ddot{\boldsymbol{x}} - \boldsymbol{b})\, dV = \int_{\partial P_{\mathrm{top}}} \boldsymbol{t}\, dA + \int_{\partial P_{\mathrm{bot}}} \boldsymbol{t}\, dA + \int_{\partial P_{\mathrm{cyl}}} \boldsymbol{t}\, dA.$$

[6] Given that much of this book was written in Minnesota and Canada, perhaps a "hockey-puck" shaped body would be a more appropriate choice of phrasing.

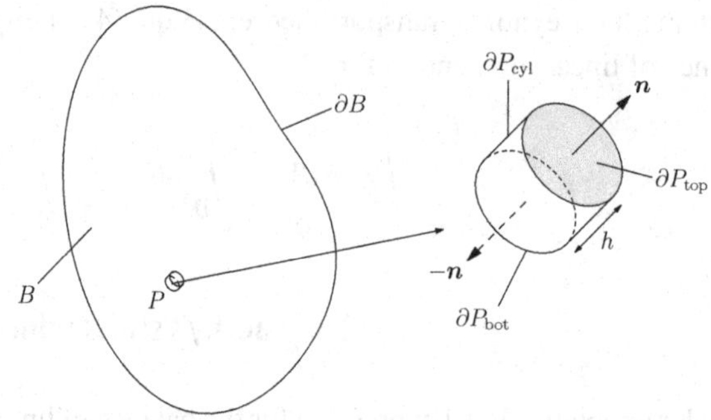

Fig. 4.3 A pillbox-shaped body P inside of a larger body B. The surfaces bounding the pillbox ($\partial P = \partial P_{\text{cyl}} \cup \partial P_{\text{bot}} \cup \partial P_{\text{top}}$), the normals to the top and bottom surfaces ($\boldsymbol{n}$ and $-\boldsymbol{n}$) and its height h are indicated.

Next, take the limit as $h \to 0$. The volume integral on the left-hand side and the surface integral on ∂P_{cyl} go to zero, while the two integrals on the top and bottom faces of the pillbox remain finite, so

$$\int_{\partial P_{\text{top}}} \boldsymbol{t}\, dA + \int_{\partial P_{\text{bot}}} \boldsymbol{t}\, dA = \boldsymbol{0}.$$

Applying the mean-value theorem,[7] this is $\boldsymbol{t}^*|_{\partial P_{\text{top}}} \Delta A + \boldsymbol{t}^*|_{\partial P_{\text{bot}}} \Delta A = \boldsymbol{0}$, where ΔA is the area of the top (or bottom) of the pillbox, $\boldsymbol{t}^* = \boldsymbol{t}(\boldsymbol{x}^*)$, and $\boldsymbol{x}^*$ is a point on ∂P_{top} or ∂P_{bot} as appropriate. In the limit that the area of the pillbox faces is taken to zero this becomes

$$\boldsymbol{t}(\boldsymbol{x})|_{\partial P_{\text{top}}} = -\, \boldsymbol{t}(\boldsymbol{x})|_{\partial P_{\text{bot}}}. \tag{4.13}$$

To continue with the derivation, let us consider $\boldsymbol{t}$ more carefully. The internal tractions are clearly a function of position and possibly time. However, since tractions are defined in terms of surfaces, they must also be related to the particulars of the surface. The only thing characterizing the surface of the pillbox is its normal[8] $\boldsymbol{n}$. Consequently, in general, we expect that $\boldsymbol{t} = \boldsymbol{t}(\boldsymbol{x}, \boldsymbol{n})$. This means that there are an infinite number of tractions (stress vectors) at each point and it is the totality of these, called the *stress state*, that characterizes the internal forces at $\boldsymbol{x}$. We have seen this idea of a vector quantity as a function of a vector before in Section 2.3. If $\boldsymbol{t}$ is a *linear* function of $\boldsymbol{n}$, then this suggests the existence of a second-order tensor. But we have still not shown that this is the case here.

Returning to the pillbox, we saw in Eqn. (4.13) that the traction on the top of the pillbox is equal to the negative of that on the bottom as the size of the pillbox is taken to zero. In

[7] The mean-value theorem for integration states that the definite integral of a continuous function over a specified domain is equal to the value of the function at some specific point within the domain multiplied by the "size" of the domain. For a surface integral, $I = \int_S f(\boldsymbol{x})\, dA$, this means that $I = f(\boldsymbol{x}^*)A$, where $\boldsymbol{x}^* \in S$ and A is the total area of S. Similarly for a volume integral, $I = \int_\Omega f(\boldsymbol{x})\, dV$, this means that $I = f(\boldsymbol{x}^*)V$, where $\boldsymbol{x}^* \in \Omega$ and V is the total volume of Ω.

[8] One may wonder whether a more general theory can be constructed where the traction depends on the surface gradient (i.e. the curvature of the surface) in addition to the normal. However, it can be shown under very general conditions that the traction can only depend on the surface normal [FV89].

terms of coordinates and normals this statement is

$$\boldsymbol{t}(\boldsymbol{x}, \boldsymbol{n}) = -\boldsymbol{t}(\boldsymbol{x}, -\boldsymbol{n}). \tag{4.14}$$

The pillbox shrinks to a single point $\boldsymbol{x}$, but the normals to the top and bottom surfaces remain opposite (Fig. 4.3). We have shown that the tractions on opposite sides of a surface are equal and opposite. This is referred to as *Cauchy's lemma*.

Another approach that leads to the same conclusion is Newton's statement of action and reaction [New62]: "To every action there is always opposed an equal reaction: as, the mutual actions of two bodies on each other are always equal and directed to contrary parts." Consider two bodies $B^{(1)}$ and $B^{(2)}$ that are in contact across some surface. The force per unit area that $B^{(1)}$ exerts on $B^{(2)}$ is $\boldsymbol{t}^{(12)}$ and the force per unit area that $B^{(2)}$ exerts on $B^{(1)}$ is $\boldsymbol{t}^{(21)}$. According to action–reaction, $\boldsymbol{t}^{(21)} = -\boldsymbol{t}^{(12)}$. This is referred to as *Newton's third law*, but since it is equivalent to Cauchy's lemma, it can be considered to be a consequence of Cauchy's stress principle.

The last use we have for the pillbox is to obtain an expression for traction boundary conditions. Consider the special case where one side of the pillbox (say the top) is on a physical surface of the body, then

$$\bar{\boldsymbol{t}}(\boldsymbol{x}) = -\boldsymbol{t}(\boldsymbol{x}, -\boldsymbol{n}) \equiv \boldsymbol{t}(\boldsymbol{x}, \boldsymbol{n}).$$

This equation relates the external applied tractions to the internal stress state. In fact, it shows that the external tractions are boundary conditions for the internal tractions:

$$\boldsymbol{t}(\boldsymbol{x}, \boldsymbol{n}) = \bar{\boldsymbol{t}}(\boldsymbol{x}) \quad \text{on} \quad \partial B. \tag{4.15}$$

4.2.4 Cauchy stress tensor

We have introduced the idea of a stress state, i.e. the fact that the internal forces at a point are characterized by an infinite set of tractions $\boldsymbol{t}(\boldsymbol{n})$ (the explicit dependence on $\boldsymbol{x}$ has been dropped for notational simplicity) for the infinite set of planes passing through the point. We have also shown that $\boldsymbol{t}(\boldsymbol{n}) = -\boldsymbol{t}(-\boldsymbol{n})$, but this just tells us that $\boldsymbol{t}$ is an odd function of $\boldsymbol{n}$. To find the functional relation between $\boldsymbol{t}$ and $\boldsymbol{n}$, we follow Cauchy and consider a small tetrahedron T of height h with one corner at $\boldsymbol{x}$, three faces ∂T_i with normals equal to $-\boldsymbol{e}_i$ and the fourth face ∂T_{n} with normal $\boldsymbol{n}$, such that $n_i > 0$ (Fig. 4.4). We denote the areas of the four faces as $\Delta A_1, \Delta A_2, \Delta A_3$ and ΔA_{n}. By simple geometric projection, we have

$$\Delta A_i = \Delta A_{\mathrm{n}}(\boldsymbol{n} \cdot \boldsymbol{e}_i) = \Delta A_{\mathrm{n}} n_i. \tag{4.16}$$

The global balance of linear momentum for T is

$$\begin{aligned}\int_T \rho(\ddot{\boldsymbol{x}} - \boldsymbol{b})\, dV &= \int_{\partial T} \boldsymbol{t}\, dA \\ &= \int_{\partial T_1} \boldsymbol{t}\, dA + \int_{\partial T_2} \boldsymbol{t}\, dA + \int_{\partial T_3} \boldsymbol{t}\, dA + \int_{\partial T_{\mathrm{n}}} \boldsymbol{t}\, dA.\end{aligned}$$

Applying the mean-value theorem (see footnote 7 on page 114), we have

$$\rho^*(\ddot{\boldsymbol{x}}^* - \boldsymbol{b}^*)\left(\frac{1}{3}h\Delta A_{\mathrm{n}}\right) = \boldsymbol{t}(-\boldsymbol{e}_1)^*\Delta A_1 + \boldsymbol{t}(-\boldsymbol{e}_2)^*\Delta A_2 + \boldsymbol{t}(-\boldsymbol{e}_3)^*\Delta A_3 + \boldsymbol{t}(\boldsymbol{n})^*\Delta A_{\mathrm{n}},$$

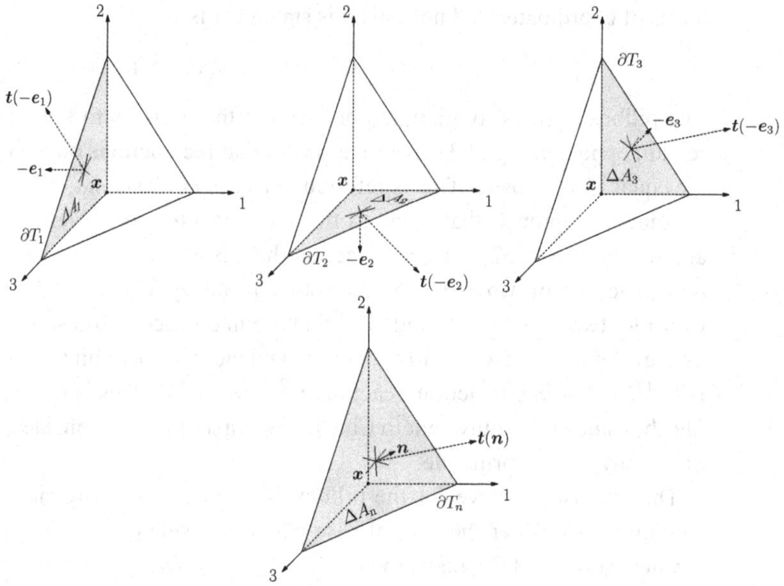

Fig. 4.4 Cauchy's tetrahedron. The four faces of the tetrahedron are indicated; ∂T_i are perpendicular to e_i $(i = 1, 2, 3)$ and have areas ΔA_i; ∂T_{n} is perpendicular to n and has area ΔA_{n}.

where we have used the expression for the volume of a tetrahedron, $\Delta V = \frac{1}{3} h \Delta A_{\text{n}}$, and the * superscript indicates that the function is evaluated in the volume or on the relevant surface as appropriate. Dividing through by ΔA_{n} and using Eqn. (4.16), we have

$$\frac{1}{3} h \rho^* (\ddot{x}^* - b^*) = t(-e_1)^* n_1 + t(-e_2)^* n_2 + t(-e_3)^* n_3 + t(n)^*.$$

Substituting in $t(-e_i) = -t(e_i)$ (Eqn. (4.14)) and shrinking the tetrahedron to a point by taking the limit as its height h goes to zero gives

$$t(n) = t_1 n_1 + t_2 n_2 + t_3 n_3 = t_j n_j, \tag{4.17}$$

where we have defined $t_j \equiv t(e_j)$. To obtain the component form of this relation, we dot both sides with e_i,

$$t_i(n) = (e_i \cdot t_j) n_j.$$

The expression in the parenthesis on the right-hand side is the ith component of the vector t_j. We denote these components by σ_{ij}, i.e. $\sigma_{ij} \equiv e_i \cdot t_j$. The traction–normal relation then takes the form

$$t_i(n) = \sigma_{ij} n_j \quad \Leftrightarrow \quad t(n) = \sigma n. \tag{4.18}$$

This important equation is referred to as *Cauchy's relation*. We now claim that σ_{ij} are the components of a second-order tensor σ called the *Cauchy stress tensor*. The proof is straightforward.

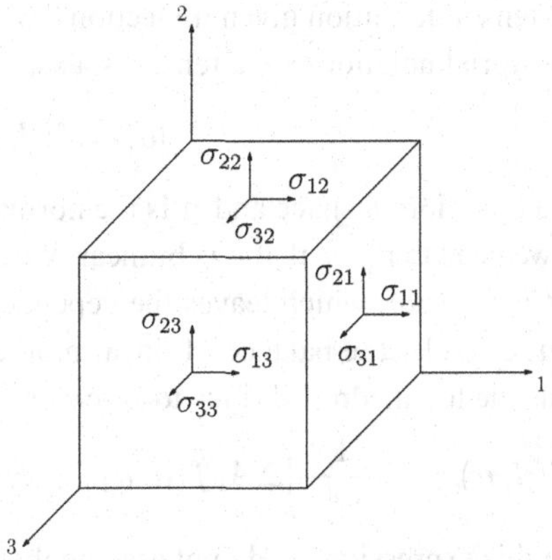

Fig. 4.5 Components of the Cauchy stress tensor. The components on the faces not shown are oriented in the reverse directions to those shown.

Proof $\boldsymbol{t}$ and $\boldsymbol{n}$ are vectors so they transform according to

$$t'_i = Q_{\beta i} t_\beta, \quad n'_j = Q_{\beta j} n_\beta, \quad \mathbf{Q}^T \mathbf{Q} = \boldsymbol{I}.$$

These vectors are related through

$$t'_i = \sigma'_{ij} n'_j.$$

Substituting in the transformation relations, we have

$$Q_{\beta i} t_\beta = \sigma'_{ij} Q_{\beta j} n_\beta.$$

Multiplying both sides by $Q_{\alpha i}$, the left-hand side becomes $Q_{\alpha i} Q_{\beta i} t_\beta = \delta_{\alpha\beta} t_\beta = t_\alpha$, so

$$t_\alpha = (Q_{\alpha i} Q_{\beta j} \sigma'_{ij}) n_\beta.$$

But we also have $t_\alpha = \sigma_{\alpha\beta} n_\beta$, so $\sigma_{\alpha\beta} = Q_{\alpha i} Q_{\beta j} \sigma'_{ij}$. Thus $\boldsymbol{\sigma}$ is a second-order tensor. □

The physical significance of the components of $\boldsymbol{\sigma}$ becomes apparent when considering a cube of material oriented along the basis vectors (see Fig. 4.5). σ_{ij} is the component[9] of the traction (i.e. the stress) acting in the direction $\boldsymbol{e}_i$ on the face normal to $\boldsymbol{e}_j$. The diagonal components σ_{11}, σ_{22}, σ_{33} are normal (tensile/compressive) stresses. The off-diagonal components $\sigma_{12}, \sigma_{13}, \sigma_{23}, \ldots$ are shear stresses.

4.2.5 An alternative ("tensorial") derivation of the stress tensor

Rather than the physical derivation of the Cauchy stress tensor given above, a more direct tensorial derivation is possible. This elegant approach due to Leigh [Lei68] is in the same

[9] We note that in some books the stress tensor is defined as the transpose of the definition given here. Thus they define $\tilde{\boldsymbol{\sigma}} = \boldsymbol{\sigma}^T$. Both definitions are equally valid as long as they are used consistently. We prefer our definition of $\boldsymbol{\sigma}$, since it leads to the Cauchy relation in Eqn. (4.18), which is consistent with the linear algebra idea that the stress tensor operates on the normal to give the traction. With the transposed definition of the stress, the Cauchy relation would be $\tilde{\boldsymbol{\sigma}}^T \boldsymbol{n}$, which is less transparent. Of course, this distinction becomes moot if the stress tensor is symmetric, which as we will see later is the case for nonpolar continua.

spirit as the tensor definition given in Section 2.3. We begin with Cauchy's stress principle. Adopting tensorial notation (i.e. a tensor is a scalar-valued function of vectors), we write

$$\boldsymbol{t}[\boldsymbol{d}] = \boldsymbol{f}(\boldsymbol{d}, \boldsymbol{n}),$$

where $\boldsymbol{d}$ is a direction in space and $\boldsymbol{n}$ is the normal to a plane. The function $\boldsymbol{f}$ looks like a tensor, but we need to prove that it is bilinear. We already know that it is linear with respect to $\boldsymbol{d}$ since $\boldsymbol{t}$ is a vector, which leaves the dependence on $\boldsymbol{n}$. To demonstrate linearity with respect to $\boldsymbol{n}$, consider the balance of linear momentum for Cauchy's tetrahedron. We saw above that as the tetrahedron shrinks to zero

$$\boldsymbol{f}(\boldsymbol{d}, \boldsymbol{n}) = -\frac{1}{\Delta A_{\mathrm{n}}}\left[\Delta A_1 \boldsymbol{f}(\boldsymbol{d}, \boldsymbol{n}_1) + \Delta A_2 \boldsymbol{f}(\boldsymbol{d}, \boldsymbol{n}_2) + \Delta A_3 \boldsymbol{f}(\boldsymbol{d}, \boldsymbol{n}_3)\right]. \tag{4.19}$$

Note that in this expression we do not assume that the tetrahedron faces are oriented along $\boldsymbol{e}_i$. This is a more general case than that assumed above and will result in a general proof. Now from the divergence theorem (Eqn. (2.108)), it is easy to show that any closed surface S satisfies

$$\int_S \boldsymbol{n}\, dA = \boldsymbol{0}.$$

Thus for Cauchy's tetrahedron we have

$$\boldsymbol{n} = -\frac{1}{\Delta A_{\mathrm{n}}}\left(\Delta A_1 \boldsymbol{n}_1 + \Delta A_2 \boldsymbol{n}_2 + \Delta A_3 \boldsymbol{n}_3\right). \tag{4.20}$$

Substituting Eqn. (4.20) into Eqn. (4.19) gives

$$\boldsymbol{f}\left(\boldsymbol{d}, -\sum_{i=1}^{3} \frac{\Delta A_i}{\Delta A_{\mathrm{n}}} \boldsymbol{n}_i\right) = -\sum_{i=1}^{3} \frac{\Delta A_i}{\Delta A_{\mathrm{n}}} \boldsymbol{f}(\boldsymbol{d}, \boldsymbol{n}_i).$$

This proves that $\boldsymbol{f}$ is linear with respect to $\boldsymbol{n}$. We can therefore write

$$\boldsymbol{t}[\boldsymbol{d}] = \boldsymbol{\sigma}[\boldsymbol{d}, \boldsymbol{n}], \tag{4.21}$$

where $\boldsymbol{\sigma}$ is a second-order tensor that we called the Cauchy stress tensor. In this tensorial approach, the components of the stress tensor in the basis $\boldsymbol{e}_i \otimes \boldsymbol{e}_j$ (see Eqn. (2.61)) are *defined* as

$$\sigma_{ij} \equiv \boldsymbol{\sigma}[\boldsymbol{e}_i, \boldsymbol{e}_j].$$

To obtain Cauchy's relation, we substitute $\boldsymbol{d} = d_i \boldsymbol{e}_i$ and $\boldsymbol{n} = n_j \boldsymbol{e}_j$ into Eqn. (4.21),

$$\begin{aligned}
\boldsymbol{t}[d_i \boldsymbol{e}_i] &= \boldsymbol{\sigma}[d_i \boldsymbol{e}_i, n_j \boldsymbol{e}_j] \\
d_i \boldsymbol{t}[\boldsymbol{e}_i] &= d_i n_j \boldsymbol{\sigma}[\boldsymbol{e}_i, \boldsymbol{e}_j] \\
d_i t_i &= d_i n_j \sigma_{ij} \\
(t_i - \sigma_{ij} n_j) d_i &= 0.
\end{aligned}$$

This must be true for any direction $\boldsymbol{d}$, therefore

$$t_i = \sigma_{ij} n_j,$$

which is Cauchy's relation.

4.2.6 Stress decomposition

A commonly employed additive decomposition of the Cauchy stress tensor is

$$\sigma_{ij} = s_{ij} - p\delta_{ij} \quad \Leftrightarrow \quad \boldsymbol{\sigma} = \boldsymbol{s} - p\boldsymbol{I}, \tag{4.22}$$

where

$$p = -\frac{1}{3}\sigma_{kk} \quad \Leftrightarrow \quad p = -\frac{1}{3}\operatorname{tr}\boldsymbol{\sigma} \tag{4.23}$$

is the *hydrostatic stress* or *pressure*, and

$$s_{ij} = \sigma_{ij} + p\delta_{ij} \quad \Leftrightarrow \quad \boldsymbol{s} = \boldsymbol{\sigma} + p\boldsymbol{I} \tag{4.24}$$

is the *deviatoric* part of the Cauchy stress tensor. Note that $\operatorname{tr}\boldsymbol{s} = \operatorname{tr}\boldsymbol{\sigma} + (\operatorname{tr}\boldsymbol{I})p = \operatorname{tr}\boldsymbol{\sigma} + 3p = 0$, thus $\boldsymbol{s}$ only includes information on shear stress. Consequently any material phenomenon that is insensitive to hydrostatic pressure, such as plastic flow in metals, depends only on $\boldsymbol{s}$. A stress state with $\boldsymbol{s} = \boldsymbol{0}$ is called *spherical* or sometimes *hydrostatic* because this is the only possible stress state for static fluids [Mal69]. In this case all directions are principal directions (see Section 2.5.3).

4.2.7 Local form of the balance of linear momentum

We are now ready to derive the local form of the balance of linear momentum in the spatial description. Recall the global form of the balance of linear momentum for a body B in Eqn. (4.12),

$$\int_B \rho(\ddot{\boldsymbol{x}} - \boldsymbol{b})\, dV = \int_{\partial B} \bar{\boldsymbol{t}}\, dA.$$

As discussed in Section 4.2.3, we now assume that we can rewrite the balance of linear momentum for an arbitrary subbody E internal to B and replace the external traction $\bar{\boldsymbol{t}}$ with the internal traction $\boldsymbol{t}$,

$$\begin{aligned}\int_E \rho(\ddot{\boldsymbol{x}} - \boldsymbol{b})\, dV &= \int_{\partial E} \boldsymbol{t}\, dA \\ &= \int_{\partial E} \boldsymbol{\sigma}\boldsymbol{n}\, dA = \int_E (\operatorname{div}\boldsymbol{\sigma})\, dV.\end{aligned}$$

To pass from the first to the second line, we substitute in Cauchy's relation (Eqn. (4.18)) and then apply the divergence theorem (Eqn. (2.108)). Gathering terms and substituting in Eqn. (3.31) we have

$$\int_E [\operatorname{div}\boldsymbol{\sigma} + \rho\boldsymbol{b} - \rho\boldsymbol{a}]\, dV = \boldsymbol{0}.$$

This must be true for any subbody E and therefore the integrand must be zero, which gives the local spatial form of the balance of linear momentum:

$$\sigma_{ij,j} + \rho b_i = \rho a_i \quad \Leftrightarrow \quad \operatorname{div} \boldsymbol{\sigma} + \rho \boldsymbol{b} = \rho \boldsymbol{a} \quad \boldsymbol{x} \in B. \tag{4.25}$$

An alternative form of the balance of linear momentum is obtained by substituting Eqn. (4.4) into the right-hand side of Eqn. (4.25):

$$\sigma_{ij,j} + \rho b_i = \frac{\partial(\rho v_i)}{\partial t} + (\rho v_i v_j)_{,j} \quad \Leftrightarrow \quad \operatorname{div} \boldsymbol{\sigma} + \rho \boldsymbol{b} = \frac{\partial(\rho \boldsymbol{v})}{\partial t} + \operatorname{div}(\rho \boldsymbol{v} \otimes \boldsymbol{v}). \tag{4.26}$$

Equation (4.26) is correct only if the continuity equation is satisfied (since it is used in the derivation of Eqn. (4.4)). It is therefore called the *continuity momentum equation*. It plays an important role in the statistical mechanics derivation of the microscopic stress tensor as shown in Section 8.2 of [TM11].

Finally, for static problems the balance of linear momentum simplifies to

$$\sigma_{ij,j} + \rho b_i = 0 \quad \Leftrightarrow \quad \operatorname{div} \boldsymbol{\sigma} + \rho \boldsymbol{b} = \boldsymbol{0} \quad \boldsymbol{x} \in B. \tag{4.27}$$

These relations are called the *stress equilibrium equations*.

4.3 Balance of angular momentum

In addition to requiring a balance of linear momentum, we must also require that the system be balanced with respect to angular momentum. The *balance of angular momentum* states that the change in angular momentum of a system is equal to the resultant moment applied to it. This is also called the *moment of momentum principle*. In mathematical form this is

$$\frac{D}{Dt} \boldsymbol{H}_0 = \boldsymbol{M}_0^{\text{ext}}, \tag{4.28}$$

where $\boldsymbol{H}_0$ is the angular momentum or moment of momentum of the system about the origin and $\boldsymbol{M}_0^{\text{ext}}$ is the total external moment about the origin.

For a system of N particles,

$$\boldsymbol{H}_0 = \sum_{\alpha=1}^{N} \boldsymbol{r}^\alpha \times (m^\alpha \dot{\boldsymbol{r}}^\alpha), \qquad \boldsymbol{M}_0^{\text{ext}} = \sum_{\alpha=1}^{N} \boldsymbol{r}^\alpha \times \boldsymbol{f}^{\text{ext},\alpha},$$

where $f^{\mathrm{ext},\alpha}$ is the external force acting on particle α and r^{α} is the particle's position. We assume that internal forces resulting from the interaction between particles can be written as a sum over terms aligned with the vectors connecting the particles, and therefore do not contribute to the moment resultant.[10] These expressions readily generalize to a continuum. For a subbody E we have

$$\boldsymbol{H}_0(E) = \int_E \boldsymbol{x} \times (dm\dot{\boldsymbol{x}}) = \int_E \boldsymbol{x} \times (\rho\dot{\boldsymbol{x}})\, dV,$$

$$\boldsymbol{M}_0^{\mathrm{ext}}(E) = \int_E \boldsymbol{x} \times (\rho\boldsymbol{b})\, dV + \int_{\partial E} \boldsymbol{x} \times \boldsymbol{t}\, dA.$$

Note that for a multipolar theory $\boldsymbol{M}_0^{\mathrm{ext}}(E)$ would also include contributions from distributed body couples and corresponding hypertractions. Substituting $\boldsymbol{H}_0(E)$ and $\boldsymbol{M}_0^{\mathrm{ext}}(E)$ into Eqn. (4.28) gives

$$\frac{D}{Dt}\int_E \boldsymbol{x} \times (\rho\dot{\boldsymbol{x}})\, dV = \int_E \boldsymbol{x} \times (\rho\boldsymbol{b})\, dV + \int_{\partial E} \boldsymbol{x} \times \boldsymbol{t}\, dA, \tag{4.29}$$

or in indicial notation

$$\frac{D}{Dt}\int_E \epsilon_{ijk} x_j \dot{x}_k \rho\, dV = \int_E \epsilon_{ijk} x_j b_k \rho\, dV + \int_{\partial E} \epsilon_{ijk} x_j t_k\, dA.$$

Applying the Reynolds transport theorem (Eqn. (4.5)) to the first term and using Cauchy's relation (Eqn. (4.18)) followed by the divergence theorem (Eqn. (2.108)) on the last term gives

$$\int_E \epsilon_{ijk}(\dot{x}_j \dot{x}_k + x_j \ddot{x}_k)\rho\, dV = \int_E \epsilon_{ijk} x_j b_k \rho\, dV + \int_E [\epsilon_{ijk} x_j \sigma_{km}]_{,m}\, dV.$$

The first term in the parenthesis in the left-hand expression cancels since $\epsilon_{ijk}\dot{x}_j\dot{x}_k = [\dot{\boldsymbol{x}} \times \dot{\boldsymbol{x}}]_i = 0$. Then carrying through the differentiation on the right-hand term and rearranging gives

$$\int_E \epsilon_{ijk} x_j \left[\rho\ddot{x}_k - \rho b_k - \sigma_{km,m}\right] dV = \int_E \epsilon_{ijk}\sigma_{kj}\, dV.$$

The expression in the square brackets on the left-hand side is zero due to the balance of linear momentum (Eqn. (4.25)), so that

$$\int_E \epsilon_{ijk}\sigma_{kj}\, dV = 0.$$

This must be satisfied for any subbody E, so it must be satisfied pointwise,

$$\epsilon_{ijk}\sigma_{kj} = 0.$$

This is a system of three equations relating the components of the stress tensor:

$$\sigma_{32} - \sigma_{23} = 0, \quad \sigma_{31} - \sigma_{13} = 0, \quad \sigma_{21} - \sigma_{12} = 0.$$

[10] This condition is *always* satisfied for a system of atoms interacting through a classical force field (see Section 5.8.1 of [TM11]).

The conclusion is that the balance of angular momentum implies that the Cauchy stress tensor is symmetric:[11]

$$\sigma_{ij} = \sigma_{ji} \quad \Leftrightarrow \quad \boldsymbol{\sigma} = \boldsymbol{\sigma}^T. \tag{4.30}$$

4.4 Material form of the momentum balance equations

The derivation of the balance equations in the previous sections is complete. However, it is often computationally more convenient (see Chapter 9) to solve the balance equations in a Lagrangian description. The convenience stems from the fact that in the reference coordinates the boundary of the body ∂B_0 is a constant, whereas in the spatial coordinates the boundary ∂B depends on the motion, which is usually what we are trying to solve for. Thus, we must obtain the material form (or referential form) of the balance of linear and angular momentum. In the process of obtaining these relations we will identify the first and second Piola–Kirchhoff stress tensors (and the related Kirchhoff stress tensor) that play important roles in the material description formulation.

4.4.1 Material form of the balance of linear momentum

To derive the material form of the balance of linear momentum, we begin with the global spatial form for an arbitrary subbody E:

$$\int_E \rho \boldsymbol{a}\, dV = \int_E \rho \boldsymbol{b}\, dV + \int_{\partial E} \boldsymbol{\sigma n}\, dA, \tag{4.31}$$

where $\boldsymbol{a} = \ddot{\boldsymbol{x}}$ is the acceleration. We rewrite each integral in the referential description replacing the spatial fields with their material descriptions. The first integral is

$$\int_E \rho a_i\, dV = \int_{E_0} \breve{\rho} \breve{a}_i J\, dV_0 = \int_{E_0} \rho_0 \breve{a}_i\, dV_0, \tag{4.32}$$

where we have used $J\breve{\rho} = \rho_0$ (Eqn. (4.1)). Similarly the second integral is

$$\int_E \rho b_i\, dV = \int_{E_0} \rho_0 \breve{b}_i\, dV_0. \tag{4.33}$$

The third integral is a bit trickier. To obtain the material form of the surface integral we must use Nanson's formula (Eqn. (3.9)),

$$\int_{\partial E} \sigma_{ij} n_j\, dA = \int_{\partial E_0} (J \breve{\sigma}_{ij} F^{-1}_{Jj}) N_J\, dA_0 = \int_{\partial E_0} P_{iJ} N_J\, dA_0, \tag{4.34}$$

[11] Note that in a multipolar theory with couple stresses, $\boldsymbol{\sigma}$ would not be symmetric since Eqn. (4.29) would include an additional volume integral over body couples and a surface integral over the applied hypertractions. The balance of angular momentum would then supply a set of three equations relating the Cauchy stress tensor to the *couple stress tensor* (see Exercise 4.7). The existence of a couple stress tensor can be derived in a manner similar to that used for the Cauchy stress tensor.

where we have defined the *first Piola–Kirchhoff stress tensor*[12]

$$P_{iJ} = J\sigma_{ij}F^{-1}_{Jj} \quad \Leftrightarrow \quad \boldsymbol{P} = J\boldsymbol{\sigma}\boldsymbol{F}^{-T}, \tag{4.35}$$

which is a two-point tensor. Equation (4.35) is referred to as the *Piola transformation*. The inverse relation is

$$\sigma_{ij} = \frac{1}{J}P_{iJ}F_{jJ} \quad \Leftrightarrow \quad \boldsymbol{\sigma} = \frac{1}{J}\boldsymbol{P}\boldsymbol{F}^{T}. \tag{4.36}$$

Another stress variable (in the deformed configuration) that can be defined at this point is the *Kirchhoff stress tensor* $\boldsymbol{\tau}$:

$$\tau_{ij} = J\sigma_{ij} \quad \Leftrightarrow \quad \boldsymbol{\tau} = J\boldsymbol{\sigma}, \tag{4.37}$$

so that $\boldsymbol{P} = \boldsymbol{\tau}\boldsymbol{F}^{-T}$. Thus, we see that the first Piola–Kirchhoff stress is a pull-back of the Kirchhoff stress (see Section 3.4.4).

It should be emphasized that the first Piola–Kirchhoff stress tensor is defined purely so that the left- and right-hand sides of Eqn. (4.34) have completely analogous symbolic forms. In this sense $\boldsymbol{P}$ is just another mathematical representation (defined for convenience) of the Cauchy stress and does not represent a new physical quantity. To see that this definition is consistent with the physical origin of Cauchy's relation, start with its spatial form, $t_i = \sigma_{ij}n_j$, and substitute in $t_i = df_i/dA$ and Eqn. (4.36):

$$df_i = \frac{1}{J}P_{iJ}F_{jJ}n_j\, dA.$$

Next, substitute in Nanson's formula (Eqn. (3.9)) and rearrange to obtain

$$\frac{df_i}{dA_0} = P_{iJ}N_J.$$

Finally, define the *nominal traction* as $T_i \equiv df_i/dA_0$ (the traction $\boldsymbol{t}$ can then be called the *true traction*) to obtain the material form of Cauchy's relation:

$$T_i = P_{iJ}N_J \quad \Leftrightarrow \quad \boldsymbol{T} = \boldsymbol{P}\boldsymbol{N}. \tag{4.38}$$

We see that $\boldsymbol{P}$ operates on the unit normal in the reference configuration in exact analogy to $\boldsymbol{\sigma}$ acting in the deformed configuration. The first Piola–Kirchhoff stress corresponds to

[12] This stress tensor is named after the Italian mathematician Gabrio Piola and the German physicist Gustav Kirchhoff who independently derived the balance of linear momentum in the material form. Piola published his results in 1832 [Pio32] and Kirchhoff in 1852 [Kir52]. For a discussion of the connection between the work of these two researchers in relation to the stress tensor named after them, see [CR07].

what is commonly called the *engineering stress* or *nominal stress*, because it is the force per unit area in the *reference* configuration. The Cauchy stress, on the other hand, is the *true stress* because it is the force per unit area in the *deformed* configuration. The fact that there are different stress measures with different meanings is often something that is not appreciated by nonexperts in mechanics, especially since these differences vanish if the deformation is small (i.e. when the deformation gradient $\boldsymbol{F} \approx \boldsymbol{I}$).

Returning to the derivation of the balance of linear momentum, substituting the three integrals in Eqns. (4.32)–(4.34) into Eqn. (4.31) gives the global material form of the balance of linear momentum

$$\int_{E_0} \rho_0 (\breve{a}_i - \breve{b}_i)\, dV_0 = \int_{\partial E_0} P_{iJ} N_J \, dA_0.$$

Applying the divergence theorem (Eqn. (2.108)) to the right-hand side, combining terms and recalling that the resulting volume integral must be true for any sub-body E_0 gives the local material form of the balance of linear momentum:

$$P_{iJ,J} + \rho_0 \breve{b}_i = \rho_0 \breve{a}_i \quad \Leftrightarrow \quad \operatorname{Div} \boldsymbol{P} + \rho_0 \breve{\boldsymbol{b}} = \rho_0 \breve{\boldsymbol{a}}, \quad \boldsymbol{X} \in B_0. \tag{4.39}$$

Again, we note that the definitions of the reference mass density and, more particularly, the first Piola–Kirchhoff stress have been chosen in such a way that the spatial form of the balance of linear momentum (Eqn. (4.25)) and the material form (Eqn. (4.39)) have perfectly analogous symbolic forms. Further, we note that although Eqn. (4.39) is called the material form of the balance of linear momentum, the equation is of a mixed nature. It describes how a set of *spatial fields* $\breve{a}_i$, $\breve{b}_i$ and $[\operatorname{Div} \boldsymbol{P}]_i$ (note the lower-case, spatial vector index) must vary from material particle to material particle. That is, how the fields must depend on $X_I \in B_0$ (note the upper-case, material coordinate index).

4.4.2 Material form of the balance of angular momentum

To obtain the material form of the balance of angular momentum, we start with its global form in the spatial description:

$$\int_E \epsilon_{ijk} \sigma_{kj} \, dV = 0,$$

where E is an arbitrary subbody. Transforming to the referential description and using Eqn. (4.36) gives

$$\int_{E_0} \epsilon_{ijk} \left(\frac{1}{J} P_{kM} F_{jM} \right) J \, dV_0 = \int_{E_0} \epsilon_{ijk} P_{kM} F_{jM} \, dV_0 = 0.$$

This must be true for any E_0, therefore

$$\epsilon_{ijk} P_{kM} F_{jM} = 0,$$

which implies that $P_{kM}F_{jM}$ is symmetric with respect to the indices jk:

$$P_{kM}F_{jM} = P_{jM}F_{kM} \quad \Leftrightarrow \quad \boldsymbol{P}\boldsymbol{F}^T = \boldsymbol{F}\boldsymbol{P}^T. \tag{4.40}$$

Note, however, that the first Piola–Kirchhoff stress tensor itself is, in general, *not* symmetric (i.e. $\boldsymbol{P} \neq \boldsymbol{P}^T$).

4.4.3 Second Piola–Kirchhoff stress

As we saw earlier, stress comes from the definition of traction as the force per unit area acting on a body. The force is a tensor defined in the deformed configuration. The area can be measured in either the deformed or the reference configuration. This leads to the two stress fields that we have encountered so far: the Cauchy stress that is defined as a mapping of a spatial vector to a spatial vector and the first Piola–Kirchhoff stress that is a two-point tensor that maps a material vector to a spatial vector. It turns out to be mathematically advantageous to define a third stress field, which is a tensor entirely in the reference configuration, by pulling the force back to the reference configuration as if it were a kinematic quantity. This stress is called the *second Piola–Kirchhoff stress tensor*.

We begin with the force–traction relation that defines the first Piola–Kirchhoff stress:

$$d\boldsymbol{f} = \boldsymbol{T}\, dA_0 = (\boldsymbol{P}\boldsymbol{N})\, dA_0.$$

We pull back $d\boldsymbol{f}$ to the reference configuration and substitute in the nominal traction definition to obtain

$$d\boldsymbol{f}_0 = \boldsymbol{F}^{-1}d\boldsymbol{f} = \boldsymbol{F}^{-1}\boldsymbol{T}\, dA_0 = \boldsymbol{F}^{-1}(\boldsymbol{P}\boldsymbol{N})\, dA_0 = \boldsymbol{S}\boldsymbol{N}\, dA_0,$$

where

$$S_{IJ} = F_{Ii}^{-1}P_{iJ} \quad \Leftrightarrow \quad \boldsymbol{S} = \boldsymbol{F}^{-1}\boldsymbol{P} \tag{4.41}$$

is the *second Piola–Kirchhoff stress tensor*. The relation between $\boldsymbol{\sigma}$ and $\boldsymbol{S}$ is obtained by using the Piola transformation in Eqn. (4.35):

$$\sigma_{ij} = \frac{1}{J}F_{iI}S_{IJ}F_{jJ} \quad \Leftrightarrow \quad \boldsymbol{\sigma} = \frac{1}{J}\boldsymbol{F}\boldsymbol{S}\boldsymbol{F}^T. \tag{4.42}$$

Inverting this relation gives $\boldsymbol{S} = J\boldsymbol{F}^{-1}\boldsymbol{\sigma}\boldsymbol{F}^{-T}$, from which it is clear that $\boldsymbol{S}$ is symmetric since $\boldsymbol{\sigma}$ is symmetric. (This can also be seen by substituting Eqn. (4.41) into the material form of the balance of angular momentum in Eqn. (4.40).)

The second Piola–Kirchhoff stress, $\boldsymbol{S}$, has no direct physical significance, but since it is symmetric it can be more convenient to work with than $\boldsymbol{P}$. The balance of linear momentum in terms of the second Piola–Kirchhoff stress follows from Eqn. (4.39) as

$$(F_{iI}S_{IJ})_{,J} + \rho_0 \breve{b}_i = \rho_0 \breve{a}_i \quad \Leftrightarrow \quad \operatorname{Div}(\boldsymbol{F}\boldsymbol{S}) + \rho_0 \breve{\boldsymbol{b}} = \rho_0 \breve{\boldsymbol{a}} \quad \boldsymbol{X} \in \mathcal{B}_0. \tag{4.43}$$

The difference between $\boldsymbol{\sigma}$, $\boldsymbol{P}$ and $\boldsymbol{S}$ is demonstrated by the following simple example.

Example 4.1 (Stretching of a bar) A bar made of an incompressible material is loaded by a force $\boldsymbol{R} = R\boldsymbol{e}_1$, where $\boldsymbol{e}_1$ is the bar's axis. The bar is uniform along its length and unconstrained in the 2- and 3-directions. The stretch in the 1-direction is α. Assume the responses in the 2- and 3-directions are the same and that no shearing deformation (with respect to the Cartesian coordinate system) takes place in the bar as a result of the uniaxial loading. The cross-sectional area of the bar when it is not loaded is A_0. Determine the 11-component of the Cauchy stress tensor (i.e. σ_{11}) and of the first and second Piola–Kirchhoff stress tensors in the bar.

Solution: Since there is no shearing and no difference between the 2- and 3-directions due to the assumed symmetry, we expect a deformation gradient of the form

$$[\boldsymbol{F}] = \begin{bmatrix} \alpha & 0 & 0 \\ 0 & \alpha^* & 0 \\ 0 & 0 & \alpha^* \end{bmatrix}.$$

The material is incompressible and so $J = \det \boldsymbol{F} = \alpha(\alpha^*)^2 = 1$, which means that $\alpha^* = 1/\sqrt{\alpha}$. The 11-component of the Cauchy stress is $\sigma_{11} = R/A$, where A is the deformed cross-sectional area. We find A from Nanson's formula (Eqn. (3.9)):

$$\boldsymbol{n}\, dA = J\boldsymbol{F}^{-T}\boldsymbol{N}\, dA_0$$

$$\begin{bmatrix} 1 \\ 0 \\ 0 \end{bmatrix} dA = \begin{bmatrix} 1/\alpha & 0 & 0 \\ 0 & \sqrt{\alpha} & 0 \\ 0 & 0 & \sqrt{\alpha} \end{bmatrix} \begin{bmatrix} 1 \\ 0 \\ 0 \end{bmatrix} dA_0 = \begin{bmatrix} 1 \\ 0 \\ 0 \end{bmatrix} \frac{1}{\alpha}\, dA_0.$$

Thus, $dA = dA_0/\alpha$ and, since α is constant within the cross-section, $A = A_0/\alpha$. Therefore $\sigma_{11} = \alpha R/A_0$. The 11-component of the first Piola–Kirchhoff stress is simply $P_{11} = R/A_0$, and the second Piola–Kirchhoff stress is obtained from the relation

$$[\boldsymbol{S}] = \left[\boldsymbol{F}^{-1}\right][\boldsymbol{P}] = \begin{bmatrix} 1/\alpha & 0 & 0 \\ 0 & \sqrt{\alpha} & 0 \\ 0 & 0 & \sqrt{\alpha} \end{bmatrix} \begin{bmatrix} R/A_0 & 0 & 0 \\ 0 & 0 & 0 \\ 0 & 0 & 0 \end{bmatrix} = \frac{R}{\alpha A_0} \begin{bmatrix} 1 & 0 & 0 \\ 0 & 0 & 0 \\ 0 & 0 & 0 \end{bmatrix}.$$

These results illustrate clearly that the first Piola–Kirchhoff stress and Cauchy stress are equivalent to what are commonly referred to in undergraduate mechanics courses as the "engineering stress" and "true stress," respectively. The former is easy to obtain from a tensile test, because there is no need to measure the changing cross-sectional area. However, the true stress experienced by the material at each stage of the test is the Cauchy stress. This example is pursued further in Exercise 4.8.

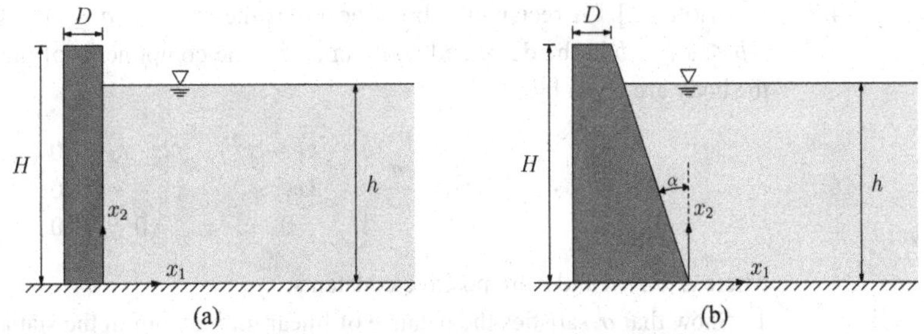

Fig. 4.6 Two configurations of dams (dark gray) with water (light gray) on the right. The dams and water are surrounded by air at atmospheric pressure. The dimensions of the dam and the level of the water are indicated. The width of the dams in the out-of-plane direction is W.

Exercises

4.1 [SECTION 4.1] Show that the continuity equation (Eqn. (4.2)) is identically satisfied for any deformation of the form

$$x_1 = \alpha_1(t)X_1, \qquad x_2 = \alpha_2(t)X_2, \qquad x_3 = \alpha_3(t)X_3,$$

where $\alpha_i(t)$ are differentiable scalar functions of time. The mass density field in the reference configuration is $\rho_0(\boldsymbol{X})$.

4.2 [SECTION 4.2] Figure 4.6 shows two configurations of dams (dimensions are shown in the figure). The width of the dams in the out-of-plane direction is W. The dams are subjected to hydrostatic pressure due to the water on the right, atmospheric pressure p_{at} due to the surrounding air, and gravity which acts downwards. The density of the water is ρ_{w} and the density of the dam material is ρ_{d}. Compute the total force (body and surface, not including the reactions where the ground supports the dams) acting on the dam for both configurations. **Hint:** The hydrostatic pressure increases linearly with depth below the water surface and is proportional to $\rho_{\text{w}} g$, where g is the gravitational acceleration.

4.3 [SECTION 4.2] In an ideal nonviscous fluid there can be no shear stress. Hence, the Cauchy stress tensor is entirely hydrostatic, $\sigma_{ij} = -p\delta_{ij}$. Show that this leads to the following form, known as Euler's equation of motion for a frictionless fluid:

$$-\frac{1}{\rho}\nabla p + \boldsymbol{b} = \frac{\partial \boldsymbol{v}}{\partial t} + (\nabla \boldsymbol{v})\boldsymbol{v}.$$

4.4 [SECTION 4.2] The rectangular Cartesian components of a particular Cauchy stress tensor are given by

$$[\boldsymbol{\sigma}] = \begin{bmatrix} a & 0 & d \\ 0 & b & e \\ d & e & c \end{bmatrix}.$$

1. Determine the unit normal $\boldsymbol{n}$ of a plane parallel to the x_3 axis (i.e. $n_3 = 0$) on which the traction vector is tangential to the plane. What are the constraints on a and b necessary to ensure a solution?
2. If a, b, c, d and e are functions of x_1 and x_2, find the most general forms for these functions that satisfy stress equilibrium in Eqn. (4.27) in the absence of body forces.

4.5 [SECTION 4.2] A rectangular body occupies the region $-a \le x_1 \le a$, $-a \le x_2 \le a$ and $-b \le x_3 \le b$ in the deformed configuration. The components of the Cauchy stress tensor in the body are given by

$$[\boldsymbol{\sigma}] = \frac{c}{a^2}\begin{bmatrix} -(x_1^2 - x_2^2) & 2x_1x_2 & 0 \\ 2x_1x_2 & x_1^2 - x_2^2 & 0 \\ 0 & 0 & 0 \end{bmatrix},$$

where a, $b > a$ and c are positive constants.

1. Show that $\boldsymbol{\sigma}$ satisfies the balance of linear momentum in the static case (Eqn. (4.27)) with no body force.
2. Determine the tractions that must be applied to the six faces of the body in order for the body to be in equilibrium.
3. Calculate the traction distribution on the sphere $x_1^2 + x_2^2 + x_3^2 = a^2$.
4. The principal values (eigenvalues) of the stress tensor (principal stresses) are denoted σ_i ($i = 1, 2, 3$), such that $\sigma_1 \ge \sigma_2 \ge \sigma_3$. These give the (algebraically) maximum and minimum normal stresses at a point. It can be shown that the maximum shear stress is given by $\tau_{\max} = (\sigma_1 - \sigma_3)/2$. Calculate the principal stresses of $\boldsymbol{\sigma}$ as a function of position. Then find the maximum value of $\tau_{\max}$ in the domain of the body.

4.6 [SECTION 4.2] A state of *plane stress* is one where the out-of-plane components of the stress tensor are zero, i.e. $\sigma_{31} = \sigma_{32} = \sigma_{33} = 0$. Show that for this case if

$$\sigma_{11} = \frac{\partial^2\phi}{\partial x_2^2}, \qquad \sigma_{22} = \frac{\partial^2\phi}{\partial x_1^2}, \qquad \sigma_{12} = -\frac{\partial^2\phi}{\partial x_1 \partial x_2},$$

where $\phi(x_1, x_2)$ is the *Airy stress function*, an unknown function to be determined, then the static equilibrium equations are satisfied identically in the absence of body forces.

4.7 [SECTION 4.3] A material is subjected to a *distributed moment* field such that every infinitesimal element (with volume $dV = dx_1\,dx_2\,dx_3$) is subjected to a moment $\mu\,dV$ (where μ is the moment per unit volume) about an axis parallel to the $\mathbf{e}_1$-direction. How will this affect the symmetry of the stress tensor $\boldsymbol{\sigma}$? Find an explicit expression for the relation between the shear components of $\boldsymbol{\sigma}$ in a Cartesian coordinate system.

4.8 [SECTION 4.4] For the stretched bar in Example 4.1, do the following:

1. Determine the plane of maximum shear stress in the deformed configuration and the value of the Cauchy shear stress on this plane.
2. Determine the material plane in the reference configuration corresponding to the plane of maximum shear stress found above. Plot the angle Θ between the normal to this plane and the horizontal axis as a function of the stretch in the 1-direction, α. Which plane does this tend to as $\alpha \to \infty$?

5 Thermodynamics

Thermodynamics is typically defined as a theory dealing with the flow of heat and energy between material systems. This definition is certainly applicable here, however, Callen provides (in his excellent book on the subject [Cal85]) an alternative definition that highlights another role that thermodynamics plays in continuum mechanics: "Thermodynamics is the study of the restrictions on the possible properties of matter that follow from the symmetry properties of the fundamental laws of physics." In this chapter (and the next), we address both of these aspects of thermodynamic theory in the context of continuum mechanics.

The theory of thermodynamics boils down to three fundamental laws, deduced from empirical observation, that all physical systems are assumed to obey. The *zeroth law of thermodynamics* is related to the concept of thermal equilibrium. The *first law of thermodynamics* is a statement of the conservation of energy. The *second law of thermodynamics* deals with the directionality of thermodynamic processes. We will discuss each of these laws in detail, but first we describe the basic concepts in which thermodynamics is phrased.

For the purposes of thermodynamic analysis, the universe is divided into two parts: the *system* whose behavior is of particular interest, and the system's *surroundings* (everything else). The behavior of the surroundings is of interest only insofar as is necessary to characterize its interactions with the system. In thermodynamics these interactions can include *mechanical interactions* in which the surroundings do work on the system, *thermal interactions* in which the surroundings transfer heat to the system and *particle transfer interactions* in which particles are transferred between the surroundings and the system. Any change in the system's surroundings which results in work, heat or particles being exchanged with the system is referred to as an *external perturbation*.[1] A perturbation can be time dependent, although for our purposes it is limited to a finite duration after which the properties of the system's surroundings remain fixed.

As an example, consider a cylinder with a movable piston containing a compressed gas situated inside a laboratory. We can take the thermodynamic system of interest to be all the gas particles inside the cylinder. Then the system's surroundings include the cylinder, the piston, the laboratory itself and, indeed, the rest of the universe. The system can interact with its surroundings which can do work on it (the piston can be moved in order to change the system's volume), the atmosphere in the laboratory can transfer heat to the gas in the cylinder (assuming the piston and cylinder allow such transfers) and it is even possible that some molecules from the air in the laboratory can diffuse through the piston (if the piston

[1] Note that we do not mean to imply by the use of the term "perturbation" that the change suffered by a system's surroundings during a perturbation is necessarily small in anyway.

is permeable) and become part of the system. A fixed change to the surroundings related to any of these modes of interaction would constitute a perturbation to the system.

In many cases, it is not necessary to consider the entire universe when studying a thermodynamic system. Often a system may interact so weakly with its surroundings that such interactions are negligible. Other times it is possible to identify a larger system that contains the system of interest such that all interactions between this larger system and the remainder of the universe may be ignored. We call such a system *isolated*. Specifically, we define an isolated system as one that the external universe is unable to do work on and to which it cannot transfer heat or particles. For example, our cylinder of gas could be put into a sealed environmental chamber which does not allow external mechanical, thermal or particle transfer interactions. Then what happens outside the chamber has no influence on the behavior of the system and can be ignored.

Extensive observation of our universe has led to two realizations. First, all macroscopic systems subjected to an external perturbation respond by undergoing a process that ultimately tends towards a simple terminal state which is quiescent and spatially homogeneous. Remarkably, these terminal states can be described by a very small number of quantities. Second, when a system already in such a terminal state is subjected to an external perturbation it transitions to another terminal state in a predictable and repeatable way that is completely characterized by a knowledge of the initial state and the external perturbation. The identification of the macroscopic quantities that characterize terminal states and perturbations, and the laws that allow predictions based on their knowledge are the goals of the theory of thermodynamics.[2]

5.1 Macroscopic observables, thermodynamic equilibrium and state variables

To begin we must identify the quantities with which the theory of thermodynamics is concerned. We know that all systems are composed of discrete particles that (to a good approximation, see Section 5.2 in [TM11]) satisfy Newton's laws of motion. Thus, to have a complete understanding of a system it is necessary to determine the number of particles N that make up the system and their positions and momenta (a total of $6N$ quantities).

[2] We do not presume to be able to provide in this short chapter a comprehensive treatise on thermodynamic theory. Indeed, the difficulty of creating a clear and precise presentation of the subject is highlighted by the following quote, attributed to the German atomic and quantum physicist Arnold Sommerfeld: "Thermodynamics is a funny subject. The first time you go through it, you don't understand it at all. The second time you go through it, you think you understand it, except for one or two small points. The third time you go through it, you know you don't understand it, but by that time you are so used to it, it doesn't bother you any more." A similar sentiment was expressed by Clifford Truesdell: "There are many who claim to understand thermodynamics, but it is best for them by common consent to avoid the topic in conversation with one another, since it leads to consequences such as can be expected from arguments over politics, religion, or the canons of female beauty. Honesty compels me to confess that in several attempts, made over decades, I have never been able to understand the subject, not only in what others have written on it, but also in my own earlier presentations" [Tru66b]. Keeping these confessions from great men in mind, our goal is to present the theory as accurately as we can, while at the same time pointing out and clarifying common pitfalls that lead to much confusion in the literature.

However, as described further below, this is a hopeless task. Thus, we must make do with a much smaller set; but what quantities will prove to be most useful? To answer this question we first must consider the nature of macroscopic observation.

5.1.1 Macroscopically observable quantities

Fundamentally, a thermodynamic system is composed of some number of particles N, where N is huge (on the order of 10^{23} for a cubic centimeter of material). The *microscopic kinematics*[3] of such a system are described by a (time-dependent) vector in a $6N$-dimensional vector space, called *phase space*, corresponding to the complete list of particle positions and momenta,

$$\boldsymbol{y} = (\boldsymbol{r}^1, \ldots, \boldsymbol{r}^N, m^1\dot{\boldsymbol{r}}^1, \ldots, m^N\dot{\boldsymbol{r}}^N),$$

where $(m^1, \ldots, m^N)$ are the masses of the particles.[4]

Although scientific advances now allow researchers to image individual atoms, we can certainly never hope (nor wish) to record the time-dependent positions and velocities of *all* atoms in a macroscopic thermodynamic system. This would seem to suggest that there is no hope of obtaining a deep understanding of the behavior of such systems. However, for hundreds of years mankind has interrogated these systems using only a relatively crude set of tools, and nevertheless we have been able to develop a sophisticated theory of their behavior. The first tools that were used for measuring kinematic quantities likely involved things such as measuring sticks and lengths of string. Later, we developed laser extensometers and laser interferometers. All of these devices have two important characteristics in common. First, they have very limited spatial resolution relative to typical interparticle distances (which are on the order of 10^{-10} m). Indeed, the spatial resolution of measuring sticks is typically on the order of 10^{-4} m and that of interferometry is on the order of 10^{-6} m (a micron). Second, these devices have very limited temporal resolution relative to characteristic atomic time scales, which are on the order of 10^{-13} s (for the oscillation period of an atom in a crystal). The temporal resolution of measuring sticks and interferometers relies on the device used to record measurements. The human eye is capable of resolving events spaced no less than 10^{-2} s apart. If a camera is used, then the shutter speed – typically on the order of 10^{-3} or 10^{-4} s – sets the temporal resolution. Clearly, these tools provide only very coarse measurements that correspond to some type of temporal and spatial averaging of the positions of the particles in the system.[5] Accordingly the only quantities these devices are capable of measuring are those that are essentially uniform in space (over lengths up to their spatial resolution) and nearly constant in time (over spans of time up to their temporal resolution). We say that these quantities are *macroscopically observable*. The fact that such

[3] See the definition of "kinematics" at the start of Chapter 3.

[4] Depending on the nature of the material there may be additional quantities that have to be known, such as the charges of the particles or their magnetic moments. Here we focus on purely thermomechanical systems for which the positions and momenta are sufficient. See Section 7.1 of [TM11] for more on the idea of a phase space.

[5] See Section 1.1 in [TM11] for a discussion of spatial and temporal scales in materials.

quantities exist is a deep truth of our universe, the discussion of which is outside the scope of our book.[6]

The measurement process described above replaces the $6N$ microscopic kinematic quantities with a dramatically smaller number of *macroscopic kinematic quantities*, such as the total volume of the system and the position of its center of mass. In addition there are also nonkinematic quantities that are macroscopically observable, such as the total linear momentum, the total number of particles in the system and its total mass.

If the volume of a thermodynamic system is large compared with the volumetric resolution of our measurement device (for the interferometers mentioned above this would be approximately one cubic micron or 10^{-18} m^3), then we are able to observe these quantities for subsystems[7] of the system. The collection of all such measurements is what we refer to when we speak of *macroscopic fields* which capture the spatial and temporal variation of the macroscopic quantities such as the *mass density* field (mass per unit volume) or conversely the *specific volume* field (volume per unit mass). Further, the arrangement of the subsystems' positions gives rise to additional macroscopic quantities that we call the shape, orientation and angular velocity of a macroscopic system. For example, consider the case where we restrict the shape of our thermodynamic system to a parallelepiped. Thus as we have seen in Section 3.4.6, the shape and volume of the system may be characterized by six independent kinematic quantities. For example,[8] the parallelepiped's three side lengths, ℓ_1, ℓ_2, ℓ_3, and three interior angles, ϕ_1, ϕ_2 and ϕ_3. Now if we choose a set of reference values, such as $L_1 = L_2 = L_3 = 1.0$ m, and $\Phi_1 = \Phi_2 = \Phi_3 = 90°$, then we can use the Lagrangian strain tensor $\boldsymbol{E}$ to describe the shape and volume of the system relative to this reference. Thus, for macroscopic systems there are macroscopic quantities that describe global, total properties of the system and there are macroscopic fields that describe how these total values are spatially distributed between the system's subsystems.

It turns out that not all of a system's macroscopic observables are relevant to the theory of thermodynamics. In particular, in most formulations of thermodynamics the total linear and angular momenta of a system are assumed to be zero.[9] We will also adopt this convention. Additionally, the position (of the center of mass) and orientation of the system are assumed to be irrelevant.[10] From now on when we refer to "macroscopic observables," we mean only those macroscopic observables not explicitly excluded in the above list.

[6] We encourage the reader to refer to Chapter 1 of Callen's book [Cal85] for a more extensive introduction, similar to the above, and to Chapter 21 of [Cal85] for a discussion of the deep fundamental reason for the existence of macroscopically observable quantities (i.e. broken symmetry and Goldstone's theorem).

[7] The idea of a thermodynamic subsystem is related to the concept of a *continuum particle* introduced in Section 3.1.

[8] One could also consider, in addition to the side lengths and interior angles, the lengths of the parallelepiped's face-diagonals. However, once the three side lengths and three interior angles are prescribed all of the face-diagonal lengths are determined. Thus, we say that there are only six independent kinematic quantities that determine the shape and volume of a parallelepiped.

[9] In fact, one can develop a version of the theory where the energy and the total linear and angular momenta all play equally important roles. See [Cal85, Part III] for further discussion and references on this point.

[10] The argument is based on the presumed symmetry of the laws of physics under spacetime translation and rotation. Again, see [Cal85, Part III] for more details.

5.1.2 Thermodynamic equilibrium

When a system experiences an external perturbation it undergoes a *dynamical process* in which its microscopic kinematic vector and, in general, its macroscopic observables, change as a function of time. As mentioned above, it is empirically observed that all systems tend to evolve to a *quiescent and spatially homogeneous* (at the macroscopic length scale) terminal state where the system's macroscopic observables have constant limiting values. Also any fields, like density or strain, must be constant since the terminal state is spatially homogeneous. Once the system reaches this terminal condition it is said to be in a state of *thermodynamic equilibrium*. In general, even once a system reaches thermodynamic equilibrium, its microscopic kinematic quantities continue to change with time. However, these quantities are of no explicit concern to thermodynamic theory.

As you might imagine, thermodynamic equilibrium can be very difficult to achieve. We may need to wait an infinite amount of time for the dynamical process to obtain the limiting equilibrium values of the macroscopic observables. Thus, most systems never reach a true state of thermodynamic equilibrium. Those that do not, however, do exhibit a characteristic "two-stage dynamical process" in which the macroscopic observables first evolve at a high rate during and immediately after an external perturbation. These values then further evolve at a rate that is many orders of magnitude smaller than in the first stage of the dynamical process. This type of system is said to be in a state of *metastable thermodynamic equilibrium* once the first part of its two-stage dynamical process is complete.[11]

An example of a system in metastable equilibrium is a single crystal of metal in a container in the presence of gravity. The crystal is not in thermodynamic equilibrium since, given enough time, the metal would flow like a fluid in order to conform to the shape of its container as its atoms preferentially diffuse towards the container's bottom. However, the time required for this to occur at room temperature is so long as to be irrelevant for typical engineering applications. Thus, for all intents and purposes the crystal *is* in thermodynamic equilibrium, which is what is meant by metastable equilibrium.

5.1.3 State variables

We have already eliminated certain macroscopic observables from consideration using physical and symmetry-based arguments. We now further reduce the set of observables of interest to those which directly affect the behavior of the thermodynamic system. We refer to these special macroscopic observables as *state variables* and define them as follows.

> The macroscopic observables that are well defined and single-valued when the system is in a state of thermodynamic equilibrium are called *state variables*. Those state variables which are related to the kinematics of the system (volume, strain, etc.) are called *kinematic state variables*.

[11] The issue of metastable equilibrium within the context of statistical mechanics is discussed in Section 11.1 of [TM11].

To explore the concept of state variables, consider the case of an ideal gas enclosed in a rigid, thermally-insulated box. Let us assume this system to be in its terminal state of thermodynamic equilibrium. Then on macroscopic time scales the atoms making up the gas will have time to explore the entire container, flying past each other and bouncing off the walls. If each atom were a point of light and we took a time-lapsed photograph of the box, we would just see a uniform bright light filling the volume. Taking this view, we can say that the positions of the atoms at the macroscale are simply characterized by the shape and volume V of the box. The momenta of the atoms manifest themselves at the macroscale in two ways: the temperature T (which is related to the kinetic energy of the atoms), and the pressure p on the box walls (coming from the momentum transfer during the collisions).[12] Formal macroscopic definitions for these quantities will have to wait until Sections 5.2 and 5.5.5. Based on empirical observation we know that the shape of the container plays no role in characterizing the equilibrium state of the gas. This means that the shape (quantified by the shear part of the strain tensor) is *not* a state variable of the gas system since it is not single-valued at equilibrium. Thus, we say that the equilibrium state of the gas is associated with (at least) four state variables: the number of particles N, the volume V, the pressure p and the temperature T.

It turns out that the above conclusions for the gas apply to any system in *true*[13] thermodynamic equilibrium. This is because given an infinite amount of time all systems are fluid-like in the sense that their atoms can fully explore the available phase space. A consequence of this is that *any* system in thermodynamic equilibrium, not just a gas, depends on only one kinematic state variable – the volume V of the system.[14]

The identification of all *kinematic* state variables is more difficult when one considers states of *metastable* thermodynamic equilibrium. Consider again the single crystal of metal in a container described in the previous section. While it is true that given unlimited time the metal would flow, our attention span is more limited so that over the hundreds or even thousands of years that we watch it, the metal may remain largely unchanged. Over such time scales the shape of the metal (quantified by the strain) is certainly necessary to characterize its behavior. How then can we determine which macroscopic kinematic observables affect the behavior of a system? To make this determination one can perform the following test. Start with the system in a state of metastable thermodynamic equilibrium. Thermally isolate the system and fix the number of particles as well as all independent kinematic quantities except for one. Now, very slowly change the free kinematic quantity.[15] If work is performed by the system as a result of changing the kinematic quantity, then that quantity is a kinematic

[12] See Chapter 7 of [TM11].

[13] We use the term "true thermodynamic equilibrium" for systems that strictly satisfy the definition of thermodynamic equilibrium as opposed to those that are in a state of metastable equilibrium.

[14] See Section 7.4.5 of [TM11] for a proof, based on statistical mechanics theory, that in the limit of an infinite number of particles (keeping the density fixed) the equilibrium properties of a system do not depend on any kinematic state variables other than the system's volume. Also see Chapter 11 of [TM11] for a detailed discussion of the metastable nature of solids.

[15] How slowly the quantity must be changed depends on the system under consideration. This illustrates the difficulty (and self-referential nature) of carefully defining the concept of metastable thermodynamic equilibrium. One must always use qualifiers, as we have done here, which implicitly refer to the laws of equilibrium thermodynamics. That is, the kinematic quantity must be changed slowly enough that the resulting identified kinematic state variables for the system satisfy all of the standard laws of thermodynamics.

state variable for the system, otherwise it is irrelevant to the system under consideration. In other words, a kinematic quantity is a state variable if the system produces a force of resistance (which does work) in response to the change of its kinematic quantity. Such a "force of resistance" is referred to as a *thermodynamic tension.*[16] For example, varying the shape of a gas's container at constant volume will not generate an opposing force (stress), but there would be a stress generated if we deformed a solid by changing any one of the six components of the Lagrangian strain tensor while holding the other five fixed. Thus, the kinematic state variables associated with (metastable) equilibrium states of solid systems include the full Lagrangian strain tensor, whereas gases only require the volume.

State variables can be divided into two categories: *intensive* and *extensive*. Intensive state variables are quantities whose values are independent of amount. Examples include the temperature and pressure (or stress) of a thermodynamic system. In contrast, extensive variables are ones whose value depends on amount. Suppose we have two identical systems which have the same values for their state variables. The extensive variables are those whose values are exactly doubled when we treat the two systems as a single composite system. Kinematic variables like volume are naturally extensive. For example, if the initial systems both have volume V, then the composite system has total volume $2V$. Strain, which is also a kinematic variable, is intensive. However, we can define a new extensive quantity, "volume strain" as $V_0 \boldsymbol{E}$, where V_0 is the reference volume. The kinematics of a system can therefore always be characterized by extensive variables. In general, we write:

A system in thermodynamic (or metastable) equilibrium is characterized by a set of n^{Γ} independent extensive kinematic state variables, which we denote generically as $\boldsymbol{\Gamma} = (\Gamma_1, \dots, \Gamma_{n^{\Gamma}})$. For a gas, $n^{\Gamma} = 1$ and $\Gamma_1 = V$. For a metastable solid, $n^{\Gamma} = 6$ and $\boldsymbol{\Gamma}$ contains the six independent components of the Lagrangian strain tensor multiplied by the reference volume.

Other important extensive variables include the number of particles making up the system and its mass. A special extensive quantity which we have not encountered yet is the total internal energy of the system $\mathcal{U}$. Later we will find that most extensive state variables are associated with corresponding intensive quantities, which play an equally important role. (In fact, these are the thermodynamic tensions mentioned above.) Table 5.1 presents a list of the extensive and intensive state variables that we will encounter (not all of which have been mentioned yet), indicating the pairings between them.

To summarize, thermodynamics deals with quantities that are macroscopically observable, well defined and single-valued at equilibrium. Such quantities are referred to as *state variables*. When the state variables relate to the motion of the system (positions and shape), we refer to them as *kinematic state variables*. The adjectives *extensive* and *intensive* indicate whether or not a state variable scales with system size. Finally, for systems that are large relative to the spatial and/or temporal resolution of the measuring device, it is also possible to record position- and time-dependent fields of the state variables. Not all state variables

[16] Thermodynamic tensions are discussed further in Section 5.5.5.

Table 5.1. Extensive and intensive state variables. Kinematic state variables are indicated with a $*$

Extensive	Intensive
internal energy ($\mathcal{U}$)	–
mass (m)	–
number of particles (N)	chemical potential (μ)
volume (V)*	pressure (p)
(Lagrangian) volume strain ($V_0 \boldsymbol{E}$)*	elastic part of the (second Piola–Kirchhoff) stress ($\boldsymbol{S}^{(e)}$)
entropy (S)	temperature (T)

are independent. Our next task is to determine the minimum number of state variables that must be fixed in order to explicitly determine all of the remaining values.

5.1.4 Independent state variables and equations of state

A system in thermodynamic equilibrium can have many state variables but not all can be specified independently. Consider again the case of an ideal gas enclosed in a rigid, thermally-insulated box as discussed in the previous section. We identified four state variables with this system: N, V, p and T. However, based on empirical observation, we know that not all four of these state variables are thermodynamically independent. Any three will determine the fourth. We will see this later in Section 5.5.5 where we discuss the ideal gas law. In fact, it turns out that *any* system in *true* thermodynamic equilibrium is fully characterized by a set of three independent state variables since, as explained above, all systems are fluid-like on the infinite time scale of thermodynamic equilibrium. For a system in metastable equilibrium the number of thermodynamically independent state variables is equal to $n^\Gamma + 2$, where n^Γ is the number of independent kinematic state variables characterizing the system, as described in the previous section. The two state variables required in addition to the kinematic state variables account for the *internal energy* and the *entropy* which we will encounter in Section 5.3 and Section 5.5.1, respectively.

We adopt the following notation. Let B be a system in thermodynamic (or metastable) equilibrium and $\mathcal{B} = (\mathcal{B}_1, \mathcal{B}_2, \dots, \mathcal{B}_{\nu^\mathsf{B}}, \mathcal{B}_{\nu^\mathsf{B}+1}, \dots)$ be the set of all state variables, where $\nu^\mathsf{B} = n^\Gamma + 2$ is the number of independent properties. The nonindependent properties are related to the independent properties through *equations of state*[17]

$$\mathcal{B}_{\nu^\mathsf{B}+j} = f_j(\mathcal{B}_1, \dots, \mathcal{B}_{\nu^\mathsf{B}}), \quad j = 1, 2, \dots.$$

As examples, we will see the equations of state for an ideal gas in Sections 5.3.2 and 5.5.5.

[17] Equations of state are closely related to *constitutive relations* which are described in Chapter 6. Typically, the term "equation of state" refers to a relationship between state variables that characterize the entire thermodynamic system, whereas "constitutive relations" relate density variables defined locally at continuum points.

5.2 Thermal equilibrium and the zeroth law of thermodynamics

Up to this point we have referred to temperature without defining it, relying on you, our reader, for an intuitive sense of this concept. We now see how temperature can be defined in a more rigorous fashion.

5.2.1 Thermal equilibrium

Our sense of touch provides us with the feeling that an object is "hotter than" or "colder than" our bodies, and thus, we have developed an intuitive sense of temperature. But how can this concept be made more explicit? We start by defining the notion of thermal equilibrium between two systems.

> Two systems A and B in thermodynamic equilibrium are said to be in *thermal equilibrium* with each other, denoted $\mathrm{A} \sim \mathrm{B}$, if they remain in thermodynamic equilibrium after being brought into thermal contact while keeping their kinematic state variables and their particle numbers fixed.

Thus, heat is allowed to flow between the two systems but they are not allowed to transfer particles or perform work. Here, heat is taken as a primitive concept similar to force. Later, when we discuss the first law of thermodynamics, we will discover that heat is simply a form of energy.

A practical test for determining whether two systems, already in thermodynamic equilibrium, are in thermal equilibrium, can be performed as follows: (1) thermally isolate both systems from their common surroundings; (2) for each system, fix its number of particles and all but one of its kinematic state variables and arrange for the systems' surroundings to remain constant; (3) bring the two systems into thermal contact; (4) wait until the two systems are again in thermodynamic equilibrium. If the free kinematic state variable in each system remains unchanged in stage (4), then the two systems were, in fact, in thermal equilibrium when they were brought into contact.[18]

As an example, consider the two cylinders of compressed gas with frictionless movable pistons shown in Fig. 5.1. In Fig. 5.1(a) the cylinders are separated and thermally isolated from their surroundings. The forces F^{A} and F^{B} are mechanical boundary conditions applied by the surroundings to the system. Both systems are in a state of thermodynamic equilibrium. Since the systems are already thermally isolated and the only extensive kinematic quantity for a gas is its volume, steps (1)–(3) of the procedure are achieved by arranging for F^{A} and F^{B} to remain constant and bringing the two systems into thermal contact. Thus, in Fig. 5.1(b) the systems are shown in thermal contact through a *diathermal partition*, which is a partition that allows only thermal interactions (heat flow) across it but is otherwise

[18] Of course, at the end of stage (4) the systems will be in thermal equilibrium regardless of whether or not they were so in the beginning. However, the purpose of the test is to determine whether the systems were in thermal equilibrium when first brought into contact.

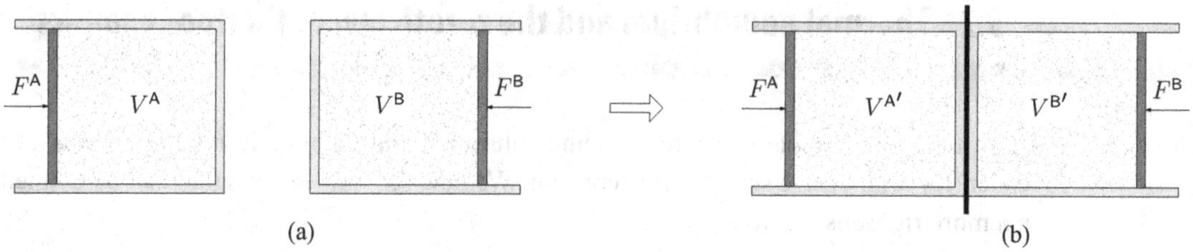

Fig. 5.1 Two cylinders of compressed gas, A and B, with movable frictionless pistons. (a) The cylinders are separated; each is in thermodynamic equilibrium. (b) The cylinders are brought into contact via a diathermal partition.

impermeable and rigid. If the volumes remain unchanged, $V^{A'} = V^A$ and $V^{B'} = V^B$, then A and B are in thermal equilibrium.

The *zeroth law of thermodynamics* is a statement about the relationship between bodies in thermal equilibrium:

Zeroth law of thermodynamics Given three thermodynamic systems, A, B and C, each in thermodynamic equilibrium, then if A $\sim$ B and B $\sim$ C it follows that A $\sim$ C.

The concept of thermal equilibrium leads to a definition for temperature.[19] If A $\sim$ B, we say that the temperature of A is the same as that of B. Otherwise, we say that the hotter system has a higher temperature.

5.2.2 Empirical temperature scales

In addition to defining temperature, thermal equilibrium also suggests an empirical approach for defining temperature scales. The idea is to calibrate temperature using a thermodynamic system that has only one independent kinematic state variable. Thus, its temperature is in one-to-one correspondence with the value of its kinematic state variable. For example, the old-fashioned mercury-filled glass thermometer is characterized by the height (volume) of the liquid mercury in the thermometer. Denote the calibrating system as Θ and its single kinematic state variable as θ. Now consider two systems, A and B. For each of these systems, there will be values θ^A and θ^B for which $\Theta \sim$ A and $\Theta \sim$ B, respectively. Then, according to the zeroth law, A $\sim$ B, if and only if $\theta^A = \theta^B$. This introduces an *empirical temperature*

[19] For those readers who are excited about the mathematics of formal logic and set theory, we note that, mathematically, thermal equilibrium is an equivalence relation. An equivalence relation $\sim$ is a binary relation between elements of a set A which satisfies the following three properties: (1) reflexivity, i.e. if $a \in A$, then $a \sim a$; (2) symmetry, i.e. if $a, b \in A$, then $a \sim b$ implies $b \sim a$; and (3) transitivity, i.e. if $a, b, c \in A$, then $a \sim b$ and $b \sim c$ implies $a \sim c$. If we consider two systems that are not in thermal equilibrium A $\not\sim$ B and put them in thermal contact, heat will flow from one system to the other. Suppose it is observed that heat flows from B to A, then we say that A is *colder than* B, A $<$ B. Thus, the thermal equilibrium equivalence relation $\sim$ and the colder than $<$ relation define a "preordering" of systems in thermodynamic equilibrium. A preordered set A is a set with binary relation $\leq$ such that for every $a, b, c \in A$ the following two properties hold: (1) reflexivity, i.e. $a \leq a$ and (2) transitivity, i.e. if $a \leq b$ and $b \leq c$, then $a \leq c$. If one adds the property of antisymmetry, i.e. $a \leq b$ and $b \leq a$ implies $a = b$, then the relation is a "partial ordering." However, this is not the case here. For example, suppose A and B are two systems in thermodynamic equilibrium with the same temperature. Then, A $\leq$ B and B $\leq$ A but A $\neq$ B. This preordering is what we call *temperature*.

scale. Different temperature scales can be defined by setting $T = f(\theta)$, where $f(\theta)$ is a monotonic function. In our example of the mercury-filled glass thermometer, the function $f(\theta)$ corresponds to the markings on the side of the thermometer that identify the spacing between specified values of the temperature T. The condition for thermal equilibrium between two systems A and B is then

$$T^{\mathsf{A}} = T^{\mathsf{B}}. \tag{5.1}$$

In fact, we will find that there exists a uniquely defined, fundamental temperature scale called the *thermodynamic temperature* (or *absolute temperature*).[20] The thermodynamic temperature scale is defined for nonnegative values only, $T \geq 0$, and the state of *zero temperature* (which can be approached but never actually obtained by any real system) is uniquely defined by the general theory. Thus, the only unambiguous part of the scale is the unit of measure for temperature. In 1954, following a procedure originally suggested by Lord Kelvin, this ambiguity was removed by the international community's establishment of the *kelvin* temperature unit K at the Tenth General Conference of Weights and Measures. The kelvin unit is defined by setting the temperature at the triple point of water (the point at which ice, water and water vapor coexist, which occurs at 0.01°C) to 273.16 K. (For a detailed explanation of empirical temperature scales, see [Adk83, Section 2].)

5.3 Energy and the first law of thermodynamics

The zeroth law introduced the concepts of thermal equilibrium and temperature. The first law establishes the fact that heat is actually just a form of energy and leads to the idea of internal energy.

5.3.1 First law of thermodynamics

Consider a thermodynamic system that is in a state of thermodynamic equilibrium (characterized by its temperature, the kinematic state variables and a *fixed* number of particles); call it state 1. Now imagine that the system is perturbed by mechanical and thermal interactions with its environment. Mechanical interaction results from tractions applied to its surfaces and body forces applied to the bulk. Thermal interactions result from heat flux in and out of the system through its surfaces and internal heat sources distributed through the body. Due to this perturbation, the system undergoes a dynamical process and eventually reaches a new state of thermodynamic equilibrium; call it state 2. During this process mechanical work $\Delta\mathcal{W}^{\mathrm{ext}}_{12}$ is performed *on the system* and heat $\Delta\mathcal{Q}_{12}$ is transferred into the system. Next consider a second perturbation that takes the system from state 2 to a third state, state 3. This perturbation is characterized by the total work $\Delta\mathcal{W}^{\mathrm{ext}}_{23}$ done on the system and heat $\Delta\mathcal{Q}_{23}$ transferred to the system. Now suppose we have the special case where state 3 coincides with state 1. In other words, the second perturbation returns the system to its original

[20] The theoretical foundation for the absolute temperature scale and its connection to the behavior of ideal gases is discussed in Section 5.5.5

state (original values of temperature and kinematic state variables) and also to the original values of total linear and angular momentum. In this case the total external work is called the *work of deformation*[21] $\Delta\mathcal{W}^{\mathrm{def}} = \Delta\mathcal{W}^{\mathrm{ext}}_{12} + \Delta\mathcal{W}^{\mathrm{ext}}_{21}$. This set of processes is called a thermodynamic cycle, since the system is returned to its original state. Through a series of exhaustive experiments in the nineteenth century, culminating with the work of English amateur scientist James Prescott Joule,[22] it was observed that in any thermodynamic cycle the amount of mechanical work performed on the system is always in constant proportion to the amount of heat *expelled by the system*:

$$\Delta\mathcal{W}^{\mathrm{def}} = -\mathcal{J}\Delta\mathfrak{Q}.$$

Here $\Delta\mathcal{W}^{\mathrm{def}}$ is the work (of deformation) *performed on the system* during the cycle, $\Delta\mathfrak{Q}$ is the external heat *supplied to the system* during the cycle and $\mathcal{J}$ is *Joule's mechanical equivalent of heat*, which expresses the constant of proportionality between work and heat.[23] Accordingly, we can define a new heat quantity that has the same units as work $\mathcal{Q} = \mathcal{J}\mathfrak{Q}$, and then Joule's observation can be rearranged to express a conservation principle for any thermodynamic system subjected to a cyclic process:

$$\Delta\mathcal{W}^{\mathrm{def}} + \Delta\mathcal{Q} = 0 \qquad \text{for any thermodynamic cycle.}$$

This implies the existence of a function that we call the *internal energy* $\mathcal{U}$ of a system in thermodynamic equilibrium.[24] The change of internal energy in going from one equilibrium state to another is therefore given by

$$\Delta\mathcal{U} = \Delta\mathcal{W}^{\mathrm{def}} + \Delta\mathcal{Q}. \tag{5.2}$$

If we consider the possibility of changes in the total linear and angular momentum of the system, we need to account for changes in the associated macroscopic kinetic energy $\mathcal{K}$. This is accomplished by the introduction of the *total energy* $\mathcal{E} \equiv \mathcal{K} + \mathcal{U}$. Then the total external work performed on a system consists of two parts: one that goes toward

[21] Our definition of a thermodynamic equilibrium state involved the number of particles and macroscopically observable state variables. The total linear and angular momentum was assumed to be zero. However, for our discussion of the first law of thermodynamics we, temporally, relax this condition and allow nonzero (constant) values of total linear and angular momentum. Thus, in the described cyclic process no external work goes toward a change in linear or angular momentum and $\Delta\mathcal{W}^{\mathrm{ext}} = \Delta\mathcal{W}^{\mathrm{def}}$, the work of deformation.

[22] Throughout most of his scientific career, Joule worked in his family's brewery. Much of his research was motivated by his desire to understand and improve the machines in the factory.

[23] Due to the success of Joule's discovery that heat and work are just different forms of energy, the constant bearing his name has fallen into disuse because independent units for heat (such as the calorie) are no longer part of the standard unit systems used by scientists.

[24] To see this consider any two thermodynamic equilibrium states, 1 and 2. Suppose $\Delta\mathcal{U}_{1\to2} = \Delta\mathcal{W} + \Delta\mathcal{Q}$ for one given process taking the system from 1 to 2. Now, let $\Delta\mathcal{U}_{2\to1}$ be the corresponding quantity for a process that takes the system from 2 to 1. The conservation principle requires that $\Delta\mathcal{U}_{2\to1} = -\Delta\mathcal{U}_{1\to2}$. In fact, this must be true for *all* processes that take the system from 2 to 1. The argument may be reversed to show that all processes that take the system from 1 to 2 must have the same value for $\Delta\mathcal{U}_{1\to2}$. We have found that the change in internal energy for any process depends only on the beginning and ending states of thermodynamic equilibrium. Thus, we can write $\Delta\mathcal{U}_{1\to2} = \mathcal{U}_2 - \mathcal{U}_1$, where $\mathcal{U}_1$ is the internal energy of state 1 and $\mathcal{U}_2$ is the internal energy of state 2.

a change in macroscopic kinetic energy and the work of deformation that goes toward a change in internal energy: $\Delta\mathcal{W}^{\text{ext}} = \Delta\mathcal{K} + \Delta\mathcal{W}^{\text{def}}$. With these definitions, Eqn. (5.2) may alternatively be given as

$$\Delta\mathcal{E} = \Delta\mathcal{W}^{\text{ext}} + \Delta\mathcal{Q}. \tag{5.3}$$

Equation (5.3) (or equivalently Eqn. (5.2)) is called the *first law of thermodynamics*. In words it is stated:

First law of thermodynamics The total energy of a thermodynamic system and its surroundings is conserved.

Mechanical and thermal energy transferred to the system (and lost by the surrounding medium) is retained in the system as part of its total energy, which consists of *kinetic energy* associated with motion of the system's particles (which includes the system's gross motion) and *potential energy* associated with deformation. In other words, energy can change form, but its amount is conserved. Two useful conclusions can be drawn from the above discussion:

1. Equation (5.2) implies that the value of $\mathcal{U}$ depends only on the state of thermodynamic equilibrium. This means that it does not depend on the details of how the system arrived at any given state, but only on the values of the independent state variables that characterize the system. It is therefore a state variable itself. For example, taking the independent state variables to be the number of particles, the values of the kinematic state variables and the temperature, we have that[25] $\mathcal{U} = \widehat{\mathcal{U}}(N, \mathbf{\Gamma}, T)$. Further, we note that the internal energy is extensive.
2. Joule's relation between work and heat implies that, although the internal energy is a state variable, the work of deformation and heat transfer are not. Their values depend on the process that occurs during a change of state. In other words, $\Delta\mathcal{W}^{\text{def}}$ and $\Delta\mathcal{Q}$ are measures of energy transfer, but associated functions $\mathcal{W}^{\text{def}}$ and $\mathcal{Q}$ (similar to the internal energy $\mathcal{U}$) do not exist. Once heat and work are absorbed into the energy of the system they are no longer separately identifiable.

   Another way of looking at this is that mechanical work and heat are just *conduits* for transmitting energy. Consider the analogy depicted in Fig. 5.2. Two containers of water are connected by two pipes, W and Q, with valves that control the flow of water. The water flows from container A to container B. The total amount of water in both containers is conserved, therefore the amount of water that flows through *both* pipes is

[25] The symbol $\widehat{\mathcal{U}}$ is used to indicate the particular functional form where the energy is determined by the values of the number of particles, kinematic state variables and temperature.

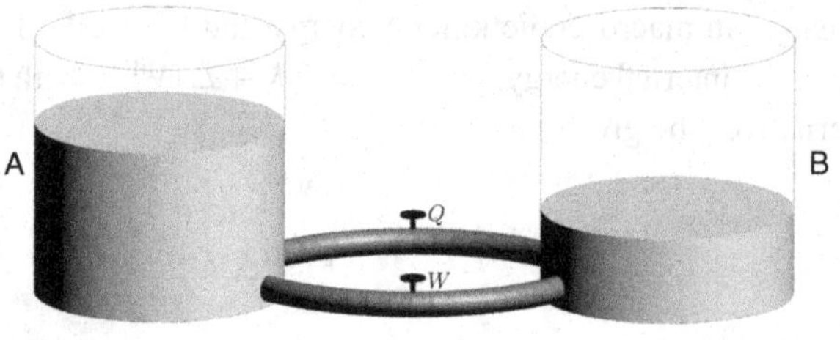

Fig. 5.2 Analogy for the first law of thermodynamics. Two water containers, A and B, are connected by pipes, W and Q, with valves. See text for explanation.

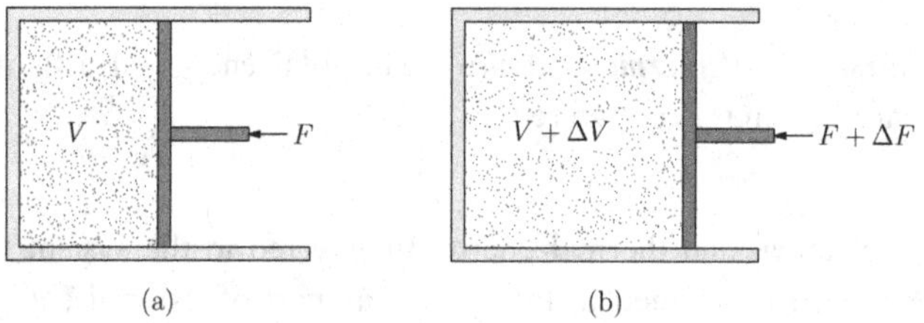

Fig. 5.3 Compressed gas in a cylinder with externally applied force equal to (a) F and (b) $F + \Delta F$.

exactly equal to the amount of water lost by container A and gained by container B. (We are assuming that no water is left in the pipes.) This is exactly what the first law states, where the water represents energy, container A represents the surroundings, container B represent the thermodynamic system and the pipes represent mechanical work and heat transfer. Once the water from the two pipes has flowed into container B there is no way to distinguish which water came through W and which water came through Q. For this reason the amount of water transferred through one of the pipes, say W (representing mechanical work, which we denoted $\Delta\mathcal{W}^{\text{def}}$), is not the difference of a function associated with B. This function, if it existed, would be the amount of water in container B that flowed into it through pipe W. But there is no way to identify this "special water" in system B after it has mixed in with the rest.

The following example demonstrates how the first law is applied to a physical system.

Example 5.1 (First law applied to a compressed gas) Consider a gas with a fixed number of N particles in a thermally-isolated cylinder compressed by a frictionless piston with applied external force $F > 0$ as shown in Fig. 5.3. The cylinder has cross-sectional area A. Initially the gas is in thermodynamic equilibrium and has a volume V (Fig. 5.3(a)). Then, the applied force is suddenly changed from F to $F + \Delta F$, after which it is held constant (Fig. 5.3(b)). This perturbation causes the gas to undergo a dynamical process. As part of this process the piston moves and oscillates, but eventually the system again reaches thermodynamic equilibrium with the new constant volume $V + \Delta V$.

The first law tells us that the change of internal energy is equal to the external heat transferred to the system plus the total work of deformation delivered to the system. Here, the system is thermally isolated, so only the work of deformation contributes. The force of $F + \Delta F$ does an amount of work given by

$$\Delta \mathcal{W}^{\text{def}} = -(F + \Delta F)\frac{\Delta V}{A}.$$

Thus, the change of the internal energy of the gas is given by

$$\Delta \mathcal{U} = \Delta \mathcal{W}^{\text{def}} = -(F + \Delta F)\frac{\Delta V}{A}.$$

If we assume that the volume decreases ($\Delta V < 0$) in response to an increase in force ($\Delta F > 0$), or vice versa – as we would expect from our physical experiences – then we see that the internal energy increases during the process. However, it is interesting to note that the first law tells us nothing about how ΔV is related to ΔF. In particular, nothing we have said so far prohibits the volume of the gas from increasing when the external force is increased (in this case the internal energy would decrease in accordance with the first law). In fact, we will have to wait until we introduce the second law of thermodynamics in order to obtain a complete description of a thermodynamic system. Once such a complete description is available we will be able to determine not only the direction in which the piston will move, but also the distance it will move when the applied force is changed from F to $F + \Delta F$.

5.3.2 Internal energy of an ideal gas

It is instructive to demonstrate the laws of thermodynamics with a simple material model. Perhaps the simplest model is the *ideal gas*, where the atoms are treated as particles of negligible radius which do not interact except when they elastically bounce off each other.[26] This idealization becomes more and more accurate as the pressure of a gas is reduced.[27] The reason for this is that the density of a gas goes to zero along with its pressure. At very low densities the size of an atom relative to the volume it occupies becomes negligible. Since the atoms in the gas are far apart most of the time, the interaction forces between them also become negligible.

Insight into the internal energy of an ideal gas was gained from Joule's experiments mentioned earlier. Joule studied the free expansion of a thermally-isolated gas (also called "Joule expansion") from an initial volume to a larger volume and measured the temperature change. The experiment is performed by rapidly removing a partition that confines the gas to the smaller volume and allowing it to expand. Since no mechanical work is performed on the gas ($\Delta \mathcal{W}^{\text{def}} = 0$) and no heat is transferred to it ($\Delta \mathcal{Q} = 0$), the first law (Eqn. (5.2)) is simply $\Delta \mathcal{U} = 0$, i.e. the internal energy is constant in any such experiment.

[26] The idea of noninteracting particles that can still bounce off each other may appear baffling to some readers. The key property of an ideal gas is that its particles do not interact. However, collisions between particles are necessary to randomize the velocity distribution (see, for example, [Les74]). The combination of these incompatible behaviors is the idealization we refer to as an "ideal gas."

[27] In Section 5.5.5 we give the formal definition of pressure and other intensive state variables that arise naturally as part of thermodynamic theory.

Now, recall that volume is the only kinematic state variable for a gas, the total differential of internal energy associated with an infinitesimal change of state is thus[28]

$$
\begin{aligned}
d\mathcal{U} &= \left.\frac{\partial \widehat{\mathcal{U}}}{\partial N}\right|_{V,T} dN + \left.\frac{\partial \widehat{\mathcal{U}}}{\partial V}\right|_{N,T} dV + \left.\frac{\partial \widehat{\mathcal{U}}}{\partial T}\right|_{N,V} dT \\
&= \left.\frac{\partial \widehat{\mathcal{U}}}{\partial N}\right|_{V,T} dN + \left.\frac{\partial \widehat{\mathcal{U}}}{\partial V}\right|_{N,T} dV + nC_v\, dT,
\end{aligned} \tag{5.4}
$$

where $n = N/N_{\mathrm{A}}$ is the number of moles of gas (with Avogadro's constant $N_{\mathrm{A}} = 6.022 \times 10^{23}\ \mathrm{mol}^{-1}$) and

$$
C_v = \frac{1}{n} \left.\frac{\partial \widehat{\mathcal{U}}}{\partial T}\right|_{N,V} \tag{5.5}
$$

is the *molar heat capacity at constant volume*.[29] The molar heat capacity of an ideal gas is a universal constant. For a *monoatomic* ideal gas it is $C_v = \frac{3}{2} N_{\mathrm{A}} k_{\mathrm{B}} = 12.472\ \mathrm{J \cdot K^{-1} \cdot mol^{-1}}$, where $k_{\mathrm{B}} = 1.3807 \times 10^{-23}$ J/K is Boltzmann's constant (see Exercise 7.8 in [TM11] for a derivation of C_v for an ideal gas based on statistical mechanics). For a real gas, C_v is a material property which can depend on the equilibrium state.

For a Joule expansion corresponding to an infinitesimal increase of volume dV at constant mole number, the first law requires $d\mathcal{U} = 0$. Joule's experiments showed that the temperature of the gas remained constant as it expanded ($dT = 0$), therefore the first and third terms of

28 The "vertical bar" notation $\partial\square/\partial T|_X$ is common in treatments of thermodynamics. It is meant to explicitly indicate which state variables (X) are to be held constant when determining the value of the partial derivative. For example $\partial\mathcal{U}/\partial T|_{N,V} \equiv \partial\widehat{\mathcal{U}}(N,V,T)/\partial T$. However, $\partial\mathcal{U}/\partial T|_{N,p}$ is completely different. It is the partial derivative of the internal energy as a function of the number of particles, the pressure and temperature: $\mathcal{U} = \widetilde{\mathcal{U}}(N,p,T)$. That is, $\partial\mathcal{U}/\partial T|_{N,p} \equiv \partial\widetilde{\mathcal{U}}(N,p,T)/\partial T$. The main advantage of the notation is that it allows for the use of a single symbol ($\mathcal{U}$) to represent the value of a state variable. Thus, it avoids the use of individual symbols to indicate the particular functional form used to obtain the quantity's value: $\mathcal{U} = \widehat{\mathcal{U}}(N,V,T) = \widetilde{\mathcal{U}}(N,p,T)$. However, we believe this leads to a great deal of confusion, obscures the mathematical structure of the theory and often results in errors by students and researchers who are not vigilant in keeping track of which particular functional form they are using. In this book, we have decided to keep the traditional notation while also using distinct symbols to explicitly indicate the functional form being used. Thus, the vertical bar notation is, strictly, redundant and can be ignored if so desired.

29 Formally, the molar heat capacity of a gas at constant volume is defined as

$$
C_v = \frac{1}{n}\frac{\Delta Q_V}{\Delta T},
$$

where ΔQ_V is the heat transferred under conditions of constant volume and n is the constant number of moles of gas. This is the amount of heat required to change the temperature of 1 mole of material by 1 degree. For a fixed amount of gas at constant volume, the first law reduces to $\Delta\mathcal{U} = \Delta Q$ (since no mechanical work is done on the gas), therefore the molar heat capacity is also

$$
C_v = \frac{1}{n} \left.\frac{\partial \widehat{\mathcal{U}}}{\partial T}\right|_{N,V}.
$$

Similar properties can be defined for a change due to temperature at constant pressure and changes due to other state variables. See [Adk83, Section 3.6] for a full discussion.

the differential in Eqn. (5.4) drop out and we have[30]

$$\left.\frac{\partial \widehat{\mathcal{U}}}{\partial V}\right|_{N,T} = 0. \tag{5.6}$$

This is an important result, since it indicates that the internal energy of an ideal gas does not depend on volume:[31]

$$\mathcal{U} = \widehat{\mathcal{U}}(n, V, T) = n\mathcal{U}_0 + nC_v T. \tag{5.7}$$

Here the number of moles n has been used to specify the amount of gas (instead of the number of particles N) and $\mathcal{U}_0$ is the molar internal energy of an ideal gas at zero temperature. Equation (5.7) is called *Joule's law*. It is exact for ideal gases, by definition, and provides a good approximation for real gases at low pressures.

Joule's law is an example of an equation of state as defined in Section 5.1.4. Of course, other choices for the independent state variables could be made. For example, instead of n, V and T, we can choose to work with n, V and $\mathcal{U}$, as the independent variables, in which case the equation of state for the ideal gas would be

$$T = \widehat{T}(n, V, \mathcal{U}) = (\mathcal{U} - n\mathcal{U}_0)/nC_v.$$

Another possibility is to use n, p and T as the independent state variables, where p is the pressure – the thermodynamic tension associated with the volume as described in Section 5.5.5. In this case the internal energy would be expressed as $\mathcal{U} = \widetilde{\mathcal{U}}(n, p, T)$. It is important to understand that in this case the internal energy would *not* be given by Eqn. (5.7). It would depend explicitly on the pressure. See Section 7.3.5 in [TM11] for a derivation of the equations of state for an ideal gas using statistical mechanics.

We now turn to two examples demonstrating how the first law can be used to compute the change in temperature of a gas.

Example 5.2 (Heating of a gas) In Example 5.1 we saw that a change in the external force compressing a gas in a thermally-isolated cylinder with a frictionless piston caused the gas to undergo a change of state that led to a change of its volume and its internal energy. Now, suppose the gas is argon, which is well approximated as an ideal gas, for which the molar heat capacity is $C_v = 12.472\,\mathrm{J\cdot K^{-1}\cdot mol^{-1}}$, and that in the initial state there are $n = 2$ mol at a temperature of $T = 300$ K, with initial volume $V = 0.5\,\mathrm{m}^3$. The piston cross-sectional area is $A = 0.01\,\mathrm{m}^2$. The force is increased from $F = 100$ N by $\Delta F = 30$ N and as a result of the ensuing dynamical process the system changes its volume

30 Actually, the temperature of a real gas does change in free expansion. However, the effect is weak and Joule's experiments lacked the precision to detect it. For an ideal gas, the change in temperature is identically zero. See Section 7.3.5 of [TM11].

31 This form for the internal energy may be obtained as follows. First, we use Joule's result to obtain $\widehat{\mathcal{U}}(n, V, T) = f(n, T)$. Second, we note that Eqn. (5.5) gives $\partial f/\partial T = nC_v$, where C_v is a constant. Third, we integrate this expression to obtain $f = nC_v T + g(n)$. Finally, we note that since $\mathcal{U}$, n and V are extensive and T is intensive we must have $M\widehat{\mathcal{U}}(n, V, T) = \widehat{\mathcal{U}}(Mn, MV, T)$, where M is a positive real number. This implies that g must be a first-order homogeneous function (equivalently, a linear function), i.e. $g(n) = n\mathcal{U}_0$, where $\mathcal{U}_0$, is a constant.

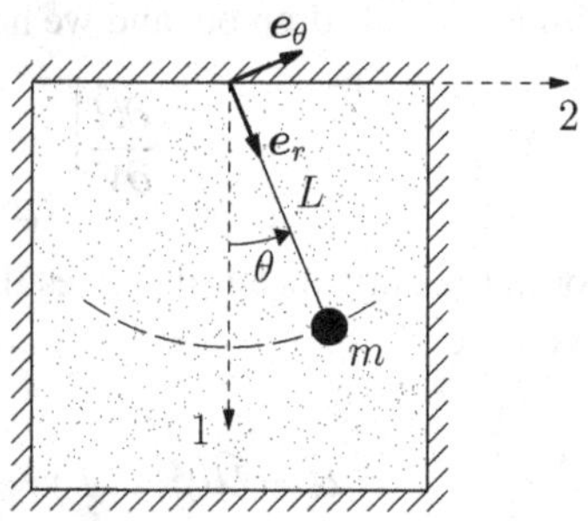

Fig. 5.4 A pendulum allowed to swing in a thermally-isolated fixed volume compartment containing an ideal gas.

by $\Delta V = -0.07\ \text{m}^3$ (a value that depends on the original state of equilibrium and ΔF). We are interested in computing the new temperature of the gas.

From the solution to Example 5.1 we find that the change of internal energy is

$$\Delta \mathcal{U} = -(F + \Delta F)\frac{\Delta V}{A} = 910\ \text{J}.$$

Using Eqn. (5.7) we find that $\Delta T = \Delta \mathcal{U}/nC_v = 910\ \text{J}/(2.0\ \text{mol} \cdot 12.47\ \text{J} \cdot \text{K}^{-1} \cdot \text{mol}^{-1}) = 36.5\ \text{K}$, and finally we find the new temperature is $T = 336.5\ \text{K}$.

Example 5.3 (Heating of a gas by a swinging pendulum) Figure 5.4 shows a pendulum swinging due to gravity in a gas, which (for small amplitude oscillations) is an example of a damped harmonic oscillator. The pendulum is thermally isolated, so that no heat is transferred to it, and has length L and mass m. Suppose the pendulum is initially at rest at an angle θ_0 and the system is in thermodynamic equilibrium. Then the pendulum is released and after a dynamical process, where the gas interacts with the pendulum as it swings back and forth, the pendulum eventually comes to rest at $\theta = 0$. As a result of this process, the gas undergoes a change of state and its temperature increases.

We will treat the gas and pendulum as a single system. The container does not change volume and no heat is transferred to the system. However, gravity acts on the pendulum, and therefore, does work on the system. Therefore, the first law (Eqn. (5.2)) reduces to $\Delta \mathcal{U} = \Delta \mathcal{W}^{\text{def}}$. The work done by gravity in moving the pendulum from $\theta = \theta_0$ to $\theta = 0$ must then be equal to the change in the internal energy of the gas

$$\Delta \mathcal{U} = \frac{1}{2}mgL(1 - \cos\theta_0) \approx \frac{1}{2}mgL\theta_0^2, \tag{5.8}$$

where we have assumed that $\theta_0 \ll 1$ so that $\cos\theta_0 \approx 1 - \theta_0^2$. Using Eqn. (5.7), the change in internal energy is $\Delta \mathcal{U} = nC_v\Delta T$, where n is the number of moles of gas. Equating this relation with Eqn. (5.8) gives

$$\Delta T = \frac{mg}{2nC_v}L\theta_0^2. \tag{5.9}$$

As a numerical example, assume the following parameters for the pendulum: $m = 1$ kg, $g = 9.81\ \text{m/s}^2$, $L = 1$ m, $\theta_0 = 0.1$. Take the gas to be air at normal room temperature and pressure for which $C_v = 20.85\ \text{J} \cdot \text{K}^{-1} \cdot \text{mol}^{-1}$ and $\rho = 1.29\ \text{kg/m}^3$. If the container is 2 m $\times$ 2 m $\times$ 2 m, then the mass of the gas is $m_{\text{gas}} = 8\rho = 10.32$ kg. The molar mass of air is $M = 28.97 \times 10^{-3}$ kg/mol. Thus, the number of moles of air in the container is $n = m_{\text{air}}/M = 356.2$ mol. Substituting the above values into Eqn. (5.9), the result is that the gas will heat by $\Delta T = 6.60 \times 10^{-6}$ K.

5.4 Thermodynamic processes

Equilibrium states are of great interest, but the true power of the theory of thermodynamics is its ability to predict the state to which a system will transition when it is perturbed from equilibrium. In fact, it is often of interest to predict an entire series of equilibrium states that will occur when a system is subjected to a series of perturbations.

5.4.1 General thermodynamic processes

We define a *thermodynamic process* as an ordered set or sequence of equilibrium states. This set need not correspond to any actual series followed by a real system. It is simply a string of possible equilibrium states. For system B with independent state variables $\mathcal{B} = (\mathcal{B}_1, \mathcal{B}_2, \dots, \mathcal{B}_{\nu^{\mathsf{B}}})$, a thermodynamic process containing M states is denoted by

$$\mathfrak{B} = (\mathcal{B}^{(1)}, \mathcal{B}^{(2)}, \dots, \mathcal{B}^{(M)}),$$

where $\mathcal{B}^{(i)} = (\mathcal{B}_1^{(i)}, \mathcal{B}_2^{(i)}, \dots, \mathcal{B}_{\nu^{\mathsf{B}}}^{(i)})$ is the ith state in the thermodynamic process. The behavior of the dependent state variables follows through the appropriate equations of state. Examples 5.2 and 5.3 above concern thermodynamic systems that undergo a "two-stage" ($M = 2$) thermodynamic process. If $M = 3$ and $\mathcal{B}^{(1)} = \mathcal{B}^{(3)}$, then we have a cyclic three-stage process such as described in Section 5.3. A general thermodynamic process can have any number of states M and there is no requirement that consecutive states in the process are close to each other. That is, the values of the independent state variables for stages i and $i+1$, $\mathcal{B}_\alpha^{(i)}$ and $\mathcal{B}_\alpha^{(i+1)}$, respectively, need not be related in any way.

5.4.2 Quasistatic processes

Although the laws of thermodynamics apply equally to all thermodynamic processes, those processes that involve a sequence of small increments to the independent state variables are of particular interest. In the limit, as the increments become infinitesimal, the process becomes a continuous path in the *thermodynamic state space* (the ν^{B}-dimensional space of independent state variables):

$$\mathfrak{B} = \mathcal{B}(s), \qquad s \in [0, 1].$$

Here functional notation is used to indicate the continuous variation of the independent state variables and s is used as a convenient variable to measure the "location" along the process.[32] Such a process is called *quasistatic*.[33]

Quasistatic processes are singularly useful within the theory of thermodynamics for two reasons. First, such processes can be associated with phenomena in the real world where small perturbations applied to a system (such as infinitesimal increments of the independent

[32] The choice of domain for s is arbitrary and the unit interval used here bears no special significance.

[33] In Section 5.5.5 we will consider the system's interaction with its surroundings as it undergoes a quasistatic process. There we will find that every such process is always to be associated with a specific amount of work and a separate specific amount of heat (and not just a total amount of energy) that are transferred to the system.

state variables) occur on a time scale that is significantly slower than that required for the system to reach equilibrium. In the limit as the perturbation rate becomes infinitely slower than the equilibration rate, the thermodynamic process becomes quasistatic. Technically, no real phenomena are quasistatic since the time required for a system to reach true equilibrium is infinite. However, in many cases the dynamical processes that lead to equilibrium are sufficiently fast for the thermodynamic process to be approximately quasistatic. This is particularly the case if we relax the condition for thermodynamic equilibrium and accept metastable equilibrium instead. Indeed, the world is replete with examples of physical phenomena that can be accurately analyzed within thermodynamic theory when they are approximated as quasistatic processes.

Second, general results of thermodynamic theory are best expressed in terms of infinitesimal changes of state. These results may then be integrated along any quasistatic process in order to obtain predictions of the theory for finite changes of state. The expressions associated with such finite changes of state are almost always considerably more complex than their infinitesimal counterparts and often are only obtainable in explicit form once the equations of state for a particular material are introduced.

5.5 The second law of thermodynamics and the direction of time

The first law of thermodynamics speaks of the conservation of energy during thermodynamic processes, but it tells us nothing about the *direction* of such processes. How is it that if we watch a movie of a shattered glass leaping onto a table and reassembling, we immediately know that it is being played in reverse? The first law provides no answer – it can be satisfied for any process. Consider the following scenario:

1. A rigid hollow sphere filled with an ideal gas is placed inside of a larger, otherwise empty, sealed box that is thermally isolated from its surroundings.
2. A hole is opened in the sphere.
3. The gas quickly expands to fill the box.
4. After some time, the gas spontaneously returns, through the hole, to occupy only its original volume within the sphere.

This scenario is perfectly legal from the perspective of the first law. In fact, we showed in our discussion of Joule's experiments in Section 5.3.2 that the internal energy of an ideal gas remains unchanged by the free expansion in step 3. It is therefore clearly *not* a violation of the first law for the gas to return to its initial state. However, our instincts, based on our familiarity with the world, tell us that this process of "reverse expansion" will never happen.

The thermodynamic process discussed in Examples 5.1 and 5.2 is another illustration of this type of scenario. If one starts with the system in the initial equilibrium state and then perturbs it by incrementing the applied force by a fixed finite amount, the system will transition to a particular final equilibrium state. However, if one starts with the system in the "final" equilibrium state and perturbs it by *decrementing* the applied force, we know

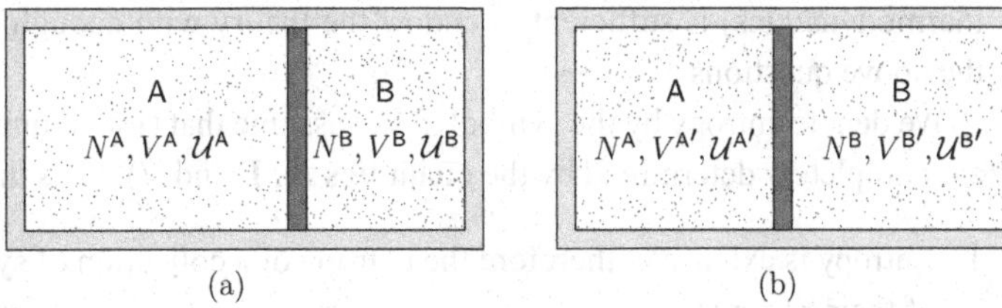

Fig. 5.5 An isolated system consisting of a rigid, sealed and thermally isolated cylinder of total volume V; an internal frictionless, impermeable piston; and two subsystems A and B containing ideal gases. (a) Initially the piston is fixed and thermally insulating and the gases are in thermodynamic equilibrium. (b) The new states of thermodynamic equilibrium obtained following a dynamical process once the piston becomes diathermal and is allowed to move.

from observation that the system will not transition to the original "initial" state. Instead it transitions to a third state that is distinct from the previous two. In other words, for *finite* increments to the force, the two-stage thermodynamic process of Examples 5.1 and 5.2 has a unique *direction*.[34] The same can be said for the above scenario of an ideal gas undergoing free expansion. In fact, we can relate this directionality of thermodynamic processes to our concept of time and why we perceive that time always evolves from the "present" to the "future" and never from the "present" to the "past." Clearly, something in addition to the first law is necessary to describe the directionality of thermodynamic processes.

5.5.1 Entropy

Suppose we have a rigid, sealed and thermally-isolated cylinder of volume V with a frictionless and impermeable internal piston that divides it into two compartments, A and B of initial volumes V^{A} and $V^{B} = V - V^{A}$, respectively, as shown in Fig. 5.5(a). Initially, the piston is fixed in place and thermally isolating. Compartment A is filled with N^{A} particles of an ideal gas with internal energy $\mathcal{U}^{A}$ and compartment B is filled with N^{B} particles of another ideal gas with internal energy $\mathcal{U}^{B}$. Thus, the composite system's total internal energy is $\mathcal{U} = \mathcal{U}^{A} + \mathcal{U}^{B}$. As long as the piston remains fixed and thermally insulating, A and B are isolated systems. If we consider the entire cylinder as a single isolated thermodynamic system consisting of two subsystems, the piston represents a set of *internal constraints*. We are interested in answering the following questions. If we release the constraints by allowing the piston to move and to transmit heat, in what direction will the piston move? How far will it move? And, why is the reverse process never observed, i.e. why does the piston never return to its original position? Since nothing in our theory so far is able to provide the answers to these questions, we postulate the existence of a new state variable, related to the direction of thermodynamic processes, that we call *entropy*.[35] We will show below that requiring this variable to satisfy a simple extremum principle (the second law of

[34] We will see later that in the limit of an *infinitesimal* increment of force the process, in principle, can occur in either direction.

[35] The word *entropy* was coined in 1865 by the German physicist Rudolf Clausius as a combination of the Greek words *en-* meaning *in* and *tropē* meaning *change or turn.*

thermodynamics) is sufficient to endow the theory with enough structure to answer all of the above questions.

We denote entropy by the symbol $\mathcal{S}$ and assume that (for all uniform systems whose state are completely determined by the quantities N, $\mathbf{\Gamma}$ and $\mathcal{U}$) it has the following properties:[36]

1. Entropy is extensive, therefore the entropy of a collection of systems is equal to the sum of their entropies:

$$\mathcal{S}^{\mathrm{A+B+C+\ldots}} = \mathcal{S}^{\mathrm{A}} + \mathcal{S}^{\mathrm{B}} + \mathcal{S}^{\mathrm{C}} + \cdots. \tag{5.10}$$

2. Entropy is a *monotonically increasing* function of the internal energy $\mathcal{U}$, when the system's independent state variables are chosen to be the number of particles N, the extensive kinematic state variables $\mathbf{\Gamma}$ and the internal energy $\mathcal{U}$

$$\mathcal{S} = \bar{\mathcal{S}}(N, \mathbf{\Gamma}, \mathcal{U}). \tag{5.11}$$

   Here $\bar{\mathcal{S}}(\cdot,\cdot,\cdot)$ indicates the functional dependence of $\mathcal{S}$ on its arguments.[37] Note that this monotonicity condition only applies to the function $\bar{\mathcal{S}}(N, \mathbf{\Gamma}, \cdot)$, where N and $\mathbf{\Gamma}$ are held constant. Thus, this condition does not restrict, in any way, how the entropy depends on N and $\mathbf{\Gamma}$.
3. $\bar{\mathcal{S}}(\cdot,\cdot,\cdot)$ is a continuous and differentiable function of its arguments. This assumption and the assumption of monotonicity imply that Eqn. (5.11) is invertible, i.e.

$$\mathcal{U} = \bar{\mathcal{U}}(N, \mathbf{\Gamma}, \mathcal{S}). \tag{5.12}$$

In Eqn. (5.12), we are using the number of particles, extensive kinematic state variables and the entropy as the independent state variables to identify any given state of thermodynamic equilibrium.

5.5.2 The second law of thermodynamics

The direction of physical processes can be expressed as a constraint on the way entropy can change during any process. This is what the *second law of thermodynamics* is about. There are many equivalent ways that this law can be stated. We choose the statement attributed to Rudolf Clausius, which we find to be physically most transparent:

> *Second law of thermodynamics* An isolated system in thermodynamic equilibrium adopts the state that has the maximum entropy of all states consistent with the imposed kinematic constraints.

[36] At this stage, these assumptions are nothing more than educated guesses which can be taken to be axioms. However, we will see below that with these properties entropy can be used to predict the direction of physical processes.

[37] This monotonicity condition should not be confused with the second law of thermodynamics. As we will see in Section 5.5.4, the physical reason for requiring the monotonicity condition is that it ensures that the temperature is always positive. However, it has nothing to do with the second law, which we see in the next section is a statement about how the entropy function of an isolated system depends on N and $\mathbf{\Gamma}$.

Let us see how the second law is applied to the cylinder with an internal piston shown in Fig. 5.5 and introduced in the previous section. The second law tells us that once the internal constraints are removed and the piston is allowed to move and to transmit heat, the system will evolve in order to maximize its entropy as shown in Fig. 5.5(b). At the end of this process, the subsystems A and B are again in thermodynamic equilibrium with state variables $(N^{\mathsf{A}}, V^{\mathsf{A}'}, \mathcal{U}^{\mathsf{A}'})$ and $(N^{\mathsf{B}}, V^{\mathsf{B}'}, \mathcal{U}^{\mathsf{B}'})$. We assume that the piston is impermeable so that the numbers of atoms do not change (i.e. $N^{\mathsf{A}'} = N^{\mathsf{A}}$ and $N^{\mathsf{B}'} = N^{\mathsf{B}}$). Since the composite system is isolated, its total volume and internal energy must be conserved and this implies that $V^{\mathsf{B}'} = V - V^{\mathsf{A}'}$ and $\mathcal{U}^{\mathsf{B}'} = \mathcal{U} - \mathcal{U}^{\mathsf{A}'}$. Thus, the equilibrium value of the entropy $\mathcal{S}'$ for the isolated composite system is

$$\mathcal{S}' = \max_{\substack{0 \le V^{\mathsf{A}'} \le V, \\ \mathcal{U}^{\mathsf{A}'} \in \mathbb{R}}} \left[\bar{\mathcal{S}}^{\mathsf{A}}(N^{\mathsf{A}}, V^{\mathsf{A}'}, \mathcal{U}^{\mathsf{A}'}) + \bar{\mathcal{S}}^{\mathsf{B}}(N^{\mathsf{B}}, V - V^{\mathsf{A}'}, \mathcal{U} - \mathcal{U}^{\mathsf{A}'}) \right],$$

where $\bar{\mathcal{S}}^{\mathsf{A}}(\cdot,\cdot,\cdot)$ and $\bar{\mathcal{S}}^{\mathsf{B}}(\cdot,\cdot,\cdot)$ are the entropy functions for the ideal gases of A and B, respectively.[38] The value of $V^{\mathsf{A}'}$ obtained from the above maximization problem determines the final position of the piston, and thus provides the answers to the questions posed earlier in this section. In particular, we see that any change of the volume of A away from the equilibrium value $V^{\mathsf{A}'}$ must necessarily result in a decrease of the total entropy. As we will see next, this would violate the second law of thermodynamics. This violation of the maximum entropy law shows us why any real thermodynamic process (and therefore time) has a unique direction and is never observed to occur in reverse.

It is useful to rephrase the second law in an alternative manner:

Second law of thermodynamics (alternative statement) The entropy of an isolated system can never decrease in any process. It can only increase or stay the same.

Mathematically this statement is

$$\Delta \mathcal{S} \ge 0, \tag{5.13}$$

for any isolated system that transitions from one equilibrium state to another in response to the release of an internal constraint. It is trivial to show that the Clausius statement of the second law leads to this conclusion. Consider a process that begins in state 1 and ends in state 2. The Clausius statement of the second law tells us that $\mathcal{S}^{(2)} \ge \mathcal{S}^{(1)}$, therefore $\Delta\mathcal{S} = \mathcal{S}^{(2)} - \mathcal{S}^{(1)} \ge 0$, which is exactly Eqn. (5.13).

Note that the statements of the second law given above have been careful to stress that the law only holds for *isolated* systems. The entropy of a system that is not isolated can and often does decrease in a process. We will see this later.

It is worth emphasizing a subtle feature of the above discussion. In order to complete the theory, we introduced a new state variable – the entropy – which exists for every thermodynamic system, but whose value can be used to determine the direction and final

[38] Note that, although the internal energy is extensive, it is not required to be positive. In fact, in principle $\mathcal{U}^{\mathsf{A}'}$ may take on any value as long as $\mathcal{U}^{\mathsf{B}'}$ is then chosen to ensure conservation of energy. Thus, the maximization with respect to energy considers all possible values of $\mathcal{U}^{\mathsf{A}'}$.

result only of processes that occur in *isolated systems*. Isolated systems are special in the sense that *all* of their *extensive state variables*, except entropy, must be conserved (fixed) during any process. In particular, all the kinematic state variables must be fixed. If this were not the case, when any kinematic state variable changed work would be done on the system by the external universe and the system would cease to be isolated. Thus, it is important to use only conserved state variables in the set of independent state variables. Accordingly, above we have introduced the entropy equation of state $\mathcal{S} = \bar{\mathcal{S}}(\cdot,\cdot,\cdot)$ as a function of the number of particles N, the kinematic state variables $\mathbf{\Gamma}$ and the internal energy $\mathcal{U}$. This function is the one to which it is appropriate to apply the extremum principle.

5.5.3 Stability conditions associated with the second law

Our discussion of equilibrium has so far been limited to spatially homogeneous states. We now consider the conditions that the entropy function, $\bar{\mathcal{S}}(N, \mathbf{\Gamma}, \mathcal{U})$, must satisfy to ensure the stability of the homogeneous state.

Consider an isolated composite system in thermodynamic equilibrium with $N' = 2N$ particles, kinematic state variables $\mathbf{\Gamma}' = 2\mathbf{\Gamma}$ and internal energy $\mathcal{U}' = 2\mathcal{U}$, consisting of two identical subsystems with N particles, kinematic state variables $\mathbf{\Gamma}$ and internal energy $\mathcal{U}$ each. Then the total entropy of the two subsystems is

$$\mathcal{S} = \bar{\mathcal{S}}(N, \mathbf{\Gamma}, \mathcal{U}) + \bar{\mathcal{S}}(N, \mathbf{\Gamma}, \mathcal{U}) = 2\bar{\mathcal{S}}(N, \mathbf{\Gamma}, \mathcal{U}).$$

The second law tells us that the entropy is maximized in this state of equilibrium, where both subsystems are in identical states. In other words, since the two systems are the same, the composite system is spatially homogeneous. However, in general the spatially homogeneous state need not maximize the entropy.

To see this, we consider what happens if some amount of energy $\Delta\mathcal{U}$ is transferred from one subsystem to the other. The total energy must be conserved because the composite system is isolated. For such an energy transfer, the total entropy becomes

$$\mathcal{S} = \bar{\mathcal{S}}(N, \mathbf{\Gamma}, \mathcal{U} + \Delta\mathcal{U}) + \bar{\mathcal{S}}(N, \mathbf{\Gamma}, \mathcal{U} - \Delta\mathcal{U}).$$

The properties, given in Section 5.5.1, for the entropy are not sufficient to determine the sign of the entropy change $\Delta\mathcal{S} = \bar{\mathcal{S}}(N, \mathbf{\Gamma}, \mathcal{U} + \Delta\mathcal{U}) + \bar{\mathcal{S}}(N, \mathbf{\Gamma}, \mathcal{U} - \Delta\mathcal{U}) - 2\bar{\mathcal{S}}(N, \mathbf{\Gamma}, \mathcal{U})$. If the entropy increases ($\Delta\mathcal{S} > 0$) due to energy transfers between subsystems a *phase transition* occurs and the system becomes a spatially inhomogeneous mixture of two distinct equilibrium states. This is an example of a *material instability*, and we say that the equilibrium state $\mathcal{B}$ of the system (identified by[39] $\mathcal{B} = (N, \mathbf{\Gamma}, \mathcal{U})$) is *unstable* with respect to changes of internal energy. An example of this is when a system of water vapor is cooled to its dew point. When this occurs, some of the water transitions from vapor to liquid and the previously spatially homogeneous vapor system splits into two subsystems: one subsystem in the liquid phase and the other in the vapor phase.

[39] Which is the same as that for $\mathcal{U}' = 2\mathcal{U}$, $\mathbf{\Gamma}' = 2\mathbf{\Gamma}$, $N' = 2N$, or, in fact, any multiple of these values since the state variables are extensive.

The alternative case is where the entropy decreases ($\Delta\mathcal{S} < 0$) when the energy transfer occurs such that

$$\overline{\mathcal{S}}(N,\boldsymbol{\Gamma},\mathcal{U}+\Delta\mathcal{U}) + \overline{\mathcal{S}}(N,\boldsymbol{\Gamma},\mathcal{U}-\Delta\mathcal{U}) \leq 2\overline{\mathcal{S}}(N,\boldsymbol{\Gamma},\mathcal{U}), \quad \text{for all } \Delta\mathcal{U}. \tag{5.14}$$

In this case we say that the *equilibrium state* $\mathcal{B}$ is *stable* with respect to changes of internal energy. A necessary condition for stability in this sense is that the entropy function be concave at $\mathcal{B}$, i.e. the second partial derivative of the entropy function with respect to the internal energy must be nonpositive:

$$\left.\frac{\partial^2\overline{\mathcal{S}}}{\partial\mathcal{U}^2}\right|_{N,\boldsymbol{\Gamma}} \leq 0. \tag{5.15}$$

This can be obtained from Eqn. (5.14) by moving all terms to the left-hand side of the inequality, dividing by $(\Delta\mathcal{U})^2$ and taking the limit as $\Delta\mathcal{U}$ goes to zero. However, it is important to note that this is not sufficient for stability. Although it ensures that Eqn. (5.14) is satisfied for infinitesimal values of $\Delta\mathcal{U}$ (i.e. $d\mathcal{U}$) it does not guarantee that it is satisfied *for all* values of $\Delta\mathcal{U}$. If a material's entropy function satisfies Eqn. (5.14) for fixed, but arbitrary, values of N, $\boldsymbol{\Gamma}$ and $\mathcal{U}$, and for all $\Delta\mathcal{U}$, i.e. *every* equilibrium state is stable with respect to changes of internal energy, then we say that the *material* is stable with respect to changes of internal energy. The entropy function of such a material is concave everywhere with respect to internal energy.

In the above discussion we have considered transfers of energy between the two subsystems, but there is nothing special about the energy; We could have instead considered transfers of particles or transfers of any one of the kinematic state variables. The same arguments can be carried out in each of these cases and similar results are obtained.[40] Thus, if a material's entropy function is concave everywhere with respect to N, then we say that the material is stable with respect to particle transfers and similarly for changes of the kinematic state variables. Finally, we can consider simultaneous transfers of two (or more) quantities, e.g. a transfer of particles and volume between the subsystems. Again, similar results are obtained. Thus, if a material's entropy function is concave everywhere with respect to all variables,[41] then we say that the material is stable.

5.5.4 Thermal equilibrium from an entropy perspective

In order to see the connection between entropy and the other thermodynamic state variables whose physical significance is more clear to us (e.g. temperature, volume and internal energy), we revisit the conditions of thermal equilibrium between two subsystems of an arbitrary isolated thermodynamic system discussed earlier in Section 5.2.

Let C be an isolated thermodynamic system made up of two subsystems, A and B, that are composed of (possibly different) stable materials. We take the independent state variables

[40] Again, we emphasize that these constraints apply to the particular functional form $\overline{\mathcal{S}}(N,\boldsymbol{\Gamma},\mathcal{U})$. If different independent state variables are used, then the functional form for entropy changes, and accordingly the constraints take different functional forms as well.

[41] Note that this implies that the matrix of all second-order partial derivatives of $\overline{\mathcal{S}}$ is everywhere negative semi-definite. Thus, it is necessary for stability (but not sufficient) that a relation such as Eqn. (5.15) is satisfied for each of the arguments of $\overline{\mathcal{S}}$.

for each system to be the number of particles N, extensive kinematic state variables $\mathbf{\Gamma}$ and the internal energy $\mathcal{U}$. Since C is isolated, according to the first law its internal energy is conserved, i.e. $\mathcal{U}^{\mathsf{C}} = \mathcal{U}^{\mathsf{A}} + \mathcal{U}^{\mathsf{B}} =$ constant. This means that any change in internal energy of subsystem A must be matched by an equal and opposite change in B:

$$\Delta\mathcal{U}^{\mathsf{A}} + \Delta\mathcal{U}^{\mathsf{B}} = 0. \tag{5.16}$$

Like the internal energy, entropy is also extensive and therefore the total entropy of the composite system is $\mathcal{S}^{\mathsf{C}} = \mathcal{S}^{\mathsf{A}} + \mathcal{S}^{\mathsf{B}}$. However, entropy is generally not constant in a change of state of an isolated system. The total entropy is a function of the two subsystems' state variables N^{A}, $\mathbf{\Gamma}^{\mathsf{A}}, \mathcal{U}^{\mathsf{A}}$, N^{B}, $\mathbf{\Gamma}^{\mathsf{B}}$ and $\mathcal{U}^{\mathsf{B}}$. The first differential of the total entropy is then[42]

$$\begin{aligned} d\mathcal{S}^{\mathsf{C}} = {} & \left.\frac{\partial\bar{\mathcal{S}}^{\mathsf{A}}}{\partial N^{\mathsf{A}}}\right|_{\mathbf{\Gamma}^{\mathsf{A}},\mathcal{U}^{\mathsf{A}}} dN^{\mathsf{A}} + \sum_{\alpha} \left.\frac{\partial\bar{\mathcal{S}}^{\mathsf{A}}}{\partial\Gamma^{\mathsf{A}}_{\alpha}}\right|_{N^{\mathsf{A}},\mathcal{U}^{\mathsf{A}}} d\Gamma^{\mathsf{A}}_{\alpha} + \left.\frac{\partial\bar{\mathcal{S}}^{\mathsf{A}}}{\partial\mathcal{U}^{\mathsf{A}}}\right|_{N^{\mathsf{A}},\mathbf{\Gamma}^{\mathsf{A}}} d\mathcal{U}^{\mathsf{A}} \\ & + \left.\frac{\partial\bar{\mathcal{S}}^{\mathsf{B}}}{\partial N^{\mathsf{B}}}\right|_{\mathbf{\Gamma}^{\mathsf{B}},\mathcal{U}^{\mathsf{B}}} dN^{\mathsf{B}} + \sum_{\beta} \left.\frac{\partial\bar{\mathcal{S}}^{\mathsf{B}}}{\partial\Gamma^{\mathsf{B}}_{\beta}}\right|_{N^{\mathsf{B}},\mathcal{U}^{\mathsf{B}}} d\Gamma^{\mathsf{B}}_{\beta} + \left.\frac{\partial\bar{\mathcal{S}}^{\mathsf{B}}}{\partial\mathcal{U}^{\mathsf{B}}}\right|_{N^{\mathsf{B}},\mathbf{\Gamma}^{\mathsf{B}}} d\mathcal{U}^{\mathsf{B}}. \end{aligned}$$

Suppose we fix the values of A's kinematic state variables $\mathbf{\Gamma}^{\mathsf{A}}$ and its number of particles N^{A} (then the corresponding values for B are determined by constraints imposed by C's isolation), but allow for energy (heat) transfer between A and B. Then the terms involving the increments of the extensive kinematic state variables and the increments of the particle numbers drop out. Further, since C is isolated, the internal energy increments must satisfy Eqn. (5.16), so likewise $d\mathcal{U}^{\mathsf{A}} = -d\mathcal{U}^{\mathsf{B}}$. All of these considerations lead to the following expression for the differential of the entropy of system C:

$$d\mathcal{S}^{\mathsf{C}} = \left[\left.\frac{\partial\bar{\mathcal{S}}^{\mathsf{A}}}{\partial\mathcal{U}^{\mathsf{A}}}\right|_{N^{\mathsf{A}},\mathbf{\Gamma}^{\mathsf{A}}} - \left.\frac{\partial\bar{\mathcal{S}}^{\mathsf{B}}}{\partial\mathcal{U}^{\mathsf{B}}}\right|_{N^{\mathsf{B}},\mathbf{\Gamma}^{\mathsf{B}}} \right] d\mathcal{U}^{\mathsf{A}}. \tag{5.17}$$

Now, according to our definition in Section 5.2, A and B are in thermal equilibrium if they remain in equilibrium when brought into thermal contact. This implies that the composite system C, subject to the above conditions, is in thermodynamic equilibrium when A and B are in thermal equilibrium. Thus, according to the second law of thermodynamics, the first differential of the entropy, Eqn. (5.17), must be zero for all $d\mathcal{U}^{\mathsf{A}}$ in this case (since the entropy is at a maximum). This leads to

$$\left.\frac{\partial\bar{\mathcal{S}}^{\mathsf{A}}}{\partial\mathcal{U}^{\mathsf{A}}}\right|_{N^{\mathsf{A}},\mathbf{\Gamma}^{\mathsf{A}}} = \left.\frac{\partial\bar{\mathcal{S}}^{\mathsf{B}}}{\partial\mathcal{U}^{\mathsf{B}}}\right|_{N^{\mathsf{B}},\mathbf{\Gamma}^{\mathsf{B}}} \tag{5.18}$$

as the condition for thermal equilibrium between A and B in terms of their entropy functions. Now recall from Eqn. (5.1) that thermal equilibrium requires $T^{\mathsf{A}} = T^{\mathsf{B}}$ or equivalently $1/T^{\mathsf{A}} = 1/T^{\mathsf{B}}$. (Here, we are referring explicitly to the thermodynamic temperature scale.) Comparing these with the equation above it is clear that $\partial\bar{\mathcal{S}}/\partial\mathcal{U}$ is either[43] T or $1/T$. To

[42] The notation $\partial\bar{\mathcal{S}}/\partial\Gamma_{\alpha}|_{N,\mathcal{U}}$ refers to the partial derivative of the function $\bar{\mathcal{S}}(N,\mathbf{\Gamma},\mathcal{U})$ with respect to the αth component of $\mathbf{\Gamma}$ (while holding all other components of $\mathbf{\Gamma}$, N, and $\mathcal{U}$ fixed). We leave out the remaining components of $\mathbf{\Gamma}$ from the list at the bottom of the bar in order to avoid extreme notational clutter.

[43] Instead of T or $1/T$ any monotonically increasing or decreasing functions of T would do. We discuss this further below.

decide which is the correct definition, we recall that the concept of temperature also included the idea of "hotter than." Thus, we must test which of the above options is consistent with our definition that if $T^{\mathsf{A}} > T^{\mathsf{B}}$, then heat (energy) will spontaneously flow from A to B when they are put into thermal contact.

To do this, consider the same combination of systems as before, and now assume that initially A has a higher temperature than B, i.e. $T^{\mathsf{A}} > T^{\mathsf{B}}$. Since the composite system is isolated, our definition of temperature and the first law of thermodynamics imply that heat will flow from A to B which will result in a decrease of $\mathcal{U}^{\mathsf{A}}$ and a correspondingly equal increase of $\mathcal{U}^{\mathsf{B}}$. However, the second law of thermodynamics says that such a change of state can only occur if it increases the total entropy of the isolated composite system. Thus, we must have that

$$d\mathcal{S}^{\mathsf{C}} = \left[\left.\frac{\partial \bar{\mathcal{S}}^{\mathsf{A}}}{\partial \mathcal{U}^{\mathsf{A}}}\right|_{N^{\mathsf{A}},\Gamma^{\mathsf{A}}} - \left.\frac{\partial \bar{\mathcal{S}}^{\mathsf{B}}}{\partial \mathcal{U}^{\mathsf{B}}}\right|_{N^{\mathsf{B}},\Gamma^{\mathsf{B}}} \right] d\mathcal{U}^{\mathsf{A}} > 0.$$

Since we expect $d\mathcal{U}^{\mathsf{A}} < 0$, this implies that

$$\left.\frac{\partial \bar{\mathcal{S}}^{\mathsf{A}}}{\partial \mathcal{U}^{\mathsf{A}}}\right|_{N^{\mathsf{A}},\Gamma^{\mathsf{A}}} < \left.\frac{\partial \bar{\mathcal{S}}^{\mathsf{B}}}{\partial \mathcal{U}^{\mathsf{B}}}\right|_{N^{\mathsf{B}},\Gamma^{\mathsf{B}}}. \tag{5.19}$$

The derivatives in Eqn. (5.19) are required to be nonnegative by the monotonically increasing nature of the entropy (see property 2 on page 150). Therefore since $T^{\mathsf{A}} > T^{\mathsf{B}}$, the definition that satisfies Eqn. (5.19) is[44]

$$\left.\frac{\partial \bar{\mathcal{S}}}{\partial \mathcal{U}}\right|_{N,\Gamma} = \frac{1}{T}, \tag{5.20}$$

where $\mathcal{S}, \mathcal{U}$ and T refer to either system A or system B. The inverse relation is

$$\left.\frac{\partial \bar{\mathcal{U}}}{\partial \mathcal{S}}\right|_{N,\Gamma} = T. \tag{5.21}$$

Equations (5.20) and (5.21) provide the key link between entropy, temperature and the internal energy.

To ensure that the extremum point at which $d\mathcal{S}^{\mathsf{C}} = 0$ is a maximum, we must also require $d^2\mathcal{S}^{\mathsf{C}} \leq 0$. Physically, this means that the system is in a state of stable equilibrium. Let

[44] As noted above, any monotonically decreasing function would do here, i.e. $\partial\bar{\mathcal{S}}/\partial\mathcal{U} = f_{-}(T)$. The choice of a particular function can be interpreted in many ways. From the above point of view the choice defines what entropy is in terms of the temperature. From another point of view, where we apply the inverse function to obtain $f_{-}^{-1}(\partial\bar{\mathcal{S}}/\partial\mathcal{U}) = T$, it defines the temperature scale in terms of the entropy. It turns out that the definition selected here provides a clear physical significance to both the thermodynamic temperature and the entropy. When viewed from a microscopic perspective, as is done in Section 7.3.4 of [TM11], this definition of entropy has a natural physical interpretation. When viewed from the macroscopic perspective the thermodynamic temperature scale is naturally related to the behavior of ideal gases as is shown in Section 5.5.5.

us explore the physical restrictions imposed by this requirement. The second differential follows from Eqn. (5.17) as

$$d^2\mathcal{S}^{\mathsf{C}} = \left[\left.\frac{\partial^2 \bar{\mathcal{S}}^{\mathsf{A}}}{\partial(\mathcal{U}^{\mathsf{A}})^2}\right|_{N^{\mathsf{A}},\mathbf{\Gamma}^{\mathsf{A}}} + \left.\frac{\partial^2 \bar{\mathcal{S}}^{\mathsf{B}}}{\partial(\mathcal{U}^{\mathsf{B}})^2}\right|_{N^{\mathsf{B}},\mathbf{\Gamma}^{\mathsf{B}}} \right] (d\mathcal{U}^{\mathsf{A}})^2. \tag{5.22}$$

In Section 5.5.3, we established that for a stable material $\partial^2\bar{\mathcal{S}}/\partial\mathcal{U}^2 \leq 0$. Therefore, we immediately see that $d^2\mathcal{S}^{\mathsf{C}} \leq 0$ and we have confirmed that the state of thermal equilibrium between A and B satisfies the second law of thermodynamics.

At this stage, it is interesting to note the following identity:

$$\left.\frac{\partial^2 \bar{\mathcal{S}}}{\partial\mathcal{U}^2}\right|_{N,\mathbf{\Gamma}} = \left.\frac{\partial}{\partial\mathcal{U}}\right|_{N,\mathbf{\Gamma}} \left.\frac{\partial\bar{\mathcal{S}}}{\partial\mathcal{U}}\right|_{N,\mathbf{\Gamma}} = \left.\frac{\partial}{\partial\mathcal{U}}\right|_{N,\mathbf{\Gamma}} \left(\frac{1}{T}\right) = -\frac{1}{T^2}\left.\frac{\partial\widehat{T}}{\partial\mathcal{U}}\right|_{N,\mathbf{\Gamma}} = -\frac{1}{T^2 n C_v} \leq 0, \tag{5.23}$$

where we have used Eqns. (5.15) and (5.20), C_v is the molar heat capacity at constant volume defined in Eqn. (5.5) and n is the number of moles. This shows that all stable materials, necessarily, have $C_v > 0$.

The introduction of entropy almost seems like the sleight of hand of a talented magician. This variable was introduced without any physical indication of what it could be. It was then tied to the internal energy and temperature through the thought experiment described above. However, this does not really provide a greater sense of what entropy actually is. An answer to that question is outside the scope of this book. However, it is discussed in detail within the context of statistical mechanics in Chapter 7 of the companion book to this one [TM11], where we make a connection between the dynamics of the atoms making up a physical system and the thermodynamic state variables introduced here. In particular, in Section 7.3.4 of [TM11], we show that entropy has a very clear and, in retrospect, almost obvious significance. It is a measure of the number of microscopic kinematic vectors (microscopic states) that are consistent with a given set of macroscopic state variables. Equilibrium is therefore simply the macroscopic state that has the most microscopic states associated with it and is therefore most likely to be observed. This is what entropy is measuring.

5.5.5 Internal energy and entropy as fundamental thermodynamic relations

The entropy function $\bar{\mathcal{S}}(N,\mathbf{\Gamma},\mathcal{U})$ and the closely related internal energy function $\bar{\mathcal{U}}(N,\mathbf{\Gamma},\mathcal{S})$ are known as *fundamental relations* for a thermodynamic system. From them we can obtain all possible information about the system when it is in any state of thermodynamic equilibrium. In particular, we can obtain all of the equations of state for a system from the internal energy fundamental relation. As we saw in the previous section, the temperature is given by the derivative of the internal energy with respect to the entropy. This can, in fact, be viewed as the definition of the temperature, and in a similar manner we can *define* a state variable associated with each argument of the internal energy function. These are the intensive state variables that were introduced in Section 5.1.3. Thus, we have:

1. Absolute temperature

$$T = \overline{T}(N, \boldsymbol{\Gamma}, \mathcal{S}) \equiv \left.\frac{\partial \overline{\mathcal{U}}}{\partial \mathcal{S}}\right|_{N,\boldsymbol{\Gamma}}. \tag{5.24}$$

2. Thermodynamic tensions

$$\gamma_\alpha = \overline{\gamma}_\alpha(N, \boldsymbol{\Gamma}, \mathcal{S}) \equiv \left.\frac{\partial \overline{\mathcal{U}}}{\partial \Gamma_\alpha}\right|_{N,\mathcal{S}}, \quad \alpha = 1, 2, \dots, n^{\Gamma}. \tag{5.25}$$

A special case is where the volume is the kinematic state variable of interest, say $\Gamma_1 = V$. In this case we introduce a negative sign and give the special name, *pressure*, and symbol, $p \equiv -\gamma_1$, to the associated thermodynamic tension. The negative sign is introduced so that, in accordance with our intuitive understanding of the concept, the pressure is positive and increases with decreasing volume. Thus, the definition of the pressure is

$$p = \overline{p}(N, \boldsymbol{\Gamma}, \mathcal{S}) \equiv -\left.\frac{\partial \overline{\mathcal{U}}}{\partial V}\right|_{N,\mathcal{S}},$$

where all kinematic state variables, except the volume, are held constant during the partial differentiation. In general, we refer to the entire set of thermodynamic tensions with the symbol $\boldsymbol{\gamma}$.

3. Chemical potential

$$\mu = \overline{\mu}(N, \boldsymbol{\Gamma}, \mathcal{S}) \equiv \left.\frac{\partial \overline{\mathcal{U}}}{\partial N}\right|_{\boldsymbol{\Gamma},\mathcal{S}}. \tag{5.26}$$

It is clear that each of the above defined quantities is intensive because each is given by the ratio of two extensive quantities. Thus, the dependence on amount cancels and we obtain a quantity that is independent of amount.

Fundamental relation for an ideal gas and the ideal gas law Recall that in Section 5.3.2, we found the internal energy of an ideal gas as a function of the mole number, the volume and the temperature (see Eqn. (5.7)):

$$\mathcal{U} = \widehat{\mathcal{U}}(n, V, T) = n\mathcal{U}_0 + nC_v T, \tag{5.27}$$

where $\mathcal{U}_0$ is the energy per mole of the gas at zero temperature. However, this equation is not a fundamental relation because it is not given in terms of the correct set of independent state variables. It is easy to see this. For instance, the derivative of this function with respect to the volume is zero. Clearly the pressure is not zero for all equilibrium states of an ideal gas. In order to obtain all thermodynamic information about an ideal gas we need the internal energy expressed as a function of the number of particles (or equivalently the mole number), the volume and the entropy. This functional form can be obtained from the statistical mechanics derivation in Section 7.3.5 of [TM11] or the classic thermodynamic approach in [Cal85, Section 3.4]. Taking the arbitrary datum of energy to be such that $\mathcal{U}_0 = 0$, we can write

$$\mathcal{U} = \overline{\mathcal{U}}(n, V, \mathcal{S}) = nK \exp\left(\frac{\mathcal{S}}{nC_v}\right)\left(\frac{V}{n}\right)^{-R_g/C_v}, \tag{5.28}$$

where K is a constant and $R_g = k_B N_A$ is the universal gas constant. Here, $k_B = 8.617 \times 10^{-5}$ eV/K $= 1.3807 \times 10^{-23}$ J/K is Boltzmann's constant and $N_A = 6.022 \times 10^{23}$ mol^{-1} is Avogadro's constant. From this fundamental relation we can obtain all of the equations of state for the intensive state variables:

1. chemical potential

$$\mu = \bar{\mu}(n, V, \mathcal{S}) = \frac{\partial \bar{\mathcal{U}}}{\partial n} = K \exp\left(\frac{\mathcal{S}}{nC_v}\right)\left(\frac{V}{n}\right)^{-R_g/C_v}\left[1 + \frac{R_g}{C_v} - \frac{\mathcal{S}}{nC_v}\right];$$

2. pressure

$$p = \bar{p}(n, V, \mathcal{S}) = -\frac{\partial \bar{\mathcal{U}}}{\partial V} = K \exp\left(\frac{\mathcal{S}}{nC_v}\right)\left(\frac{V}{n}\right)^{-(\frac{R_g}{C_v}+1)}\frac{R_g}{C_v};$$

3. temperature

$$T = \bar{T}(n, V, \mathcal{S}) = \frac{\partial \bar{\mathcal{U}}}{\partial \mathcal{S}} = nK \exp\left(\frac{\mathcal{S}}{nC_v}\right)\left(\frac{V}{n}\right)^{-R_g/C_v}\frac{1}{nC_v}.$$

We may now recover from these functions the original internal energy function and the *ideal gas law* by eliminating the entropy from the equations for the pressure and the temperature. First, notice that the temperature contains a factor which is equal to the internal energy in Eqn. (5.28), giving $T = \mathcal{U}/nC_v$. From this we may solve for the internal energy and immediately obtain Eqn. (5.27) (where we recall that we have chosen $\mathcal{U}_0 = 0$ as the energy datum). Next we recognize that the equation for the pressure can be written $p = \mathcal{U}(1/V)(R_g/C_v)$. Substituting the relation we just obtained for the internal energy in terms of the temperature, we find that $p = nR_gT/V$ or

$$pV = nR_gT, \tag{5.29}$$

which is the *ideal gas law* that is familiar from introductory physics and chemistry courses.

From the ideal gas law we can obtain a physical interpretation of the thermodynamic temperature scale referred to in Section 5.2.2 and defined in Eqn. (5.24). Since all gases behave like ideal gases as the pressure goes to zero,[45] gas thermometers provide a unique temperature scale at low pressure [Adk83]:

$$T = \lim_{p \to 0} \frac{pV}{nR_g}.$$

The value of the ideal gas constant R_g appearing in this relation (and by extension, the value of Boltzmann's constant k_B) is set by defining the thermodynamic temperature $T = 273.16$ K to be the triple point of water (see page 139).

[45] See also Section 5.3.2 where ideal gases are defined and discussed.

5.5.6 Entropy form of the first law

The above definitions for the intensive state variables allow us to obtain a very useful interpretation of the first law of thermodynamics in the context of a quasistatic process. Consider the first differential of internal energy

$$d\mathcal{U} = \left.\frac{\partial \bar{\mathcal{U}}}{\partial N}\right|_{\Gamma,\mathcal{S}} dN + \sum_\alpha \left.\frac{\partial \bar{\mathcal{U}}}{\partial \Gamma_\alpha}\right|_{N,\mathcal{S}} d\Gamma_\alpha + \left.\frac{\partial \bar{\mathcal{U}}}{\partial \mathcal{S}}\right|_{N,\Gamma} d\mathcal{S}.$$

Substituting in Eqns. (5.24), (5.25) and (5.26), we obtain the result

$$d\mathcal{U} = \mu dN + \sum_\alpha \gamma_\alpha d\Gamma_\alpha + T d\mathcal{S}. \tag{5.30}$$

Restricting our attention to the case where the number of particles is fixed, the first term in the differential drops out and we find

$$d\mathcal{U} = \sum_\alpha \gamma_\alpha d\Gamma_\alpha + T d\mathcal{S}. \tag{5.31}$$

If we compare the above equation with the first law in Eqn. (5.2), it is natural to associate the first term which depends on the kinematic variables with the mechanical work $\Delta\mathcal{W}^{\text{def}}$ and the second term which depends on the temperature with the heat $\Delta\mathcal{Q}$. Therefore[46]

$$đ\mathcal{W}^{\text{def}} = \sum_\alpha \gamma_\alpha d\Gamma_\alpha, \qquad đ\mathcal{Q} = T d\mathcal{S}, \tag{5.32}$$

which are increments of *quasistatic work* and *quasistatic heat*, respectively. We will take Eqn. (5.32) as an additional defining property of quasistatic processes. An important special case is that of a thermally isolated system undergoing a quasistatic process. In this situation there is no heat transferred to the system, $d\mathcal{Q} = 0$. Since the temperature will generally not be zero, the only way that this can be true is if $d\mathcal{S} = 0$ for the system. Thus, we have found that when a thermally-isolated system undergoes a quasistatic process its entropy remains constant, and we say the process is *adiabatic*.

Based on the identification of work as the product of a thermodynamic tension with its associated kinematic state variable, it is common to refer to these quantities as *work conjugate* or simply *conjugate* pairs. Thus, we say that γ_α and Γ_α are work conjugate, or that the pressure is conjugate to the volume.

Equations (5.30) and (5.31) are called the *entropy form of the first law* of thermodynamics and, as discussed above, they identify the work performed on the system and the heat transferred to the system as it undergoes a quasistatic process. Thus, when a system's surroundings change in such a way that they cause it to undergo a quasistatic process

[46] Here we use the notation $đ$ in $đ\mathcal{W}^{\text{def}}$ and $đ\mathcal{Q}$ to explicitly indicate that these quantities are not the differentials of functions $\mathcal{W}^{\text{def}}$ and $\mathcal{Q}$. This will serve to remind us that the heat and work transferred to a system generally depend on the process being considered. See Fig. 5.2 and the associated discussion on page 141.

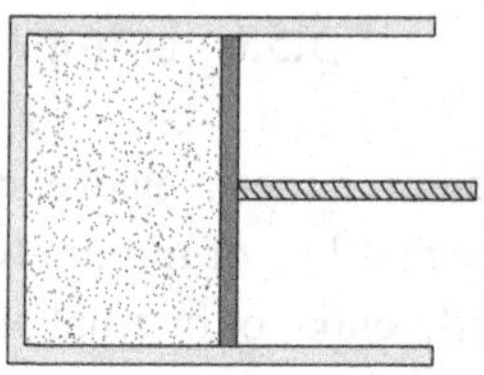

Fig. 5.6 A container with a screw press piston and n moles of an ideal gas.

$\mathfrak{B} = \mathcal{B}(s)$ beginning at state $\mathcal{B}(0)$ and ending in sate $\mathcal{B}(1)$ while keeping the same number of particles ($N(s) = N$), we say that the surroundings perform *quasistatic work* on the system such that

$$\Delta\mathcal{W}^{\mathrm{def}} = \sum_{\alpha} \int_0^1 \left[\bar{\gamma}_\alpha(N, \mathbf{\Gamma}(s), \mathcal{S}(s))\, \dot{\Gamma}_\alpha(s) \right] ds, \tag{5.33}$$

where $\dot{\Gamma}_\alpha \equiv d\Gamma_\alpha/ds$ is the rate of change of Γ_α along the quasistatic process. Similarly, in the same process the system's surroundings will perform a *quasistatic heat transfer* to the system equal to

$$\Delta\mathcal{Q} = \int_0^1 \left[\bar{T}(N, \mathbf{\Gamma}(s), \mathcal{S}(s))\, \dot{\mathcal{S}}(s) \right] ds, \tag{5.34}$$

where $\dot{\mathcal{S}} \equiv d\mathcal{S}/ds$ is the rate of entropy change along the process.

Example 5.4 (Quasistatic work and heat) In Fig. 5.6 we see a container with n moles of an ideal gas that is initially at temperature T_0. The container has a screw press piston which is used to change the gas's volume quasistatically from its initial value of V_0 to V_1. We will consider two scenarios: (1) the container is thermally isolated, i.e. the process is adiabatic, and (2) the container is diathermal and its surroundings are maintained at the initial temperature T_0.

(1) Adiabatic volume changes We will determine the pressure, temperature and total amount of quasistatic work performed by the screw press for any point along the quasistatic adiabatic process. We start by considering the differential relations that must be satisfied along the quasistatic process that occurs as the volume is changed from V_0 to V_1 and then integrate the results. Since the container is thermally isolated, the system's entropy remains constant, and the first law gives $d\mathcal{U} = đ\mathcal{W}^{\mathrm{def}}$. Using Eqn. (5.32)$_1$ for the quasistatic work, the ideal gas law Eqn. (5.29) and Eqn. (5.7) (in its differential form) for the internal energy of an ideal gas, we obtain $nC_v\,dT = -(nR_gT/V)dV$. This can be integrated by separation of variables, and the ideal gas law can be used to obtain the temperature and pressure as functions of the volume

$$T = \mathring{T}(V) = T_0 \left(\frac{V}{V_0}\right)^{-R_g/C_v}, \qquad p = \mathring{p}(V) = p_0 \left(\frac{V}{V_0}\right)^{-(\frac{R_g}{C_v}+1)},$$

where the ideal gas law has also been used to identify the initial pressure, p_0. The total quasistatic work performed by the screw press is obtained from Eqn. (5.33) by recognizing that since the mole number and entropy are constant during the process the integral may be written as

$$\Delta \mathcal{W}^{\text{def}} = \Delta \overset{\circ}{\mathcal{W}}{}^{\text{def}}(V) = \int_{V_0}^{V} -\overset{\circ}{p}(V)dV = -\int_{V_0}^{V} p_0 \left(\frac{V}{V_0}\right)^{-\left(\frac{R_g}{C_v}+1\right)} dV$$
$$= nC_v T_0 \left[\left(\frac{V}{V_0}\right)^{-R_g/C_v} - 1\right].$$

It is interesting to note that once we realized that n and $\mathcal{S}$ remain constant during the quasistatic process, we could have immediately obtained these results from the equations of state for an ideal gas given in Section 5.5.5 without having to invoke the differential form of the first law.

(2) Volume changes at constant temperature In this case, the gas exchanges heat with its surroundings and the quasistatic process proceeds with the gas always at a constant temperature equal to that of its surroundings T_0. We will obtain the pressure in the gas, the total amount of quasistatic work performed by the screw press and the total amount of quasistatic heat transferred to the gas by the surroundings. We can obtain the pressure from the ideal gas law:

$$p = \overset{\circ}{p}(V) = \frac{nR_g T_0}{V}.$$

The quasistatic work then follows as

$$\Delta \mathcal{W}^{\text{def}} = \Delta \overset{\circ}{\mathcal{W}}{}^{\text{def}}(V) = -nR_g T_0 \ln\left(\frac{V}{V_0}\right).$$

In order to obtain the quasistatic heat transferred to the system we need to compute the total change of the system's entropy so that we can use Eqn. (5.34). From Eqn. (5.7) we know that at constant temperature the ideal gas's internal energy is independent of its volume, and thus, constant. The differential form of the first law for the quasistatic process then tells us that $d\mathcal{U} = d\mathcal{W}^{\text{def}} + d\mathcal{Q} = 0$. Using this together with Eqn. $(5.32)_2$ gives

$$T_0 d\mathcal{S} = pdV.$$

Substituting the expression we just found for the pressure and integrating we find the system's entropy as a function of its volume

$$\mathcal{S} = \overset{\circ}{\mathcal{S}}(V) = \mathcal{S}_0 + nR_g \ln\left(\frac{V}{V_0}\right),$$

where $\mathcal{S}_0$ is the entropy of the gas at the beginning of the process. The quasistatic heat transfer is then given by

$$\Delta \mathcal{Q} = \Delta \overset{\circ}{\mathcal{Q}}(V) = nR_g T_0 \ln\left(\frac{V}{V_0}\right).$$

Notice that this confirms that the gas's internal energy remains constant for the *entire* process, since we find that $\Delta\mathcal{U} = \Delta\mathcal{W}^{\text{def}} + \Delta\mathcal{Q} = 0$.

5.5.7 Reversible and irreversible processes

According to the statement of the second law of thermodynamics in Eqn. (5.13), the entropy of an isolated system cannot decrease in any process, rather it must remain constant or else increase. Clearly, if an isolated system undergoes a process in which its entropy increases, then the reverse process can never occur. We say that such a process is *irreversible*. However, if the process leaves the system's entropy unchanged, then the reverse process is also possible and we say that the process is *reversible*. Next, we will explore the differences between these two fundamental types of processes.

We start by considering a general thermodynamic process $\mathfrak{C}$ as defined in Section 5.4. Suppose the isolated system of interest is a composite system C made up of some finite number of subsystems containing stable materials and that internal constraints between the subsystems exist. The process begins at a state of thermodynamic equilibrium for the constrained composite system. Next, one or more internal constraints between the subsystems are released. Because the subsystems are stable, it is easy to show that the isolated composite system is stable with respect to variations of its unconstrained state variables. Thus, there are three possibilities for the initial state of the process after the internal constraints are released: (1) The initial state is a generic point on the total entropy hypersurface (taken as a function of the unconstrained state variables, including those associated with the released internal constraints). (2) The initial state is *the* maximum point on the entropy hypersurface. (3) The initial state is one of a continuum of maxima along the entropy hypersurface, i.e. the hypersurface has a flat region of constant maximum entropy. We will consider each of these cases in turn.

1. In this case the entropy is not at its maximum value and the system, starting from the initial state $\mathcal{C}^{(1)}$, undergoes a dynamical process that eventually ends in state $\mathcal{C}^{(2)}$ which is finitely removed from $\mathcal{C}^{(1)}$. That is the entropy and the unconstrained state variables undergo finite changes. Due to the stability (concavity) of the system, this necessarily corresponds to an *increase* of the total entropy. Thus, the two-stage process $\mathfrak{C} = (\mathcal{C}^{(1)}, \mathcal{C}^{(2)})$ is irreversible. Similarly, any such general thermodynamic process with any number of stages will also be irreversible.
2. The initial state corresponds to the entropy maximum for the system even after the constraints are released. Thus, nothing changes and there is, in fact, no process.
3. In this case the system finds that it can take on any of a continuum of states contained in the hypersurface, all of which have the same value of total entropy. In particular, starting from the initial state on this hypersurface $\mathcal{C}(0)$, the system can change its state in a continuous and arbitrary way along any path $\mathcal{C}(s)$ (for $s \in [0, 1]$) on the hypersurface, such that it ends up in state $\mathcal{C}(1)$. Thus, because every such process consists of a continuous variation of the state variables, it is quasistatic. Further, since the entropy is constant everywhere along the process, it is reversible.[47]

From the above discussion we have learned two important things. First, we see that if a process is reversible, then it must also be quasistatic. (However, the converse is not true.) Second, we infer that most thermodynamic processes are irreversible. This is because the probability of the initial state corresponding to case 1 is much higher than that of case 2

[47] Note that in this and the previous items we have consistently used the notation introduced in Section 5.4. In the first item, the process is found to be a general thermodynamic process, and thus, its discrete states are labeled with superscript parenthesized integers, as in $\mathcal{C}^{(1)}$ and $\mathcal{C}^{(2)}$. In the last item, the process is found to be quasistatic, and thus, its states are given by the *functions* $\mathcal{C}(s)$, $s \in [0, 1]$.

which is much higher than that of case 3.[48] In fact, due to the stability conditions, case 3 – where a flat region exists in system C's entropy function at the state $\mathcal{C}(0)$ – can occur only if two or more of its subsystems have flat regions in their respective entropy functions for the appropriate equilibrium state corresponding to $\mathcal{C}(0)$. This is so unlikely that it is fair to say that no real process is ever truly reversible. However, it is possible, in theory, to construct (very special) isolated systems with processes that are arbitrarily close to being reversible. In order to understand exactly how to do so, let us explore the differences between reversible and irreversible quasistatic processes.

Let C be an isolated composite system with two subsystems A and B. Since C is isolated, knowledge of A's kinematic state variables implies knowledge of the corresponding values for B due to the extensive nature of the $\mathbf{\Gamma}^{\mathsf{C}}$ variables (i.e.[49] $\mathbf{\Gamma}^{\mathsf{B}} = \mathbf{\Gamma}^{\mathsf{C}} - \mathbf{\Gamma}^{\mathsf{A}}$). We may therefore take $\mathbf{\Gamma}^{\mathsf{A}}$ as the unconstrained state variables for C. We suppose that C undergoes a reversible quasistatic process $\mathfrak{C} = \mathcal{C}(s)$ in which C's unconstrained state variables vary continuously.[50] The process is quasistatic, so we will study it by considering the differential forms of the laws of thermodynamics for an arbitrary increment ds along the process. The differential form of the extensivity relation between A and B's kinematic state variables is $d\mathbf{\Gamma}^{\mathsf{B}} = -d\mathbf{\Gamma}^{\mathsf{A}}$. Since the process is reversible we must also have that $d\mathcal{S}^{\mathsf{C}} = d\mathcal{S}^{\mathsf{A}} + d\mathcal{S}^{\mathsf{B}} = 0$, which gives $d\mathcal{S}^{\mathsf{B}} = -d\mathcal{S}^{\mathsf{A}}$ and satisfies the second law. Finally because C is isolated and the process is quasistatic, the subsystems exchange equal amounts of quasistatic work, $\text{đ}\mathcal{W}^{\text{def},\mathsf{B}} = -\text{đ}\mathcal{W}^{\text{def},\mathsf{A}}$, and quasistatic heat, $\text{đ}\mathcal{Q}^{\mathsf{B}} = -\text{đ}\mathcal{Q}^{\mathsf{A}}$. This automatically satisfies the first law and using the definitions for quasistatic work and heat in Eqn. (5.32) gives

$$\sum_\alpha \gamma_\alpha^{\mathsf{A}} d\Gamma_\alpha^{\mathsf{A}} = -\sum_\alpha \gamma_\alpha^{\mathsf{B}} d\Gamma_\alpha^{\mathsf{B}}, \qquad T^{\mathsf{A}} d\mathcal{S}^{\mathsf{A}} = -T^{\mathsf{B}} d\mathcal{S}^{\mathsf{B}}.$$

Introducing the above differential relations connecting the increments of the kinematic state variables and the entropy and rearranging, we find

$$\sum_\alpha \left(\gamma_\alpha^{\mathsf{A}} - \gamma_\alpha^{\mathsf{B}}\right) d\Gamma_\alpha^{\mathsf{A}} = 0, \qquad \left(T^{\mathsf{A}} - T^{\mathsf{B}}\right) d\mathcal{S}^{\mathsf{A}} = 0. \tag{5.35}$$

Since the system is free to explore increments of each individual $d\Gamma_\alpha^{\mathsf{A}}$ and $d\mathcal{S}^{\mathsf{A}}$ separately, these equations imply that A and B must be in equilibrium. That is, we must have $\gamma^{\mathsf{B}} = \gamma^{\mathsf{A}}$ and $T^{\mathsf{B}} = T^{\mathsf{A}}$. Our analysis is valid for an arbitrary state along the quasistatic process, and thus its results must hold for *every* state in the process.

It should now be clear why a reversible quasistatic process is such a special process. The subsystems must undergo changes of state, by exchanging heat and work, in such a way

[48] This can be seen by realizing that case 2 requires the special condition that $d\mathcal{S} = 0$ in the initial state and that case 3 requires *two* special conditions: $d\mathcal{S} = 0$ and $d^2\mathcal{S} = 0$. However, case 1 has no such special requirements and is therefore the most likely situation to be encountered.

[49] Here we are considering a special case where A and B have the same set of kinematic state variables and interact with each other so that the described constraint is correct. Other scenarios are similar and follow as variations of the case discussed here. For example, suppose A and B are ideal gases in two cylindrical containers of different radius (say R^{A} and R^{B}, respectively) and that the movable pistons containing the gases are connected by a rigid rod. Then, a change of system A's volume ΔV^{A} will correspond to a linear displacement of its piston equal to $\Delta x = \Delta V^{\mathsf{A}}/(\pi(R^{\mathsf{A}})^2)$. Accordingly, system B's piston will experience an equal and opposite displacement which leads to a change of its volume $\Delta V^{\mathsf{B}} = -\Delta x \pi (R^{\mathsf{B}})^2 = -(R^{\mathsf{B}}/R^{\mathsf{A}})^2 \Delta V^{\mathsf{A}}$.

[50] We will assume here that the particle numbers of A and B remain constant.

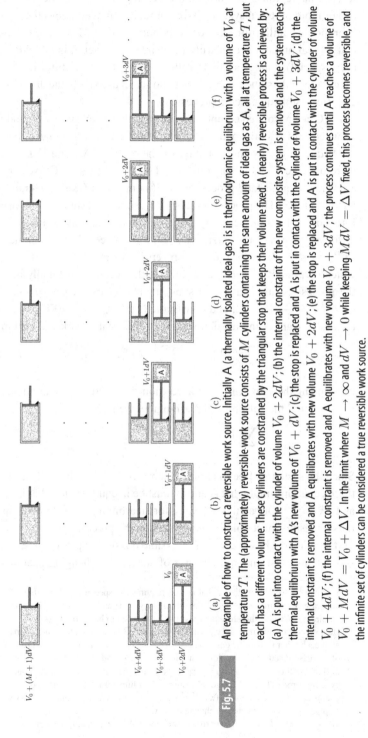

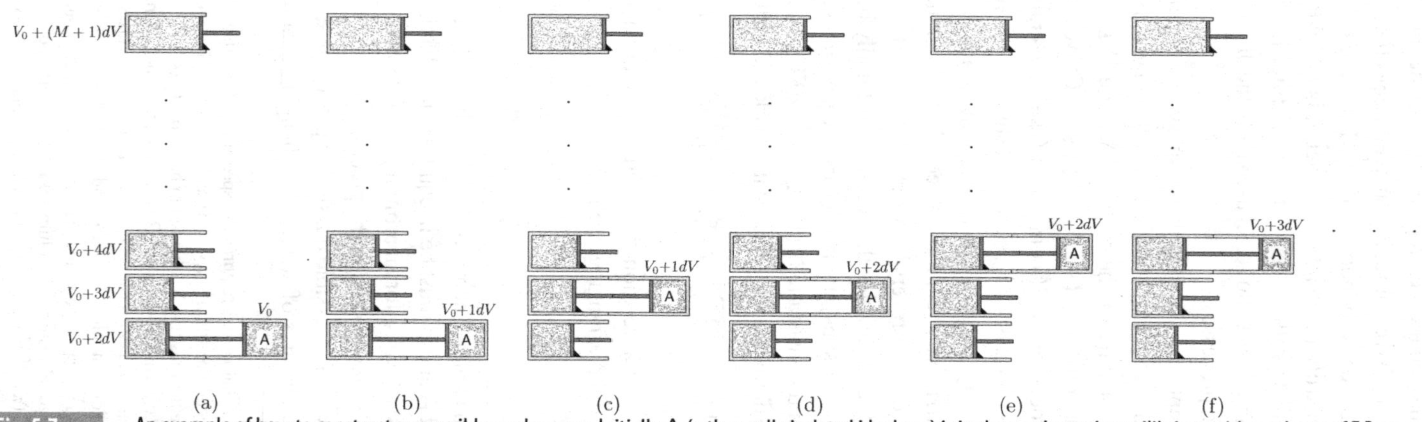

Fig. 5.7 An example of how to construct a reversible work source. Initially A (a thermally isolated ideal gas) is in thermodynamic equilibrium with a volume of V_0 at temperature T. The (approximately) reversible work source consists of M cylinders containing the same amount of ideal gas as A, all at temperature T, but each has a different volume. These cylinders are constrained by the triangular stop that keeps their volume fixed. A (nearly) reversible process is achieved by: (a) A is put into contact with the cylinder of volume $V_0 + 2dV$; (b) the internal constraint of the new composite system is removed and the system reaches thermal equilibrium with A's new volume of $V_0 + dV$; (c) the stop is replaced and A is put in contact with the cylinder of volume $V_0 + 3dV$; (d) the internal constraint is removed and A equilibrates with new volume $V_0 + 2dV$; (e) the stop is replaced and A is put in contact with the cylinder of volume $V_0 + 4dV$; (f) the internal constraint is removed and A equilibrates with new volume $V_0 + 3dV$; the process continues until A reaches a volume of $V_0 + MdV = V_0 + \Delta V$. In the limit where $M \to \infty$ and $dV \to 0$ while keeping $MdV = \Delta V$ fixed, this process becomes reversible, and the infinite set of cylinders can be considered a true reversible work source.

that they remain in equilibrium at all times. This is not possible in general. For example, let us consider a hypothetical reversible process where A and B are both composed of ideal gases and they are thermally insulated from each other so that they only interact by the transfer of work. They must start the process in equilibrium. Now, imagine that the constraint keeping V^{A} and V^{B} fixed is removed and one increment along the hypothetical process occurs. Suppose this involves A expanding by an amount dV^{A}. Necessarily, B's volume will decrease by the same amount. However, at the end of this process increment the pressure in A is smaller than its original value and the pressure in B is larger. Thus, the systems are no longer in equilibrium and it is, therefore, not possible for the next increment of the process to occur reversibly.

In order to construct a quasistatic reversible process in which the ideal gas in subsystem A increases its volume by a finite amount ΔV, one would need to have an infinite number of additional subsystems B_m, $m = 1, 2, \ldots, \infty$ such that the volume in B_m is infinitesimally larger than that in B_{m-1}, i.e.[51] $V_m = V_{m-1} + 2dV$. Such a system is illustrated in Fig. 5.7. We can expand A by having it undergo an infinite series of infinitesimal processes, one with each B_m in which A performs an increment of quasistatic work, at the end of which A has reached its specified final volume and the total entropy change of the isolated system C (consisting of A and all of the B_m s) is zero. In fact, since no heat was transferred, each of the subsystems has exactly the same value of entropy at the end of the process as it did at the beginning. The composite subsystem made up of all the B_m s is called a *reversible work source*. Thus, a reversible work source supplies (or accepts) work from another system while keeping its own entropy constant. A similar procedure can be used to construct a *reversible heat source* that accepts heat from another system by undergoing a quasistatic process at constant values of its particle number and kinematic state variables. This construction is further explored in Exercise 5.10.

These idealized systems are useful because they can be used to construct reversible processes. Indeed, for any system A and for any two of its equilibrium states $\mathcal{A}$ and $\mathcal{A}'$, we can always construct an isolated composite system – consisting of a reversible heat source, a reversible work source and A as subsystems – for which there exists a reversible process in which A starts in state $\mathcal{A}$ and ends in state $\mathcal{A}'$. The second law may then be used to make statements about how the equilibrium state of any system A (not necessarily isolated) must change during a process.

For the described isolated system, there are many different processes that can occur for which A starts in state $\mathcal{A}$ and ends in state $\mathcal{A}'$. Each of these processes results in the same amount of energy being transferred from A to the rest of the system. The distinguishing factor between the processes is exactly how this total energy transfer is partitioned between the reversible work and heat sources. Since the reversible work source does not change its entropy during any of these processes, the second law tells us that the total entropy change must satisfy

$$\Delta\mathcal{S} = \Delta\mathcal{S}^{\mathsf{A}} + \Delta\mathcal{S}^{\mathrm{RHS}} \geq 0,$$

where $\Delta\mathcal{S}^{\mathrm{RHS}}$ is the change in entropy of the reversible heat source (RHS) and the equality holds only for reversible processes. Thus, we find that $\Delta\mathcal{S}^{\mathsf{A}} \geq -\Delta\mathcal{S}^{\mathrm{RHS}}$. If we consider

[51] For ideal gases this is equivalent to having the pressure infinitesimally decreasing, i.e. $p_m = p_{m-1} - dp$.

an infinitesimal change of A's state, then this becomes $d\mathcal{S}^{\mathsf{A}} \geq -d\mathcal{S}^{\mathrm{RHS}} = -đ\mathcal{Q}^{\mathrm{RHS}}/T^{\mathrm{RHS}}$, since the reversible heat source supplies heat quasistatically. Finally, if the reversible heat source accepts an amount of heat $đ\mathcal{Q}^{\mathrm{RHS}}$, then the heat transferred to A is $đ\mathcal{Q}^{\mathsf{A}} = -đ\mathcal{Q}^{\mathrm{RHS}}$. Using this relation, we find that the minus signs cancel and we obtain (dropping the subscript A to indicate that this relation is true for any system)

$$d\mathcal{S} \geq \frac{đ\mathcal{Q}}{T^{\mathrm{RHS}}}. \tag{5.36}$$

This is called the *Clausius–Planck inequality*, which is an alternative statement of the second law of thermodynamics. It is emphasized that T^{RHS} is not generally equal to the system's temperature T. Rather, T^{RHS} is the "temperature at which heat is supplied to the system."

If we define the *external entropy input* as

$$d\mathcal{S}^{\mathrm{ext}} \equiv \frac{đ\mathcal{Q}}{T^{\mathrm{RHS}}},$$

then the difference between the actual change in the system's entropy and the external entropy input is called the *internal entropy production* and is defined as

$$d\mathcal{S}^{\mathrm{int}} \equiv d\mathcal{S} - d\mathcal{S}^{\mathrm{ext}}.$$

Then, according to the Clausius–Planck inequality, $d\mathcal{S}^{\mathrm{int}} \geq 0$. We can convert this into a statement about the work performed on the system by noting that the change of internal energy is, by definition, $d\mathcal{U} = Td\mathcal{S} + \sum_\alpha \gamma_\alpha d\Gamma_\alpha$ and that the first law requires $d\mathcal{U} = đ\mathcal{Q} + đ\mathcal{W}^{\mathrm{def}}$ for all processes. Equating these two expressions for $d\mathcal{U}$, solving for $d\mathcal{S}$ and substituting the result and the definition of $d\mathcal{S}^{\mathrm{ext}}$ into the definition for the internal entropy production we obtain

$$d\mathcal{S}^{\mathrm{int}} = đ\mathcal{Q}\left(\frac{1}{T} - \frac{1}{T^{\mathrm{RHS}}}\right) + \frac{1}{T}\left(đ\mathcal{W}^{\mathrm{def}} - \sum_\alpha \gamma_\alpha d\Gamma_\alpha\right) \geq 0. \tag{5.37}$$

The equality holds only for reversible processes, in which case it is then necessary that $T = T^{\mathrm{RHS}}$. We can further note that if $đ\mathcal{Q} > 0$ then $T < T^{\mathrm{RHS}}$ and if $đ\mathcal{Q} < 0$ then $T > T^{\mathrm{RHS}}$. Either way, the first term on the right-hand side of the inequality in Eqn. (5.37) is positive. This allows us to conclude that for any irreversible process

$$đ\mathcal{W}^{\mathrm{def}} > \sum_\alpha \gamma_\alpha d\Gamma_\alpha.$$

That is, in an irreversible process, the work of deformation performed on a system is greater than it would be in a quasistatic process. The difference goes towards increasing the entropy.

Example 5.5 (Entropy production in adiabatic expansion of an ideal gas) The difference between reversible and irreversible processes and how the first and second laws apply to them can be confusing. Let us examine an irreversible process – free expansion of an ideal gas from volume V_0 into a confining box with volume V_1 – and compare it with the quasistatic expansion of an ideal gas discussed in Example 5.4. Assume that the system is insulated from its surroundings so that no heat is exchanged, i.e. the process is adiabatic. Let the initial temperature be T_0 and the initial entropy be $\mathcal{S}_0$. The pressure at the initial state follows from the ideal gas law, Eqn. (5.29),

$$p_0 = nR_g T_0/V_0.$$

Let us compute the final state. From the first law, we know that $\Delta\mathcal{U} = 0$, since there is no external work or heat input. Since $\mathcal{U}$ is unchanged, for an ideal gas the temperature is also unchanged ($T_1 = T_0$). To determine the volume the gas occupies in its final equilibrium state, we must compute the entropy as a function of volume and find its maximum. To do so, we use the entropy form of the first law, which for an ideal gas is

$$d\mathcal{U} = Td\mathcal{S} - pdV.$$

This law tells us how changes in $\mathcal{U}$, $\mathcal{S}$ and V are related along any quasistatic path. Free expansion of a gas does *not* follow a quasistatic path; however, its end states are in thermodynamic equilibrium and may be assumed to be known. We can therefore compute changes in the variables between the end states by integrating the above equation along any quasistatic path that connects the initial and final states. One option is to very slowly expand the gas by the controlled motion of a piston while maintaining a constant temperature with appropriate heating as we did in Example 5.4(2).[52] There we found

$$\mathcal{S} = \mathring{\mathcal{S}}(V) = \mathcal{S}_0 + nR_g \log \frac{V}{V_0}.$$

This function monotonically increases with V. The maximum possible value is $\mathcal{S}(V_1)$, therefore according to the second law this will be the equilibrium state.[53] The final pressure follows from the ideal gas law, $p_1 = nR_g T_1/V_1$. The pressure is reduced relative to p_0 since $T_1 = T_0$, while $V_1 > V_0$.

The difference between the irreversible free expansion process considered here and the quasistatic isothermal expansion considered in Example 5.4 is very important. In both cases the gases have the same starting and ending equilibrium states. The isothermal expansion process is reversible, assuming that the gas interacts with reversible heat and work sources. In this case, the entropy of the gas increases, but the entropy of the reversible heat source decreases by exactly the same amount so that the total change in entropy is zero. In contrast, in the case of free expansion the gas is an isolated system. Accordingly, it performs no work on and exchanges no heat with its surroundings. Since the process is adiabatic ($\Delta\mathcal{Q} = 0$) the change in entropy is entirely due to internal entropy production (Eqn. (5.37)), and the process is irreversible.

[52] The heating is the key. In this case, the process occurs at constant temperature, whereas an adiabatic expansion of the gas would result in a reduction in temperature as seen in Example 5.4(1). To maintain a constant temperature in the process it is necessary to transfer heat to the system as the gas is expanded. The transferred heat increases the entropy of the gas by increments of $d\mathcal{S} = đ\mathcal{Q}/T$. This is exactly the entropy that is generated internally in the irreversible free expansion process that we are calculating! See also Exercise 5.8 where an alternative quasistatic path is considered.

[53] This result can be viewed as a confirmation that ideal gases are stable materials, and therefore, no phase transformations – where the system splits into part gas and part liquid – can occur.

So far, our discussion of thermodynamics has been limited to homogeneous thermodynamic systems. We now make the assumption of *local thermodynamic equilibrium* and derive the continuum counterparts to the first and second laws.

5.6 Continuum thermodynamics

Our discussion of thermodynamics has led us to definitions for familiar quantities such as the pressure p and temperature T as derivatives of a system's fundamental relation. This relation describes the system only for states of thermodynamic equilibrium, which by definition are homogeneous, i.e. without spatial and temporal variation. Accordingly, it makes sense to talk about the temperature and pressure of the gas inside the rigid sphere discussed at the start of Section 5.5 *before* the hole is opened. However, the temperature and pressure are not defined for the system *while the gas expands* after the hole is opened. This may seem reasonable to you because the expansion process is so fast (relative to the rate of processes we encounter on a day-to-day basis) that it seems impossible to measure the temperature of the gas at any given spatial position. However, consider the case of a large swimming pool into which hot water is being poured from a garden hose. In this case your intuition and experience would lead you to argue that it is certainly possible to identify locations within the pool that are hotter than others. That is, we believe we can identify a spatially varying temperature field. The question we are exploring is: *Is it possible to describe real processes using a continuum theory where we replace p, V and T with fields of pressure $p(\boldsymbol{x})$, density $\rho(\boldsymbol{x})$ and temperature $T(\boldsymbol{x})$?* As the above examples suggest, the answer depends on the conditions of the experiment.

It is correct to represent state variables as spatial fields provided that the length scale over which the continuum fields vary appreciably is much larger than the microscopic length scale. In fluids, this is measured by the *Knudsen number* $Kn = \lambda/L$, where λ is the mean free path (the average distance between collisions of gas atoms) and L is a characteristic macroscopic length (such as the diameter of the rigid sphere from Section 5.5). The continuum approximation is valid as long as $Kn \ll 1$. For an ideal gas, where the velocities of the atoms are distributed according to the Maxwell–Boltzmann distribution (see Section 9.3.3 of [TM11]), the mean free path is [TM04, Section 17.5]

$$\lambda = \frac{k_B T}{\sqrt{2}\pi\delta^2 p},$$

where δ is the atom diameter. For a gas at room temperature and atmospheric pressure, $\lambda \approx 70$ nm. That means that for the gas in the rigid sphere the continuum assumption is valid as long as the diameter of the sphere is much larger than 70 nm. However, if the sphere is filled with a rarefied gas ($p \approx 1$ torr), then $\lambda \approx 0.1$ mm. This is still small relative to, say, a typical pressure gauge, but we see that we are beginning to approach the length scale where the continuum model breaks down.[54]

[54] See [Moo90] for an interesting comparison between the continuum case ($Kn \to 0$) and the free-molecular case ($Kn \to \infty$) for the expansion of a gas in vacuum.

By accepting the "continuum assumptions" and the existence of state variable fields, we are in fact accepting the *postulate of local thermodynamic equilibrium*. This postulate states that *the local and instantaneous relations between thermodynamic quantities in a system out of equilibrium are the same as for a uniform system in equilibrium.*[55] Thus although the system as a whole is not in equilibrium, the laws of thermodynamics and the equations of state developed for uniform systems in thermodynamic equilibrium are applied locally. For example, for the expanding gas, the relation between pressure, density and temperature at a point:

$$p(\boldsymbol{x}) = \frac{k_B}{m}\rho(\boldsymbol{x})T(\boldsymbol{x}),$$

follows from the ideal gas law in Eqn. (5.29) by setting $\rho = Nm/V$, where m is the mass of one atom of the gas.

In addition to the spatial dependence of continuum fields, a temporal dependence is also possible. Certainly the expansion of a gas is a time-dependent phenomenon. Again, the definitions of equilibrium thermodynamics can be stretched to accommodate this requirement provided that the rate of change of continuum field variables is slow compared to the atomistic equilibration time scale. This means that change occurs sufficiently slowly on the macroscopic scale so that all heat transfers can be approximated as quasistatic and that at each instant the thermodynamic system underlying each continuum particle has sufficient time to reach a close approximation to thermodynamic (or at least metastable) equilibrium.

Since the thermodynamic system associated with each continuum particle is not exactly in equilibrium, there is some error in the quasistatic heat transfer assumption and the use of the equilibrium fundamental relations to describe a nonequilibrium process. However, this error is small enough so that it can be accurately compensated for by introducing an irreversible viscous, or dissipative, contribution to the stress. Thus, the total stress will have an elastic contribution (corresponding to the thermodynamic tensions and determined by the equilibrium fundamental relation) and a viscous contribution.

By definition, any process that we can accurately predict as a continuum time-dependent process is one that satisfies the above requirements. Consider the following two examples.

1. Imagine placing a cold piece of metal in a hot oven. The metal will gradually heat to the ambient temperature of the oven. During this transient phase the metal will be hottest where it is in contact with the oven wall. Although the metal as a whole will not be in thermodynamic equilibrium until the end of the process, it is possible to define a temperature field in the metal and to describe the process using continuum mechanics. This will be true as long as the oven is not so hot or the metal so small that the spatial variations in the temperature field or its rate of change are too large.
2. Imagine hitting a piece of metal with a hammer. The head of the hammer striking the metal will create a compressive stress wave in the material that will expand outward from the impact site, racing through the metal, bouncing off its boundaries and gradually dissipating as heat. This problem can also be formulated as a continuum mechanics problem in terms of fields of stress and temperature. As before there are conditions. In

[55] This is the particular form of the postulate given by [LJCV08].

this case the hammer cannot be too small (so that the spatial variations are not too large) and it cannot hit too hard (with resulting high deformation rates.)

Clearly, neither of the systems in these examples is in macroscopic equilibrium, however since its solution is described in terms of fields of state variables, locally at each continuum point there must exist a thermodynamic system that is nearly in thermodynamic equilibrium at each step. These conditions will be satisfied as long as the system is "sufficiently close to equilibrium." There are no clear quantitative measures that determine when this condition is satisfied, but experience has shown that the postulate of local thermodynamic equilibrium is satisfied for a broad range of systems over a broad range of conditions [EM90c]. When it fails, there is no recourse but to turn to a more general theory of *non-equilibrium statistical mechanics* that is valid far from equilibrium. This is a very difficult subject that remains an area of active research.[56] In this book we will restrict ourselves to nonequilibrium processes that are at least approximately in local thermodynamic equilibrium.

5.6.1 Local form of the first law (energy equation)

We now turn to the derivation of the local forms of the first and second laws of thermodynamics. It is useful to introduce the rate of heat supply $\mathcal{R} \equiv đ\mathcal{Q}/dt$ and the rate of external work (also called the external power) $\mathcal{P}^{\text{ext}} \equiv đ\mathcal{W}^{\text{ext}}/dt$. Then, the first law is written in terms of three variables: total energy $\mathcal{E}$, external power $\mathcal{P}^{\text{ext}}$ and heat transfer rate $\mathcal{R}$. Let us examine these quantities more closely for a continuous medium.

Total energy $\mathcal{E}$ Consider the infinitesimal volume element shown in Fig. 4.2. This continuum particle has a macroscopic kinetic energy, $d\mathcal{K} = \frac{1}{2}\rho \|\boldsymbol{v}\|^2 \, dV$, associated with its gross motion. Any additional energy associated with the particle is called its internal energy, $d\mathcal{U} = \rho u dV$, where u is called the *specific internal energy* (i.e. internal energy per unit mass).[57] The specific internal energy includes the strain energy due to deformation of the particle, the microscopic kinetic energy associated with vibrations of the atoms making up the particle and any other energy not explicitly accounted for in the system. (See Appendix A for a heuristic microscopic derivation of the internal energy, and [AT11] for a more rigorous derivation based on nonequilibrium statistical mechanics.) Integrating the kinetic and internal energy densities over the entire body B gives the total energy,

$$\mathcal{E} = \mathcal{K} + \mathcal{U}, \tag{5.38}$$

where $\mathcal{K}$ is the total (gross) kinetic energy,

$$\mathcal{K} = \int_B \frac{1}{2}\rho \|\boldsymbol{v}\|^2 \, dV, \tag{5.39}$$

[56] See, for example, [Rue99] for a review of this field.

[57] This should not be confused with the differential of the internal energy.

and $\mathcal{U}$ is the total internal energy,

$$\mathcal{U} = \int_B \rho u \, dV. \tag{5.40}$$

The first law of thermodynamics in Eqn. (5.3) can then be written as

$$\dot{\mathcal{K}} + \dot{\mathcal{U}} = \mathcal{P}^{\text{ext}} + \mathcal{R}. \tag{5.41}$$

The rates of change of the kinetic and the internal energy are given by

$$\dot{\mathcal{K}} = \frac{D}{Dt} \int_B \frac{1}{2} \rho v_i v_i \, dV = \int_B \frac{1}{2} \rho (a_i v_i + v_i a_i) \, dV = \int_B \rho a_i v_i \, dV, \tag{5.42}$$

$$\dot{\mathcal{U}} = \frac{D}{Dt} \int_B \rho u = \int_B \rho \dot{u} \, dV, \tag{5.43}$$

respectively, where we have used the Reynolds transport theorem (Eqn. (4.5)).

External power $\mathcal{P}^{\text{ext}}$ A continuum body may be subjected to distributed body forces and surface tractions as shown in Fig. 4.2. The work per unit time transferred to the continuum by these fields is the external power,

$$\mathcal{P}^{\text{ext}} = \int_B \rho b_i v_i \, dV + \int_{\partial B} \bar{t}_i v_i \, dA, \tag{5.44}$$

where $\bar{t}$ is the external traction acting on the surfaces of the body. Focusing on the second term, we apply Cauchy's relation (Eqn. (4.18)) followed by the divergence theorem (Eqn. (2.108)):

$$\int_{\partial B} \bar{t}_i v_i \, dA = \int_{\partial B} (\sigma_{ij} n_j) v_i \, dA = \int_B (\sigma_{ij} v_i)_{,j} \, dV = \int_B (\sigma_{ij,j} v_i + \sigma_{ij} v_{i,j}) \, dV.$$

Substituting this into Eqn. (5.44) and rearranging gives

$$\mathcal{P}^{\text{ext}} = \int_B (\sigma_{ij,j} + \rho b_i) v_i \, dV + \int_B \sigma_{ij} v_{i,j} \, dV.$$

Due to the symmetry of the stress tensor, $\sigma_{ij} v_{i,j} = \sigma_{ij} d_{ij}$, where $\boldsymbol{d}$ is the rate of deformation tensor (Eqn. (3.38)). Using this together with the balance of linear momentum (Eqn. (4.25)) to simplify the first term gives

$$\mathcal{P}^{\text{ext}} = \int_B \rho a_i v_i \, dV + \int_B \sigma_{ij} d_{ij} \, dV.$$

Comparing this relation with Eqn. (5.42) we see that the first term is simply the rate of change of the kinetic energy, so that

$$\mathcal{P}^{\text{ext}} = \dot{\mathcal{K}} + \mathcal{P}^{\text{def}}, \tag{5.45}$$

where

$$\mathcal{P}^{\text{def}} = \int_B \sigma_{ij} d_{ij} \, dV \quad \Leftrightarrow \quad \mathcal{P}^{\text{def}} = \int_B \boldsymbol{\sigma} : \boldsymbol{d} \, dV \tag{5.46}$$

is the continuum form of the deformation power (corresponding to the rate of the work of deformation $đ\mathcal{W}^{\text{def}}$ we encountered in Section 5.3). This is the portion of the external power contributing to the deformation of the body with the remainder going towards kinetic energy. We note that since $\boldsymbol{d} = \dot{\boldsymbol{\epsilon}}$ (see Eqn. (3.28)), Eqn. (5.46) can also be written

$$\mathcal{P}^{\text{def}} = \int_B \sigma_{ij} \dot{\epsilon}_{ij} \, dV \quad \Leftrightarrow \quad \mathcal{P}^{\text{def}} = \int_B \boldsymbol{\sigma} : \dot{\boldsymbol{\epsilon}} \, dV. \tag{5.47}$$

Returning now to the representation of the first law in Eqn. (5.41) and substituting in Eqn. (5.45), we see that the first law can be written more concisely as

$$\dot{\mathcal{U}} = \mathcal{P}^{\text{def}} + \mathcal{R}, \tag{5.48}$$

which is similar to the form obtained previously in Eqn. (5.2).

Alternative forms for the deformation power It is also possible to obtain expressions for the deformation power in terms of other stress variables that are often useful. Starting with the definition in Eqn. (5.46), we note that

$$\mathcal{P}^{\text{def}} = \int_B \sigma_{ij} d_{ij} \, dV = \int_B \sigma_{ij} v_{i,j} \, dV = \int_{B_0} \breve{\sigma}_{ij} \frac{\partial \breve{v}_i}{\partial X_J} \frac{\partial X_J}{\partial x_j} J \, dV_0,$$

where we have used $\breve{\sigma}_{ij}$ and $\breve{v}_i$ to emphasize that the stress and velocity fields are expressed in the material description. Now use

$$\frac{\partial \breve{v}_i}{\partial X_J} = \frac{\partial}{\partial X_J}\left(\frac{\partial x_i}{\partial t}\right) = \frac{\partial}{\partial t}\left(\frac{\partial x_i}{\partial X_J}\right) = \dot{F}_{iJ}, \qquad \frac{\partial X_J}{\partial x_j} = F^{-1}_{Jj}$$

together with Eqn. (4.35) for the first Piola–Kirchhoff stress $\boldsymbol{P}$ to obtain the material form of the deformation power:

$$\mathcal{P}^{\text{def}} = \int_{B_0} P_{iJ} \dot{F}_{iJ} \, dV_0 \quad \Leftrightarrow \quad \mathcal{P}^{\text{def}} = \int_{B_0} \boldsymbol{P} : \dot{\boldsymbol{F}} \, dV_0. \tag{5.49}$$

Substituting $P_{iJ} = F_{iI} S_{IJ}$ (inverse of Eqn. (4.41)) and using the following identity,

$$S_{IJ} \dot{C}_{IJ} = S_{IJ} (\overline{F_{iI} F_{iJ}})^{\cdot} = 2 S_{IJ} F_{iI} \dot{F}_{iJ},$$

we find the material form of the deformation power in terms of the second Piola–Kirchhoff stress $\boldsymbol{S}$:

$$\mathcal{P}^{\text{def}} = \frac{1}{2} \int_{B_0} S_{IJ} \dot{C}_{IJ} \, dV_0.$$

Recalling the definition of the Lagrangian strain in Eqn. (3.23), we see that $\dot{E} = \frac{1}{2}\dot{C}$, so that

$$\mathcal{P}^{\text{def}} = \int_{\mathcal{B}_0} S_{IJ}\dot{E}_{IJ}\, dV_0 \quad \Leftrightarrow \quad \mathcal{P}^{\text{def}} = \int_{\mathcal{B}_0} \boldsymbol{S} : \dot{\boldsymbol{E}}\, dV_0. \tag{5.50}$$

Elastic and viscous (dissipative) parts of the stress As indicated at the beginning of this section, generally a continuum particle will not be in a perfect state of thermodynamic equilibrium, and so the stress will generally not be equal to the thermodynamic tensions that are work conjugate to the strain, i.e. the stress is *not* a state variable. To correct for this, continuum thermodynamic theory introduces the ideas of the *elastic part of the stress* $\boldsymbol{\sigma}^{(e)}$ and the *viscous part of the stress* [ZM67]:[58]

$$\boldsymbol{\sigma} = \boldsymbol{\sigma}^{(e)} + \boldsymbol{\sigma}^{(v)}. \tag{5.51}$$

By definition, the elastic part of the stress is given by the material's fundamental relation, and therefore it *is* a state variable. The viscous part of the stress is the part which is not associated with an equilibrium state of the material, and is therefore not a state variable. Substituting Eqn. (5.51) into the definitions for the first and second Piola–Kirchhoff stresses, we can similarly obtain the elastic and viscous parts of these stress measures.

Power conjugate variables The three equations for the deformation power, Eqns. (5.47), (5.49) and (5.50), provide three pairs of variables whose product yields a power density: $(\boldsymbol{\sigma}, \dot{\boldsymbol{\epsilon}})$, $(\boldsymbol{P}, \dot{\boldsymbol{F}})$ and $(\boldsymbol{S}, \dot{\boldsymbol{E}})$. These power conjugate variables fit the general form given in Eqn. (5.32) except that for the continuum formulation the kinematic state variables are intensive and written as rates. This allows us to use the general and convenient notation we introduced in Section 5.1.3. Thus, in general, the deformation power is written

$$\mathcal{P}^{\text{def}} = \int_{\mathcal{B}_0} \sum_{\alpha} (\gamma_\alpha + \gamma_\alpha^{(v)}) \dot{\Gamma}^{\text{i}}_{\alpha}\, dV_0, \tag{5.52}$$

where $\boldsymbol{\Gamma}^{\text{i}} = (\Gamma^{\text{i}}_1, \dots, \Gamma^{\text{i}}_{n^\Gamma})$ is a relevant set of n^Γ intensive state variables that describe the local kinematics of the continuum, and $\boldsymbol{\gamma} = (\gamma_1, \dots, \gamma_{n^\Gamma})$ and $\boldsymbol{\gamma} = (\gamma_1^{(v)}, \dots, \gamma_{n^\Gamma}^{(v)})$ are the thermodynamic tensions (work conjugate to $\boldsymbol{\Gamma}^{\text{i}}$) and their viscous counterparts, respectively, which when added together are power conjugate to $\dot{\boldsymbol{\Gamma}}^{\text{i}}$. For example, for Eqn. (5.50) we can make the assignment in Tab. 5.2, which is called *Voigt notation*.[59]

Heat transfer rate $\mathcal{R}$ The heat transfer rate $\mathcal{R}$ can be divided into two parts:

$$\mathcal{R} = \int_{\mathcal{B}} \rho r\, dV - \int_{\partial\mathcal{B}} h\, dA. \tag{5.53}$$

[58] It is not definite that an additive partitioning can always be made. In plasticity theory, for example, it is common to partition the deformation gradient into a plastic and an elastic part, instead of the stress. See [Mal69, p. 267] for a discussion of this issue.

[59] Voigt notation is a concatenated notation used for symmetric stress and strain tensors. The two coordinate indices of the tensor are replaced with a single index ranging from 1 to 6. See more details in Section 6.5.1.

Table 5.2. Power conjugate variables for a continuum system under finite strain. Representation in Voigt notation

α	$\dot{\Gamma}^{\mathrm{i}}_{\alpha}$	γ_α	$\gamma^{(v)}_\alpha$
1	$\dot{E}_{11}$	$S^{(e)}_{11}$	$S^{(v)}_{11}$
2	$\dot{E}_{22}$	$S^{(e)}_{22}$	$S^{(v)}_{22}$
3	$\dot{E}_{33}$	$S^{(e)}_{33}$	$S^{(v)}_{33}$
4	$2\dot{E}_{23}$	$S^{(e)}_{23}$	$S^{(v)}_{23}$
5	$2\dot{E}_{13}$	$S^{(e)}_{13}$	$S^{(v)}_{13}$
6	$2\dot{E}_{12}$	$S^{(e)}_{12}$	$S^{(v)}_{12}$

Here, $r = r(\boldsymbol{x}, t)$ is the strength of a distributed heat source per unit mass, and $h = h(\boldsymbol{x}, t, \boldsymbol{n})$ is the *outward* heat flux across an element of the surface of the body with normal $\boldsymbol{n}$. Substituting Eqns. (5.43), (5.46) and (5.53) into Eqn. (5.48) and combining terms gives

$$\int_B [\sigma_{ij} d_{ij} + \rho r - \rho \dot{u}] \, dV = \int_{\partial B} h(\boldsymbol{n}) \, dA. \tag{5.54}$$

It may seem that progress beyond this point would be material and environment specific since it depends on the particular form of $h(\boldsymbol{n})$. However, an explicit universal form for $h(\boldsymbol{n})$ can be obtained by following the same reasoning that Cauchy used for the traction vectors (see Section 4.2):[60]

1. Rewrite Eqn. (5.54) for a pillbox and take the height to zero. This shows that

$$h(\boldsymbol{n}) = -h(-\boldsymbol{n}). \tag{5.55}$$

2. Rewrite Eqn. (5.54) for a tetrahedron with three of its sides oriented along the Cartesian axes and take the volume to zero. Together with Eqn. (5.55) this shows that

$$h(\boldsymbol{n}) = \boldsymbol{q} \cdot \boldsymbol{n} = q_i n_i, \tag{5.56}$$

where $\boldsymbol{q}$ is called the *heat flux vector*.

Substituting Eqn. (5.56) into Eqn. (5.54), applying the divergence theorem and combining terms gives

$$\int_B [\sigma_{ij} d_{ij} + \rho r - \rho \dot{u} - q_{i,i}] \, dV = 0.$$

This can be rewritten for any arbitrary subbody E, so it must be satisfied pointwise:

$$\sigma_{ij} d_{ij} + \rho r - q_{i,i} = \rho \dot{u} \quad \Leftrightarrow \quad \boldsymbol{\sigma} : \boldsymbol{d} + \rho r - \operatorname{div} \boldsymbol{q} = \rho \dot{u}. \tag{5.57}$$

[60] This was first shown by the Irish mathematician Sir George Gabriel Stokes [Tru84].

This equation, called the *energy equation*, is the local spatial form of the first law of thermodynamics. It can be thought of as a statement of conservation of energy for an infinitesimal continuum particle. The first term in the equation ($\boldsymbol{\sigma} : \boldsymbol{d}$) is the portion of the mechanical power going towards deformation of the particle; the second term (ρr) is the internal source of heat;[61] the third term ($-\operatorname{div} \boldsymbol{q}$) is the inflow of heat through the boundaries of the particle; the term on the right-hand side ($\rho \dot{u}$) is the rate of change of internal energy. The energy equation can also be written in the material form:

$$P_{iJ} \dot{F}_{iJ} + \rho_0 r_0 - q_{0I,I} = \rho_0 \dot{u}_0 \quad \Leftrightarrow \quad \boldsymbol{P} : \dot{\boldsymbol{F}} + \rho_0 r_0 - \operatorname{Div} \boldsymbol{q}_0 = \rho_0 \dot{u}_0, \tag{5.58}$$

where r_0, $\boldsymbol{q}_0$ and u_0 are respectively the specific heat source, heat flux vector and specific internal energy defined in the reference configuration.

5.6.2 Local form of the second law (Clausius–Duhem inequality)

Having established the local form of the first law, we now turn to the second law of thermodynamics. Our objective is to obtain a local form of the second law. We begin with the Clausius–Planck inequality (Eqn. (5.36)) in its rate form:

$$\dot{\mathcal{S}} \geq \dot{\mathcal{S}}^{\mathrm{ext}} = \frac{\mathcal{R}}{T^{\mathrm{RHS}}}, \tag{5.59}$$

where T^{RHS} is the temperature of the reversible heat source from which the heat is *quasistatically* transferred to the body. We now introduce continuum variables. The entropy $\mathcal{S}$ is an extensive variable, we therefore define the entropy content of an arbitrary subbody E as a volume integral over the *specific entropy* s (i.e. the entropy per unit mass):

$$\mathcal{S}(E) = \int_E \rho s \, dV. \tag{5.60}$$

The rate of heat transfer to E is

$$\mathcal{R}(E) = \int_E \rho r \, dV - \int_{\partial E} \boldsymbol{q} \cdot \boldsymbol{n} \, dA. \tag{5.61}$$

This can be substituted into Eqn. (5.59), but to progress further we must address an important subtlety. There can be a reversible heat source associated with every point on the boundary of the body and the temperature of these sources is not, in principle, equal to the temperature of the material point at the boundary. However, in continuum thermodynamics theory, it is assumed that the boundary points are always in thermal equilibrium with their reversible heat sources. The argument is that even if the boundary of the body starts a process at a different temperature, a thin layer at the boundary heats (or cools) nearly instantaneously to the source's temperature. Also, it is assumed that the internal heat sources are always in

[61] The idea of an internal heat source is used to model interactions of the material with the external world that are like body forces but are otherwise not accounted for in the thermomechanical formulation. For example, electromagnetic interactions may cause a current to flow in the material and its natural electrical resistance will then generate heat in the material.

thermal equilibrium with their material point.[62] Accordingly, we can substitute Eqn. (5.61) into Eqn. (5.59) and take the factor of $1/T$ inside the integrals where it is treated as a function of position and obtained from the material's fundamental relation. This means that the external entropy input rate is

$$\dot{\mathcal{S}}^{\text{ext}}(E) = \int_E \frac{\rho r}{T}\, dV - \int_{\partial E} \frac{\boldsymbol{q}\cdot\boldsymbol{n}}{T}\, dA. \tag{5.62}$$

Substituting Eqns. (5.60) and (5.62) into Eqn. (5.59), we have

$$\frac{D}{Dt}\int_E \rho s\, dV \geq \int_E \frac{\rho r}{T}\, dV - \int_{\partial E} \frac{\boldsymbol{q}\cdot\boldsymbol{n}}{T}\, dA.$$

Applying the Reynolds transport theorem (Eqn. (4.5)) to the left-hand side of the equation and the divergence theorem (Eqn. (2.108)) to the surface integral on the right gives

$$\int_E \rho\dot{s}\, dV \geq \int_E \frac{\rho r}{T}\, dV - \int_E \operatorname{div}\frac{\boldsymbol{q}}{T}\, dV.$$

Note that the use of the divergence theorem requires that $\boldsymbol{q}/T$ must be continuously differentiable. Broadly speaking, this means that we are assuming there are no jumps in the temperature field.[63] Combining terms and recognizing that the inequality must hold for any subbody E, we obtain the local condition

$$\dot{s} \geq \dot{s}^{\text{ext}} = \frac{r}{T} - \frac{1}{\rho}\operatorname{div}\frac{\boldsymbol{q}}{T}, \tag{5.63}$$

where $\dot{s}^{\text{ext}}$ is the specific external entropy input rate. Equation (5.63) is called the *Clausius–Duhem inequality*. This relation can also be obtained directly by considering a thermodynamic system consisting of an infinitesimal continuum particle and accounting for the heat transfer through its surfaces and from internal sources. (See Exercise 5.11 for a one-dimensional example of this approach.) The specific internal entropy production rate, $\dot{s}^{\text{int}}$, follows as

$$\dot{s}^{\text{int}} \equiv \dot{s} - \dot{s}^{\text{ext}} = \dot{s} - \frac{r}{T} + \frac{1}{\rho}\operatorname{div}\frac{\boldsymbol{q}}{T}. \tag{5.64}$$

The Clausius–Duhem inequality is then simply

$$\dot{s}^{\text{int}} \geq 0. \tag{5.65}$$

This is the local analog to Eqn. (5.37).

This concludes our overview of thermodynamics. We have introduced the important concepts of energy, temperature and entropy that will remain with us for the rest of the book. In the next chapter we turn to the remaining piece of the continuum puzzle, the establishment of constitutive relations that govern the behavior of materials. We will see that

[62] However, some authors have argued that a different temperature should be used, see, for example, [GW66].
[63] Heat flow across such a jump would be a source of additional entropy production.

the Clausius–Duhem inequality derived above provides constraints on allowable functional forms for constitutive relations. In the process of deriving these constraints, we will also learn more about the nature of the second law for the types of materials considered in this book.

Exercises

5.1 [SECTION 5.1] Consider a two-dimensional rectangular body made from a typical engineering material. Assuming the solid can undergo only homogeneous deformation, it has three independent macroscopic kinematic quantities. These are its two side lengths, L_1 and L_2, and the angle between two adjacent sides, γ.

1. Suppose we fix L_1 and γ. Apply the test for state variables, discussed in Section 5.1.3, to determine if L_2 is a state variable. Use your intuition about the behavior of a typical solid and explain your reasoning.
2. Now, suppose we fix L_1 and L_2. Is γ a state variable? Again, use your intuition about the behavior of a typical solid and explain your reasoning.

5.2 [SECTION 5.2] Consider, again, the two-dimensional solid of the previous problem. Fix L_1 and γ. Also, consider a cylinder, similar to those in Fig. 5.1, containing an ideal gas of volume V subject to a force F. Generally speaking, when these two systems are initially in thermal equilibrium and they are brought into thermal contact, both their free kinematic state variables will remain constant. When they are not initially in equilibrium, both their free kinematic state variables will change. Now, suppose the temperature of the solid is T_s, the temperature of the gas is T_g and $T_s < T_g$. The solid and gas are put into thermal contact.

1. Do you expect L_2 and/or V to change? Why or why not? If you expect a change, use your physical intuition to describe how these quantities will change.
2. Now suppose, instead of fixing L_1 and γ, we fix L_1 and L_2. In this case, γ does not change, but V decreases. Based on the test for thermal equilibrium, this result seems to imply that the solid is in thermal equilibrium with the gas, but the gas is not in thermal equilibrium with the solid. Explain this apparent contradiction. **Hint:** Think carefully about the assumed behavior of the solid.

5.3 [SECTION 5.3] A sealed and thermally insulated container of volume $V = 1\ \text{m}^3$ contains $n = 20$ mol of an ideal gas at $T = 250$ K. A propeller is immersed in the gas and connected to a shaft that passes through the container via a sealed frictionless bearing. A cable is wound around the exterior part of the shaft and is attached to a 25 kg mass which is suspended 5 m above the ground. The mass is released and it falls to the ground under the influence of gravity, causing the cable to unwind and spin the shaft with the propeller attached. After a period of time, the propeller comes to rest and the gas in the container reaches thermodynamic equilibrium. Determine the final temperature of the gas.

5.4 [SECTION 5.4] For each of the thermodynamic processes described below, identify the process as either a *quasistatic* process or a *general* process.

1. A ball of molten steel is quenched in a bucket of ice water.
2. A glacier melts due to global warming.
3. A "solid" ball of pitch is placed in a funnel and *very* slowly drips on the floor. (See, http://en.wikipedia.org/wiki/Pitch_drop_experiment.)
4. A nail becomes hot as it is pulled out of a wooden board with the claw of a hammer.

5.5 [SECTION 5.5] Consider, again, the isolated system of Fig. 5.5, with its two subsystems A and B that exchange heat and volume in order for the composite system to reach thermodynamic equilibrium.

1. First, we will derive an identity that will be needed for the rest of the problem. The partial derivatives of the entropy function can be directly related to the thermodynamic tensions. This is accomplished by a careful application of the rules of partial differentiation. For example, show that

$$\left.\frac{\partial \bar{\mathcal{S}}}{\partial V}\right|_{N,\mathcal{U}} = \frac{1}{T}p.$$

Start by carefully noting which variables are held constant during the above partial differentiation, and then write the differential of $\bar{\mathcal{U}}$. From this expression and the definitions in Section 5.5.5, you can obtain the desired result.

2. Show that the second law implies that the two subsystems must have equal temperatures and pressures in the final state of thermodynamic equilibrium.
3. Assume that initially (when the piston is impermeable and fixed) the two subsystems are in thermodynamic equilibrium. Their states are given by the values $N^{\mathsf{A}} = N$, $N^{\mathsf{B}} = 2N$, T^{A}, T^{B}, V^{A} and V^{B}. Find the final temperature T_f, pressure p_f and volumes V_f^{A} and V_f^{B}, in terms of N, T^{A}, T^{B}, V^{A} and V^{B}.

5.6 [SECTION 5.5] Consider an isolated system consisting of two separate cylinders containing ideal gases. The first gas cylinder system is called A and has a cross-sectional area of A^{A}. The second gas cylinder system is called B and has a cross-sectional area of $A^{\mathsf{B}} < A^{\mathsf{A}}$. Initially, the two systems are mechanically and thermally isolated from each other and their initial states are given by the values $N^{\mathsf{A}} = N$, $N^{\mathsf{B}} = 2N$, T^{A}, T^{B}, V^{A} and V^{B}. The two cylinders are then allowed to interact thermally (heat may be transferred between them) and their pistons are connected (they may perform work on each other) so that when the piston of A moves by a distance d, the piston of B moves by the same amount in the opposite direction. Find the final temperature T_f, pressures p_f^{A} and p_f^{B} and volumes V_f^{A} and V_f^{B}, in terms of A^{A}, A^{B}, N, T^{A}, T^{B}, V^{A} and V^{B}.

5.7 [SECTION 5.5] Suppose the entropy function of a fixed amount of a material at a fixed value of its internal energy is given by

$$\bar{\mathcal{S}}(V) = V_0^2(V_0 - V)^2 - 2(V_0 - V)^4 + \mathcal{S}_{V_0},$$

where $\mathcal{S}_{V_0}$ is the value of the entropy at the reference volume V_0.

1. Using Eqn. (5.14), suitably modified to apply to changes of volume instead of changes of internal energy, show that for the reference volume this system is unstable. That is, find a value of ΔV for which the inequality in the modified version of Eqn. (5.14) is violated.
2. Now consider the volume $V = V_0/4$. Prove that the system is stable for this volume.
3. Describe the behavior of this system as its volume is quasistatically increased from $V = V_0/4$ to $V = 7V_0/4$ at constant internal energy. Be as quantitative as possible.

5.8 [SECTION 5.5] In Example 5.5, we considered the free expansion of an ideal gas in a container. The change in state variables was computed using a quasistatic process with the same end points where the gas is slowly expanded at constant temperature. As an alternative, consider a two-part quasistatic process where the gas is first adiabatically expanded (as in Example 5.4) and then reversibly heated to the correct temperature. Show that the same result for the change in entropy and pressure as in Example 5.5 is obtained.

5.9 [SECTION 5.5] A closed isolated cylinder of volume V contains n moles of an ideal gas. The cylinder has a removable, frictionless, piston that can be inserted at the end, quasistatically and adiabatically moved to a position where the available volume is $V/2$ and then quickly (instantaneously) removed to allow the gas to freely expand back to the full volume of the cylinder. This procedure is repeated k times. The gas has a molar heat capacity at constant volume of C_v and a reference internal energy $\mathcal{U}_0$. The gas initially has temperature T_{init}, internal energy $\mathcal{U}_{\text{init}}$, pressure p_{init}, and entropy $\mathcal{S}_{\text{init}}$.

1. Obtain expressions for the temperature $T(k)$, pressure $p(k)$, internal energy $\mathcal{U}(k)$ and entropy $\mathcal{S}(k)$ after k repetitions of the procedure.
2. Plot $T(k)/T_{\text{init}}$ and $p(k)/p_{\text{init}}$ as a function of k. Use material constants for air.

5.10 [Section 5.5] Consider an ideal gas, with molar heat capacity C_v, contained in a rigid diathermal cylinder.[64] Suppose we have $N+1$ large buckets of water with temperatures T_0, T_1,...,T_N. The ratio of successive temperatures is constant, such that

$$\frac{T_{i+1}}{T_i} = \left(\frac{T_N}{T_0}\right)^{1/N}, \qquad i = 0, \dots, N.$$

Initially, the cylinder containing the gas is in the first bucket and in thermal equilibrium. Thus, the gas has initial temperature T_0. The cylinder is then taken out of the bucket, placed into the next bucket and allowed to reach thermal equilibrium. This process is repeated until the cylinder and gas are in bucket $N+1$ at a temperature of T_N. The procedure is then reversed and ultimately the cylinder and gas return to the first bucket at temperature T_0. The cylinder, gas, and the $N+1$ buckets of water form an isolated system and no work is performed as part of the process. Assume the buckets contain enough water that any change in the value of their temperature is negligible.

1. Determine the change in the entire system's entropy that occurs between the beginning of the procedure and the end of the first part of the procedure, where the gas is at temperature T_N.
2. Determine the change in the entire system's entropy that occurs between the beginning and end of the entire procedure, where the gas has been heated from T_0 to T_N and then cooled back to T_0 again.
3. Calculate the entropy change, computed in the previous part of this problem, in the limit as $N \to \infty$, while T_0 and T_N remain fixed. **Hint:** You will need to use the fact that, for large N

$$N(x^{1/N} - 1) \approx \ln x + \frac{1}{2N}(\ln x)^2 + \cdots .$$

5.11 [Section 5.6] Consider a one-dimensional system with temperature $T(x)$, heat flux $q(x)$, heat source density $r(x)$, mass density $\rho(x)$ and entropy density $s(x)$. Construct a one-dimensional differential element and show that for a quasistatic process the balance of entropy is

$$\rho\dot{s} = \frac{\rho r}{T} - \frac{\partial}{\partial x}\left(\frac{q}{T}\right),$$

in agreement with the Clausius–Duhem inequality. **Hint:** You will need to use the following expansion: $1/(1+\delta) \approx 1 - \delta + \delta^2 - \cdots$, where $\delta = dT/T \ll 1$, and retain only first order terms.

[64] This problem is based on Problem 4.4-6 of [Cal85].

6 Constitutive relations

In the previous two chapters, we explored the physical laws that govern the behavior of continuum systems. The result was the following set of partial differential equations expressed in the deformed configuration taken from Eqns. (4.2), (4.25) and (5.57):

conservation of mass:	$\dot{\rho} + \rho(\operatorname{div} \boldsymbol{v}) = 0$	(1 equation),
balance of linear momentum:	$\operatorname{div} \boldsymbol{\sigma} + \rho \boldsymbol{b} = \rho \boldsymbol{a}$	(3 equations),
conservation of energy (first law):	$\boldsymbol{\sigma} : \boldsymbol{d} + \rho r - \operatorname{div} \boldsymbol{q} = \rho \dot{u}$	(1 equation),

along with the algebraic Eqns. (4.30) and the differential inequality (5.63):

balance of angular momentum:	$\boldsymbol{\sigma}^T = \boldsymbol{\sigma}$	(3 equations),
Clausius–Duhem inequality (second law):	$\dot{s} \geq \frac{r}{T} - \frac{1}{\rho} \operatorname{div} \frac{\boldsymbol{q}}{T}$	(1 equation).

Excluding the balance of angular momentum and the Clausius–Duhem inequality, which provide constraints on material behavior but are not governing equations, a continuum thermomechanical system is therefore governed by five differential equations. These are called the *field equations* or *governing equations* of continuum mechanics.

The independent fields entering into these equations are:

$$\rho \quad \text{(1 unknown)}, \quad \boldsymbol{\sigma} \quad \text{(6 unknowns)}, \quad u \quad \text{(1 unknown)}, \quad T \quad \text{(1 unknown)},$$
$$\boldsymbol{x} \quad \text{(3 unknowns)}, \quad \boldsymbol{q} \quad \text{(3 unknowns)}, \quad s \quad \text{(1 unknown)},$$

where we have imposed the symmetry of the stress tensor due to the constraint of the balance of angular momentum. The result is a total of sixteen unknowns. The heat source r and body force $\boldsymbol{b}$ are assumed to be known external interactions of the body with its environment. The velocity, acceleration and the rate of deformation tensor are not independent fields. They are given by

$$\boldsymbol{v} = \dot{\boldsymbol{x}}, \qquad \boldsymbol{a} = \ddot{\boldsymbol{x}}, \qquad \boldsymbol{d} = \frac{1}{2}(\nabla \dot{\boldsymbol{x}} + (\nabla \dot{\boldsymbol{x}})^T).$$

Consequently, a continuum thermomechanical system is characterized by five equations with sixteen unknowns. The missing equations are the *constitutive relations* (or *response functions*) that describe the response of the material to the mechanical and thermal loading imposed on it. Constitutive relations are required for u, T, $\boldsymbol{\sigma}$ and $\boldsymbol{q}$ [CN63]. These provide the additional eleven equations required to close the system.

Constitutive relations cannot be selected arbitrarily. They must conform to certain constraints imposed on them by physical laws and they must be consistent with the structure of the material. To derive these constraints, we will take a fresh look at the theory of continuum thermomechanical systems. Our approach will be to, temporarily, forget all of the relationships among the thermodynamic variables that we discovered in the previous chapter, and instead take the above governing equations, and the five basic principles given below, as fundamental. These principles and field equations will serve as our starting point from which we will discover the constraints on constitutive relations that we seek.

These constraints will help immensely to reduce the set of possible forms from which all constitutive equations must be chosen. However, the relations that we will obtain are still quite general. That is, there will be many possible choices of constitutive relations that satisfy the constraints. In particular, we will find that the *postulate of local thermodynamic equilibrium* described in Section 5.6 satisfies all of the constraints, and therefore is a valid and consistent choice for the constitutive relations.[1] However, it is important to note that when formulated from this point of view, the theory of continuum thermomechanical systems allows for a much broader set of possible constitutive relations than simply those associated with the postulate of local thermodynamic equilibrium.

In this chapter, we will derive restrictions on the possible functional forms of constitutive relations and present some important prototypical examples of such relations. The possibility of computing constitutive relations directly from an atomistic model is discussed in Chapter 11 of [TM11]. In such a case, one starts with an "atomistic constitutive relation," describing how individual atoms interact based on their kinematic description, and then use certain averaging techniques to obtain the continuum-level relations.

6.1 Constraints on constitutive relations

Constitutive relations are assumed to be governed by the following fundamental principles.

I *Principle of determinism*

This is a fundamental philosophical statement at the heart of science that proposes that past events determine the present. This principle was most optimistically stated in 1820 by the French mathematician Pierre-Simon de Laplace [Lap51]:

> Present events are connected with preceding ones by a tie based on the evident principle that a thing cannot occur without a cause which produces it... We ought to regard the present state of the universe as the effect of its antecedent state and as the cause of the state that is to follow. An intelligence knowing all the forces acting in nature at a given instant, as well as the momentary positions of all things in the universe, would be able to comprehend in one single formula the motions of the largest bodies as well as the lightest atoms in the world, provided that its intellect

[1] In Section 5.6 we started with the laws of equilibrium thermodynamics, assumed the postulate of local thermodynamic equilibrium and then derived the Clausius–Duhem inequality. Here, in a sense, we make the converse argument: we assume the existence of temperature and entropy fields and take the Clausius–Duhem inequality as given and fundamental. Then we derive relations that admit the postulate of local thermodynamic equilibrium as one (of many) possible constitutive relations.

were sufficiently powerful to subject all data to analysis; to it nothing would be uncertain, the future as well as the past would be present to its eyes. The perfection that the human mind has been able to give to astronomy affords but a feeble outline of such an intelligence.

The development of quantum mechanics over the following 100 years, initiated by the experiments of Gustav Kirchhoff and others with black body radiation, spoiled Laplace's triumphant mood. We no longer believe in perfect determinism as a fundamental law of nature. Nevertheless, at the macroscopic level described by continuum mechanics we still subscribe to determinism in the sense that the current value of any physical variable can be determined from the knowledge of the present and past values of other variables. For example, we assume that the stress at a material particle $\boldsymbol{X}$ in a body at time t can be determined from the history of the motion of the body, its temperature history and so on [Jau67]:

$$\boldsymbol{\sigma}(\boldsymbol{X},t) = \boldsymbol{f}(\varphi^t(\cdot), T^t(\cdot), \dots, \boldsymbol{X}, t). \tag{6.1}$$

Here, $\varphi^t(\cdot)$ and $T^t(\cdot)$ represent the time histories of the deformation mapping and temperature at all points in the body. A material that depends on the past as well as the present is called a material with memory. The explicit dependence of $\boldsymbol{f}$ on $\boldsymbol{X}$ allows for heterogeneous materials where the constitutive relation is different in different parts of the body. The explicit dependence on t allows the response of a material to change with time to account for material aging.

II *Principle of local action*
The principle of local action states that the material response at a point depends only on the conditions within an arbitrarily small region about that point.[2] We assume that a physical variable in the vicinity of particle $\boldsymbol{X}$ can be characterized by a Taylor expansion. For example, the deformation $\boldsymbol{x} = \varphi(\boldsymbol{X})$ near $\boldsymbol{X}$ is described by

$$\boldsymbol{x} + \Delta\boldsymbol{x} = \varphi(\boldsymbol{X}) + \boldsymbol{F}(\boldsymbol{X})\Delta\boldsymbol{X} + \frac{1}{2}\nabla_0\boldsymbol{F} : (\Delta\boldsymbol{X} \otimes \Delta\boldsymbol{X}) + \cdots,$$

where $\boldsymbol{F} = \nabla_0\varphi$ is the deformation gradient and ∇_0 is the gradient with respect to the material coordinates. The stress function in Eqn. (6.1), under the assumption of local action, is then

$$\boldsymbol{\sigma}(\boldsymbol{X},t) = \boldsymbol{g}(\varphi^t(\boldsymbol{X}), \boldsymbol{F}^t(\boldsymbol{X}), \dots, T^t(\boldsymbol{X}), (\nabla_0 T)^t(\boldsymbol{X}), \dots, \boldsymbol{X}, t), \tag{6.2}$$

where a dependence on a finite number of terms in the Taylor expansion is assumed. If the material has no memory, the expression simplifies to

$$\boldsymbol{\sigma}(\boldsymbol{X},t) = \boldsymbol{h}(\varphi(\boldsymbol{X},t), \boldsymbol{F}(\boldsymbol{X},t), \dots, T(\boldsymbol{X},t), \nabla_0 T(\boldsymbol{X},t), \dots, \boldsymbol{X}, t). \tag{6.3}$$

An example of such a model is the generalized Hooke's law for a hyperelastic material[3] under conditions of infinitesimal deformations, where the stress is a linear function of

[2] This definition is originally due to Noll. See [TN65, Section 26] for a detailed discussion.
[3] We define what we mean by "elastic" and "hyperelastic" materials in the next section.

the small strain tensor at a point:

$$\sigma_{ij}(\boldsymbol{X}) = c_{ijkl}(\boldsymbol{X})\epsilon_{kl}(\boldsymbol{X}).$$

Here $\boldsymbol{c}$ is the small strain elasticity tensor. It is important to point out that the principle of local action is not universally accepted. There are *nonlocal* continuum theories that reject this hypothesis. In such theories, the constitutive response at a point is obtained by integrating over the volume of the body. For example in Eringen's nonlocal continuum theory the Cauchy stress $\boldsymbol{\sigma}$ at a point is [Eri02]

$$\sigma_{ij}(\boldsymbol{X}) = \int_{B_0} K(\|\boldsymbol{X} - \boldsymbol{X}'\|) t_{ij}(\boldsymbol{X}')\, dV_0(\boldsymbol{X}'), \tag{6.4}$$

where the kernel $K(r)$ is an *influence function* (often taken to be a Gaussian and of finite support, i.e. it is identically zero for all $r > r_{\text{cut}}$ for some cutoff distance $r_{\text{cut}} > 0$) and $t_{ij} = c_{ijkl}\epsilon_{kl}$ are the usual local stresses. Alternatively, Silling has developed a nonlocal continuum theory called *peridynamics* formulated entirely in terms of forces [Sil02].

Nonlocal theories can be very useful in certain situations, such as in the presence of discontinuities; however, local constitutive relations tend to be the dominant choice due to their simplicity and their ability to adequately describe most phenomena of interest. In particular, in the context of the multiscale methods discussed in [TM11], continuum theories are applied only in regions where gradients are sufficiently smooth to warrant the local action approximation. The regions where such approximations break down are described using atomistic methods that are naturally nonlocal. For more on this see Chapter 12 of [TM11].

III *Second law restrictions*
A constitutive relation cannot violate the second law of thermodynamics, which states that the entropy of an isolated system remains constant for a reversible process and increases for an irreversible process. For example, a constitutive model for heat flux must ensure that heat flows from hot to cold regions and not vice versa. The second law for continuum thermomechanical systems takes the form of the Clausius–Duhem inequality. The application of this inequality to impose constraints on the form of constitutive relations was pioneered in the seminal 1963 paper of Coleman and Noll [CN63]. The approach outlined in that paper is referred to as the *Coleman–Noll procedure*.

IV *Principle of material frame-indifference (objectivity)*
All physical variables for which constitutive relations are required must be *objective* tensors. An objective tensor is a tensor which is physically the same in all frames of reference. For example, the relative position between two physical points is an objective vector, whereas the velocity of a physical point is not objective since it will change depending on the frame of reference in which it is measured. The condition of objectivity imposes certain constraints on the functional form of constitutive relations, which ensures that the resulting variables are objective or *material frame-indifferent*.

V *Material symmetry*
A constitutive relation must respect any symmetries that the material possesses. For example, the stress in a uniformly strained homogeneous isotropic material (i.e. a material that has the same mechanical properties in all directions at all points) is the same regardless of how the material is rotated before the strain is applied.

In addition to the five general principles described above, in this book we will restrict the discussion further to the most commonly encountered types of constitutive relations with two additional constraints:

VI *Only materials without memory and without aging are considered*
This, along with the principle of local action, means that the constitutive relations for the variables u, T, $\boldsymbol{\sigma}$ and $\boldsymbol{q}$ only depend on the local values of other state variables (including possibly a finite number of terms – higher-order gradients – from their Taylor expansion) and their time rates of change.

VII *Only materials whose internal energy depends solely on the entropy and deformation gradient are considered*
That is, we explicitly exclude the possibility of dependence on any rates of deformation as well as the higher-order gradients of the deformation. This is consistent with the thermodynamic definition in Eqn. (5.12).

In the next three sections we see the implications of the restrictions described above on allowable forms of the constitutive relations.

6.2 Local action and the second law of thermodynamics

In this section we consider the implications of the principle of local action and the second law of thermodynamics (principles II and III) along with constraints VI and VII for the functional forms of the constitutive relations for u, T, $\boldsymbol{\sigma}$ and $\boldsymbol{q}$. The implications of principles IV and V will be considered later in the chapter.

6.2.1 Specific internal energy constitutive relation

The statement of the second law introduced the concept of entropy as a state variable and the following functional dependence for the internal energy (Eqn. (5.12)):

$$\mathcal{U} = \bar{\mathcal{U}}(N, \boldsymbol{\Gamma}, \mathcal{S}),$$

where $\boldsymbol{\Gamma}$ is a set of extensive kinematic variables. We can eliminate the particle number from the list of state variables if we work with *intensive* versions of the extensive state variables. Thus dividing all extensive state variables by the total mass of the particles, we obtain the specific internal energy u (i.e. the internal energy per unit mass) as a function of the specific entropy s and the intensive versions of the kinematic state variables $\boldsymbol{\Gamma}^{\mathrm{i}}$:

$$u = \bar{u}(s, \boldsymbol{\Gamma}^{\mathrm{i}}). \tag{6.5}$$

For notational simplicity, we drop the "i" superscript on $\boldsymbol{\Gamma}^{\mathrm{i}}$ in subsequent discussion, since the extensive or intensive nature of $\boldsymbol{\Gamma}$ is clear from the context. As before, a bar or (other accent) over a variable, as in $\overline{u}$, is used to denote the response function (as opposed to the actual quantity). Considering constraint VII, we obtain the functional form for the specific internal energy constitutive relation:

$$u = \overline{u}(s, \boldsymbol{F}). \tag{6.6}$$

This is referred to as the *caloric equation of state*. A material whose constitutive relation depends on the deformation only through the history of the local value of $\boldsymbol{F}$ is called a *simple material*. A simple material without memory (depending only on the instantaneous value of $\boldsymbol{F}$) is called an *elastic simple material*.

Before continuing, we note that it is necessary for some materials to augment constraint VII to include additional *internal variables* that describe microstructural features (additional kinematic state variables) of the continuum such as dislocation density, vacancy density, impurity concentration, phase fraction, microcrack density and so on:[4]

$$u = \widetilde{u}(s, \boldsymbol{F}, \delta_1, \delta_2, \dots).$$

The inclusion of these parameters leads to additional rate equations that model their evolution [Lub72].

Another set of possible constitutive relations, which we have excluded from discussion via constraint VII, are those that include a dependence on higher-order gradients of the deformation:[5]

$$u = \widehat{u}(s, \boldsymbol{F}, \nabla_0 \boldsymbol{F}, \dots).$$

The result is a *strain gradient theory*. This approach has been successfully used to study length scale[6] dependence in plasticity [FMAH94] and localization of deformation in the form of shear bands [TA86]. See the discussion in Section 6.6. An alternative approach is the polar *Cosserat theory* in which nonuniform local deformation is characterized by associating a triad of orthonormal director vectors with each material point [Rub00]. These approaches are beyond the scope of the present discussion.

[4] In this chapter we use accents over a function's symbol to indicate differences between functional forms. For example, here we switch from $\overline{u}(s, \boldsymbol{F})$ to $\widetilde{u}(s, \boldsymbol{F}, \delta_1, \delta_2, \dots)$ to emphasize the fact that these are two distinct functional forms; however, the different accents ($\widehat{\cdot}$, $\bar{\cdot}$, $\tilde{\cdot}$ and $\breve{\cdot}$) are not associated with any particular set of functional arguments. In contrast, in Chapter 5 we used accents over a function's symbol to indicate its specific arguments (e.g. $\overline{\mathcal{U}}(\mathcal{S}, \boldsymbol{\Gamma})$ and $\widehat{\mathcal{U}}(T, \boldsymbol{\Gamma})$, where $\bar{\cdot}$ is identified with the functional arguments $\mathcal{S}$ and $\boldsymbol{\Gamma}$, and $\widehat{\cdot}$ is identified with the functional arguments T and $\boldsymbol{\Gamma}$, respectively).

[5] Interestingly, it is not possible to simply add a dependence on higher-order gradients without introducing additional variables that are conjugate with the higher-order gradient fields, and modifying the energy equation and the Clausius–Duhem inequality [Gur65]. For example, a second-gradient theory requires the introduction of couple stresses. Therefore, classical continuum thermodynamics is by necessity limited to simple materials.

[6] Each higher-order gradient introduced into the formulation is associated with a length scale. For example, a second-order gradient has units of 1/length. It must therefore be multiplied by a parameter with units of length to cancel this out in the energy expression. In contrast, the classical continuum mechanics of simple materials has no length scale. This qualitative difference has sometimes led authors to call these strain gradient theories "nonlocal." However, this terminology does not appear to be consistent with the original definition of the term "local."

6.2.2 Coleman–Noll procedure

In order to obtain functional forms for the temperature, heat flux vector and stress tensor, it is advantageous to revisit the second law of thermodynamics and concepts of reversible and irreversible processes. By doing so, we will be able to obtain the specific functional dependence of the temperature and heat flux response functions. In addition, we will show that the stress tensor can be divided into two parts: a conservative elastic part and an irreversible viscous part. The procedure followed here is due to Coleman and Noll [CN63] and Ziegler and McVean [ZM67].

We saw earlier in Eqn. (5.65) that the Clausius–Duhem inequality can be written in abbreviated form as

$$\dot{s}^{\text{int}} \equiv \dot{s} - \dot{s}^{\text{ext}} \geq 0, \tag{6.7}$$

where $\dot{s}^{\text{int}}$ is the specific internal entropy production rate and

$$\dot{s}^{\text{ext}} = \frac{r}{T} - \frac{1}{\rho}\operatorname{div}\frac{\boldsymbol{q}}{T} \tag{6.8}$$

is the specific external entropy input rate. Substituting Eqn. (6.8) into Eqn. (6.7) and expanding the divergence term, we have

$$\begin{aligned}
\dot{s}^{\text{int}} &= \dot{s} - \frac{r}{T} + \frac{1}{\rho}\operatorname{div}\frac{\boldsymbol{q}}{T} \\
&= \dot{s} - \frac{r}{T} + \frac{(\operatorname{div}\boldsymbol{q})T - \boldsymbol{q}\cdot\nabla T}{\rho T^2} \\
&= \dot{s} - \frac{1}{\rho T}\left[\rho r - \operatorname{div}\boldsymbol{q}\right] - \frac{1}{\rho T^2}\boldsymbol{q}\cdot\nabla T \geq 0.
\end{aligned}$$

Rearranging, we obtain

$$\rho T\dot{s}^{\text{int}} = \rho T\dot{s} - \left[\rho r - \operatorname{div}\boldsymbol{q}\right] - \frac{1}{T}\boldsymbol{q}\cdot\nabla T \geq 0. \tag{6.9}$$

The expression in the square brackets in Eqn. (6.9) appears in exactly the same form in the energy equation (Eqn. (5.57)):

$$\rho r - \operatorname{div}\boldsymbol{q} = \rho\dot{u} - \boldsymbol{\sigma} : \boldsymbol{d}. \tag{6.10}$$

Substituting Eqn. (6.10) into Eqn. (6.9) gives

$$\rho T\dot{s}^{\text{int}} = \rho T\dot{s} - \rho\dot{u} + \boldsymbol{\sigma} : \boldsymbol{d} - \frac{1}{T}\boldsymbol{q}\cdot\nabla T \geq 0. \tag{6.11}$$

Taking a material time derivative of Eqn. (6.6), we have

$$\dot{u} = \frac{\partial\bar{u}}{\partial s}\dot{s} + \frac{\partial\bar{u}}{\partial\boldsymbol{F}} : \dot{\boldsymbol{F}}. \tag{6.12}$$

Substituting Eqn. (6.12) into Eqn. (6.11) and rearranging gives

$$\rho\left[T - \frac{\partial\bar{u}}{\partial s}\right]\dot{s} + \left[\boldsymbol{\sigma} : \boldsymbol{d} - \rho\frac{\partial\bar{u}}{\partial\boldsymbol{F}} : \dot{\boldsymbol{F}}\right] - \frac{1}{T}\boldsymbol{q}\cdot\nabla T \geq 0. \tag{6.13}$$

Now, since $\boldsymbol{\sigma}$ is symmetric, we have $\boldsymbol{\sigma} : \boldsymbol{d} = \boldsymbol{\sigma} : \boldsymbol{l}$ (where $\boldsymbol{l}$ is the velocity gradient). Recall also that $\dot{\boldsymbol{F}} = \boldsymbol{l}\boldsymbol{F}$ (Eqn. (3.36)), therefore

$$\boldsymbol{l} = \dot{\boldsymbol{F}}\boldsymbol{F}^{-1}. \tag{6.14}$$

Replacing $\boldsymbol{\sigma} : \boldsymbol{d}$ in Eqn. (6.13) with $\boldsymbol{\sigma} : \boldsymbol{l}$ and substituting in Eqn. (6.14), we have

$$\rho\left[T - \frac{\partial \bar{u}}{\partial s}\right]\dot{s} + \left[\boldsymbol{\sigma}\boldsymbol{F}^{-T} - \rho\frac{\partial \bar{u}}{\partial \boldsymbol{F}}\right] : \dot{\boldsymbol{F}} - \frac{1}{T}\boldsymbol{q}\cdot\nabla T \geq 0. \tag{6.15}$$

The argument made by Coleman and Noll is that Eqn. (6.15) must be satisfied for *every* admissible process. By selecting special cases, insight is gained into the relation between the different continuum fields. This line of thinking is referred to as the *Coleman–Noll procedure*. We apply it below to obtain the functional forms for the constitutive relations for temperature, heat flux and stress.

Temperature constitutive relation Consider a process where the deformation is constant in time ($\dot{\boldsymbol{F}} = \boldsymbol{0}$) and the temperature is uniform across the body, so that $\nabla T = \boldsymbol{0}$. In this case, Eqn. (6.15) reduces to

$$\rho\left[T - \frac{\partial \bar{u}}{\partial s}\right]\dot{s} \geq 0. \tag{6.16}$$

The rate of change of entropy $\dot{s}$ can be assigned arbitrarily (e.g. by modifying an external heat source r (see Eqn. (5.63)). Since the sign of $\dot{s}$ is arbitrary, Eqn. (6.16) can only be satisfied for every process if

$$T = \bar{T}(s, \boldsymbol{F}) \equiv \frac{\partial \bar{u}}{\partial s}, \tag{6.17}$$

where the functional dependence follows from Eqn. (6.6). We see that the specific internal energy density has the same relation to the local temperature as the total internal energy does to temperature in a homogeneous system as given in Eqn. (5.21).

Heat flux constitutive relation Substituting Eqn. (6.17) in Eqn. (6.15), the second law inequality reduces to

$$\left[\boldsymbol{\sigma}\boldsymbol{F}^{-T} - \rho\frac{\partial \bar{u}}{\partial \boldsymbol{F}}\right] : \dot{\boldsymbol{F}} - \frac{1}{T}\boldsymbol{q}\cdot\nabla T \geq 0. \tag{6.18}$$

Again, considering a process where the deformation is constant ($\dot{\boldsymbol{F}} = \boldsymbol{0}$), we have

$$-\frac{1}{T}\boldsymbol{q}\cdot\nabla T = -\frac{1}{T}q_i T_{,i} \geq 0. \tag{6.19}$$

This inequality is consistent with our physical intuition: heat flows from hot to cold. This result does not provide an explicit form for the heat flux constitutive relation. However, since Eqn. (6.19) must be satisfied for any ∇T, the heat flux must depend on this variable. For example, if we consider the two heat flux fields ∇T and $-\nabla T$, then Eqn. (6.19) must be satisfied for both of these. This means that $\boldsymbol{q}$ must change sign in accordance with ∇T to ensure that the inequality remains valid. We can therefore state in general that $\boldsymbol{q}$ must have the following functional dependence:[7]

$$\boldsymbol{q} = \bar{\boldsymbol{q}}(s, \boldsymbol{F}, \nabla T). \tag{6.20}$$

Cauchy stress constitutive relation Returning to Eqn. (6.18), consider the case where the temperature is uniform across the body ($\nabla T = \boldsymbol{0}$). In this case, the second law inequality is

$$\left[\boldsymbol{\sigma}\boldsymbol{F}^{-T} - \rho\frac{\partial \bar{u}}{\partial \boldsymbol{F}}\right] : \dot{\boldsymbol{F}} \geq 0. \tag{6.21}$$

This equation must hold for any choice of $\dot{\boldsymbol{F}}$. This can only be satisfied for all $\dot{\boldsymbol{F}}$ if

$$\boldsymbol{\sigma}\boldsymbol{F}^{-T} - \rho\frac{\partial \bar{u}}{\partial \boldsymbol{F}} = \boldsymbol{0}. \tag{6.22}$$

Therefore, unless Eqn. (6.22) is satisfied, Eqn. (6.21) can be violated by a particular choice of $\dot{\boldsymbol{F}}$. There is a problem with this conclusion. Equation (6.22) implies that all irreversibility enters through the heat flux term in Eqn. (6.18) and consequently that no irreversibility is possible under uniform temperature conditions. This is not consistent with experimental observation. The implication of this is that the stress is *not* a state variable as anticipated by the discussion on page 173. To proceed, we partition $\boldsymbol{\sigma}$ into two (as in Eqn. (5.51)): an elastic reversible part that is a state variable and a "viscous," or dissipative, part that is irreversible:[8]

$$\boldsymbol{\sigma} = \boldsymbol{\sigma}^{(e)} + \boldsymbol{\sigma}^{(v)}. \tag{6.23}$$

Substituting Eqn. (6.23) into Eqn. (6.18) gives

$$\left[\boldsymbol{\sigma}^{(e)}\boldsymbol{F}^{-T} - \rho\frac{\partial \bar{u}}{\partial \boldsymbol{F}}\right] : \dot{\boldsymbol{F}} + \boldsymbol{\sigma}^{(v)} : \boldsymbol{l} - \frac{1}{T}\boldsymbol{q} \cdot \nabla T \geq 0. \tag{6.24}$$

If we now assume that $\boldsymbol{\sigma}^{(v)}$ represents an irreversible process, then its entropy production is always positive (the dissipated energy is converted into heat which causes the entropy to

[7] It is curious to note that if electromagnetic effects are included, the heat flux constitutive relation will generally include bilinear coupling between the temperature gradient and the electric current. This coupling gives rise to the Thomson effect whereby, through the application of a suitable electric current through a specimen, it is possible to violate Eqn. (6.19) and have heat flow from cold to hot. Do not despair, however; with the appropriate formulation of thermodynamic theory it is found that this does not, in any way, violate the second law of thermodynamics.

[8] An additive partitioning of the stress may not always be appropriate. See footnote 58 on page 173.

increase),

$$\boldsymbol{\sigma}^{(v)} : \boldsymbol{d} \geq 0, \tag{6.25}$$

where by replacing $\boldsymbol{l}$ with $\boldsymbol{d}$, we have assumed that the viscous stress is symmetric.[9] Since the last two terms in Eqn. (6.24) have a fixed sign (always positive) and (by choice of $\dot{\boldsymbol{F}}$) the first term can take on any value, the inequality can only be guaranteed to be satisfied if

$$\boldsymbol{\sigma}^{(e)} = \bar{\boldsymbol{\sigma}}^{(e)}(s, \boldsymbol{F}) \equiv \rho \frac{\partial \bar{u}}{\partial \boldsymbol{F}} \boldsymbol{F}^T, \tag{6.26}$$

or in component form

$$\sigma_{ij}^{(e)} = \rho \frac{\partial \bar{u}}{\partial F_{iJ}} F_{jJ}.$$

Furthermore, we require that the inequality in Eqn. (6.25) must be satisfied for every process. Similarly to the heat flux, this inequality on its own is not enough to obtain an explicit form for $\boldsymbol{\sigma}^{(v)}$. However, it does indicate that the viscous stress must depend on the rate of deformation tensor, therefore

$$\boldsymbol{\sigma}^{(v)} = \bar{\boldsymbol{\sigma}}^{(v)}(s, \boldsymbol{F}, \boldsymbol{d}). \tag{6.27}$$

A material for which $\boldsymbol{\sigma}^{(v)} = \boldsymbol{0}$, and for which an energy function exists, such that the stress is entirely determined by Eqn. (6.26), is called a *hyperelastic* material.

Entropy change in reversible and irreversible processes Following the definition of the elastic stress in Eqn. (6.26), the Clausius–Duhem inequality in Eqn. (6.24) is reduced to its final form:

$$\rho T \dot{s}^{\text{int}} = \boldsymbol{\sigma}^{(v)} : \boldsymbol{d} - \frac{1}{T} \boldsymbol{q} \cdot \nabla T \geq 0. \tag{6.28}$$

This relation can be used to shed some light on local entropy changes in materials whose stress can be decomposed according to Eqn. (6.23). Equating the expressions for $\rho T \dot{s}^{\text{int}}$ in Eqns. (6.28) and (6.9), we obtain

$$\dot{s} = \frac{r}{T} - \frac{1}{\rho T} \operatorname{div} \boldsymbol{q} + \frac{1}{\rho T} \boldsymbol{\sigma}^{(v)} : \boldsymbol{d}. \tag{6.29}$$

[9] We prove that the elastic stress is symmetric immediately after Eqn. (6.105) and therefore the viscous stress must also be symmetric in order for the balance of angular momentum to be satisfied. For now we treat both $\boldsymbol{\sigma}^{(e)}$ and $\boldsymbol{\sigma}^{(v)}$ as symmetric tensors.

If the process is reversible, then each of the terms in Eqn. (6.28) is zero (since a sum of two positive terms is zero only if they are both zero):

$$\boldsymbol{\sigma}^{(v)} : \boldsymbol{d} = 0, \qquad -\frac{1}{T}\boldsymbol{q} \cdot \nabla T = 0, \tag{6.30}$$

and Eqn. (6.29) reduces to

$$\dot{s} = \dot{s}_{\text{rev}} = \frac{r}{T} - \frac{1}{\rho T}\operatorname{div}\boldsymbol{q}. \tag{6.31}$$

In this case, $\dot{s}$ is exactly equal to $\dot{s}_{\text{ext}}$ in Eqn. (6.8), since $(\boldsymbol{q}/T) \cdot \nabla T = 0$. For an irreversible process (where the system interacts with reversible heat and work sources), $\dot{s} > \dot{s}_{\text{ext}}$ and the difference is exactly $\dot{s}^{\text{int}}$.

6.2.3 Onsager reciprocal relations

In Eqns. (6.20) and (6.27), we established guidelines for the constitutive forms of $\boldsymbol{q}$ and $\boldsymbol{\sigma}^{(v)}$ that appear in the entropy production expression in Eqn. (6.28). However, the actual functional forms are unknown. As a result, a phenomenological approach is normally adopted where a functional form is postulated and the parameters appearing in it are obtained by fitting to experimental measurements. The simplest possibility is to assume a linear relation between the arguments. In general, we have

$$J_i = \mathsf{L}_{ij} Y_j, \tag{6.32}$$

where $\boldsymbol{J}$ is the viscous flux vector and $\boldsymbol{Y}$ is the corresponding generalized viscous force. The entries of the matrix L coupling them are called the *phenomenological coefficients*. The identities of $\boldsymbol{J}$ and $\boldsymbol{Y}$ are somewhat arbitrary. In our case, there are two sets of flux-force pairs. We can choose $\boldsymbol{\sigma}^{(v)}$ (in concatenated Voigt form as shown in Tab. 5.2) and $\boldsymbol{q}/T$ to be the generalized forces and $\boldsymbol{d}$ (suitably concatenated) and ∇T to be the corresponding fluxes.[10] Other terms are possible when additional irreversible phenomena are considered.

The heart of the phenomenological relations is the coefficient matrix L. What can be said in general about this matrix? We established earlier that the contribution of each irreversible term to entropy production must be nonnegative. Therefore, we must have that

$$J_i Y_i = \mathsf{L}_{ij} Y_i Y_j \geq 0, \tag{6.33}$$

for all forces $\boldsymbol{Y}$. This means that the matrix L must be positive definite (or at least positive semi-definite), which imposes constraints on the coefficients L_{ij}.

A second set of constraints can be inferred from the fact that the microscopic equations of motion are symmetric with respect to time. This means that if the velocities of all atoms are instantaneously reversed the atoms will retrace their earlier trajectories. In an important theorem in nonequilibrium statistical mechanics, Lars Onsager proved that for systems close to equilibrium the phenomenological coefficients matrix must be symmetric:

$$\mathsf{L}_{ij} = \mathsf{L}_{ji}. \tag{6.34}$$

[10] The resulting linear constitutive relations are well known. The first relation describes the viscous response of a Newtonian fluid and the second is called Fourier's law of heat conduction. We explore both relations later.

This is referred to as the *Onsager reciprocal relations*.[11] For a symmetric matrix, the earlier requirement of positive definiteness is equivalent to requiring that the eigenvalues of **L** be positive.

6.2.4 Constitutive relations for alternative stress variables

Continuum formulations for solids are often expressed in a Lagrangian description, where the appropriate stress variables are the first or second Piola–Kirchhoff stress tensors. The constitutive relations for these variables can be found by suitably transforming the Cauchy stress function.

The constitutive relation for the elastic part of the first Piola–Kirchhoff stress is obtained by substituting Eqn. (6.26) into Eqn. (4.35). The result after using Eqn. (4.1) is

$$P_{iJ}^{(e)} = \rho_0 \frac{\partial \breve{u}}{\partial F_{iJ}} \quad \Leftrightarrow \quad \boldsymbol{P}^{(e)} = \rho_0 \frac{\partial \breve{u}}{\partial \boldsymbol{F}}. \tag{6.35}$$

The second Piola–Kirchhoff stress is obtained in similar fashion from Eqn. (4.41) as

$$S_{IJ}^{(e)} = \rho_0 F_{Ii}^{-1} \frac{\partial \breve{u}}{\partial F_{iJ}}. \tag{6.36}$$

We will see in Section 6.3 that due to material frame-indifference the internal energy can only depend on $\boldsymbol{F}$ through the right stretch tensor $\boldsymbol{U}$ (or equivalently through the right Cauchy–Green deformation tensor $\boldsymbol{C}$ or the Lagrangian strain tensor $\boldsymbol{E}$). We therefore rewrite Eqn. (6.36) using an alternative internal energy function, $\widetilde{u}(s, \boldsymbol{E})$, that depends on the Lagrangian strain. Thus,

$$\begin{aligned} S_{IJ}^{(e)} &= \rho_0 F_{Ii}^{-1} \frac{\partial \widetilde{u}}{\partial E_{MN}} \frac{\partial E_{MN}}{\partial F_{iJ}} \\ &= \rho_0 F_{Ii}^{-1} \frac{\partial \widetilde{u}}{\partial E_{MN}} \left(F_{iN} \delta_{MJ} + F_{iM} \delta_{NJ} \right)/2 \\ &= \rho_0 F_{Ii}^{-1} \left(\frac{\partial \widetilde{u}}{\partial E_{JN}} F_{iN} + \frac{\partial \widetilde{u}}{\partial E_{MJ}} F_{iM} \right)/2 \\ &= \rho_0 F_{Ii}^{-1} F_{iM} \frac{\partial \widetilde{u}}{\partial E_{MJ}}, \end{aligned}$$

where the symmetry of $\boldsymbol{E}$ was used in passing from the third to the fourth line. The $\boldsymbol{F}^{-1}\boldsymbol{F}$ product gives the identity, so the final result is

$$S_{IJ}^{(e)} = \rho_0 \frac{\partial \widetilde{u}}{\partial E_{IJ}} \quad \Leftrightarrow \quad \boldsymbol{S}^{(e)} = \rho_0 \frac{\partial \widetilde{u}}{\partial \boldsymbol{E}}. \tag{6.37}$$

[11] Onsager received the Nobel Prize in Chemistry in 1968 for the discovery of the reciprocal relations. See de Groot and Mazur [dGM62] for a detailed discussion of the reciprocal relations and their derivation. The application of Onsager's relations to continuum field theories is not without controversy. Truesdell [Tru84, Lecture 7] pointed to the arbitrariness of the definition of fluxes and forces and questioned Onsager's basic assumptions. In typical Truesdellian fashion, he attacked the proponents of "Onsagerism."

Equations (6.26), (6.35) and (6.37) provide the constitutive relations for the elastic parts of the Cauchy and Piola–Kirchhoff stress tensors. These expressions provide insight into the power conjugate pairs obtained earlier in the derivation of the deformation power in Section 5.6. That analysis identified three pairs of power conjugate variables: $(\boldsymbol{\sigma}, \dot{\boldsymbol{\epsilon}})$, $(\boldsymbol{P}, \dot{\boldsymbol{F}})$ and $(\boldsymbol{S}, \dot{\boldsymbol{E}})$. From Eqns. (6.35) and (6.37) we see that the elastic parts of the first and second Piola–Kirchhoff stress tensors are conservative thermodynamic tensions work conjugate with their respective kinematic variables. In contrast, the elastic part of the Cauchy stress tensor *cannot* be written as the derivative of the energy with respect to the small strain tensor $\boldsymbol{\epsilon}$. The reason is that unlike $\boldsymbol{F}$ and $\boldsymbol{E}$, the small strain tensor $\boldsymbol{\epsilon}$ is not a state variable. Rather it is an incremental deformation measure. The conclusion is that $\boldsymbol{\sigma}^{(e)}$ is not a conservative thermodynamic tension. Consequently, a calculation of the change in internal energy using the power conjugate pair $(\boldsymbol{\sigma}, \dot{\boldsymbol{\epsilon}})$ requires an integration over the time history.[12]

The constitutive relations derived above have taken the entropy and deformation gradient as the independent state variables. For example, the stress response functions correspond to the change in energy with deformation under conditions of constant entropy. Other scenarios require a transformation from the specific internal energy to other thermodynamic potentials. This is discussed next along with the physical significance of selecting different independent state variables.

6.2.5 Thermodynamic potentials and connection with experiments

The mathematical description of a process can be significantly simplified by an appropriate choice of independent state variables. A process occurring at constant entropy ($\dot{s} = 0$) is called an *isentropic process*. A process where $\boldsymbol{F}$ is controlled is subject to *displacement control*. Thus, $u = \breve{u}(s, \boldsymbol{F})$ is the appropriate energy variable for isentropic processes under displacement control. If, in addition to being isentropic, the process is also reversible, it then follows from Eqn. (6.31) that

$$\rho r - \operatorname{div} \boldsymbol{q} = 0. \tag{6.38}$$

A process satisfying this condition is called *adiabatic*. It is important to note that for continuum systems, adiabatic conditions are not ensured by thermally isolating the system from its environment, which given Eqn. (5.61), only ensures that

$$\mathcal{R}(B) = \int_B \rho r \, dV - \int_{\partial B} \boldsymbol{q} \cdot \boldsymbol{n} \, dA = \int_B [\rho r - \operatorname{div} \boldsymbol{q}] \, dV = 0. \tag{6.39}$$

This does not translate to the local requirement in Eqn. (6.38), unless Eqn. (6.39) is assumed to hold for every subbody of the body. This implies that there is no transfer of heat between different parts of the body. The assumption is that such conditions can be approximately satisfied if the loading is performed "rapidly" on time scales associated with heat transfer [Mal69]. For example, if a tension test in the elastic regime is performed in a laboratory where the sample is thermally isolated from its environment and is loaded (sufficiently fast) by applying a fixed displacement to its end, the engineering stress (i.e. the first Piola–Kirchhoff stress) measured in the experiment will be $\rho_0 \partial \breve{u}(s, \boldsymbol{F}) / \partial \boldsymbol{F}$.

[12] This has important implications for the application of constant stress boundary conditions in atomistic simulations as explained in Section 6.4.3 of [TM11].

Table 6.1. Summary of the form and properties of the thermodynamic potentials

Potential	Functional form	Independent variables	Dependent variables	
internal energy	u	$s, \mathbf{\Gamma}$	$T = \partial u/\partial s$	$\gamma = \partial u/\partial \mathbf{\Gamma}$
Helmholtz free energy	$\psi = u - Ts$	$T, \mathbf{\Gamma}$	$s = -\partial \psi/\partial T$	$\gamma = \partial \psi/\partial \mathbf{\Gamma}$
enthalpy	$h = u - \gamma \cdot \mathbf{\Gamma}$	s, γ	$T = \partial h/\partial s$	$\mathbf{\Gamma} = -\partial h/\partial \gamma$
Gibbs free energy	$g = u - Ts - \gamma \cdot \mathbf{\Gamma}$	T, γ	$s = -\partial g/\partial T$	$\mathbf{\Gamma} = -\partial g/\partial \gamma$

In many cases, the loading conditions will differ. For example, if the tension test mentioned above is performed in a temperature-controlled laboratory with an uninsulated sample, then the process is *isothermal* (i.e. it occurs at constant temperature) and the result of the test will be different. Yet another result will be observed if the device controlling the displacement of the test frame is replaced with a *load control* device that maintains a specified force. The suitable energy variable in either of these cases is not the specific internal energy. Instead, alternative thermodynamic potentials, derived below using Legendre transformations, must be used (see also Exercise 6.1). The results are summarized in Tab. 6.1. We write the expressions below in generic form for arbitrary kinematic variables $\mathbf{\Gamma}$ and thermodynamic tensions γ and then give the results for two particular choices of $\mathbf{\Gamma}$: $\boldsymbol{F}$ and $\boldsymbol{E}$.

Helmholtz free energy The *Helmholtz free energy* is the appropriate energy variable for processes where T and $\mathbf{\Gamma}$ are the independent variables. Let us derive this potential. We begin with the temperature $T = \overline{T}(s, \mathbf{\Gamma})$, which is given by $T = \partial \overline{u}/\partial s$ (Eqn. (6.17)). We seek an alternative potential $\widehat{\mathcal{P}}(T, \mathbf{\Gamma})$ which leads to the inverse relation,

$$s = \widehat{s}(T, \mathbf{\Gamma}) \equiv \frac{\partial \widehat{\mathcal{P}}(T, \mathbf{\Gamma})}{\partial T}.$$

It is straightforward to show that the correct form is given by the following transformation called a *Legendre transformation*:

$$\mathcal{P} = sT - u.$$

The proof is elementary.

Proof Let $\mathcal{P} = \widehat{\mathcal{P}}(T, \mathbf{\Gamma}) = \widehat{s}(T, \mathbf{\Gamma})T - \overline{u}(\widehat{s}(T, \mathbf{\Gamma}), \mathbf{\Gamma})$. Then,

$$\frac{\partial \widehat{\mathcal{P}}}{\partial T} = \frac{\partial \widehat{s}}{\partial T} T + \widehat{s} - \frac{\partial \overline{u}}{\partial s}\frac{\partial \widehat{s}}{\partial T} = \frac{\partial \widehat{s}}{\partial T} T + \widehat{s} - T\frac{\partial \widehat{s}}{\partial T} = \widehat{s}.$$

□

By convention, the negative of $\mathcal{P}$ is taken as the *specific Helmholtz free energy* ψ:

$$\psi = u - Ts. \tag{6.40}$$

The explicit expression showing the variable dependence is

$$\widehat{\psi}(T,\boldsymbol{\Gamma}) = \overline{u}(\widehat{s}(T,\boldsymbol{\Gamma}),\boldsymbol{\Gamma}) - T\widehat{s}(T,\boldsymbol{\Gamma}),$$

with

$$s = -\frac{\partial\widehat{\psi}(T,\boldsymbol{\Gamma})}{\partial T}, \qquad \boldsymbol{\gamma} = \frac{\partial\widehat{\psi}(T,\boldsymbol{\Gamma})}{\partial\boldsymbol{\Gamma}}.$$

The continuum stress variables at constant temperature for the two choices of $\boldsymbol{\Gamma}$ are

$$\boldsymbol{P}^{(e)} = \rho_0\frac{\partial\widehat{\psi}(T,\boldsymbol{F})}{\partial\boldsymbol{F}}, \qquad \boldsymbol{S}^{(e)} = \rho_0\frac{\partial\widetilde{\psi}(T,\boldsymbol{E})}{\partial\boldsymbol{E}}. \tag{6.41}$$

A potential closely related to the specific Helmholtz free energy is the *strain energy density function* W. This is simply the free energy per unit reference volume instead of per unit mass:

$$W = \rho_0\psi. \tag{6.42}$$

In some atomistic simulations, where calculations are performed at "zero temperature," the strain energy density is directly related to the internal energy, $W = \rho_0 u$. In this way strain energy density can be used as a catch-all for both zero temperature and finite temperature conditions. The stress variables follow as

$$\boldsymbol{P}^{(e)} = \frac{\partial\widehat{W}(T,\boldsymbol{F})}{\partial\boldsymbol{F}}, \qquad \boldsymbol{S}^{(e)} = \frac{\partial\widetilde{W}(T,\boldsymbol{E})}{\partial\boldsymbol{E}}. \tag{6.43}$$

Enthalpy The *specific enthalpy* h:

$$h = u - \boldsymbol{\gamma}\cdot\boldsymbol{\Gamma}, \tag{6.44}$$

is the appropriate energy variable for processes where s and $\boldsymbol{\gamma}$ are the independent variables. The explicit expression showing the variable dependence is

$$\widehat{h}(s,\boldsymbol{\gamma}) = \overline{u}(s,\widehat{\boldsymbol{\Gamma}}(s,\boldsymbol{\gamma})) - \boldsymbol{\gamma}\cdot\widehat{\boldsymbol{\Gamma}}(s,\boldsymbol{\gamma}),$$

with

$$T = \frac{\partial\widehat{h}(s,\boldsymbol{\gamma})}{\partial s}, \qquad \boldsymbol{\Gamma} = -\frac{\partial\widehat{h}(s,\boldsymbol{\gamma})}{\partial\boldsymbol{\gamma}}.$$

The continuum deformation measures at constant entropy are

$$F = -\rho_0 \frac{\partial \widehat{h}(s, P^{(e)})}{\partial P^{(e)}}, \qquad E = -\rho_0 \frac{\partial \widetilde{h}(s, S^{(e)})}{\partial S^{(e)}}. \tag{6.45}$$

Gibbs free energy The *specific Gibbs free energy* (or *specific Gibbs function*) g:

$$g = u - Ts - \gamma \cdot \Gamma, \tag{6.46}$$

is the appropriate energy variable for processes where T and γ are the independent variables. The explicit expression showing the variable dependence is

$$\widehat{g}(T, \gamma) = \overline{u}(\widehat{s}(T, \gamma), \widehat{\Gamma}(T, \gamma)) - T\widehat{s}(T, \gamma) - \gamma \cdot \widehat{\Gamma}(T, \gamma), \tag{6.47}$$

with

$$s = -\frac{\partial \widehat{g}(T, \gamma)}{\partial T}, \qquad \Gamma = -\frac{\partial \widehat{g}(T, \gamma)}{\partial \gamma}.$$

The continuum deformation measures at constant temperature are

$$F = -\rho_0 \frac{\partial \widehat{g}(T, P^{(e)})}{\partial P^{(e)}}, \qquad E = -\rho_0 \frac{\partial \widetilde{g}(T, S^{(e)})}{\partial S^{(e)}}. \tag{6.48}$$

6.3 Material frame-indifference

Constitutive relations provide a connection between a material's deformation and its entropy, stress and temperature. A fundamental assumption in continuum mechanics is that this response is intrinsic to the material and should therefore be independent of the frame of reference used to describe the motion of the material. This hypothesis is referred to as the *principle of material frame-indifference*. Explicitly, it states that (intrinsic) constitutive relations must be invariant with respect to changes of frame.[13]

[13] The principle of material frame-indifference has a long history (see [TN65, Section 19A] for a review). The principle was first clearly stated by James Oldroyd who wrote in 1950 [Old50, Section 1]: "The form of the completely general equations [of state] must be restricted by the requirement that the equations describe properties independent of the frame of reference." In Oldroyd's formulation, invariance is guaranteed by expressing constitutive relations in a convected coordinate system that deforms with the material and then mapping them back to a particular fixed frame of reference of interest. In this book, we follow the work of Walter Noll, where constitutive relations can be formulated in any frame of reference, but must satisfy certain constraints to ensure invariance with respect to change of frame. Noll initially used the term "principle of isotropy of space" to describe this principle. Later it was renamed the "principle of objectivity" and then again to its current name [Nol04]. An early publication that describes Noll's formulation is [Nol58].

The application of the principle of material frame-indifference to constitutive relations is a two-step process. First, it must be established how different variables transform under a change of frame of reference. Variables that are unaffected, in a certain sense, by such transformations are called *objective*. Second, variables for which constitutive relations are necessary are required to be objective. The second step imposes constraints on the allowable form of the constitutive relations. We discuss the two steps in order.

At the end of this section we briefly discuss a controversy surrounding the universality of the principle of material frame-indifference. Some authors claim that this principle is not a principle at all, but an approximation which is valid as long as macroscopic time and length scales are large relative to microscopic phenomena. We argue that the controversy is essentially a debate over semantics. Material frame-indifference is a principle for *intrinsic*[14] constitutive relations as they are defined in continuum mechanics. However, these relations are an idealization of a more complex physical reality that is not necessarily frame-indifferent.

6.3.1 Transformation between frames of reference

The description of physical events, characterized by positions in space and the times at which they occur, requires the specification of a *frame of reference*; a concept introduced in Section 2.1. A frame of reference $\mathcal{F}$ is defined as a rigid object (which may be moving), relative to which positions are measured, and a clock to measure time. Mathematically, the space associated with a frame of reference is identified with a Euclidean point space E (see Section 2.3.1).[15] An *event* in the physical world is represented in frame $\mathcal{F}$ as a point x in E and a time t in $\mathbb{R}$. The distance $d(x, y)$ between two points x and y in E is computed from the distance function of the associated inner-product vector space $\mathbb{R}^{n_d}$ (the translation space of E):[16]

$$d(x, y) = \|\boldsymbol{x} - \boldsymbol{y}\| .$$

Here $\boldsymbol{x}$ and $\boldsymbol{y}$ are the position vectors of x and y relative to an origin o (see Eqn. (2.19)).

The choice of frame of reference is not unique, of course. There is an infinite number of possible choices, each of which is associated with a different Euclidean point space and a different clock. Thus, the same event will be associated with different points and times depending on the frame of reference in which it is represented. We now consider two frames of reference, frame $\mathcal{F}$ with points $x \in E$ and times $t \in \mathbb{R}$ and frame $\mathcal{F}^+$ with $x^+ \in E^+$ and $t^+ \in \mathbb{R}$, which may be moving relative to each other. Since we are not dealing with

[14] See Section 6.3.7 for a definition of "intrinsic" constitutive relations.

[15] See [Nol04, Chapter 2] for a description of the formal process by which a Euclidean point space, which Noll calls a "frame-space," is constructed from a rigid material system.

[16] An "event" in the physical world is an abstract concept. There is no way for us to know what those events actually are, Noll calls them "atoms of experience" [Nol73]. In a classical model, all we assume is that using our senses and brains we can measure the distance between the locations of two events and the time lapse between them. This is the information used to make the connection with the mathematical representation of physical reality. In particular, the inner product of the Euclidean vector space $\mathbb{R}^{n_d}$ is constructed specifically so that the distance computed for two points x and y in E coincides with the distance measured between the physical events that x and y represent. Similarly, the difference between two times t_x and t_y in $\mathbb{R}$ equals the time lapse between the corresponding physical events.

relativistic phenomena, we can assume that it is possible for the clocks in both frames to agree on the sequence of two events and the time difference between them. This means that times in the two frames are related by

$$t^+ = t - a,$$

where a is a constant. Since a plays no role in the subsequent derivation, we simplify by setting $a = 0$, so that $t^+ = t$. Physically, this means that it is possible for measurements performed in both frames to agree that a particular event occurred at a particular instant.

The relation between the Euclidean point spaces of frames $\mathcal{F}$ and $\mathcal{F}^+$ is defined formally by the following bijective (one-to-one and onto) transformation [Mur82, Nol87]:

$$\alpha_t : E \rightarrow E^+,$$

where α_t is a linear mapping from E to E^+ at time t. This means that an event at time t, which according to frame $\mathcal{F}$ occurs at point x, is identified with point $x^+ = \alpha_t(x)$ in frame $\mathcal{F}^+$ at time t. The transformation α_t cannot be arbitrary. In order to qualify as a transformation between frames of reference it must preserve distances between points, i.e. $d(x, y) = d(x^+, y^+)$ for any x and y in E. We recall from Section 2.5 that transformations that satisfy this condition must be orthogonal. Thus, in terms of relative positions in frames $\mathcal{F}$ and $\mathcal{F}^+$, we must have at time t [Nol87, Section 33]

$$\boldsymbol{x}^+ - \boldsymbol{y}^+ = \mathcal{Q}_t(\boldsymbol{x} - \boldsymbol{y}), \tag{6.49}$$

where $\mathcal{Q}_t = \nabla \alpha_t$ is a spatially constant, time-dependent, orthogonal,[17] linear transformation from $\mathbb{R}^{n_d}$ to $\mathbb{R}^{n_d+}$ (where $\mathbb{R}^{n_d+}$ is the Euclidean vector space associated with E^+). We use a calligraphic symbol for $\mathcal{Q}_t$, to stress the fact that this transformation is different from a proper orthogonal tensor $\boldsymbol{Q}$, introduced in Section 2.5.1, which maps vectors in the translation space of a single frame to the same translation space and which will be used later to impose constraints on the functional form of constitutive relations.

To make the above discussion more concrete, consider the following example.

Example 6.1 (Different frames of reference) A two-dimensional physical world contains only a rigid cross and a rigid rectangle that are translating and rotating relative to each other. The motion is represented relative to two different frames of reference: the cross (frame $\mathcal{F}$) and the rectangle (frame $\mathcal{F}^+$) as shown in Fig. 6.1. In frame $\mathcal{F}$, the cross appears stationary and the rectangle is rotating and translating (top row of images in Fig. 6.1). In frame $\mathcal{F}^+$, the rectangle appears stationary and the cross appears to be moving (bottom row of images in Fig. 6.1). The pair of points $(\boldsymbol{x}, \boldsymbol{y})$ and $(\boldsymbol{x}^+, \boldsymbol{y}^+)$ are the representation of two points in the physical world in frames $\mathcal{F}$ and $\mathcal{F}^+$, respectively. The distance between the points is the same in both frames, but their orientation appears different in each frame and is related through Eqn. (6.49). It is important to understand that the vectors $\boldsymbol{x} - \boldsymbol{y}$ and $\boldsymbol{x}^+ - \boldsymbol{y}^+$ exist in different Euclidean vector spaces, so it is not possible (or necessary) to draw them together on the same graph.

[17] The concept of an "orthogonal" linear transformation between different Euclidean vector spaces is discussed by Noll in [Nol73, Nol06] and Murdoch in [Mur03].

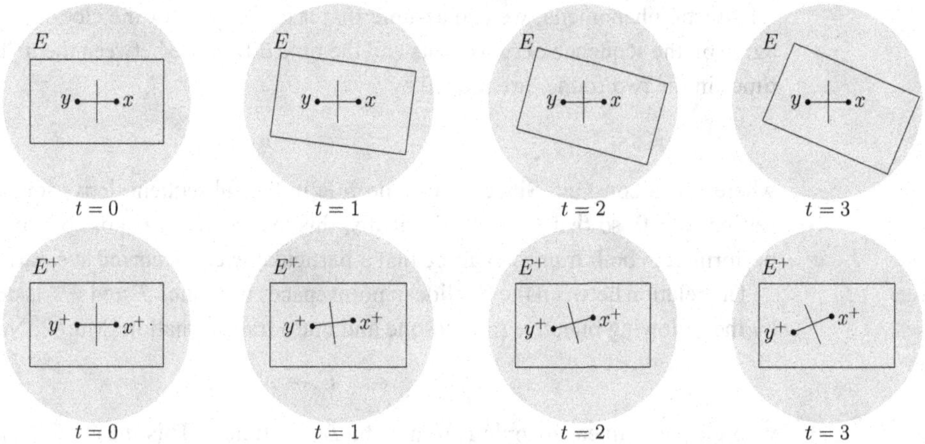

Fig. 6.1 A two-dimensional example demonstrating how the physical world is represented in two different frames of reference (see Example 6.1). In this example a cross and a rectangle are translating and rotating relative to each other. The top row is a series of snapshots in time showing the two objects in which the cross is the frame of reference ($\mathcal{F}$). The bottom row shows snapshots of the same process in which the rectangle is the frame of reference ($\mathcal{F}^+$). The gray background represents the Euclidean point spaces, E and E^+ associated with the two frames. The points x and x^+ represent the location of the same physical event in the two frames. Similarly, y and y^+ represent the same event.

Having defined the basic transformation formula between frames of reference in Eqn. (6.49), we now obtain the transformation relations for important kinematic variables that will be needed later. Equation (6.49) can be rewritten as

$$\boldsymbol{x}^+ = \boldsymbol{c}^+(t) + \mathcal{Q}_t\boldsymbol{x}, \tag{6.50}$$

where $\boldsymbol{y}^+$ has been moved to the right-hand side and $\boldsymbol{c}^+(t) = \boldsymbol{y}^+ - \mathcal{Q}_t\boldsymbol{y}$ is a vector in $\mathbb{R}^{n_d+}$. Equation (6.50) shows how the position vector of a point transforms between frames of reference. In the context of a continuum body, $\boldsymbol{x}$ represents the deformed position of a material particle $\boldsymbol{X} \in \mathcal{B}_0$ through a motion $\varphi(\boldsymbol{X}, t)$, so that $\boldsymbol{x} = \varphi(\boldsymbol{X}, t)$. Similarly, $\boldsymbol{x}^+ = \varphi^+(\boldsymbol{X}^+, t)$. The transformation of the velocity and acceleration of a particle follow by material time differentiation of Eqn. (6.50) (see Section 3.6):

$$\boldsymbol{v}^+ = \dot{\boldsymbol{c}}^+ + \dot{\mathcal{Q}}_t\boldsymbol{x} + \mathcal{Q}_t\boldsymbol{v}, \tag{6.51}$$

$$\boldsymbol{a}^+ = \ddot{\boldsymbol{c}}^+ + \ddot{\mathcal{Q}}_t\boldsymbol{x} + 2\dot{\mathcal{Q}}_t\boldsymbol{v} + \mathcal{Q}_t\boldsymbol{a}. \tag{6.52}$$

The transformation relation for the velocity gradient, $\boldsymbol{l} = \nabla\boldsymbol{v}$, is obtained by taking the spatial gradient with respect to $\boldsymbol{x}^+$ (denoted by ∇_+) of $\boldsymbol{v}^+$ in Eqn. (6.51):

$$\boldsymbol{l}^+ = \dot{\mathcal{Q}}_t\nabla_+\boldsymbol{x} + \mathcal{Q}_t\boldsymbol{l}\nabla_+\boldsymbol{x}, \tag{6.53}$$

where we have used the chain rule. From Eqn. (6.50), we have

$$\nabla_+\boldsymbol{x} = \nabla_+\left[\mathcal{Q}_t^T(\boldsymbol{x}^+ - \boldsymbol{c}^+)\right] = \mathcal{Q}_t^T. \tag{6.54}$$

Substituting Eqn. (6.54) into Eqn. (6.53) gives

$$\boldsymbol{l}^+ = \boldsymbol{\Omega}^+ + \mathcal{Q}_t \boldsymbol{l} \mathcal{Q}_t^T, \tag{6.55}$$

where $\boldsymbol{\Omega}^+ = \dot{\mathcal{Q}}_t \mathcal{Q}_t^T$ is a second-order tensor over $\mathbb{R}^{n_d+}$. This is clear since $\boldsymbol{\Omega}^+$ maps a vector in $\mathbb{R}^{n_d+}$ to another vector in $\mathbb{R}^{n_d+}$:

$$\boldsymbol{\Omega}^+ \boldsymbol{a}^+ = \dot{\mathcal{Q}}_t(\mathcal{Q}_t^T \boldsymbol{a}^+) = \dot{\mathcal{Q}}_t \boldsymbol{a} = \boldsymbol{b}^+,$$

where $\boldsymbol{a}^+, \boldsymbol{b}^+ \in \mathbb{R}^{n_d+}$ and $\boldsymbol{a} \in \mathbb{R}^{n_d}$. The tensor $\boldsymbol{\Omega}^+$ has the important property that it is antisymmetric. The proof is straightforward:

Proof Start with $\mathcal{Q}_t \mathcal{Q}_t^T = \boldsymbol{I}^+$, where $\boldsymbol{I}^+$ is the identity transformation on $\mathbb{R}^{n_d+}$ [Mur03]. Taking a material time derivative, we have, $(D/Dt)(\mathcal{Q}_t \mathcal{Q}_t^T) = \dot{\mathcal{Q}}_t \mathcal{Q}_t^T + \mathcal{Q}_t \dot{\mathcal{Q}}_t^T = \boldsymbol{0}$, and therefore $\dot{\mathcal{Q}}_t \mathcal{Q}_t^T = -\mathcal{Q}_t \dot{\mathcal{Q}}_t^T = -(\dot{\mathcal{Q}}_t \mathcal{Q}_t^T)^T$. Thus, $\dot{\mathcal{Q}}_t \mathcal{Q}_t^T$ is antisymmetric. □

We saw earlier that it is not $\boldsymbol{l}$, but rather its symmetric part, the rate of deformation tensor $\boldsymbol{d}$, that appears in constitutive relations. Substituting Eqn. (6.55) into $\boldsymbol{d}^+ = \frac{1}{2}[\boldsymbol{l}^+ + (\boldsymbol{l}^+)^T]$, the antisymmetric term drops out and we have

$$\boldsymbol{d}^+ = \mathcal{Q}_t \boldsymbol{d} \mathcal{Q}_t^T. \tag{6.56}$$

So far we have only considered spatial measures. In order to obtain relations for the transformation of reference measures, we must first consider how reference configurations defined in the different frames transform. For simplicity, we assume that both frames of reference adopt a Lagrangian description, which means that the reference configuration is the configuration that a body of interest occupies at time $t = 0$. Since we have assumed that both frames use the same clock, we have

$$\boldsymbol{X}^+ = \boldsymbol{c}^+(0) + \mathcal{Q}_0 \boldsymbol{X}, \tag{6.57}$$

where $\boldsymbol{X}$ and $\boldsymbol{X}^+$ are particles in the reference configuration in frames $\mathcal{F}$ and $\mathcal{F}^+$, respectively. The deformation gradient follows from Eqn. (6.50) as

$$\boldsymbol{F}^+ = \nabla_{0+} \boldsymbol{x}^+ = \mathcal{Q}_t \nabla_{0+} \boldsymbol{x} = \mathcal{Q}_t \boldsymbol{F} \nabla_{0+} \boldsymbol{X}, \tag{6.58}$$

where ∇_{0+} is the gradient with respect to $\boldsymbol{X}^+$ and where we have used the chain rule. The gradient $\nabla_{0+} \boldsymbol{X}$ appearing in Eqn. (6.58) can be computed from Eqn. (6.57):

$$\nabla_{0+} \boldsymbol{X} = \nabla_{0+} \left[\mathcal{Q}_0^T (\boldsymbol{X}^+ - \boldsymbol{c}^+(0))\right] = \mathcal{Q}_0^T. \tag{6.59}$$

Substituting Eqn. (6.59) into Eqn. (6.58), we have

$$\boldsymbol{F}^+ = \mathcal{Q}_t \boldsymbol{F} \mathcal{Q}_0^T. \tag{6.60}$$

The right Cauchy–Green deformation tensor, $C^+ = (F^+)^T F^+$, follows as

$$C^+ = Q_0 C Q_0^T. \tag{6.61}$$

We now turn to the definition of objective tensors which will be used later to establish the material frame-indifference constraints.

6.3.2 Objective tensors

A tensor is called *objective* if it appears the same in all frames of reference. Exactly what we mean by the "same" is discussed below. We will later argue that the variables for which constitutive relations are required must be objective. But first we define the conditions under which zeroth-, first- and second-order tensors are objective.

Objectivity condition for a scalar invariant A zeroth-order tensor (scalar invariant) s is objective if it satisfies

$$s^+ = s \tag{6.62}$$

for all mappings α_t. A zeroth-order tensor is just a real number, so objectivity simply means that this number is the same in all frames of reference. It may seem that all scalar invariants are objective since by definition they do not depend on coordinates. However, consider, for example, the speed $s = \|v\|$ at which an object is moving. $\|v\|$ is a zeroth-order tensor; however, it is not objective since the speed of an object depends on the frame of reference in which it is represented. In Fig. 6.1, the speed of points on the cross is zero in frame $\mathcal{F}$ and nonzero in frame $\mathcal{F}^+$. Thus, speed is a *subjective* variable. We postulate that physical variables, such as the mass density ρ, temperature T, entropy density s and so on are objective.

Objectivity condition for a vector A first-order tensor (vector) u is objective if it satisfies

$$u^+ = Q_t u \tag{6.63}$$

for all mappings α_t (or equivalently, for all orthogonal transformations Q_t). This definition stems from the manner in which relative vectors transform between frames of reference in Eqn. (6.49). Thus, a vector is objective if it differs only by the rotation that all relative vectors experience. Vectors satisfying this condition have the same orientation relative to actual "physical directions." To explain what we mean by this, we introduce an orthonormal basis $\{e_i\}$ in frame $\mathcal{F}$. In Fig. 6.1, the horizontal and vertical lines of the cross can be used to define basis vectors for $\mathcal{F}$, so that $e_1 = (x - y)/\|x - y\|$ and e_2 is defined in a similar manner along the vertical direction. This choice is not unique, of course. Any set of vectors obtained through a *fixed* (time-independent) rotation of these vectors is also a valid basis. This simply corresponds to a change of basis in frame $\mathcal{F}$.

Since the basis vectors are defined from relative position vectors, they transform according to Eqn. (6.49):

$$\mathrm{e}_i^+(t) = \mathcal{Q}_t \mathbf{e}_i. \tag{6.64}$$

Here $\mathrm{e}_i^+(t)$ is the unit vector in $\mathcal{F}^+$ constructed from the same events from which $\mathbf{e}_i$ is constructed. In frame $\mathcal{F}$, the vector $\mathbf{e}_i$ is constant, but in $\mathcal{F}^+$ its image, the vector $\mathrm{e}_i^+(t)$, will move according to the time-dependent transformation $\mathcal{Q}_t$, just as in Fig. 6.1 the cross is moving in the lower row of images. This means that $\mathrm{e}_i^+(t)$ is *not* a fixed basis for frame $\mathcal{F}^+$. We stress this by writing its explicit dependence on time.

We now consider an objective vector $\boldsymbol{u}$ in frame $\mathcal{F}$ with components, $u_i = \boldsymbol{u} \cdot \mathbf{e}_i$, relative to the basis $\{\mathbf{e}_i\}$. Since $\boldsymbol{u}$ is objective it transforms according to Eqn. (6.63), therefore

$$\boldsymbol{u}^+ = \mathcal{Q}_t \boldsymbol{u} = \mathcal{Q}_t(u_i \mathbf{e}_i) = u_i(\mathcal{Q}_t \mathbf{e}_i) = u_i \mathrm{e}_i^+(t), \tag{6.65}$$

where Eqn. (6.64) was used in the last step. We see that $\boldsymbol{u}$ has the same components along the vectors $\mathbf{e}_i$ and their images $\mathrm{e}_i^+(t)$, which represent the same directions in the physical world. Thus, although an objective vector appears differently in different frames it actually has the same orientation relative to events in the physical world. This is the meaning of objective.

Comparing Eqn. (6.63) with Eqns. (6.49)–(6.52), we see that the relative position between points, $\boldsymbol{x} - \boldsymbol{y}$ is objective, whereas the position, velocity and acceleration of a single point, $\boldsymbol{x}$, $\boldsymbol{v}$ and $\boldsymbol{a}$, are not. The latter is not surprising since, naturally, the position and motion of a physical point depends on the frame of reference in which it is represented. In particular, the additional terms in Eqns. (6.51)–(6.52) relative to Eqn. (6.63), reflect the motion of the frame of reference itself. This is clear if we repeat the procedure that led to Eqn. (6.65) for a nonobjective vector like the position vector, $\boldsymbol{x} = x_i \mathbf{e}_i$. We find

$$\boldsymbol{x}^+ = \mathbf{c}^+(t) + x_i \mathrm{e}_i^+(t) = (c_i^+(t) + x_i)\mathrm{e}_i^+(t).$$

So the components of $\boldsymbol{x}^+$ relative to $\mathrm{e}_i^+(t)$ are *not* the same as those of $\boldsymbol{x}$ relative to $\mathbf{e}_i$.

Objectivity condition for a second-order tensor A second-order tensor $\boldsymbol{T}$ is objective if it satisfies

$$\boldsymbol{T}^+ = \mathcal{Q}_t \boldsymbol{T} \mathcal{Q}_t^T \tag{6.66}$$

for all transformations α_t (or equivalently, for all orthogonal transformations $\mathcal{Q}_t$). This relation can be obtained from the objective vector definition in Eqn. (6.63) as follows.

Proof The tensor $\boldsymbol{T}$ is objective, if for every objective vector $\boldsymbol{a}$, the vector $\boldsymbol{b}$ defined by

$$\boldsymbol{b} = \boldsymbol{T}\boldsymbol{a} \tag{6.67}$$

is also objective. The corresponding expression in frame $\mathcal{F}^+$ is

$$b^+ = T^+ a^+. \tag{6.68}$$

Since a and b are objective, we have $a^+ = \mathcal{Q}_t a$ and $b^+ = \mathcal{Q}_t b$. Substituting this into Eqn. (6.68) gives

$$\mathcal{Q}_t b = T^+ \mathcal{Q}_t a. \tag{6.69}$$

Substituting Eqn. (6.67) into Eqn. (6.69) and rearranging gives

$$(T^+ \mathcal{Q}_t - \mathcal{Q}_t T)a = 0.$$

The above relation must be true for every objective vector a, therefore $T^+ \mathcal{Q}_t = \mathcal{Q}_t T$, from which we obtain the result in Eqn. (6.66). (An alternative approach for obtaining Eqn. (6.66) using the dyadic representation of a second-order tensor, which generalizes to higher-order tensors, is discussed in Exercise 6.4.) □

As for an objective vector, an objective tensor preserves components relative to a given basis in a transformation between frames of reference, i.e. $e_i^+(t) \cdot T^+ e_j^+(t) = e_i \cdot T e_j$. The proof is analogous to that given above for a vector.

Comparing Eqn. (6.66) with Eqns. (6.55)–(6.56), we see that the velocity gradient l is not objective due to the presence of the extra term Ω^+; however, the rate of deformation tensor d is objective. Further, considering Eqns. (6.60) and (6.61) we see that F and C are not objective.

6.3.3 Principle of material frame-indifference

The basic postulate of the principle of material frame-indifference is that all variables for which constitutive relations are required must be objective tensors. To understand the reasoning underlying this requirement, let us revisit Example 6.1 and Fig. 6.1. Imagine that a spring is connected between the physical points that are labeled x and y in frame $\mathcal{F}$. Assume that the free length of the spring ℓ_0 is shorter than the distance $\|x - y\|$ so that the spring is in tension. What is the force in the spring? In frame $\mathcal{F}$ the spring is stationary. In frame $\mathcal{F}^+$ (where its ends are located at points x^+ and y^+) the spring is translating and rotating. According to material frame-indifference the force must be the same in both cases.[18] Understand that this is the same physical spring, the only difference is the frame of reference in which its motion is represented.

In our case, we consider constitutive relations for four variables: the internal energy density u, the temperature T, the heat flux vector q and the Cauchy stress tensor σ (separated into elastic and viscous parts). We therefore require these variables to be objective.

The objectivity of the stress tensor is fundamentally tied to the objectivity of force. The Cauchy stress tensor σ is defined by the Cauchy relation, $t = \sigma n$, where t is the traction vector and n is a normal to a plane. The normal to a plane is an objective vector since it can be defined in terms of relative positions between particles, which is objective [Mur03].

[18] We discuss the subtleties associated with this requirement at the end of this section where we address the controversy surrounding material frame-indifference.

Therefore, in order for the stress to be objective the traction must be objective (see proof after Eqn. (6.66)). Now, traction is defined as force per unit area, and area (which also depends on relative positions of particles) is objective. Thus, the requirement that stress is objective translates to the basic postulate that force is objective.

The notion that the force $\boldsymbol{f}$ is an objective vector may seem at odds with Newton's second law, $\boldsymbol{f} = m\boldsymbol{a}$, since although mass is objective, acceleration is not. In fact, this is precisely the origin of the concept of inertial reference frames discussed in Section 2.1. We start with the assumption that force is objective and that we know, $\boldsymbol{f} = m\boldsymbol{a}$, in frame $\mathcal{F}$. (This is Thomson's *law of inertia* discussed on page 14.) From this we can observe that Newton's second law holds only in inertial frames of reference, say $\mathcal{F}^+$, for which (relative to $\mathcal{F}$) $\ddot{c}^+ = \boldsymbol{0}$ and $\dot{\mathcal{Q}}_t = \ddot{\mathcal{Q}}_t = \boldsymbol{0}$, and therefore according to Eqn. (6.52), $\boldsymbol{a}^+ = \mathcal{Q}_t\boldsymbol{a}$. Thus the relation $\boldsymbol{f} = m\boldsymbol{a}$, which is true in frame $\mathcal{F}$, is also satisfied in all inertial frames $\mathcal{F}^+$, i.e. $\boldsymbol{f}^+ = m\boldsymbol{a}^+$. We see that the postulate that force is objective and that $\boldsymbol{f} = m\boldsymbol{a}$ in at least one frame of reference is equivalent to the fact that Newton's laws hold in all inertial frames of reference.

6.3.4 Constraints on constitutive relations due to material frame-indifference

Constitutive relations are required for scalar invariant, vector and tensor variables:

$$s = \widehat{s}(\gamma), \qquad \boldsymbol{u} = \widehat{\boldsymbol{u}}(\gamma), \qquad \boldsymbol{T} = \widehat{\boldsymbol{T}}(\gamma). \tag{6.70}$$

Here γ represents a scalar invariant, vector or tensor argument. Of course, each constitutive relation can have different arguments and more than one. We proceed with the derivation for a single generic argument γ. The results can then be immediately extended to any specific set of arguments of an actual constitutive relation.

According to material frame-indifference, the variables s, $\boldsymbol{u}$ and $\boldsymbol{T}$ in Eqn. (6.70) must be objective. We therefore require according to Eqns. (6.62), (6.63) and (6.66) that

$$\overline{s}^+(\gamma_t^+) = \widehat{s}(\gamma_t), \qquad \overline{\boldsymbol{u}}^+(\gamma_t^+) = \mathcal{Q}_t\widehat{\boldsymbol{u}}(\gamma_t), \qquad \overline{\boldsymbol{T}}^+(\gamma_t^+) = \mathcal{Q}_t\widehat{\boldsymbol{T}}(\gamma_t)\mathcal{Q}_t^T, \tag{6.71}$$

for all motions (or equivalently, for all functions of time γ_t) and for all changes of frame. Note that we do not assume that the functional forms in the two frames are the same, only that the result is an objective variable. Further, we now indicate, with a subscript t, all quantities that explicitly depend on time.[19] Below (following [Nol06] in spirit), the frame-indifference conditions are reformulated in a single frame of reference in order to obtain constraints for a given functional form.

The variables γ_t^+ and γ_t are related through the appropriate frame of reference transformations derived earlier. We define $\mathcal{L}_t$ as the mapping taking γ_t to γ_t^+ at time t, thus

$$\gamma_t^+ = \mathcal{L}_t\gamma_t, \qquad \gamma_t = \mathcal{L}_t^{-1}\gamma_t^+. \tag{6.72}$$

For example, from Eqn. (6.56), the rate of deformation tensor is mapped according to $\boldsymbol{d}_t^+ = \mathcal{L}_t\boldsymbol{d}_t = \mathcal{Q}_t\boldsymbol{d}_t\mathcal{Q}_t^T$. We now rewrite Eqn. (6.71) explicitly accounting for the transformation

[19] Recall that we have already eliminated the possibility that the constitutive relations depend explicitly on time via constraint VI, which says that we are only considering materials *without memory and without aging*.

of the arguments:

$$\bar{s}^+(\mathcal{L}_t\gamma_t) = \widehat{s}(\gamma_t), \qquad \bar{\boldsymbol{u}}^+(\mathcal{L}_t\gamma_t) = \mathcal{Q}_t\widehat{\boldsymbol{u}}(\gamma_t), \qquad \bar{\boldsymbol{T}}^+(\mathcal{L}_t\gamma_t) = \mathcal{Q}_t\widehat{\boldsymbol{T}}(\gamma_t)\mathcal{Q}_t^T. \tag{6.73}$$

This relation must hold for all γ_t and for all frames $\mathcal{F}^+$. In particular, consider the time[20] $t = 0$. Then, Eqn. (6.73) gives

$$\bar{s}^+(\mathcal{L}_0\gamma) = \widehat{s}(\gamma), \qquad \bar{\boldsymbol{u}}^+(\mathcal{L}_0\gamma) = \mathcal{Q}_0\widehat{\boldsymbol{u}}(\gamma), \qquad \bar{\boldsymbol{T}}^+(\mathcal{L}_0\gamma) = \mathcal{Q}_0\widehat{\boldsymbol{T}}(\gamma)\mathcal{Q}_0^T, \tag{6.74}$$

which must hold for all γ, since γ_0 will range over all possible values when all motions γ_t are considered. Now, in Eqn. (6.73) write $\mathcal{L}_t\gamma_t = \mathcal{L}_0\mathcal{L}_0^{-1}\mathcal{L}_t\gamma_t = \mathcal{L}_0\gamma_t^*$ with

$$\gamma_t^* \equiv \mathcal{L}_0^{-1}\mathcal{L}_t\gamma_t \tag{6.75}$$

and apply Eqn. (6.74), i.e. $\bar{s}^+(\mathcal{L}_0\gamma_t^*) = \widehat{s}(\gamma_t^*)$, to obtain

$$\widehat{s}(\gamma_t^*) = \widehat{s}(\gamma_t), \qquad \mathcal{Q}_0\widehat{\boldsymbol{u}}(\gamma_t^*) = \mathcal{Q}_t\widehat{\boldsymbol{u}}(\gamma_t), \qquad \mathcal{Q}_0\widehat{\boldsymbol{T}}(\gamma_t^*)\mathcal{Q}_0^T = \mathcal{Q}_t\widehat{\boldsymbol{T}}(\gamma_t)\mathcal{Q}_t^T. \tag{6.76}$$

Here, γ_t^* can be interpreted as a second, different, motion measured in the frame $\mathcal{F}$. Thus, we have found that, for any given motion γ_t, the principle of material frame-indifference implies a relation between the response associated with γ_t and the response associated with the related motion γ_t^* *in a single frame*. However, Eqn. (6.76) is not the most convenient mathematical form of the relation. Transferring the $\mathcal{Q}_0$ terms to the right and substituting the definition of γ_t^*, we have

$$\widehat{s}(\mathcal{L}_0^{-1}\mathcal{L}_t\gamma_t) = \widehat{s}(\gamma_t), \qquad \widehat{\boldsymbol{u}}(\mathcal{L}_0^{-1}\mathcal{L}_t\gamma_t) = \boldsymbol{Q}_t\widehat{\boldsymbol{u}}(\gamma_t), \qquad \widehat{\boldsymbol{T}}(\mathcal{L}_0^{-1}\mathcal{L}_t\gamma_t) = \boldsymbol{Q}_t\widehat{\boldsymbol{T}}(\gamma_t)\boldsymbol{Q}_t^T, \tag{6.77}$$

where[21]

$$\boldsymbol{Q}_t = \mathcal{Q}_0^T\mathcal{Q}_t \tag{6.78}$$

is a *proper*[22] orthogonal tensor defined over the Euclidean vector space $\mathbb{R}^{n_d}$ of frame $\mathcal{F}$ (i.e. it maps vectors from $\mathbb{R}^{n_d}$ into itself). Similarly, $\mathcal{L}_0^{-1}\mathcal{L}_t\gamma_t$ is expressed in $\mathcal{F}$. This equation must be satisfied for all changes of frame and for all motions. Because $\boldsymbol{Q}_t$ and $\mathcal{L}_0^{-1}\mathcal{L}_t$ depend only on the change of frame and γ_t depends only on the motion, it is clear that Eqn. (6.77) must be satisfied for *arbitrary and independent* values of $\boldsymbol{Q}$ and γ. For this reason the choice of the particular fixed time ($t = 0$ in this case) is unimportant. Thus, we find that for all $\boldsymbol{Q}$ and for every γ the following must be true:

$$\widehat{s}(\mathcal{L}_0^{-1}\mathcal{L}_t\gamma) = \widehat{s}(\gamma), \qquad \widehat{\boldsymbol{u}}(\mathcal{L}_0^{-1}\mathcal{L}_t\gamma) = \boldsymbol{Q}\widehat{\boldsymbol{u}}(\gamma), \qquad \widehat{\boldsymbol{T}}(\mathcal{L}_0^{-1}\mathcal{L}_t\gamma) = \boldsymbol{Q}\widehat{\boldsymbol{T}}(\gamma)\boldsymbol{Q}^T, \tag{6.79}$$

[20] There is nothing special about $t = 0$. Any *fixed* time would do as we will see later.

[21] Notice the difference in font between $\boldsymbol{Q}_t$ and $\mathcal{Q}_t$, which is meant to distinguish these distinct entities. That is, $\boldsymbol{Q}_t$ is a proper orthogonal tensor mapping vectors in the translation space of frame $\mathcal{F}$ to itself; whereas $\mathcal{Q}_t$ is an orthogonal linear transformation mapping vectors in the translation space of frame $\mathcal{F}$ to vectors in the translation space of frame $\mathcal{F}^+$.

[22] A proof that $\boldsymbol{Q}_t$ is proper orthogonal is given later in this section.

where we note that the value of $\mathcal{L}_0^{-1}\mathcal{L}_t\gamma$ does not depend on the frame $\mathcal{F}^+$ or time t, but does depend on the type of the variable γ. If γ is objective, then it transforms according to Eqns. (6.62), (6.63) and (6.66), depending on whether it is a scalar invariant, a vector or a tensor. The results for these three cases are:

$$\mathcal{L}_0^{-1}\mathcal{L}_t s = s, \qquad \mathcal{L}_0^{-1}\mathcal{L}_t \boldsymbol{u} = \boldsymbol{Q}\boldsymbol{u}, \qquad \mathcal{L}_0^{-1}\mathcal{L}_t \boldsymbol{T} = \boldsymbol{Q}\boldsymbol{T}\boldsymbol{Q}^T, \tag{6.80}$$

where as before $\boldsymbol{Q} = \mathcal{Q}_0^T\mathcal{Q}_t$. In addition, we will require the transformations for the deformation gradient and the right Cauchy–Green deformation tensor. For $\boldsymbol{F}$ we have from Eqn. (6.60) that $\mathcal{L}_t\boldsymbol{F} = \mathcal{Q}_t\boldsymbol{F}\mathcal{Q}_0^T$. Consequently, $\mathcal{L}_0^{-1}\mathcal{L}_t\boldsymbol{F} = \mathcal{Q}_0^T(\mathcal{Q}_t\boldsymbol{F}\mathcal{Q}_0^T)\mathcal{Q}_0 = \boldsymbol{Q}\boldsymbol{F}$. The transformation for $\boldsymbol{C}$ is obtained in a similar manner. In summary,

$$\mathcal{L}_0^{-1}\mathcal{L}_t\boldsymbol{F} = \boldsymbol{Q}\boldsymbol{F}, \qquad \mathcal{L}_0^{-1}\mathcal{L}_t\boldsymbol{C} = \boldsymbol{C}. \tag{6.81}$$

The proof that $\boldsymbol{Q}_t$ (and thus $\boldsymbol{Q}$) is proper orthogonal is similar to a proof by Murdoch [Mur03], although the motivation and details are different. A sketch of the proof follows.

Proof Introduce the basis $\{\mathbf{e}_i^+\}$ for $\mathcal{F}^+$. Similar to the argument that led to Eqn. (6.64), we have that the basis vectors are mapped to their images in $\mathcal{F}$ by $\mathbf{e}_i(t) = \mathcal{Q}_t^T\mathbf{e}_i^+$. At time $t = 0$, the mapping is $\mathbf{e}_i(0) = \mathcal{Q}_0^T\mathbf{e}_i^+$. Extracting $\mathbf{e}_i^+$ from both expressions and equating them, we have, $\mathcal{Q}_t\mathbf{e}_i(t) = \mathcal{Q}_0\mathbf{e}_i(0)$ and so

$$\mathbf{e}_i(t) = \mathcal{Q}_t^T\mathcal{Q}_0\mathbf{e}_i(0) = (\mathcal{Q}_0^T\mathcal{Q}_t)^T\mathbf{e}_i(0) = \boldsymbol{Q}_t^T\mathbf{e}_i(0).$$

The handedness of $\mathbf{e}_i(t)$ is arbitrary; however, this handedness must be preserved over time. The reason for this is that $\mathbf{e}_i(t)$ are the images of the basis $\{\mathbf{e}_i^+\}$ that has a fixed handedness and frames $\mathcal{F}$ and $\mathcal{F}^+$ are constructed from rigid physical bodies that over time can translate and rotate relative to each other but cannot reflect. If the handedness of the triad $\{\mathbf{e}_i(t)\}$ has to be preserved, then $\mathbf{e}_i(t)$ can only differ from $\mathbf{e}_i(0)$ by a rotation and hence $\boldsymbol{Q}_t^T$ (and therefore $\boldsymbol{Q}_t$) is proper orthogonal. □

The distinction between Eqns. (6.73) and (6.79) is important. Equation (6.73) is a relationship between the constitutive relations in two different frames of reference with possibly different functional forms. The transformation $\mathcal{Q}_t$ appearing in this relation is an orthogonal mapping between two different vector spaces associated with the different frames. This is the mathematical form of material frame-indifference corresponding to the original statement: "constitutive relations must be invariant with respect to changes of frame."

In contrast, Eqn. (6.79) is a condition obtained in a single frame of reference for the same functional form. The tensor $\boldsymbol{Q}$ appearing in this relation is a *proper* orthogonal tensor defined over a single vector space. Equation (6.79) is not a direct statement of material frame-indifference. Rather it is a constraint that all constitutive relations must satisfy in their own frame of reference. It is straightforward to show that this constraint, along with the relations given by Eqn. (6.74), ensure that the mapping between any two frames obeys Eqn. (6.73) and therefore material frame-indifference is satisfied. When expressed in this form, material frame-indifference is sometimes referred to as *invariance with respect to superposed rigid-body motion.*[23]

We demonstrate the application of material frame-indifference with a simple example.

Example 6.2 (Material frame-indifference in a two-particle system[24]) In motivating the application of material frame-indifference to constitutive relations we used the example of a spring connected between two points $\boldsymbol{x}$ and $\boldsymbol{y}$ in frame $\mathcal{F}$. One can think of this as a two-particle system interacting through a force field. What constraints does material frame-indifference place on the form of the force between the particles?

Let $\boldsymbol{f}$ be the force on particle $\boldsymbol{x}$ due to particle $\boldsymbol{y}$. In the most general case, the force can depend on the position of both particles, $\boldsymbol{f} = \widetilde{\boldsymbol{f}}(\boldsymbol{x}, \boldsymbol{y})$. The requirement of material frame-indifference for this function according to Eqn. $(6.79)_2$ is

$$\widetilde{\boldsymbol{f}}(\mathcal{L}_0^{-1}\mathcal{L}_t\boldsymbol{x}, \mathcal{L}_0^{-1}\mathcal{L}_t\boldsymbol{y}) = \boldsymbol{Q}\widetilde{\boldsymbol{f}}(\boldsymbol{x}, \boldsymbol{y}). \tag{6.82}$$

First, we must determine the form of $\mathcal{L}_0^{-1}\mathcal{L}_t$ for position vectors. From Eqn. (6.50), we have, $\mathcal{L}_t\boldsymbol{x} = \boldsymbol{c}^+(t) + \mathcal{Q}_t\boldsymbol{x}$. Therefore $\mathcal{L}_0\boldsymbol{x} = \boldsymbol{c}^+(0) + \mathcal{Q}_0\boldsymbol{x}$, with an inverse $\mathcal{L}_0^{-1}\boldsymbol{x}^+ = \mathcal{Q}_0^T(\boldsymbol{x}^+ - \boldsymbol{c}^+(0))$. The composition $\mathcal{L}_0^{-1}\mathcal{L}_t$ follows as

$$\mathcal{L}_0^{-1}\mathcal{L}_t\boldsymbol{x} = \mathcal{Q}_0^T\left([\boldsymbol{c}^+(t) + \mathcal{Q}_t\boldsymbol{x}] - \boldsymbol{c}^+(0)\right) = \mathcal{Q}_0^T\mathcal{Q}_t\boldsymbol{x} + \mathcal{Q}_0^T\left(\boldsymbol{c}^+(t) - \boldsymbol{c}^+(0)\right) = \boldsymbol{Q}\boldsymbol{x} + \boldsymbol{c}, \tag{6.83}$$

where $\boldsymbol{Q} = \mathcal{Q}_0^T\mathcal{Q}_t$ is a proper orthogonal tensor and $\boldsymbol{c}$ is an arbitrary vector in $\mathbb{R}^{n_d}$. Substituting Eqn. (6.83) into Eqn. (6.82) gives

$$\widetilde{\boldsymbol{f}}(\boldsymbol{Q}\boldsymbol{x} + \boldsymbol{c}, \boldsymbol{Q}\boldsymbol{y} + \boldsymbol{c}) = \boldsymbol{Q}\widetilde{\boldsymbol{f}}(\boldsymbol{x}, \boldsymbol{y}). \tag{6.84}$$

This condition must be true for all $\boldsymbol{Q} \in SO(n_d)$ and for all $\boldsymbol{c} \in \mathbb{R}^{n_d}$, therefore it must also be true for the special case, $\boldsymbol{Q} = \boldsymbol{I}$ and $\boldsymbol{c} = \boldsymbol{d} - \boldsymbol{y}$, where $\boldsymbol{d}$ is an arbitrary point. Substituting these values into Eqn. (6.84) gives $\widetilde{\boldsymbol{f}}(\boldsymbol{d} + \boldsymbol{x} - \boldsymbol{y}, \boldsymbol{d}) = \widetilde{\boldsymbol{f}}(\boldsymbol{x}, \boldsymbol{y})$. The only way this can be satisfied for any point $\boldsymbol{d}$ is if $\boldsymbol{f}$ only depends on $\boldsymbol{x} - \boldsymbol{y}$, i.e.

$$\boldsymbol{f} = \widetilde{\boldsymbol{f}}(\boldsymbol{x}, \boldsymbol{y}) = \bar{\boldsymbol{f}}(\boldsymbol{x} - \boldsymbol{y}).$$

We have shown that the force that one particle exerts on another in a two-particle system can only depend on the *difference* between their positions. We can say more, though.

The condition for material frame-indifference for the new functional form, $\bar{\boldsymbol{f}}(\boldsymbol{x} - \boldsymbol{y})$, follows from Eqn. $(6.79)_2$ together with Eqn. $(6.80)_2$:

$$\bar{\boldsymbol{f}}(\boldsymbol{Q}\boldsymbol{u}) = \boldsymbol{Q}\bar{\boldsymbol{f}}(\boldsymbol{u}), \tag{6.85}$$

[23] Equation (6.79) is the one normally given in textbooks, but since most books write the relation from the start in a single frame of reference it is necessary to make the additional "assumptions" that the functional form of constitutive relations is the same in different frames (this is sometimes referred to as *form invariance*) and that $\boldsymbol{Q}$ is proper orthogonal. These assumptions are normally introduced without clear physical motivation. We see here that these additional assumptions are unnecessary and are merely a consequence of the mathematical "short-cut" of working in a single frame.

[24] This example is based on a derivation in [Nol04, page 18].

where $\boldsymbol{u} = \boldsymbol{x} - \boldsymbol{y}$ is the relative position vector. Equation (6.85) must be satisfied for all $\boldsymbol{Q} \in SO(n_d)$, so it must also be satisfied for the special case $\boldsymbol{Q}^*$ defined through the relation, $\boldsymbol{Q}^*\boldsymbol{u} = \boldsymbol{u}$, i.e. $\boldsymbol{Q}^*$ is a rotation about $\boldsymbol{u}$. Substituting $\boldsymbol{Q}^*$ into Eqn. (6.85) gives $\bar{\boldsymbol{f}}(\boldsymbol{u}) = \boldsymbol{Q}^*\bar{\boldsymbol{f}}(\boldsymbol{u})$. This is only satisfied if $\bar{\boldsymbol{f}}(\boldsymbol{u})$ is oriented along $\boldsymbol{u}$, so that

$$\boldsymbol{f} = \bar{\boldsymbol{f}}(\boldsymbol{u}) = \bar{\varphi}(\boldsymbol{u})\frac{\boldsymbol{u}}{\|\boldsymbol{u}\|}, \tag{6.86}$$

where $\bar{\varphi}(\boldsymbol{u})$ is a scalar function. Substituting Eqn. (6.86) into Eqn. (6.85), we obtain the material frame-indifference constraint on $\bar{\varphi}$:

$$\bar{\varphi}(\boldsymbol{Q}\boldsymbol{u}) = \bar{\varphi}(\boldsymbol{u}), \tag{6.87}$$

for all $\boldsymbol{Q} \in SO(n_d)$. Now, consider another vector $\boldsymbol{v} = \|\boldsymbol{u}\|\,\boldsymbol{e}$, where $\boldsymbol{e}$ is an arbitrary unit vector. Equation (6.87) must hold for any vector $\boldsymbol{u}$, so it must also hold for the new vector $\boldsymbol{v}$. Therefore,

$$\bar{\varphi}(\boldsymbol{Q}\,\|\boldsymbol{u}\|\,\boldsymbol{e}) = \bar{\varphi}(\|\boldsymbol{u}\|\,\boldsymbol{e}). \tag{6.88}$$

Equation (6.88) must hold for all $\boldsymbol{Q} \in SO(n_d)$, so it must also hold for the special case defined by $\boldsymbol{Q}\boldsymbol{e} = \boldsymbol{u}/\|\boldsymbol{u}\|$. Substituting this into Eqn. (6.88) gives $\bar{\varphi}(\boldsymbol{u}) = \bar{\varphi}(\|\boldsymbol{u}\|\,\boldsymbol{e})$. This must be true for all unit vectors $\boldsymbol{e}$, which is only possible if $\bar{\varphi}(\boldsymbol{u}) = \widehat{\varphi}(\|\boldsymbol{u}\|)$. Combining this result with the form in Eqn. (6.86), we have

$$\boldsymbol{f} = \widehat{\varphi}(\|\boldsymbol{u}\|)\frac{\boldsymbol{u}}{\|\boldsymbol{u}\|}. \tag{6.89}$$

Thus the force on a particle in a two-particle system can only depend on the *distance* between the particles and must be oriented along the line connecting them. This result implies that a two-particle system is conservative, since Eqn. (6.89) can be rewritten as the (negative) gradient of a scalar (energy) function:

$$\boldsymbol{f} = -\nabla_{\boldsymbol{x}}\widehat{e}(\|\boldsymbol{u}\|) = -\frac{\partial}{\partial \boldsymbol{x}}\widehat{e}(\|\boldsymbol{x}-\boldsymbol{y}\|) = -\widehat{e}'(\|\boldsymbol{x}-\boldsymbol{y}\|)\frac{\boldsymbol{x}-\boldsymbol{y}}{\|\boldsymbol{x}-\boldsymbol{y}\|},$$

so that $\widehat{\varphi}(\|\boldsymbol{u}\|) = -\widehat{e}'(\|\boldsymbol{u}\|)$. An extension to systems of more than two particles is discussed in Section 5.3.2 of [TM11] and in Appendix A of [TM11].

6.3.5 Reduced constitutive relations

We have established in Eqn. (6.79) together with Eqns. (6.80) and (6.81), the general framework for imposing constraints on constitutive relations. We now apply these constraints to the specific constitutive relation forms obtained earlier in Section 6.2. The resulting functional forms are called *reduced constitutive relations*.

Reduced internal energy density function The functional form of the internal energy density, given in Eqn. (6.6), is

$$u = \breve{u}(s, \boldsymbol{F}).$$

We require u to be objective, therefore according to Eqn. $(6.79)_1$ together with Eqns. $(6.80)_1$ and $(6.81)_1$ for the arguments s and $\boldsymbol{F}$, we have

$$\breve{u}(s, \boldsymbol{Q}\boldsymbol{F}) = \breve{u}(s, \boldsymbol{F}), \qquad \forall \boldsymbol{Q} \in SO(3). \tag{6.90}$$

Equation (6.90) places a constraint on the way that the function $\overline{u}$ can depend on $\boldsymbol{F}$. In fact, we can show that Eqn. (6.90) is satisfied if and only if the dependence of $\overline{u}$ on $\boldsymbol{F}$ is through the right stretch tensor, i.e.

$$u = \widehat{u}(s, \boldsymbol{U}). \tag{6.91}$$

Proof Assume u is objective, so that Eqn. (6.90) is satisfied. Substitute the right polar decomposition of $\boldsymbol{F}$ (Eqn. (3.10)), $\boldsymbol{F} = \boldsymbol{R}\boldsymbol{U}$, into the left-hand side of Eqn. (6.90):

$$\overline{u}(s, \boldsymbol{Q}\boldsymbol{R}\boldsymbol{U}) = \overline{u}(s, \boldsymbol{F}), \qquad \forall \boldsymbol{Q} \in SO(3). \tag{6.92}$$

Since Eqn. (6.92) is true for all $\boldsymbol{Q} \in SO(3)$, it must also be true for $\boldsymbol{Q} = \boldsymbol{R}^T$, since $\boldsymbol{R}$ is proper orthogonal. Substituting this into Eqn. (6.92) and noting that $\boldsymbol{R}^T\boldsymbol{R} = \boldsymbol{I}$, we have

$$\overline{u}(s, \boldsymbol{U}) = \overline{u}(s, \boldsymbol{F}).$$

This shows that the value of the internal energy density is determined by the value of $\boldsymbol{U}$ and implies the existence of the function $\widehat{u}(s, \boldsymbol{U})$. Using Eqn. (3.11), we can write the original function $\overline{u}$ in terms of this new function $\widehat{u}$:

$$\overline{u}(s, \boldsymbol{F}) \equiv \widehat{u}(s, \boldsymbol{R}^T\boldsymbol{F}) = \widehat{u}(s, \sqrt{\boldsymbol{F}^T\boldsymbol{F}}). \tag{6.93}$$

□

In the above proof we are careful to distinguish between the functional forms $\overline{u}(s, \boldsymbol{F})$ and $\widehat{u}(s, \boldsymbol{U})$. Although the two energy functions have the same value when evaluated at any given symmetric second-order tensor, the derivatives of the two functions with respect to their kinematic arguments are not equal. The former gives the first Piola–Kirchhoff stress tensor which is work conjugate to the deformation gradient, while the latter is a symmetric tensor that is work conjugate to the right stretch tensor. This indicates that they are, in fact, two distinct functional forms. We demonstrate this in the following example.

Example 6.3 (Energy constitutive relations) Let us consider a simple example, ignoring for the moment the dependence on the entropy density. Suppose

$$\widehat{u}(\boldsymbol{U}) = (\boldsymbol{L} : \boldsymbol{U}^2) : \boldsymbol{U}^2,$$

where $\boldsymbol{L}$ is a fourth-order tensor. Then according to Eqn. (6.93),

$$\overline{u}(\boldsymbol{F}) = (\boldsymbol{L} : (\boldsymbol{F}^T\boldsymbol{F})) : (\boldsymbol{F}^T\boldsymbol{F}).$$

We will show that $\overline{u}(\boldsymbol{U}^*) = \widehat{u}(\boldsymbol{U}^*)$, and $\partial\overline{u}/\partial\boldsymbol{F}|_{\boldsymbol{F}=\boldsymbol{U}^*} \neq \partial\widehat{u}/\partial\boldsymbol{U}|_{\boldsymbol{U}=\boldsymbol{U}^*}$ for any symmetric $\boldsymbol{U}^*$.
First, we may simply evaluate the functions at the value of $\boldsymbol{U}^*$:

$$\overline{u}(\boldsymbol{U}^*) = \left(\boldsymbol{L} : \left[(\boldsymbol{U}^*)^T\boldsymbol{U}^*\right]\right) : \left[(\boldsymbol{U}^*)^T\boldsymbol{U}^*\right] = (\boldsymbol{L} : (\boldsymbol{U}^*)^2) : (\boldsymbol{U}^*)^2 = \widehat{u}(\boldsymbol{U}^*),$$

where we have used the symmetry of $\boldsymbol{U}^*$ in going from the first to the second equality.

Second, we start by computing the derivatives of $\overline{u}$ and $\widehat{u}$. In indicial notation, we have

$$\begin{aligned}\frac{\partial \overline{u}}{\partial F_{iJ}} &= \frac{\partial}{\partial F_{iJ}}(L_{ABCD}(F_{aC}F_{aD})(F_{bA}F_{bB})) \\ &= L_{ABJD}F_{iD}F_{bA}F_{bB} + L_{ABCJ}F_{iC}F_{bA}F_{bB} \\ &\quad + L_{JBCD}F_{aC}F_{aD}F_{iB} + L_{AJCD}F_{aC}F_{aD}F_{iA}\end{aligned}$$

and

$$\begin{aligned}\frac{\partial \widehat{u}}{\partial U_{IJ}} &= \frac{\partial}{\partial U_{IJ}}(L_{ABCD}U_{CE}U_{ED}U_{AF}U_{FB}) \\ &= \frac{1}{2}L_{ABID}U_{JD}U_{AF}U_{FB} + \frac{1}{2}L_{ABJD}U_{ID}U_{AF}U_{FB} \\ &\quad + \frac{1}{2}L_{ABCJ}U_{CI}U_{AF}U_{FB} + \frac{1}{2}L_{ABCI}U_{CJ}U_{AF}U_{FB} \\ &\quad + \frac{1}{2}L_{IBCD}U_{CE}U_{ED}U_{JB} + \frac{1}{2}L_{JBCD}U_{CE}U_{ED}U_{IB} \\ &\quad + \frac{1}{2}L_{AJCD}U_{CE}U_{ED}U_{AI} + \frac{1}{2}L_{AICD}U_{CE}U_{ED}U_{AJ}.\end{aligned}$$

Above, we have used the facts that

$$\frac{\partial F_{iJ}}{\partial F_{kL}} = \delta_{ik}\delta_{JL} \quad \Leftrightarrow \quad \frac{\partial \boldsymbol{F}}{\partial \boldsymbol{F}} = \boldsymbol{I},$$

where $\boldsymbol{I}$ is the fourth-order identity tensor, and that

$$\frac{\partial U_{IJ}}{\partial U_{KL}} = \frac{1}{2}(\delta_{IK}\delta_{JL} + \delta_{IL}\delta_{JK}) \quad \Leftrightarrow \quad \frac{\partial \boldsymbol{U}}{\partial \boldsymbol{U}} = \boldsymbol{I}_{(s)},$$

where $\boldsymbol{I}_{(s)}$ is the "fourth-order symmetric identity" tensor (which accounts for the symmetry of $\boldsymbol{U}$). Now, it is easy to see, upon substituting $\boldsymbol{F} = \boldsymbol{U}^*$ and $\boldsymbol{U} = \boldsymbol{U}^*$ into the above equations, that the derivatives are not equal unless the tensor $\boldsymbol{L}$ has certain special symmetries.

We established above that u depends on deformation through the right stretch tensor. Since $\boldsymbol{U}$ is uniquely related to the right Cauchy–Green deformation tensor, $\boldsymbol{C} = \boldsymbol{U}^2$, and therefore also to the Lagrangian strain tensor, $\boldsymbol{E} = \frac{1}{2}(\boldsymbol{C} - \boldsymbol{I})$, we can also write

$$u = \widetilde{u}(s, \boldsymbol{C}) \quad \text{or} \quad u = \breve{u}(s, \boldsymbol{E}), \tag{6.94}$$

which are often more convenient forms to use in practice. The different accents over u indicate different functional forms.

The temperature follows from Eqn. (6.17) as

$$T = \frac{\partial \widehat{u}(s, \boldsymbol{U})}{\partial s} \quad \text{or} \quad T = \frac{\partial \widetilde{u}(s, \boldsymbol{C})}{\partial s} \quad \text{or} \quad T = \frac{\partial \breve{u}(s, \boldsymbol{E})}{\partial s}. \tag{6.95}$$

Reduced heat flux vector function The functional form of the heat flux vector, given in Eqn. (6.20), is

$$\boldsymbol{q} = \bar{\boldsymbol{q}}(s, \boldsymbol{F}, \boldsymbol{\tau}), \tag{6.96}$$

where $\boldsymbol{\tau} = \nabla T$ is the temperature gradient vector. We require $\boldsymbol{q}$ to be objective, therefore according to Eqn. $(6.79)_2$ together with Eqns. $(6.80)_1$, $(6.81)_1$ and $(6.80)_2$ for the arguments s, $\boldsymbol{F}$ and $\boldsymbol{\tau}$, we have

$$\bar{\boldsymbol{q}}(s, \boldsymbol{QF}, \boldsymbol{Q\tau}) = \boldsymbol{Q}\bar{\boldsymbol{q}}(s, \boldsymbol{F}, \boldsymbol{\tau}) \qquad \forall \boldsymbol{Q} \in SO(3).$$

Substituting $\boldsymbol{F} = \boldsymbol{RU}$ on the left, premultiplying by $\boldsymbol{Q}^T$ and rearranging, we have

$$\bar{\boldsymbol{q}}(s, \boldsymbol{F}, \boldsymbol{\tau}) = \boldsymbol{Q}^T \bar{\boldsymbol{q}}(s, \boldsymbol{QRU}, \boldsymbol{Q\tau}) \qquad \forall \boldsymbol{Q} \in SO(3). \tag{6.97}$$

Now, as we did for the internal energy density above, select the special case $\boldsymbol{Q} = \boldsymbol{R}^T$, then Eqn. (6.97) becomes

$$\bar{\boldsymbol{q}}(s, \boldsymbol{F}, \boldsymbol{\tau}) = \boldsymbol{R}\bar{\boldsymbol{q}}(s, \boldsymbol{U}, \boldsymbol{R}^T\boldsymbol{\tau}).$$

This indicates that we can identify a function $\widehat{\boldsymbol{q}}(s, \boldsymbol{U}, \boldsymbol{\tau})$ such that we have

$$\boldsymbol{q} = \boldsymbol{R}\widehat{\boldsymbol{q}}(s, \boldsymbol{U}, \boldsymbol{R}^T\nabla T). \tag{6.98}$$

This relation shows that the functional form for the heat flux vector has to depend in a very specific manner on the finite rotation part $\boldsymbol{R}$ of the polar decomposition of $\boldsymbol{F}$. Further progress can be made by assuming a linear relation between the temperature gradient and heat flux vector as suggested in Section 6.2.3. Consider, for example, the simplest linear heat flux relation (Fourier's law), $\boldsymbol{q} = -k\nabla T$, where k is the thermal conductivity of the material. Fourier's law is a special case of Eqn. (6.98), since

$$\boldsymbol{q} = -k\nabla T = \boldsymbol{R}(-k\boldsymbol{R}^T\nabla T). \tag{6.99}$$

In this case, the dependence on $\boldsymbol{R}$ drops out. However, for more general constitutive forms it does not. Finally, as for the internal energy density, Eqn. (6.98) can be more conveniently rewritten in terms of $\boldsymbol{C}$ or $\boldsymbol{E}$ instead of $\boldsymbol{U}$:

$$\boldsymbol{q} = \boldsymbol{R}\widetilde{\boldsymbol{q}}(s, \boldsymbol{C}, \boldsymbol{R}^T\nabla T) \qquad \text{or} \qquad \boldsymbol{q} = \boldsymbol{R}\check{\boldsymbol{q}}(s, \boldsymbol{E}, \boldsymbol{R}^T\nabla T). \tag{6.100}$$

Reduced elastic stress function The functional form of the elastic part of the stress tensor, given in Eqn. (6.26), is

$$\boldsymbol{\sigma}^{(e)} = \bar{\boldsymbol{\sigma}}^{(e)}(s, \boldsymbol{F}). \tag{6.101}$$

According to Eqn. $(6.79)_3$, in order for $\boldsymbol{\sigma}^{(e)}$ to be objective it must satisfy

$$\bar{\boldsymbol{\sigma}}^{(e)}(s, \boldsymbol{QF}) = \boldsymbol{Q}\bar{\boldsymbol{\sigma}}^{(e)}(s, \boldsymbol{F})\boldsymbol{Q}^T, \tag{6.102}$$

where we have used Eqns. $(6.80)_1$ and $(6.81)_1$ for the arguments s and $\boldsymbol{F}$ on the left-hand side. For the elastic part of the stress tensor $\boldsymbol{\sigma}^{(e)}$, the Coleman–Noll procedure does more than just identify the variables on which the constitutive relation depends. Equation (6.26) provides a specific functional form for $\boldsymbol{\sigma}^{(e)}$ in terms of the internal energy density:

$$\bar{\boldsymbol{\sigma}}^{(e)}(s, \boldsymbol{F}) = \rho \frac{\partial \bar{u}}{\partial \boldsymbol{F}} \boldsymbol{F}^T = \frac{\rho_0}{\det \boldsymbol{F}} \frac{\partial \bar{u}}{\partial \boldsymbol{F}} \boldsymbol{F}^T, \tag{6.103}$$

where $\rho = \rho_0 / \det \boldsymbol{F}$ due to conservation of mass (Eqn. (4.1)). Let us verify that $\boldsymbol{\sigma}^{(e)}$ defined in this way is objective.

Proof We need to show that the functional form of $\boldsymbol{\sigma}^{(e)}$ defined in Eqn. (6.103) satisfies Eqn. (6.102). Substituting Eqn. (6.103) into the left-hand side of Eqn. (6.102), we have

$$\bar{\boldsymbol{\sigma}}^{(e)}(s, \boldsymbol{QF}) = \frac{\rho_0}{\det \boldsymbol{QF}} \frac{\partial \bar{u}(s, \boldsymbol{QF})}{\partial (\boldsymbol{QF})} (\boldsymbol{QF})^T = \frac{\rho_0}{\det \boldsymbol{F}} \frac{\partial \bar{u}(s, \boldsymbol{QF})}{\partial (\boldsymbol{QF})} \boldsymbol{F}^T \boldsymbol{Q}^T, \tag{6.104}$$

where we have used the fact that $\det \boldsymbol{QF} = \det \boldsymbol{F}$, since $\boldsymbol{Q}$ is proper orthogonal. Now, from Eqn. (6.90), we have the following identity:

$$\frac{\partial \bar{u}(s, \boldsymbol{F})}{\partial \boldsymbol{F}} = \frac{\partial \bar{u}(s, \boldsymbol{QF})}{\partial \boldsymbol{F}} = \boldsymbol{Q}^T \frac{\partial \bar{u}(s, \boldsymbol{QF})}{\partial (\boldsymbol{QF})},$$

where the chain rule was used in the last step. Inverting this relation and substituting into Eqn. (6.104), we have

$$\bar{\boldsymbol{\sigma}}^{(e)}(s, \boldsymbol{QF}) = \boldsymbol{Q} \left(\frac{\rho_0}{\det \boldsymbol{F}} \frac{\partial \bar{u}(s, \boldsymbol{F})}{\partial \boldsymbol{F}} \boldsymbol{F}^T \right) \boldsymbol{Q}^T = \boldsymbol{Q} \bar{\boldsymbol{\sigma}}^{(e)}(s, \boldsymbol{F}) \boldsymbol{Q}^T,$$

which shows that Eqn. (6.102) is satisfied. □

Next, we demonstrate that $\boldsymbol{\sigma}^{(e)}$ is symmetric. We showed in Eqn. (6.91) that the dependence of $\bar{u}$ on $\boldsymbol{F}$ must be through $\boldsymbol{U}$ (or equivalently through $\boldsymbol{C}$). We therefore rewrite Eqn. (6.103) in terms of $\breve{u}(s, \boldsymbol{C})$ and apply the chain rule:

$$\begin{aligned}
\sigma_{ij}^{(e)} = \rho \frac{\partial \bar{u}}{\partial F_{iJ}} F_{jJ} &= \rho \frac{\partial \breve{u}}{\partial C_{MN}} \frac{\partial C_{MN}}{\partial F_{iJ}} F_{jJ} \\
&= \rho \frac{\partial \breve{u}}{\partial C_{MN}} \frac{\partial (F_{kM} F_{kN})}{\partial F_{iJ}} F_{jJ} \\
&= \rho \frac{\partial \breve{u}}{\partial C_{MN}} \left[\delta_{MJ} F_{iN} + \delta_{NJ} F_{iM} \right] F_{jJ} = 2\rho F_{iM} \frac{\partial \breve{u}}{\partial C_{MJ}} F_{jJ},
\end{aligned}$$

where in the last step we used the symmetry of $\boldsymbol{C}$. In direct notation the result is

$$\boldsymbol{\sigma}^{(e)} = 2\rho \boldsymbol{F} \frac{\partial \breve{u}(s, \boldsymbol{C})}{\partial \boldsymbol{C}} \boldsymbol{F}^T = \breve{\boldsymbol{\sigma}}^{(e)}(s, \boldsymbol{C}). \tag{6.105}$$

Equation (6.105) shows that the elastic part of the stress tensor is symmetric, since

$$\left(\boldsymbol{\sigma}^{(e)}\right)^T = 2\rho \left(\boldsymbol{F} \frac{\partial \breve{u}}{\partial \boldsymbol{C}} \boldsymbol{F}^T \right)^T = 2\rho \boldsymbol{F} \left(\frac{\partial \breve{u}}{\partial \boldsymbol{C}} \right)^T \boldsymbol{F}^T = 2\rho \boldsymbol{F} \frac{\partial \breve{u}}{\partial \boldsymbol{C}} \boldsymbol{F}^T = \boldsymbol{\sigma}^{(e)},$$

where the symmetry of $\boldsymbol{C}$ has been used. This result also establishes the symmetry of the viscous part of the stress, $\boldsymbol{\sigma}^{(v)}$, since the total stress, $\boldsymbol{\sigma} = \boldsymbol{\sigma}^{(e)} + \boldsymbol{\sigma}^{(v)}$, is symmetric due to the balance of angular momentum. We assumed the symmetry of the viscous stress earlier in the derivation of Eqn. (6.25). Our result here verifies the correctness of that assumption.

The stress expression derived above was obtained for an isentropic process where the internal energy density is the appropriate energy variable. More commonly, experiments are performed under isothermal conditions for which the specific Helmholtz free energy ψ must be used, or more conveniently the strain energy density W defined in Eqn. (6.42). Replacing u with ψ in Eqn. (6.105) and using Eqn. (4.1), we have

$$\boldsymbol{\sigma}^{(e)} = \frac{2}{J}\boldsymbol{F}\frac{\partial \widetilde{W}(T,\boldsymbol{C})}{\partial \boldsymbol{C}}\boldsymbol{F}^T, \tag{6.106}$$

where $J = \det \boldsymbol{F}$ is the Jacobian of the deformation. The reference stress variables follow from Eqns. (4.35) and (4.41) as

$$\boldsymbol{P}^{(e)} = 2\boldsymbol{F}\frac{\partial \widetilde{W}(T,\boldsymbol{C})}{\partial \boldsymbol{C}} \quad \text{or} \quad \boldsymbol{S}^{(e)} = 2\frac{\partial \widetilde{W}(T,\boldsymbol{C})}{\partial \boldsymbol{C}}. \tag{6.107}$$

Reduced viscous stress function The functional form of the viscous part of the stress tensor, given in Eqn. (6.27), is

$$\boldsymbol{\sigma}^{(v)} = \bar{\boldsymbol{\sigma}}^{(v)}(s, \boldsymbol{F}, \boldsymbol{d}). \tag{6.108}$$

We require $\boldsymbol{\sigma}^{(v)}$ to be objective, therefore according to Eqn. $(6.79)_3$ together with Eqns. $(6.80)_1$, $(6.81)_1$ and $(6.80)_3$ for the arguments s, $\boldsymbol{F}$ and $\boldsymbol{d}$, we have

$$\bar{\boldsymbol{\sigma}}^{(v)}(s, \boldsymbol{QF}, \boldsymbol{QdQ}^T) = \boldsymbol{Q}\bar{\boldsymbol{\sigma}}^{(v)}(s, \boldsymbol{F}, \boldsymbol{d})\boldsymbol{Q}^T.$$

Following an analogous procedure to the one used in deriving the reduced heat flux constitutive relation, we find

$$\boldsymbol{\sigma}^{(v)} = \boldsymbol{R}\hat{\boldsymbol{\sigma}}^{(v)}(s, \boldsymbol{U}, \boldsymbol{R}^T\boldsymbol{dR})\boldsymbol{R}^T. \tag{6.109}$$

Thus, as before, the constitutive relation involves a function with arbitrary dependence on its arguments together with an explicit dependence on $\boldsymbol{R}$. The simplest constitutive relation that satisfies this relation is a linear response model where the components of $\boldsymbol{\sigma}^{(v)}$ are proportional to those of $\boldsymbol{d}$ (Newtonian fluid) as suggested in Section 6.2.3. In this case the $\boldsymbol{R}$ terms cancel out. In more complex models the explicit dependence on $\boldsymbol{R}$ remains.

Finally, more convenient forms of Eqn. (6.109) in terms of C and E are

$$\sigma^{(v)} = R\tilde{\sigma}^{(v)}(s, C, R^T dR)R^T \quad \text{and} \quad \sigma^{(v)} = R\breve{\sigma}^{(v)}(s, E, R^T dR)R^T. \tag{6.110}$$

6.3.6 Continuum field equations and material frame-indifference

The complete set of field equations that a continuum body must satisfy are summarized on page 180 at the start of this chapter. It can be shown that the continuity equation, balance of angular momentum and energy equation are all frame indifferent (see, for example, [Cha99, Section 4.3]) and can therefore be written in any frame of reference. However, we know from Section 2.1 that the balance of linear momentum,

$$\operatorname{div} \sigma + \rho b = \rho a, \tag{6.111}$$

where $a = \ddot{x}$, only holds in an inertial frame of reference. Let us assume that $\mathcal{F}$ is an inertial frame and transform the balance of linear momentum to a noninertial frame $\mathcal{F}^+$. Since σ is objective, we can show that the first term in Eqn. (6.111) is an objective vector:

$$\operatorname{div}_+ \sigma^+ = \operatorname{div}_+ \mathcal{Q}_t \sigma \mathcal{Q}_t^T = \mathcal{Q}_t \left(\frac{\partial \sigma}{\partial x} \frac{\partial x}{\partial x^+} \right) \mathcal{Q}_t = \mathcal{Q}_t \left(\frac{\partial \sigma}{\partial x} \mathcal{Q}_t^T \right) \mathcal{Q}_t = \mathcal{Q}_t \operatorname{div} \sigma, \tag{6.112}$$

where we have used Eqns. (6.54) and (6.66). Applying $\mathcal{Q}_t$ from the left to Eqn. (6.111) and using Eqns. (6.52) and (6.112) and $\rho = \rho^+$ gives

$$\operatorname{div}_+ \sigma^+ + \rho^+ \mathcal{Q}_t b = \rho^+ (a^+ - \ddot{c}^+ - \ddot{\mathcal{Q}}_t x - 2\dot{\mathcal{Q}}_t v).$$

This equation can be made to look like Eqn. (6.111),

$$\operatorname{div}_+ \sigma^+ + \rho^+ b^* = \rho^+ a^+ \tag{6.113}$$

where $a^+ = \ddot{x}^+$, by defining a special body force

$$b^* = \mathcal{Q}_t b + \ddot{c}^+ + \ddot{\mathcal{Q}}_t x + 2\dot{\mathcal{Q}}_t v. \tag{6.114}$$

The first term is the external body force and the remaining terms are "fictitious" forces resulting from the motion of the frame of reference. Thus, the motion of a body in a noninertial frame can be represented as motion in an inertial frame where the additional fictitious forces are treated as though they were real.[25]

6.3.7 Controversy regarding the principle of material frame-indifference

In a study in 1972, Ingo Müller [Mül72] demonstrated that constitutive relations derived from the kinetic theory of gases violated the principle of material frame-indifference.

[25] These fictitious forces are analogous to the Coriolis forces that appear in the analysis of rigid-body motion in a rotating (noninertial) frame of reference.

Briefly, the kinetic theory of gases is a microscopic theory that describes a gas in terms of a distribution function $\hat{f}(\boldsymbol{x}, \boldsymbol{v}, t)$, where $\hat{f}(\boldsymbol{x}, \boldsymbol{v}, t) d\boldsymbol{x} d\boldsymbol{v}$ is the probability of finding an atom with a velocity within $d\boldsymbol{v}$ of $\boldsymbol{v}$ and a position within $d\boldsymbol{x}$ of $\boldsymbol{x}$ at time t. The evolution of $\hat{f}$ is governed by the Boltzmann equation that explicitly considers collisions between gas particles. The kinetic theory leads to expressions for stress and heat flux, which depend on the rotation of the gas and are therefore "frame dependent." We do not discuss further the kinetic theory of gases in this book.

Müller's study sparked off a long series of articles arguing whether his results constitute a failure of the principle of material frame-indifference. There were more calculations based on kinetic theory [EM73, Wan75, Mül76, Tru76, Söd76, HM83, Woo83, Mur83, Ban84, Duf84, Spe87, SK95] and on molecular dynamics [HMML81, EH89], as well as a series of theoretical discussions (only some of which are cited here) [Lum70, AF73, BM80, Spe81, Mur82, BdGH83, Rys85, Eu85, LBCJ86, Eu86, AK86, Mat86, Kem89, SH96, SB99, Mur03, Liu04, Mur05, Liu05, MR08, Fre09]. In our opinion, Müller's results do not constitute a failure of material frame-indifference, but rather a more basic failure of continuum mechanics itself, and in particular the idea that it is possible to describe the response of a material using constitutive relations that are *intrinsic* to the material, where "intrinsic" means that the relations are not affected by whether the motion of the material is represented in an inertial frame or not.

The basic issue is that materials, which are composed of a large number of particles undergoing dynamical motion (not to mention the electromagnetic fields associated with the electronic structure of the material), are inherently not frame-indifferent. The constitutive relations of continuum mechanics provide an idealized description of materials. It is reasonable to require that these constitutive relations should be frame-indifferent, but there is no reason to expect that the real material shares this property.

As an example, let us consider the simplest case of a spring connected between two points in physical space located at $\boldsymbol{x}$ and $\boldsymbol{y}$ in frame $\mathcal{F}$ (as shown in Fig. 6.1). We used this example to motivate the idea that material frame-indifference leads to constraints on constitutive relations. In Example 6.2, we showed that the constitutive relation for the spring, i.e. the force in the spring based on the position of its end points, must be proportional to its length and oriented along it. Clearly, a constitutive relation phrased in this way is independent of the motion of the frame in which it is expressed. However, an actual spring is not a function that returns a force based on the distance between its ends, but rather a material consisting of a huge number of atoms arranged in a coiled structure that vibrate about their positions with a mean kinetic energy determined by the temperature of the spring. The dynamical motion of these atoms is governed by Newton's second law which is not frame-indifferent since acceleration is not an objective variable.

From the above discussion it is clear that there are no grounds for expecting a real material to be frame-indifferent. In that sense the term "material frame-indifference" is a bit misleading. A better term might be "constitutive frame-indifference." However, normally, the macroscopic motions associated with continuum deformation are so slow relative to the microscopic scales that this effect is negligible and continuum constitutive relations provide an excellent model for the behavior of a real material irrespective of the frame of reference. Let us consider, though, what a failure of material frame-indifference means.

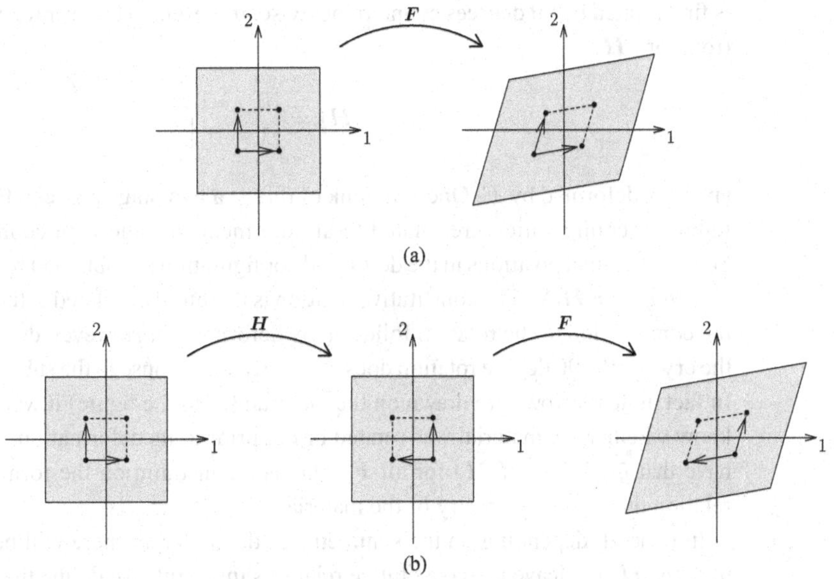

Fig. 6.2 A two-dimensional example of material symmetry. A material with a square lattice structure is (a) subjected to a homogeneous deformation $\boldsymbol{F}$, or (b) first subjected to a rotation $\boldsymbol{H}$ by 90 degrees in the counterclockwise direction and then deformed by $\boldsymbol{F}$. The constitutive response of the material is the same in both cases due to the symmetry of the crystal structure.

In this case, the basic separation between the continuum balance laws and the constitutive relations that describe material response breaks down. The underlying dynamical behavior of the material itself becomes important and must be solved together with the dynamics of the overall system. This constitutes a basic failure of the entire continuum mechanics framework. In that sense, one can argue that the principle of material frame-indifference, being part and parcel of the continuum world view, cannot fail within this context.

6.4 Material symmetry

Most materials possess certain symmetries which are reflected by their constitutive relations. Consider, for example, the deformation of a material with a two-dimensional square lattice structure as shown in Fig. 6.2.[26] The unit cell and lattice vectors of the crystal are shown. In Fig. 6.2(a), the material is uniformly deformed with a deformation gradient $\boldsymbol{F}$, so that a particle $\boldsymbol{X}$ in the reference configuration is mapped to $\boldsymbol{x} = \boldsymbol{F}\boldsymbol{X}$ in the deformed configuration. The response of the material to the deformation is given by a constitutive relation, $\bar{g}(\boldsymbol{F})$, where $\bar{g}$ can be the internal energy density function $\bar{u}$, the temperature function $\bar{T}$, etc. Now consider a second scenario, illustrated in Fig. 6.2(b), where the material

[26] The concepts of lattice vectors and crystal structures are discussed extensively in Chapter 3 of [TM11]. Here we will assume the reader has some basic familiarity with these concepts.

is first rotated by 90 degrees counterclockwise, represented by the proper orthogonal tensor (rotation) $\boldsymbol{H}$,

$$[\boldsymbol{H}] = \begin{bmatrix} 0 & -1 \\ 1 & 0 \end{bmatrix},$$

and then deformed by $\boldsymbol{F}$. One can think of this as a two-stage process. First, particles in the reference configuration are rotated to an intermediate stage with coordinates $\boldsymbol{y} = \boldsymbol{H}\boldsymbol{X}$. Second, the final positions in the deformed configuration are obtained by applying $\boldsymbol{F}$, so that $\boldsymbol{x} = \boldsymbol{F}\boldsymbol{y} = \boldsymbol{F}\boldsymbol{H}\boldsymbol{X}$. The constitutive relation is therefore evaluated at the deformation $\boldsymbol{F}\boldsymbol{H}$, the composition of the rotation followed by deformation. However, due to the symmetry of the crystal, the 90 degree rotation does not affect its response to the subsequent deformation. In fact, unless arrows are drawn on the material (as in the figure) it would be impossible to know whether the material was rotated or not prior to its deformation. Therefore, we must have that $\breve{g}(\boldsymbol{F}) = \breve{g}(\boldsymbol{F}\boldsymbol{H})$ for all $\boldsymbol{F}$. This is a constraint on the form of the constitutive relation due to the symmetry of the material.

In general, depending on the symmetry of the material, there will be multiple transformations $\boldsymbol{H}$ that leave the constitutive relations invariant. We define the *material symmetry group* G of a material as the set of uniform density-preserving changes of its reference configuration that leave all of its constitutive relations unchanged [CN63]. Thus, G is the set of all second-order tensors $\boldsymbol{H}$ for which $\det \boldsymbol{H} = 1$ (density-preserving) and for which

$$\begin{aligned} \breve{u}(s, \boldsymbol{F}) &= \breve{u}(s, \boldsymbol{F}\boldsymbol{H}), \\ \breve{T}(s, \boldsymbol{F}) &= \breve{T}(s, \boldsymbol{F}\boldsymbol{H}), \\ \breve{q}(s, \boldsymbol{F}, \nabla T) &= \breve{q}(s, \boldsymbol{F}\boldsymbol{H}, \nabla T), \\ \breve{\boldsymbol{\sigma}}^{(e)}(s, \boldsymbol{F}) &= \breve{\boldsymbol{\sigma}}^{(e)}(s, \boldsymbol{F}\boldsymbol{H}), \\ \breve{\boldsymbol{\sigma}}^{(v)}(s, \boldsymbol{F}, \boldsymbol{d}) &= \breve{\boldsymbol{\sigma}}^{(v)}(s, \boldsymbol{F}\boldsymbol{H}, \boldsymbol{d}), \end{aligned} \tag{6.115}$$

for all s, ∇T, $\boldsymbol{d}$ and $\boldsymbol{F}$ (i.e. all second-order tensors with positive determinants). Note that the symmetry relations for mixed and material tensors take slightly different forms than those shown in Eqn. (6.115). For example, the relations for the elastic part of the first and second Piola–Kirchhoff stress tensors are

$$\breve{\boldsymbol{P}}^{(e)}(s, \boldsymbol{F}) = \breve{\boldsymbol{P}}^{(e)}(s, \boldsymbol{F}\boldsymbol{H})\boldsymbol{H}^T \quad \text{and} \quad \breve{\boldsymbol{S}}^{(e)}(s, \boldsymbol{F}) = \boldsymbol{H}\breve{\boldsymbol{S}}^{(e)}(s, \boldsymbol{F}\boldsymbol{H})\boldsymbol{H}^T.$$

These may be obtained directly from Eqn. (6.115) by substituting Eqns. (4.36) and (4.42).[27]

The concept of a *group* was defined on page 32. It is straightforward to prove that G constitutes a group. We need to show that G is closed with respect to tensor multiplication and that it satisfies the properties of associativity, existence of an identity and existence of an inverse element. We show the proof below for a generic constitutive relation $\breve{g}(\boldsymbol{F})$, which represents any of the relations in Eqn. (6.115).

[27] When substituting on the right-hand side of Eqn. (6.115) do not forget that Eqns. (4.36) and (4.42) relate $\breve{\boldsymbol{\sigma}}(s, \boldsymbol{F})$ to $\breve{\boldsymbol{P}}(s, \boldsymbol{F})$ and $\breve{\boldsymbol{S}}(s, \boldsymbol{F})$, respectively.

Proof For G to be closed with respect to tensor multiplication, we need to show that $\forall \boldsymbol{H}_1, \boldsymbol{H}_2 \in G$ we have $\boldsymbol{H}_1\boldsymbol{H}_2 \in G$. This means that we need to show that:

(i) $\det(\boldsymbol{H}_1\boldsymbol{H}_2) = 1$,
(ii) $\bar{g}(\boldsymbol{F}) = \bar{g}(\boldsymbol{F}\boldsymbol{H}_1\boldsymbol{H}_2)$.

For (i), by definition, $\det \boldsymbol{H}_1\boldsymbol{H}_2 = \det \boldsymbol{H}_1 \det \boldsymbol{H}_2 = 1 \cdot 1 = 1$. For (ii), denote $\boldsymbol{K} \equiv \boldsymbol{F}\boldsymbol{H}_1$. Note that $\boldsymbol{K}$ has a positive determinant and therefore the material symmetry operations must hold for $\boldsymbol{K}$ just as for $\boldsymbol{F}$. Therefore $\bar{g}(\boldsymbol{K}\boldsymbol{H}_2) = \bar{g}(\boldsymbol{K})$ since $\boldsymbol{H}_2 \in G$. Substituting in the definition of $\boldsymbol{K}$ gives $\bar{g}(\boldsymbol{F}\boldsymbol{H}_1\boldsymbol{H}_2) = \bar{g}(\boldsymbol{F}\boldsymbol{H}_1) = \bar{g}(\boldsymbol{F})$, where the last equality follows since $\boldsymbol{H}_1 \in G$. Thus, we have shown that $\boldsymbol{H}_1\boldsymbol{H}_2 \in G$.

The remaining three properties are also satisfied. Associativity is satisfied because tensor multiplication is associative. The identity element is $\boldsymbol{I}$. The inverse element is guaranteed to exist $\forall \boldsymbol{H} \in G$ since $\det \boldsymbol{H} \neq 0$. Further, it belongs to G since $\det \boldsymbol{H}^{-1} = (\det \boldsymbol{H})^{-1} = 1$ and, denoting $\boldsymbol{L} \equiv \boldsymbol{F}\boldsymbol{H}^{-1}$, we have $\bar{g}(\boldsymbol{L}) = \bar{g}(\boldsymbol{L}\boldsymbol{H})$ since $\boldsymbol{H} \in G$. Substituting in the definition of $\boldsymbol{L}$ gives $\bar{g}(\boldsymbol{F}\boldsymbol{H}^{-1}) = \bar{g}(\boldsymbol{F}\boldsymbol{H}^{-1}\boldsymbol{H}) = \bar{g}(\boldsymbol{F})$. Thus, indeed, $\boldsymbol{H}^{-1} \in G$. □

The largest possible material symmetry group is the *proper unimodular group* $SL(3)$ (also called the *special linear group*), which is the set of all tensors with determinant equal to $+1$. This material symmetry group describes simple fluids which can be subjected to any density-preserving deformation without a change to their constitutive response. Note that if a tensor is proper unimodular this does not imply that it is also proper orthogonal. For example, in two dimensions, the tensor with components

$$[\boldsymbol{A}] = \begin{bmatrix} 1 & 2 \\ 3 & 7 \end{bmatrix}$$

is proper unimodular, since $\det \boldsymbol{A} = 1$, but it is not proper orthogonal, since $\boldsymbol{A}^T \neq \boldsymbol{A}^{-1}$.

An important material symmetry group for solids is the *proper orthogonal group* $SO(3)$ already encountered in Section 2.5.1. A member of this group represents a rigid-body rotation of the material. Materials possessing this symmetry are *isotropic*. They have the same constitutive response regardless of how they are rotated before being deformed.

The smallest possible material symmetry group is the set that contains only the identity tensor $\boldsymbol{I}$. This is the case for a material that possesses no symmetries. Crystals with this property are called *triclinic*. Other crystals lie between the triclinic and isotropic limiting cases. See Section 6.5.1 for the effect of symmetry on the elastic constants of crystals. Also, see Chapter 3 of [TM11] for a detailed discussion of the symmetries associated with different crystal structures.

Below, we apply the symmetry constraints together with material frame-indifference for the two special cases of a simple fluid and an isotropic elastic solid, and derive simplified stress constitutive relations. In the linear limit, these relations reduce to the well-known linear viscosity relation (Newton's law) for a Newtonian fluid and Hooke's law for a linear elastic isotropic solid.

6.4.1 Simple fluids

A *simple fluid* is a simple material whose material symmetry group coincides with the full proper unimodular group, $G = SL(3)$. As noted above, this means that a simple fluid can undergo any density-preserving change to its reference configuration without affecting its constitutive response. For example, fish swimming in a level aquarium or one that has been tilted are expected to report the same constitutive experience. This is consistent with our concept of a fluid.

Our objective is to derive the most general form for the constitutive relation for the Cauchy stress of a simple fluid. We treat the elastic and viscous parts of the stress separately. The material symmetry condition for the elastic stress is

$$\bar{\boldsymbol{\sigma}}^{(e)}(\boldsymbol{F}) = \bar{\boldsymbol{\sigma}}^{(e)}(\boldsymbol{F}\boldsymbol{H}), \tag{6.116}$$

for all $\boldsymbol{H} \in SL(3)$. Note that to reduce clutter, we have dropped the explicit dependence of $\bar{\boldsymbol{\sigma}}^{(e)}$ on other variables which do not play a role in this derivation. Equation (6.116) must hold for all $\boldsymbol{H} \in SL(3)$, therefore it must also hold for the following particular choice:

$$\boldsymbol{H}^* = J^{1/3}\boldsymbol{F}^{-1}, \tag{6.117}$$

where $J = \det F$ is the Jacobian. This is a valid choice since $\det \boldsymbol{H}^* = \det(J^{1/3}\boldsymbol{F}^{-1}) = J \det \boldsymbol{F}^{-1} = J/\det \boldsymbol{F} = J/J = 1$ so $\boldsymbol{H}^* \in SL(3)$. Substituting Eqn. (6.117) into Eqn. (6.116) gives

$$\bar{\boldsymbol{\sigma}}^{(e)}(\boldsymbol{F}) = \bar{\boldsymbol{\sigma}}^{(e)}(\boldsymbol{F}\boldsymbol{H}^*) = \bar{\boldsymbol{\sigma}}^{(e)}(\boldsymbol{F}J^{1/3}\boldsymbol{F}^{-1}) = \bar{\boldsymbol{\sigma}}^{(e)}(J^{1/3}\boldsymbol{I}). \tag{6.118}$$

We see that $\widehat{\boldsymbol{\sigma}}^{(e)}$ can only depend on $\boldsymbol{F}$ through its scalar invariant J. This makes sense, since J is unaffected by density-preserving transformations.

Equation (6.118) places a strong constraint on the form of the elastic stress function for a simple fluid. We can go even further, though, by considering the implications of material frame-indifference for this case. The material frame-indifference constraint for the elastic stress tensor is $\boldsymbol{Q}\bar{\boldsymbol{\sigma}}^{(e)}(\boldsymbol{F})\boldsymbol{Q}^T = \bar{\boldsymbol{\sigma}}^{(e)}(\boldsymbol{Q}\boldsymbol{F})$ for all $\boldsymbol{Q} \in SO(3)$ (Eqn. (6.102)). Substituting in the functional dependence implied by Eqn. (6.118), we have

$$\boldsymbol{Q}\bar{\boldsymbol{\sigma}}^{(e)}(J^{1/3}\boldsymbol{I})\boldsymbol{Q}^T = \bar{\boldsymbol{\sigma}}^{(e)}((\det \boldsymbol{Q}\boldsymbol{F})^{1/3}\boldsymbol{I}) = \bar{\boldsymbol{\sigma}}^{(e)}(J^{1/3}\boldsymbol{I}), \tag{6.119}$$

where we have used $\det \boldsymbol{Q} = 1$. Equation (6.119) is satisfied for all proper orthogonal $\boldsymbol{Q}$, which means that $\bar{\boldsymbol{\sigma}}^{(e)}(J^{1/3}\boldsymbol{I})$ is an isotropic tensor. We showed in Section 2.5.6 that a second-order isotropic tensor is proportional to the identity tensor. Therefore, the elastic stress of a simple fluid must have the following form:

$$\boldsymbol{\sigma}^{(e)} = \bar{f}(J)\boldsymbol{I}, \tag{6.120}$$

where the $1/3$ power is absorbed into the arbitrary functional form $\bar{f}(J)$. According to the material form of the conservation of mass in Eqn. (4.1), $J = \rho_0/\rho$, so Eqn. (6.120) can be written in the generally more convenient form:

$$\boldsymbol{\sigma}^{(e)} = \widehat{f}(\rho)\boldsymbol{I}. \tag{6.121}$$

We can draw two important conclusions from Eqn. (6.121) for the special case of a simple *elastic* fluid, i.e. one that does not support viscous stresses so that $\boldsymbol{\sigma} = \boldsymbol{\sigma}^{(e)}$:

1. Simple elastic fluids are incapable of sustaining shear stresses, since

$$\sigma_{ij}^{(e)} = 0 \qquad \forall i \neq j.$$

2. The pressure p in a simple elastic fluid depends on the local density,

$$p = -\frac{1}{3}\operatorname{tr}\boldsymbol{\sigma}^{(e)} = -\widehat{f}(\rho).$$

In fact, we see that $\widehat{f}(\rho)$ is just the (negative) pressure function.

Next, we turn to the viscous part of the stress of a simple fluid. A similar procedure to the one followed for the elastic stress leads to the following constraint due to material symmetry and material frame-indifference on the form of the viscous stress:

$$\bar{\boldsymbol{\sigma}}^{(v)}(J^{1/3}\boldsymbol{I}, \boldsymbol{Q}\boldsymbol{d}\boldsymbol{Q}^T) = \boldsymbol{Q}\bar{\boldsymbol{\sigma}}^{(v)}(J^{1/3}\boldsymbol{I}, \boldsymbol{d})\boldsymbol{Q}^T. \tag{6.122}$$

Focusing on the dependence on $\boldsymbol{d}$, we see that

$$\bar{\boldsymbol{\sigma}}^{(v)}(\boldsymbol{Q}\boldsymbol{d}\boldsymbol{Q}^T) = \boldsymbol{Q}\bar{\boldsymbol{\sigma}}^{(v)}(\boldsymbol{d})\boldsymbol{Q}^T.$$

A function satisfying this condition is called an *isotropic tensor function*. Note that this is different from the condition in Eqn. (6.119) which defines an isotropic tensor. It can be shown that an isotropic tensor function can be represented in the following form [Gur81, Section 37]:

$$\bar{\boldsymbol{\sigma}}^{(v)}(\boldsymbol{d}) = \eta_0\boldsymbol{I} + \eta_1\boldsymbol{d} + \eta_2\boldsymbol{d}^2,$$

where η_i are arbitrary scalar functions of the principal invariants of $\boldsymbol{d}$,

$$I_1^{\boldsymbol{d}} = \operatorname{tr}\boldsymbol{d}, \qquad I_2^{\boldsymbol{d}} = \frac{1}{2}\left[(\operatorname{tr}\boldsymbol{d})^2 - \operatorname{tr}\boldsymbol{d}^2\right], \qquad I_3^{\boldsymbol{d}} = \det\boldsymbol{d}.$$

Given the dependence of $\bar{\boldsymbol{\sigma}}^{(v)}$ on the Jacobian, the scalar functions in the representation of the viscous stress can also depend on J or equivalently on the density ρ. So, in general,

$$\boldsymbol{\sigma}^{(v)} = \widehat{\varphi}_0(\rho, I_i^{\boldsymbol{d}})\boldsymbol{I} + \widehat{\varphi}_1(\rho, I_i^{\boldsymbol{d}})\boldsymbol{d} + \widehat{\varphi}_2(\rho, I_i^{\boldsymbol{d}})\boldsymbol{d}^2. \tag{6.123}$$

Adding the elastic and viscous parts of the stress in Eqns. (6.121) and (6.123), we have

$$\boldsymbol{\sigma} = \left[\widehat{f}(\rho) + \widehat{\varphi}_0(\rho, I_i^{\boldsymbol{d}})\right] \boldsymbol{I} + \widehat{\varphi}_1(\rho, I_i^{\boldsymbol{d}})\boldsymbol{d} + \widehat{\varphi}_2(\rho, I_i^{\boldsymbol{d}})\boldsymbol{d}^2. \qquad (6.124)$$

Equation (6.124) is the most general possible form for the constitutive relation of a simple fluid. A fluid of this type is called a *Reiner–Rivlin fluid*. When the dependence on $\boldsymbol{d}$ is linear, we obtain as a special case a *Newtonian fluid*, for which

$$\boldsymbol{\sigma} = \widehat{f}(\rho)\boldsymbol{I} + \left[\kappa(\rho) - \frac{2}{3}\mu(\rho)\right](\text{tr}\,\boldsymbol{d})\boldsymbol{I} + 2\mu(\rho)\boldsymbol{d}, \qquad (6.125)$$

where κ is the *bulk viscosity* and μ is the *shear viscosity*, which are material parameters that in general can depend on density (see Exercise 6.9). Equation (6.125) provides an adequate approximation for many fluids including water and air. For the even simpler case of an incompressible fluid, Eqn. (6.125) reduces to *Newton's law*,

$$\boldsymbol{\sigma} = 2\mu\boldsymbol{d} - p\boldsymbol{I},$$

which is the functional form that Isaac Newton proposed in 1687, giving this type of fluid its name [TG06]. Here, p is an undetermined hydrostatic pressure whose value can only be obtained as part of the solution to a boundary-value problem. (For an example of such a procedure in the case of incompressible solids, see Section 8.2.)

When Reiner [Rei45] and Rivlin [Riv47] derived the constitutive form in Eqn. (6.124) in the mid-1940s it was hoped that this form could provide a general framework for complex fluids that could not be adequately described as Newtonian fluids. However, experimental studies have shown that the Reiner–Rivlin form is inadequate and that in fact all fluids that can be described in that form are actually Newtonian [Cha99]. This indicates that memory effects play an important role in the flow of complex fluids. The Reiner–Rivlin material is based on the simple viscous stress assumption in Eqn. (6.27) that depends on the rate of deformation tensor, but not its history. More sophisticated models account for memory effects either by including a dependence on the time derivatives of the rate of deformation tensor (analogous to the approach used in the strain gradient theories described in Section 6.2.1) or by having $\boldsymbol{\sigma}$ directly dependent on the history of the deformation, for example,

$$\boldsymbol{\sigma}(t) = \int_{-\infty}^{t} G(t - t')\boldsymbol{d}(t')\,dt',$$

where t is the current time and $G(t - t')$ is called the *relaxation modulus*. This approach is analogous to spatially nonlocal constitutive relations such as Eringen's model in Eqn. (6.4). For a more detailed discussion of constitutive relations for complex fluids, see, for example, [TG06, Section 3.3].

6.4.2 Isotropic solids

A *simple isotropic material* is a simple material (see page 185) whose material symmetry group coincides with the proper orthogonal group, $G = SO(3)$. As noted above, this means that an arbitrary rigid-body rotation can be applied to the reference configuration without affecting the constitutive response of the material. Crystalline materials are not isotropic at the level of a single crystal, however, at the continuum level many materials appear isotropic since the response at a point represents an average over a large number of randomly oriented single crystals (or grains). We focus here on simple elastic materials,[28] where $\boldsymbol{\sigma} = \boldsymbol{\sigma}^{(e)}$. For this reason we drop the superscript on the stress terms in the following derivation.

The material symmetry condition for the stress of an isotropic elastic solid is

$$\overline{\boldsymbol{\sigma}}(\boldsymbol{F}) = \overline{\boldsymbol{\sigma}}(\boldsymbol{F}\boldsymbol{H}), \tag{6.126}$$

for all $\boldsymbol{H} \in SO(3)$. Substituting the left polar decomposition (Eqn. (3.10)), $\boldsymbol{F} = \boldsymbol{V}\boldsymbol{R}$, into the right-hand side of Eqn. (6.126) gives

$$\overline{\boldsymbol{\sigma}}(\boldsymbol{F}) = \overline{\boldsymbol{\sigma}}(\boldsymbol{V}\boldsymbol{R}\boldsymbol{H}).$$

This relation is true for all $\boldsymbol{H} \in SO(3)$, so it is also true for $\boldsymbol{H} = \boldsymbol{R}^T$, since $\boldsymbol{R} \in SO(3)$. Therefore,

$$\overline{\boldsymbol{\sigma}}(\boldsymbol{F}) = \overline{\boldsymbol{\sigma}}(\boldsymbol{V}). \tag{6.127}$$

We see that the stress can only depend on $\boldsymbol{F}$ through the left stretch tensor. This makes sense, since $\boldsymbol{V}$ is insensitive to rotations of the reference configuration. Equation (6.127) implies the existence of a function $\widehat{\boldsymbol{\sigma}}(\boldsymbol{B})$ that depends only on the left Cauchy–Green tensor, $\boldsymbol{B} = \boldsymbol{F}\boldsymbol{F}^T = \boldsymbol{V}^2$. Thus, we can write

$$\overline{\boldsymbol{\sigma}}(\boldsymbol{F}) = \widehat{\boldsymbol{\sigma}}(\boldsymbol{F}\boldsymbol{F}^T) = \widehat{\boldsymbol{\sigma}}(\boldsymbol{B}). \tag{6.128}$$

Equation (6.128) constitutes a constraint on the form of the stress function due to the isotropy of the material. A more explicit functional form is obtained by considering the material frame-indifference condition for $\widehat{\boldsymbol{\sigma}}(\boldsymbol{B})$:

$$\widehat{\boldsymbol{\sigma}}(\boldsymbol{Q}\boldsymbol{B}\boldsymbol{Q}^T) = \boldsymbol{Q}\widehat{\boldsymbol{\sigma}}(\boldsymbol{B})\boldsymbol{Q}^T \qquad \forall \boldsymbol{Q} \in SO(3). \tag{6.129}$$

Equation (6.129) shows that $\widehat{\boldsymbol{\sigma}}$ is an isotropic tensor function (as explained above for the viscous part of the stress of a simple fluid) and can therefore be represented as

$$\boldsymbol{\sigma} = \eta_0 \boldsymbol{I} + \eta_1 \boldsymbol{B} + \eta_2 \boldsymbol{B}^2, \tag{6.130}$$

where η_i are arbitrary scalar-valued functions of the principal invariants of $\boldsymbol{B}$:

$$I_1^{\boldsymbol{B}} = \operatorname{tr} \boldsymbol{B}, \qquad I_2^{\boldsymbol{B}} = \frac{1}{2}\left[(\operatorname{tr} \boldsymbol{B})^2 - \operatorname{tr} \boldsymbol{B}^2\right], \qquad I_3^{\boldsymbol{B}} = \det \boldsymbol{B}. \tag{6.131}$$

[28] Elastic materials are defined on page 185.

Equation (6.130) is the most general form for the stress function of a simple elastic isotropic solid.

Hyperelastic solids As defined on page 189, a simple hyperelastic material is one which possesses a strain energy density $W(\boldsymbol{F})$ from which the stress may be obtained from Eqn. (6.43). A procedure similar to that used to obtain Eqn. (6.130) shows that the most general form of the strain energy density function for a simple hyperelastic isotropic solid is

$$W = W(I_1^{\boldsymbol{B}}, I_2^{\boldsymbol{B}}, I_3^{\boldsymbol{B}}), \tag{6.132}$$

where $I_i^{\boldsymbol{B}}$ are the principal invariants of the left Cauchy–Green deformation tensor $\boldsymbol{B}$ given in Eqn. (6.131). In order to use Eqn. (6.43) to obtain the stress from the strain energy density function, we will require expressions for certain derivatives of the principal invariants. First, the derivative of $\boldsymbol{B}$ with respect to $\boldsymbol{F}$ is

$$\frac{\partial B_{ij}}{\partial F_{kL}} = \delta_{ik} F_{jL} + \delta_{jk} F_{iL}. \tag{6.133}$$

Second, the derivatives of the principal invariants of $\boldsymbol{B}$ with respect to $\boldsymbol{B}$ and $\boldsymbol{F}$ are:

$$\frac{\partial I_1^{\boldsymbol{B}}}{\partial B_{ij}} = \delta_{ij}, \qquad \frac{\partial I_1^{\boldsymbol{B}}}{\partial F_{iJ}} = 2F_{iJ}, \tag{6.134}$$

$$\frac{\partial I_2^{\boldsymbol{B}}}{\partial B_{ij}} = I_1^{\boldsymbol{B}} \delta_{ij} - B_{ij}, \qquad \frac{\partial I_2^{\boldsymbol{B}}}{\partial F_{iJ}} = 2(I_1^{\boldsymbol{B}} F_{iJ} - B_{ik} F_{kJ}), \tag{6.135}$$

$$\frac{\partial I_3^{\boldsymbol{B}}}{\partial B_{ij}} = I_3^{\boldsymbol{B}} B_{ji}^{-1}, \qquad \frac{\partial I_3^{\boldsymbol{B}}}{\partial F_{iJ}} = I_3^{\boldsymbol{B}} F_{Ji}^{-1}. \tag{6.136}$$

Using these expressions, we can write the stress in terms of the strain energy density. The first Piola–Kirchhoff stress is

$$\boldsymbol{P} = 2\left[W_{,I_1^{\boldsymbol{B}}} + I_1^{\boldsymbol{B}} W_{,I_2^{\boldsymbol{B}}}\right] \boldsymbol{F} - 2W_{,I_2^{\boldsymbol{B}}} \boldsymbol{B}\boldsymbol{F} + I_3^{\boldsymbol{B}} W_{,I_3^{\boldsymbol{B}}} \boldsymbol{F}^{-T},$$

where $W_{,I_k^{\boldsymbol{B}}} = \partial W / \partial I_k^{\boldsymbol{B}}$. The second Piola–Kirchhoff stress follows from Eqn. (4.41) as

$$\boldsymbol{S} = 2\left[W_{,I_1^{\boldsymbol{B}}} + I_1^{\boldsymbol{B}} W_{,I_2^{\boldsymbol{B}}}\right] \boldsymbol{F} - 2W_{,I_2^{\boldsymbol{B}}} \boldsymbol{C} + I_3^{\boldsymbol{B}} W_{,I_3^{\boldsymbol{B}}} \boldsymbol{C}^{-1}.$$

Using Eqn. (4.36), the Cauchy stress has the form

$$\boldsymbol{\sigma} = \frac{1}{I_3^{\boldsymbol{B}}} \left(I_3^{\boldsymbol{B}} W_{,I_3^{\boldsymbol{B}}} \boldsymbol{I} + 2\left[W_{,I_1^{\boldsymbol{B}}} + I_1^{\boldsymbol{B}} W_{,I_2^{\boldsymbol{B}}}\right] \boldsymbol{B} - 2W_{,I_2^{\boldsymbol{B}}} \boldsymbol{B}^2 \right).$$

Notice that this expression is a special case of Eqn. (6.130) (as it must be). That is, an isotropic hyperelastic material (which has a strain energy density function) is an isotropic elastic material. However, the converse is not true. (Not all isotropic elastic materials with constitutive relations of the form Eqn. (6.130) have a strain energy density function). Next, we give an example of a constitutive relation for an isotropic hyperelastic material.

Blatz-Ko materials In [BK62] Blatz and Ko developed a series of isotropic material models for foamed rubbers based on experimental tests. These tests showed that the stress–strain behavior of such materials is (nearly) independent of I_1^B. One of the most common of these models has the following strain energy density function:[29]

$$\overline{W}(I_2^B, I_3^B) = c_1 \left[\frac{I_2^B}{I_3^B} + 2\sqrt{I_3^B} - 5 \right], \tag{6.137}$$

where c_1 is a constant. The first Piola–Kirchhoff stress tensor is

$$\boldsymbol{P} = c_1 \left[(\sqrt{I_3^B} - I_2^B/I_3^B)\boldsymbol{F}^{-T} + 2\frac{I_1^B}{I_3^B}\boldsymbol{F} - \frac{2}{I_3^B}\boldsymbol{BF} \right], \tag{6.138}$$

and the corresponding Cauchy stress is

$$\boldsymbol{\sigma} = c_1 \left[\left(\frac{1}{\sqrt{I_3^B}} - \frac{I_2^B}{(I_3^B)^2} \right) \boldsymbol{I} + 2\frac{I_1^B}{(I_3^B)^2}\boldsymbol{B} - \frac{2}{(I_3^B)^2}\boldsymbol{B}^2 \right].$$

Blatz and Ko showed that this model was capable of accurately predicting the behavior of foamed polyurethane rubber under isothermal conditions for strains of up to 140%.

Constrained solids: incompressibility An important special class of isotropic hyperelastic materials are those which are *incompressible*. Many materials are approximately incompressible and the study of ideal incompressible materials has been an important factor in the rigorous and complete development of the theory of continuum mechanics (see Chapter 8 for more on this). Incompressible materials, by definition, cannot change their volume, and we must have that any admissible deformation φ satisfies the constraint $\det \boldsymbol{F} = \det(\nabla_0 \varphi) = 1$ everywhere in $\mathcal{B}_0$. Since the material is incompressible it is possible to apply an arbitrary hydrostatic pressure without deforming the material. This means that the material's constitutive relation does not uniquely determine the hydrostatic part of the stress. The pressure must be obtained as part of the solution to a particular boundary-value problem.[30] Since $\det \boldsymbol{F} = 1$ implies $I_3^B = 1$, the strain energy density for isotropic incompressible materials is only a function of I_1^B and I_2^B. Below, we provide some common examples of nonlinear constitutive laws for isotropic incompressible simple materials. For more discussion on many of these models see [Ogd84].

Neo-Hookean materials One of the simplest possible incompressible constitutive relations, the neo-Hookean material model, has been extensively used in theoretical studies where the focus is more on developing an understanding of general continuum mechanics principles rather than obtaining results for a particular material. Motivated by experiments that show the constitutive behavior of rubber to be nearly independent of I_2^B, the neo-Hookean strain energy density is defined as

$$\overline{W}(I_1^B) = c_1(I_1^B - 3). \tag{6.139}$$

[29] This is a simplified version of a more general form given in [BK62] which contains three parameters: (1) the shear modulus μ (above we use the symbol c_1), (2) Poisson's ratio ν and (3) a parameter f which is more difficult to describe in physical terms. Equation (6.137) results from the more general form when one takes the parameter values $f = 0$ and $\nu = 1/4$ (motivated by the experiments of Blatz and Ko). The shear modulus and Poisson's ratio are discussed in Section 6.5.1.

[30] In general, the value of the hydrostatic pressure part of the stress will vary from point to point within the body. See Section 8.2 for a practical example.

The first Piola–Kirchhoff stress tensor, given by Eqn. (6.43), for a neo-Hookean incompressible material is

$$\boldsymbol{P} = 2c_1\boldsymbol{F} - c_0\boldsymbol{F}^{-T}, \tag{6.140}$$

where the final term accounts for the undetermined part of the hydrostatic pressure c_0. Using Eqns. (4.36) and (6.140) we find the Cauchy stress to be

$$\boldsymbol{\sigma} = 2c_1\boldsymbol{B} - c_0\boldsymbol{I},$$

where we have used $J = 1$ (which is due to the incompressibility condition). Notice that, in general, the pressure $p = -\operatorname{tr}\boldsymbol{\sigma}/3 = c_0 - 2c_1 I_1^{\boldsymbol{B}}/3$ has a contribution from $\overline{W}$ in addition to the undetermined contribution c_0. For more on the stability of neo-Hookean materials, see Example 7.1.

Moony–Rivlin materials This incompressible material model includes a dependence on $I_2^{\boldsymbol{B}}$ and has a strain energy density given by

$$\overline{W}(I_1^{\boldsymbol{B}}, I_2^{\boldsymbol{B}}) = c_1(I_1^{\boldsymbol{B}} - 3) + c_2(I_2^{\boldsymbol{B}} - 3). \tag{6.141}$$

The first Piola–Kirchhoff stress for a Moony–Rivlin material is given by

$$\boldsymbol{P} = 2(c_1 + c_2 I_1^{\boldsymbol{B}})\boldsymbol{F} - 2c_2\boldsymbol{B}\boldsymbol{F} - c_0\boldsymbol{F}^{-T}, \tag{6.142}$$

and the corresponding Cauchy stress is

$$\boldsymbol{\sigma} = 2(c_1 + c_2 I_1^{\boldsymbol{B}})\boldsymbol{B} - 2c_2\boldsymbol{B}^2 - c_0\boldsymbol{I}.$$

The neo-Hookean material is a special case of the Moony–Rivlin model (for $c_2 = 0$).

Ogden materials In his book [Ogd84], Ogden describes a general class of incompressible material models for which the strain energy density is given by a power-series:

$$\overline{W}(I_1^{\boldsymbol{B}}, I_2^{\boldsymbol{B}}) = \sum_{p,q=0}^{\infty} c_{pq}(I_1^{\boldsymbol{B}} - 3)^p(I_2^{\boldsymbol{B}} - 3)^q. \tag{6.143}$$

It is easy to see that the Moony–Rivlin and neo-Hookean models are special cases of the Ogden model. The first Piola–Kirchhoff stress is

$$\boldsymbol{P} = \sum_{p,q=0}^{\infty} 2(I_1^{\boldsymbol{B}}-3)^p(I_2^{\boldsymbol{B}}-3)^q\left[(p+1)c_{(p+1)q}\boldsymbol{F} + (q+1)c_{p(q+1)}(I_1^{\boldsymbol{B}}\boldsymbol{F} - \boldsymbol{B}\boldsymbol{F})\right] - c_0\boldsymbol{F}^{-T}, \tag{6.144}$$

and the Cauchy stress is

$$\boldsymbol{\sigma} = \sum_{p,q=0}^{\infty} 2(I_1^{\boldsymbol{B}}-3)^p(I_2^{\boldsymbol{B}}-3)^q\left[(p+1)c_{(p+1)q}\boldsymbol{B} + (q+1)c_{p(q+1)}(I_1^{\boldsymbol{B}}\boldsymbol{B} - \boldsymbol{B}^2)\right] - c_0\boldsymbol{I}.$$

Gent materials In [Gen96] Gent, using the experimental observation that the stress appears to go to infinity as $I_1^{\boldsymbol{B}}$ asymptotically approaches a value I_m, proposed the following incompressible strain energy density function:

$$\overline{W}(I_1^{\boldsymbol{B}}) = -\frac{c_1 I_m}{2}\ln\left(1 - \frac{I_1^{\boldsymbol{B}} - 3}{I_m - 3}\right). \tag{6.145}$$

Here c_1 is a constant and I_m is the limiting value that I_1^B is allowed to approach. The first Piola–Kirchhoff stress and the Cauchy stress are, respectively,

$$\boldsymbol{P} = \frac{c_1}{1 - I_1^B/I_m}\boldsymbol{F} - c_0\boldsymbol{F}^{-T}, \qquad \boldsymbol{\sigma} = \frac{c_1}{1 - I_1^B/I_m}\boldsymbol{B} - c_0\boldsymbol{I}. \tag{6.146}$$

Beyond isotropy As a prelude to our study of anisotropic linearized constitutive relations in the next section, we now present an example of an anisotropic, (geometrically) nonlinear material model.

Saint Venant–Kirchhoff materials These materials have strain energy density functions that are simply quadratic in the Lagrangian strain $\boldsymbol{E}$:

$$\widetilde{W}(\boldsymbol{E}) = \frac{1}{2}(\mathcal{C} : \boldsymbol{E}) : \boldsymbol{E}. \tag{6.147}$$

Here $\mathcal{C}$ is a constant fourth-order tensor with both *minor* and *major* symmetries (see Eqns. (6.152) and (6.153) in the next section). The second Piola–Kirchhoff stress is found, from Eqn. (6.43), to be

$$\boldsymbol{S} = \mathcal{C} : \boldsymbol{E}.$$

Thus, we see that the second Piola–Kirchhoff stress is linearly related to the Lagrangian strain for Saint Venant–Kirchhoff materials. The first Piola–Kirchhoff stress and the Cauchy stress follow as, respectively

$$\boldsymbol{P} = \boldsymbol{F}(\mathcal{C} : \boldsymbol{E}), \qquad \boldsymbol{\sigma} = \frac{1}{J}\boldsymbol{F}(\mathcal{C} : \boldsymbol{E})\boldsymbol{F}^T. \tag{6.148}$$

For more on the stability of Saint Venant–Kirchhoff materials, see Examples 7.2 and 7.3.

6.5 Linearized constitutive relations for anisotropic hyperelastic solids

An *anisotropic* material has different properties along different directions and therefore has less symmetry than the isotropic materials discussed above. The term *hyperelastic*, defined on page 189, means that the material has no dissipation and that an energy function exists for it. The stress then follows as the gradient of the energy function with respect to a conjugate strain variable. For example, the Piola–Kirchhoff stress tensors for a hyperelastic material are given in Eqn. (6.43) and reproduced here for convenience (dropping the functional dependence on T for notational simplicity):

$$\boldsymbol{S}^{(e)} = \frac{\partial\widetilde{W}(\boldsymbol{E})}{\partial\boldsymbol{E}}, \qquad \boldsymbol{P}^{(e)} = \frac{\partial\widehat{W}(\boldsymbol{F})}{\partial\boldsymbol{F}}. \tag{6.149}$$

Additional constraints on these functional forms can be obtained by considering material symmetry (as done above in Section 6.4.2). This together with carefully planned experiments can then be used to construct *phenomenological* (i.e. fitted) models for the nonlinear material response such as the examples given in the last section (see also, for example, [Hol00] for a discussion of phenomenological constitutive relations). Alternatively, $\widetilde{\boldsymbol{S}}(\boldsymbol{E})$ can be computed directly from an atomistic model as explained in Chapter 11 of [TM11]. A third possibility that is often used in numerical solutions to continuum boundary-value problems is an incremental approach, where the equations are linearized. This requires the calculation of *linearized constitutive relations* for the material which involve the definition of elasticity tensors. When the linearization is about the reference configuration of the material this approach leads to the well-known generalized Hooke's law.

The linearized form of Eqn. $(6.149)_1$ relates the increment of the second Piola–Kirchhoff stress $d\boldsymbol{S}$ to the increment of the Lagrangian strain $d\boldsymbol{E}$ and is given by

$$dS_{IJ} = C_{IJKL} dE_{KL}, \quad \Leftrightarrow \quad d\boldsymbol{S} = \boldsymbol{C} : d\boldsymbol{E}, \tag{6.150}$$

where

$$C_{IJKL} = \frac{\partial \widetilde{S}_{IJ}(\boldsymbol{E})}{\partial E_{KL}} = \frac{\partial^2 \widetilde{W}(\boldsymbol{E})}{\partial E_{IJ} E_{KL}} \quad \Leftrightarrow \quad \boldsymbol{C} = \frac{\partial \widetilde{\boldsymbol{S}}(\boldsymbol{E})}{\partial \boldsymbol{E}} = \frac{\partial^2 \widetilde{W}(\boldsymbol{E})}{\partial \boldsymbol{E}^2}, \tag{6.151}$$

is a fourth-order tensor called the *material elasticity tensor*.[31] Due to the symmetry of $\boldsymbol{S}$ and $\boldsymbol{E}$, the tensor $\boldsymbol{C}$ has the following symmetries:

$$C_{IJKL} = C_{JIKL} = C_{IJLK}. \tag{6.152}$$

These are called the *minor* symmetries of $\boldsymbol{C}$. In addition, hyperelastic materials have the following additional *major* symmetry:

$$C_{IJKL} = C_{KLIJ}, \tag{6.153}$$

due to the fact that $\boldsymbol{C}$ is the second derivative of an energy with respect to strain and the order of differentiation is unimportant.

Similarly, we may obtain the relationship between increments of the first Piola–Kirchhoff stress $d\boldsymbol{P}$ and the deformation gradient $d\boldsymbol{F}$ by linearizing Eqn. $(6.149)_2$:

$$dP_{iJ} = D_{iJkL} dF_{kL}, \quad \Leftrightarrow \quad d\boldsymbol{P} = \boldsymbol{D} : d\boldsymbol{F}, \tag{6.154}$$

[31] There should be no confusion between the fourth-order material elasticity tensor and the second-order right Cauchy–Green tensor which are denoted by the same symbol $\boldsymbol{C}$.

where $\boldsymbol{D}$ is the *mixed elasticity tensor* given by

$$D_{iJkL} = \frac{\partial \widehat{P}_{iJ}(\boldsymbol{F})}{\partial F_{kL}} = \frac{\partial^2 \widehat{W}(\boldsymbol{F})}{\partial F_{iJ} F_{kL}} \quad \Leftrightarrow \quad \boldsymbol{D} = \frac{\partial \widehat{\boldsymbol{P}}(\boldsymbol{F})}{\partial \boldsymbol{F}} = \frac{\partial^2 \widehat{W}(\boldsymbol{F})}{\partial \boldsymbol{F}^2}. \tag{6.155}$$

$\boldsymbol{D}$ does not have the minor symmetries that $\boldsymbol{C}$ possesses since $\boldsymbol{P}$ and $\boldsymbol{F}$ are not symmetric. However, for a hyperelastic material it does possess the major symmetry, $D_{iJkL} = D_{kLiJ}$, due to invariance with respect to the order of differentiation. We can obtain the relation between $\boldsymbol{D}$ and $\boldsymbol{C}$. First, we use Eqn. (3.23) to find the incremental relation

$$d\boldsymbol{E} = \frac{1}{2}(d\boldsymbol{F}^T \boldsymbol{F} + \boldsymbol{F}^T d\boldsymbol{F}). \tag{6.156}$$

Next, we use Eqn. (4.41) to obtain the incremental relation

$$d\boldsymbol{S} = \boldsymbol{F}^{-1} d\boldsymbol{P} - \boldsymbol{F}^{-1} d\boldsymbol{F} \boldsymbol{S},$$

where we have also used the identity $d\boldsymbol{F}^{-1} = -\boldsymbol{F}^{-1} d\boldsymbol{F} \boldsymbol{F}^{-1}$, which is obtained in a similar fashion to Eqn. (3.53). Finally, we substitute these relations into Eqn. (6.150), simplify (taking advantage of the symmetries of $\boldsymbol{C}$ and $\boldsymbol{S}$) and compare the result with Eqn. (6.154) in order to find that

$$D_{iJkL} = C_{IJKL} F_{iI} F_{kK} + \delta_{ik} S_{JL}. \tag{6.157}$$

For practical reasons, it is often useful to treat the deformed configuration as a new reference configuration and then consider increments of deformation and stress measured from this configuration. Suppose the deformed configuration is given by $\boldsymbol{x} = \boldsymbol{\varphi}(\boldsymbol{X})$ and define the new reference configuration as $\boldsymbol{X}^* \equiv \boldsymbol{\varphi}(\boldsymbol{X})$. Now we consider an additional deformation to a "new deformed configuration" which we can represent as $\boldsymbol{x}^* = \boldsymbol{\varphi}^*(\boldsymbol{X}^*) = \boldsymbol{\varphi}^*(\boldsymbol{\varphi}(\boldsymbol{X}))$. The deformation gradients measured from the new and original reference configurations are

$$\boldsymbol{F}^* = \frac{\partial \boldsymbol{x}^*}{\partial \boldsymbol{X}^*} \equiv \nabla_* \boldsymbol{\varphi}^*,$$

and

$$\boldsymbol{F}^0 = \frac{\partial \boldsymbol{x}^*}{\partial \boldsymbol{X}} = (\nabla_* \boldsymbol{\varphi}^*)(\nabla_0 \boldsymbol{\varphi}) = \boldsymbol{F}^* \boldsymbol{F},$$

respectively. Using these expressions and Eqn. (3.23) we find that the Lagrangian strain $\boldsymbol{E}^0$ measured from the original reference configuration can be written in terms of the Lagrangian strain $\boldsymbol{E}^*$ measured from the new reference configuration and the Lagrangian strain $\boldsymbol{E}$ relating the original and new reference configurations as

$$\boldsymbol{E}^0 = \boldsymbol{F}^T \boldsymbol{E}^* \boldsymbol{F} + \boldsymbol{E}.$$

This allows us to define the strain energy density function measured from the new reference configuration as

$$W^*(\boldsymbol{E}^*) \equiv \frac{W(\boldsymbol{F}^T\boldsymbol{E}^*\boldsymbol{F}+\boldsymbol{E})}{J}, \tag{6.158}$$

where we have divided by the Jacobian to ensure that W^* is the energy per unit volume in the new reference configuration. The associated second Piola–Kirchhoff stress is given by $\boldsymbol{S}^* = \partial W^*/\partial \boldsymbol{E}^*$ and the linearized form of this relation is

$$dS^*_{IJ} = C^*_{IJKL}dE^*_{KL}, \tag{6.159}$$

where $\boldsymbol{C}^* = \partial^2 W^*/\partial(\boldsymbol{E}^*)^2$. From Eqn. (4.42) we know that $J^*\boldsymbol{\sigma} = \boldsymbol{F}^*\boldsymbol{S}^*(\boldsymbol{F}^*)^T$. Taking the full differential of this equation and solving for the differential of the second Piola–Kirchhoff stress $d\boldsymbol{S}^*$ we find that

$$d\boldsymbol{S}^* = J^*(\boldsymbol{F}^*)^{-1}\left[d\boldsymbol{\sigma} - d\boldsymbol{F}^*(\boldsymbol{F}^*)^{-1}\boldsymbol{\sigma} - \boldsymbol{\sigma}(\boldsymbol{F}^*)^{-T}d(\boldsymbol{F}^*)^T + \boldsymbol{\sigma}(\boldsymbol{F}^*)^{-T} : d\boldsymbol{F}^*\right](\boldsymbol{F}^*)^{-T},$$

where we have also used the fact that $dJ^* = J^*(\boldsymbol{F}^*)^{-T} : d\boldsymbol{F}^*$. If we now consider the above increments to be associated with dynamic motion, then we can divide by an increment of time dt and take the limit to obtain the stress rate relation

$$\begin{aligned}\dot{\boldsymbol{S}}^* &= J^*(\boldsymbol{F}^*)^{-1}\left[\dot{\boldsymbol{\sigma}} - \dot{\boldsymbol{F}}^*(\boldsymbol{F}^*)^{-1}\boldsymbol{\sigma} - \boldsymbol{\sigma}(\boldsymbol{F}^*)^{-T}(\dot{\boldsymbol{F}}^*)^T + \boldsymbol{\sigma}(\boldsymbol{F}^*)^{-T} : \dot{\boldsymbol{F}}^*\right](\boldsymbol{F}^*)^{-T} \\ &= J^*(\boldsymbol{F}^*)^{-1}\left[\dot{\boldsymbol{\sigma}} - \boldsymbol{l}\boldsymbol{\sigma} - \boldsymbol{\sigma}\boldsymbol{l}^T + \boldsymbol{\sigma}\,\mathrm{tr}\,\boldsymbol{l}\right](\boldsymbol{F}^*)^{-T} \\ &= J^*(\boldsymbol{F}^*)^{-1}\mathring{\boldsymbol{\sigma}}(\boldsymbol{F}^*)^{-T},\end{aligned} \tag{6.160}$$

where we have used Eqn. (3.36) and

$$\mathring{\boldsymbol{\sigma}} \equiv \dot{\boldsymbol{\sigma}} - \boldsymbol{l}\boldsymbol{\sigma} - \boldsymbol{\sigma}\boldsymbol{l}^T + \boldsymbol{\sigma}\,\mathrm{tr}\,\boldsymbol{l} \tag{6.161}$$

is the objective *Truesdell stress rate* of the Cauchy stress tensor [Hol00].[32] Also note that the rate of Lagrangian strain is given by $\dot{\boldsymbol{E}}^* = \frac{1}{2}(\boldsymbol{F}^*)^T(\boldsymbol{l}+\boldsymbol{l}^T)\boldsymbol{F}^* = (\boldsymbol{F}^*)^T\dot{\boldsymbol{\epsilon}}\boldsymbol{F}^*$. Substituting these expressions into Eqn. (6.159), evaluating at the new reference configuration (where $\boldsymbol{F}^* = \boldsymbol{I}$, $J^* = 1$) and simplifying we find

$$\mathring{\sigma}_{ij} = C^*_{IJKL}\delta_{Ii}\delta_{Jj}\delta_{Kk}\delta_{Ll}\dot{\epsilon}_{kl}, \tag{6.162}$$

or

$$\mathring{\sigma}_{ij} = c_{ijkl}\dot{\epsilon}_{kl}, \tag{6.163}$$

[32] See Exercise 6.6 for a discussion of objective stress rates.

where $\mathbf{c} = \mathbf{C}^*$ is the *spatial elasticity tensor*.[33] Note that $\mathbf{c}$ has the same minor and major symmetries as its material counterpart:

$$c_{ijkl} = c_{jikl} = c_{ijlk} = c_{klij}. \tag{6.164}$$

Using Eqn. (6.158) and the definitions of $\mathbf{c}$, $\mathbf{C}^*$ and $\mathbf{C}$ we can obtain the relation

$$c_{ijkl} = J^{-1} F_{iI} F_{jJ} F_{kK} F_{lL} C_{IJKL}, \tag{6.165}$$

where it is understood that $\mathbf{C}$ is evaluated at the deformed configuration corresponding to the new reference configuration. Similarly, the relation between $\mathbf{c}$ and $\mathbf{D}$ is

$$c_{ijkl} = J^{-1} \left(F_{jJ} F_{lL} D_{iJkL} - \delta_{ik} F_{lL} P_{jL} \right). \tag{6.166}$$

6.5.1 Generalized Hooke's law and the elastic constants

When the new reference configuration considered above is taken to be the same as the original reference configuration (which is assumed to be stress free), then we can again start with Eqn. (6.159) and follow a procedure similar to the one used to obtain Eqn. (6.163). However, instead of dividing the expression for the increment of second Piola–Kirchhoff stress by dt, we simply evaluate it at the stress-free reference configuration (corresponding to the values $J^* = 1$, $\mathbf{F}^* = \mathbf{I}$, $\boldsymbol{\sigma} = \mathbf{0}$) to obtain $d\mathbf{S}^* = d\boldsymbol{\sigma}$. Evaluating Eqn. (6.156) in the same manner, we find $d\mathbf{E} = d\boldsymbol{\epsilon}$. Next, we notice from Eqn. (6.165) that for the case considered here $\mathbf{c} = \mathbf{C}^*$. Finally, since the reference configuration is stress-free we can identify $d\boldsymbol{\sigma}$ with $\boldsymbol{\sigma}$ and $d\boldsymbol{\epsilon}$ with $\boldsymbol{\epsilon}$ to obtain

$$\sigma_{ij} = c_{ijkl}\epsilon_{kl} \quad \Leftrightarrow \quad \boldsymbol{\sigma} = \mathbf{c} : \boldsymbol{\epsilon}, \tag{6.167}$$

which is valid for small strains. This is called the *generalized Hooke's law*.[34] The fourth-order tensor $\mathbf{c}$ is the *elasticity tensor*. (The epithet "spatial" is dropped since all elasticity tensors are the same in this case. The term "small strain elasticity tensor" is also used.)

[33] Note that some authors use an alternative definition, $\mathbf{c} = J\mathbf{C}^*$, for the spatial elasticity tensor. As a result, the corresponding expressions relating $\mathbf{c}$ with the material and mixed elasticity tensors will be slightly different than the ones derived here.

[34] For a discussion of the origin of Hooke's law, see footnote 45 on page 235.

Hooke's law can also be inverted to relate strain to stress:

$$\epsilon_{ij} = s_{ijkl}\sigma_{kl} \quad \Leftrightarrow \quad \boldsymbol{\epsilon} = \boldsymbol{s} : \boldsymbol{\sigma}, \tag{6.168}$$

where $\boldsymbol{s}$ is the *compliance tensor*. The corresponding strain energy density function, W, is

$$W = \frac{1}{2}\sigma_{ij}\epsilon_{ij} = \frac{1}{2}c_{ijkl}\epsilon_{ij}\epsilon_{kl} = \frac{1}{2}s_{ijkl}\sigma_{ij}\sigma_{kl}. \tag{6.169}$$

The strain energy density expression in terms of strain can also be written in terms of the displacement gradient:

$$W = \frac{1}{2}c_{ijkl}u_{i,j}u_{k,l}, \tag{6.170}$$

since the contraction of the antisymmetric part of $\nabla \boldsymbol{u}$ with $\boldsymbol{c}$ is zero due to the symmetry properties of the elasticity tensor (see Section 2.5.2). In the above relations, we assumed a stress-free reference configuration. If this is not the case, then an additional constant stress term $\boldsymbol{\sigma}^0$ is added to Eqn. (6.167), $\boldsymbol{\sigma}$ is replaced by $\boldsymbol{\sigma} - \boldsymbol{\sigma}^0$ in Eqn. (6.168) and the energy expression has an additional term linear[35] in strain, $(\boldsymbol{\sigma}^0 : \boldsymbol{\epsilon})/2$. In addition, a constant reference strain energy density W_0 can always be added to W.

Due to the symmetry of the stress and strain tensors, it is convenient to write Eqn. (6.167) in a contracted matrix notation referred to as *Voigt notation*, where pairs of indices in the tensor notation are replaced with a single index in the matrix notation (see also Tab. 5.2):

tensor indices ij:	11	22	33	23, 32	13, 31	12, 21
matrix index m:	1	2	3	4	5	6

Using this notation, the generalized Hooke's law (Eqn. (6.167)) is

$$\begin{bmatrix} \sigma_{11} \\ \sigma_{22} \\ \sigma_{33} \\ \sigma_{23} \\ \sigma_{13} \\ \sigma_{12} \end{bmatrix} = \begin{bmatrix} \mathsf{c}_{11} & \mathsf{c}_{12} & \mathsf{c}_{13} & \mathsf{c}_{14} & \mathsf{c}_{15} & \mathsf{c}_{16} \\ \mathsf{c}_{21} & \mathsf{c}_{22} & \mathsf{c}_{23} & \mathsf{c}_{24} & \mathsf{c}_{25} & \mathsf{c}_{26} \\ \mathsf{c}_{31} & \mathsf{c}_{32} & \mathsf{c}_{33} & \mathsf{c}_{34} & \mathsf{c}_{35} & \mathsf{c}_{36} \\ \mathsf{c}_{41} & \mathsf{c}_{42} & \mathsf{c}_{43} & \mathsf{c}_{44} & \mathsf{c}_{45} & \mathsf{c}_{46} \\ \mathsf{c}_{51} & \mathsf{c}_{52} & \mathsf{c}_{53} & \mathsf{c}_{54} & \mathsf{c}_{55} & \mathsf{c}_{56} \\ \mathsf{c}_{61} & \mathsf{c}_{62} & \mathsf{c}_{63} & \mathsf{c}_{64} & \mathsf{c}_{65} & \mathsf{c}_{66} \end{bmatrix} \begin{bmatrix} \epsilon_{11} \\ \epsilon_{22} \\ \epsilon_{33} \\ 2\epsilon_{23} \\ 2\epsilon_{13} \\ 2\epsilon_{12} \end{bmatrix}, \tag{6.171}$$

where c is the *elasticity matrix*.[36] The entries c_{mn} of the elasticity matrix are referred to as the *elastic constants*. Therefore c is also called the "elastic constants matrix." The stress and

[35] Note, however, that the resulting stress–strain relations are no longer linear. Thus, in this case the principle of superposition is not valid for solutions to boundary-value problems that use this type of stress–strain relation.

[36] Note that we use a sans serif font for the elasticity matrix. This stresses the fact that the numbers that constitute this 6×6 matrix are not the components of a second-order tensor in a six-dimensional space and therefore do not transform according to standard tensor transformation rules.

strain tensors can also be expressed in compact notation by defining the column matrices,

$$\boldsymbol{\sigma} = [\sigma_{11}, \sigma_{22}, \sigma_{33}, \sigma_{23}, \sigma_{13}, \sigma_{12}]^T, \qquad \boldsymbol{\epsilon} = [\epsilon_{11}, \epsilon_{22}, \epsilon_{33}, 2\epsilon_{23}, 2\epsilon_{13}, 2\epsilon_{12}]^T.$$

Hooke's law is then

$$\sigma_m = \mathsf{c}_{mn}\epsilon_n \qquad \text{or} \qquad \epsilon_m = \mathsf{s}_{mn}\sigma_n, \tag{6.172}$$

where $\mathsf{s} = \mathsf{c}^{-1}$ is the *compliance matrix*.[37]

The minor symmetries of c_{ijkl} (and s_{ijkl}) are automatically accounted for in c_{mn} (and s_{mn}) by the Voigt notation. The major symmetry of c_{ijkl} (and s_{ijkl}) implies that c_{mn} (and s_{mn}) are symmetric, i.e. $\mathsf{c}_{mn} = \mathsf{c}_{nm}$ (and $\mathsf{s}_{mn} = \mathsf{s}_{nm}$). Therefore in the most general case a material can have 21 independent elastic constants.

The material symmetry condition for the elastic stress tensor is given in Eqn. (6.126). For the linear elastic case considered here this translates to the following set of constraints on the elasticity tensor [FV96]:

$$c_{ijkl} = Q_{ip}Q_{jq}Q_{kr}Q_{ls}c_{pqrs} \qquad \forall \boldsymbol{Q} \in G \subset SO(3), \tag{6.173}$$

where G is the material symmetry group of the material, which for a solid is a subgroup of the set of rotations[38] $SO(3)$. As an example, let us consider the simplest case where the material has a symmetry plane normal to the 3-direction. We use the "direct inspection method" described in [Nye85, pp. 118–120]. The symmetry reflection operation is represented by the following transformation:

$$[Q] = \begin{bmatrix} 1 & 0 & 0 \\ 0 & 1 & 0 \\ 0 & 0 & -1 \end{bmatrix}.$$

This takes any point $\boldsymbol{X} = [X_1, X_2, X_3]^T$ to $\boldsymbol{x} = [X_1, X_2, -X_3]^T$. Substituting this into Eqn. (6.173) gives the following relations between the elastic constants in Voigt matrix notation:

$$\begin{bmatrix} \mathsf{c}_{11} & \mathsf{c}_{12} & \mathsf{c}_{13} & \mathsf{c}_{14} & \mathsf{c}_{15} & \mathsf{c}_{16} \\ & \mathsf{c}_{22} & \mathsf{c}_{23} & \mathsf{c}_{24} & \mathsf{c}_{25} & \mathsf{c}_{26} \\ & & \mathsf{c}_{33} & \mathsf{c}_{34} & \mathsf{c}_{35} & \mathsf{c}_{36} \\ & & & \mathsf{c}_{44} & \mathsf{c}_{45} & \mathsf{c}_{46} \\ & \text{sym} & & & \mathsf{c}_{55} & \mathsf{c}_{56} \\ & & & & & \mathsf{c}_{66} \end{bmatrix} = \begin{bmatrix} \mathsf{c}_{11} & \mathsf{c}_{12} & \mathsf{c}_{13} & -\mathsf{c}_{14} & -\mathsf{c}_{15} & \mathsf{c}_{16} \\ & \mathsf{c}_{22} & \mathsf{c}_{23} & -\mathsf{c}_{24} & -\mathsf{c}_{25} & \mathsf{c}_{26} \\ & & \mathsf{c}_{33} & -\mathsf{c}_{34} & -\mathsf{c}_{35} & \mathsf{c}_{36} \\ & & & \mathsf{c}_{44} & \mathsf{c}_{45} & -\mathsf{c}_{46} \\ & \text{sym} & & & \mathsf{c}_{55} & -\mathsf{c}_{56} \\ & & & & & \mathsf{c}_{66} \end{bmatrix}.$$

[37] Note, however, that the fourth-order tensor $\boldsymbol{s} \neq \boldsymbol{c}^{-1}$. This is because, strictly speaking, $\boldsymbol{c}$ is not invertible, since $\boldsymbol{c} : \boldsymbol{w} = \boldsymbol{0}$, where $\boldsymbol{w} = -\boldsymbol{w}^T$ is any antisymmetric second-order tensor. This indicates that $\boldsymbol{w}$ is an "eigentensor" of $\boldsymbol{c}$ associated with the eigenvalue 0, and further implies that $\boldsymbol{c}$ is not invertible. However, if $\boldsymbol{c}$ and $\boldsymbol{s}$ are viewed as linear mappings from the space of all *symmetric* second-order tensors to itself (as opposed to the space of *all* second-order tensors), then $\boldsymbol{c} : \boldsymbol{w}$ is not a valid operation. In this sense, $\boldsymbol{c}$ is invertible and only then do we have that $\boldsymbol{s} = \boldsymbol{c}^{-1}$.

[38] Strictly speaking we should include *lattice invariant shears* in the material symmetry group G of crystalline solids. These are shear deformations that carry all the atoms in an infinite crystal to other atomic positions leaving the crystal unchanged. Such deformations do not affect the symmetry properties of the elasticity tensor and therefore need not be considered here. For more on the importance of lattice invariant shears see [Eri77].

We see by inspection that $c_{14} = -c_{14}$, which means that $c_{14} = 0$. Similarly, we see that $c_{15} = c_{24} = c_{25} = c_{34} = c_{35} = c_{46} = c_{56} = 0$. The most general form for **c** for this symmetry is therefore

$$\mathbf{c} = \begin{bmatrix} c_{11} & c_{12} & c_{13} & 0 & 0 & c_{16} \\ & c_{22} & c_{23} & 0 & 0 & c_{26} \\ & & c_{33} & 0 & 0 & c_{36} \\ & & & c_{44} & c_{45} & 0 \\ & \text{sym} & & & c_{55} & 0 \\ & & & & & c_{66} \end{bmatrix}.$$

This form corresponds to the *monoclinic* symmetry class. We see that the number of distinct elastic constants has been reduced from 21 to 13.

An interesting question is: how many distinct symmetry classes exist? Originally, a crystallographic approach was taken to answer this question going back to the work of Woldemar Voigt published in his 1910 *magnum opus* [Voi10]. The idea was to painstakingly go through all of the crystal classes and to identify by brute-force inspection (along the lines of the above example) the resulting distinct elasticity matrices. For example, Wallace [Wal72, p. 28] classifies the symmetry classes according to the 11 crystallographic Laue groups,[39] since "all [crystal] classes in a given group have a common array of elastic constants." When limited to second-order elastic constants (i.e. the elasticity matrix), the number of distinct symmetry classes is reduced to nine. Adding to this the isotropy group gives the classical result that there are ten distinct symmetry classes for the elasticity tensor.[40] This is the result cited in many books including the classical book on the subject by Nye [Nye85].

The crystallographic approach seems reasonable, but its conclusions are incorrect. The modern approach is to pose the question mathematically by directly identifying the equivalence classes corresponding to Eqn. (6.173) without considering crystallography at all. This is a far more general approach since many materials of interest are not crystalline (an important example is composite materials). Interestingly, despite the generality of the approach, the conclusion to emerge from these studies is that there are in fact only *eight* distinct symmetry classes. This was first conclusively shown by Forte and Vianello in 1996 [FV96] (although there were partial indications of this result earlier as noted in the interesting historical review in this paper).

Forte and Vianello's proof is based on harmonic and Cartan decomposition techniques. Since then several additional proofs have been advanced including a simple one based on the idea of mirror symmetry planes due to Chadwick *et al.* [CVC01]. In this paper, the authors were able to connect their symmetry plane argument with Forte and Vianello's classification

[39] Crystallographically, there are 32 unique point groups, of which only 11 are centrosymmetric. These form 11 unique diffraction patterns. The diffraction patterns of the remaining noncentrosymmetric crystal structures are each indistinguishable from one of the 11 centrosymmetric crystals, and thus we can organize the 32 point groups into 11 distinct classes based on their diffraction patterns. These 11 classes are called the Laue classes. It is interesting that crystals sharing the same diffraction pattern also share elastic symmetry.

[40] The ten classes are called triclinic, monoclinic, orthotropic, hexagonal (7), hexagonal (6), tetragonal (7), tetragonal (6), cubic, transversely isotropic and isotropic. Following each name in parenthesis is the number of distinct elastic constants for this class. See, for example, [CM87, Table 1].

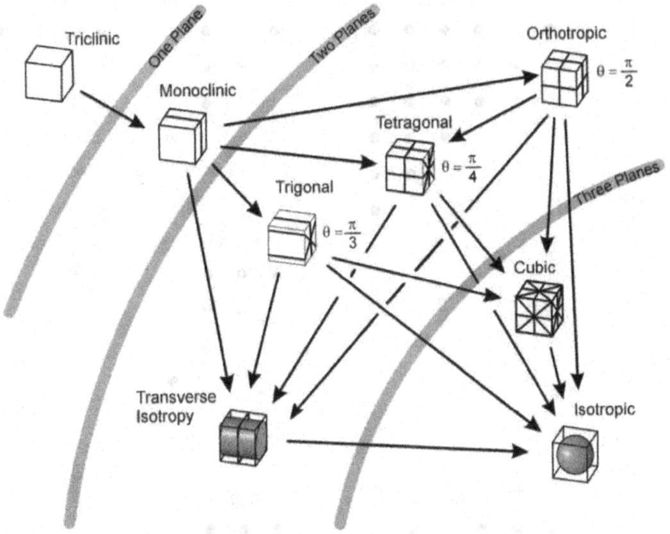

Fig. 6.3 The eight distinct symmetry classes of the elasticity tensor. Note that 'orthotropic' is also called orthorhombic and 'transverse isotropy' is also called hexagonal. Reprinted from [CVC01], with permission from Elsevier. A similar figure also appears in [BBS04].

and in this manner identify the distinct symmetry classes with the earlier crystallographic categories. The names they give the symmetry classes are borrowed from the traditional seven crystal systems (see Section 3.4 of [TM11]) based on the symmetry classes in which these systems fall: triclinic, monoclinic, orthorhombic (the term 'orthotropic' is also used), tetragonal, trigonal, hexagonal (the term 'transverse isotropy' is also used), cubic and isotropic.[41] The relation between the different symmetry classes is illustrated in Fig. 6.3. The arrows indicate how one symmetry class is obtained from another through the addition of symmetry planes.

In addition to knowing the number of symmetry classes, it is of course also of interest to know the number of independent elastic constants in each case and the structure of the elasticity matrix as shown above for the special case of monoclinic symmetry. Figure 6.4 provides this information for the eight symmetry classes,[42] where the number in parentheses after the name of the class is the number of distinct elastic constants. The diagrams are based on the notation introduced by Nye [Nye85] (see the caption for an explanation). The most general material belongs to the triclinic class with 21 independent constants,[43] and

[41] The relationship between the new terms and the traditional ten classes listed in footnote 40 above is as follows (new = old): triclinic = triclinic, monoclinic = monoclinic, orthorhombic = orthotropic, trigonal = hexagonal (6), tetragonal = tetragonal (6), cubic = cubic, hexagonal = transversely isotropic, isotropic = isotropic. We see that the two classes that were dropped are hexagonal (7) and tetragonal (7).

[42] The structures of the elasticity matrices given in Fig. 6.4 are the simple forms associated with basis vectors that are suitably aligned with the crystallographic axes. For arbitrary basis vector orientation, the matrices can be full, with all entries being functions of the independent elastic constants for the relevant symmetry class.

[43] Actually, the maximum number of independent elastic constants is 18. The reason is that it is always possible to orient the coordinate system in such a way that three of the constants are zeroed. Similarly, the number of elastic constants for monoclinic symmetry can be reduced from 13 to 12. See, for example, [CVC01].

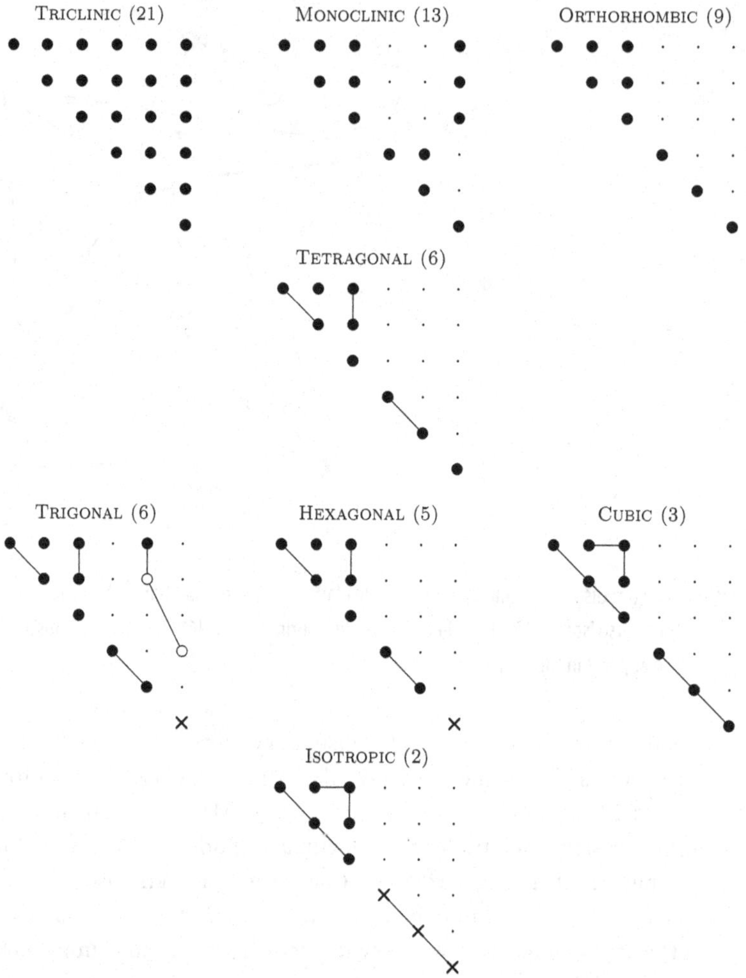

Fig. 6.4 Symmetry classes of the elasticity matrix. Following the name of the class in parentheses is the number of distinct elastic constants for this class. The arrays of dots, circles and $\times$-symbols under the name are the elements of the 6×6 elasticity matrix for the symmetry class. Only half the matrix is shown since it is symmetric. Small dots correspond to zero elements. Circles and $\times$-symbols are nonzero elements. Elements are equal when connected by a line. A white filled circle is equal to the negative of the black element to which it is connected. Elements marked with an $\times$ are equal to $\frac{1}{2}(c_{11} - c_{12})$.

this number is reduced with increasing symmetry of the material. Additional symmetry in the monoclinic and orthorhombic classes implies that certain constants must be zero (appearing as dots in the diagram) and therefore the number of distinct constants is reduced. In the tetragonal class, symmetry also dictates that some constants must be equal (shown connected by lines). We see that $c_{11} = c_{22}$, $c_{13} = c_{23}$, and $c_{44} = c_{55}$. There are therefore six distinct constants: c_{11}, c_{12}, c_{13}, c_{33}, c_{44} and c_{66}. In the trigonal class, there are two new features. The constants c_{15} and c_{25} are constrained to have opposite signs, i.e. $c_{25} = -c_{15}$, which is indicated by the black and white circles. The constant c_{66} is equal to $\frac{1}{2}(c_{11} - c_{12})$ (indicated by the $\times$-symbol). The remaining classes can be seen in the diagram. We note that for isotropic symmetry the elasticity tensor has two independent constants. This special case is described below.

Hooke's law for isotropic linear elastic materials The elasticity tensor for isotropic materials can be written as

$$c_{ijkl} = \lambda\delta_{ij}\delta_{kl} + \mu(\delta_{ik}\delta_{jl} + \delta_{il}\delta_{jk}), \tag{6.174}$$

where $\lambda = c_{12}$ and $\mu = c_{44} = (c_{11} - c_{12})/2$ are called the *Lamé constants* (μ is also called the *shear modulus*). (Note that there is no connection between the Lamé constants introduced here and the viscosity coefficients in Newton's law.) Substituting Eqn. (6.174) into Eqn. (6.167), we obtain *Hooke's law* for an isotropic linear elastic solid:

$$\boldsymbol{\sigma} = \lambda(\operatorname{tr}\boldsymbol{\epsilon})\boldsymbol{I} + 2\mu\boldsymbol{\epsilon}, \tag{6.175}$$

This relation can be inverted, in which case it is more conveniently expressed in terms of two other material parameters, *Young's modulus*[44], E, and *Poisson's ratio*, ν:

$$\boldsymbol{\epsilon} = -\frac{\nu}{E}(\operatorname{tr}\boldsymbol{\sigma})\boldsymbol{I} + \frac{1+\nu}{E}\boldsymbol{\sigma}. \tag{6.176}$$

The two sets of material parameters are related through

$$\mu = \frac{E}{2(1+\nu)}, \quad \lambda = \frac{\nu E}{(1+\nu)(1-2\nu)} \quad \text{or} \quad E = \frac{\mu(3\lambda + 2\mu)}{\lambda + \mu}, \quad \nu = \frac{\lambda}{2(\lambda+\mu)}. \tag{6.177}$$

Equation (6.176) can be reduced to one dimension by setting all stresses to zero, except $\sigma_{11} = \sigma$, and solving for the strains. The result is the one-dimensional Hooke's law:[45]

$$\sigma = E\epsilon, \tag{6.178}$$

where $\epsilon = \epsilon_{11}$ is the strain in the 1-direction (see Exercise 6.13).

44 "Young's modulus" is named after the English polymath Thomas Young and is often attributed to an article that he published in 1807. Actually, as pointed out by Truesdell [Tru68], the "modulus of extension" was introduced by Euler 100 years before Young. In fact, Young defined his modulus as the ratio of *force* to strain (rather than stress to strain as Euler did). Young's definition does not constitute a material property since it depends on the geometry of the structure for which it is defined.

45 Robert Hooke's original law was published in 1676 in the form of the anagram "ceiiinosssttuv," which unscrambles to the Latin "ut tensio sic vis" or in English "as the extension so the force." The anagram, which appeared at the end of an unrelated paper, was a way of establishing precedence without divulging the details of the theory which was published several years later. In his work, Hooke was referring to the constitutive relation for a linear spring. He had no understanding of the concepts of stress or strain. It was actually James Bernoulli in 1704 who provides the first instance of a true stress–strain constitutive relation [Tru68, p. 103].

It is important to stress in closing this section that the symmetry classes and corresponding elasticity matrices are only valid for infinitesimal perturbations about the reference state. Once the deformations become "large" (or finite) the original symmetries of the reference structure are lost (except for special loadings that are consistent with the symmetry of the structure) and the linear elastic constants no longer adequately describe the response of the material.

6.6 Limitations of continuum constitutive relations

In our discussion of constitutive relations, we have made the assumption of local action (see Section 6.2), according to which the strain energy density is a pointwise function of the deformation gradient, $W = \widehat{W}(\boldsymbol{F})$. Since real materials are not continuous, it is clear that the response at a "point" represents an average over a small domain surrounding this point. We noted this at the very start of the discussion of continuum mechanics when we introduced the notion of a "continuum particle" in Section 3.1. This begs the physical question: just how *large* must this particle be for the continuum assumptions to work?

The answer depends on what we want to model, or more precisely on the characteristic length scales of the structure or body relative to the characteristic length scales of the material. It is also only easy to answer this question *a posteriori*, as it also depends on the length scales over which the deformation gradient itself varies. Once we choose a constitutive law of the form $\overline{W}(\boldsymbol{F})$, the solutions we obtain will not respect any notion of a material length scale. Instead, we will have to quantify the variations in $\boldsymbol{F}$ in any obtained solutions and decide whether our assumptions about the size of a continuum particle remain valid. We saw an example of this with the Knudsen number for a fluid in Section 5.6. We seek a similar criterion for a solid under static conditions. Imagine that for a certain displacement field, we can identify a sphere of radius r_ϵ such that

$$\|\boldsymbol{F}(\boldsymbol{X}+\boldsymbol{\xi}) - \boldsymbol{F}(\boldsymbol{X})\| < \epsilon \tag{6.179}$$

for all $\boldsymbol{X}$ in the body and for all $\|\boldsymbol{\xi}\| < r_\epsilon$, where ϵ is the tolerance that defines some limit of a "negligible" variability in $\boldsymbol{F}$. The choice of a norm in Eqn. (6.179) is arbitrary since all norms are equivalent in a finite-dimensional space (see footnote 20 on page 26). A standard norm for second-order tensors is the scalar contraction operation defined at the end of Section 2.4.5, according to which Eqn. (6.179) is

$$[(\boldsymbol{F}(\boldsymbol{X}+\boldsymbol{\xi}) - \boldsymbol{F}(\boldsymbol{X})) : (\boldsymbol{F}(\boldsymbol{X}+\boldsymbol{\xi}) - \boldsymbol{F}(\boldsymbol{X}))]^{1/2} < \epsilon. \tag{6.180}$$

In words, this relation implies that the deformation field is such that $\boldsymbol{F}$ can be considered constant within any sphere of radius r_ϵ.

The radius r_ϵ can now be compared to our material length scales in the context of the constitutive assumptions we have made. For this purpose, we define a representative volume element (RVE) of the material as a sphere of radius r_{RVE}. The RVE must be large enough

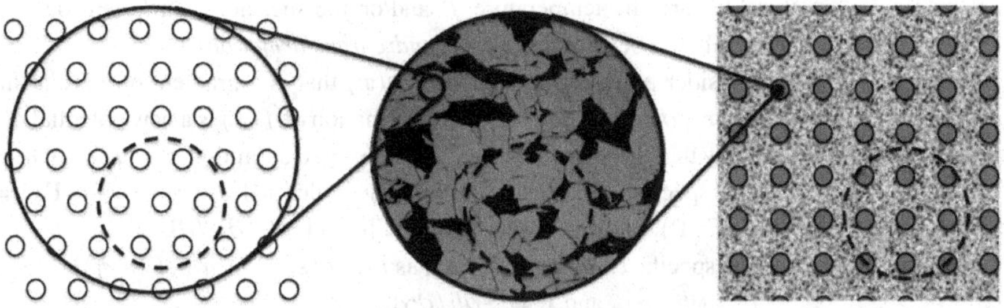

Fig. 6.5 Examples of how the representative volume element shown by the dashed circles, depends on the scale of the material of interest. On the left is a single crystal of bcc Fe, in the center is the microstructure of a single bar of steel and on the right is concrete reinforced by an array of steel bars.

that its response to a globally applied uniform $\boldsymbol{F}$ is the same as the response of any larger volume of the same material.[46]

Figure 6.5 shows several examples. Imagine that we are interested in building a constitutive model based on the assumption of a uniform, elastic, material response. A single crystal of bcc Fe, represented on the left of Fig. 6.5 by an array of Fe atoms, might be adequately modeled using an RVE whose size is on the order of the unit cell size of the lattice. However, to study the macroscopic response of steel shown in the center of Fig. 6.5, we need to consider not only a large number of randomly oriented and sized bcc Fe grains, but grains of other phases as well; steel contains a complex mixture of different crystal phases. As a result, the RVE may need to be as large as several microns in this case. Finally, we might want to model an entire bridge made of concrete that is reinforced with steel bars. In that case the RVE will need to contain one or more entire steel bars and the concrete that surrounds them, as shown on the right in Fig. 6.5.

If the radius of homogeneous deformation, r_ϵ is less than the radius of the RVE, we expect that our constitutive model assumptions will not hold; the deformation varies on a scale that is finer than the material length scale and we must refine our constitutive description. On the other hand, if the deformation gradient can be assumed constant over the scale of the RVE, we can now trust a constitutive law based on that premise. In situations where Eqn. (6.179) breaks down, it is necessary to resort to multiscale methods that combine lower-level microscopic models with continuum models. Methods of this type are discussed in Part IV of [TM11].

Exercises

6.1 [SECTION 6.2] The specific internal energy, $u = \widehat{u}(s, \boldsymbol{\Gamma})$, is a function of the specific entropy s and kinematic variables $\boldsymbol{\Gamma}$. Depending on the thermal and mechanical loading conditions, it is often more convenient to work with other thermodynamic potentials, where the control

[46] Here we assume we have a homogeneous material for which the energy density function does not depend explicitly on $\boldsymbol{X}$.

variables are the temperature T and/or the thermodynamic tensions γ. These alternative potentials can be obtained via *Legendre transformations*.

1. Consider a vector function $\boldsymbol{y} = \boldsymbol{y}(\boldsymbol{x})$ that is a gradient of a scalar field $f(\boldsymbol{x})$, i.e. $y_i = \partial f(\boldsymbol{x})/\partial x_i$. The Legendre transformation of $f(\boldsymbol{x})$ is a new potential $g(\boldsymbol{y}) = \boldsymbol{x} \cdot \boldsymbol{y} - f(\boldsymbol{x})$. Show that this function provides the inverse definition $\boldsymbol{x} = \boldsymbol{x}(\boldsymbol{y})$, where $x_i = \partial g(\boldsymbol{y})/\partial y_i$.
2. The specific *Helmholtz free energy* is defined as $\psi = \widehat{\psi}(T, \boldsymbol{\Gamma}) = \widehat{u}(\widehat{s}(T, \boldsymbol{\Gamma}), \boldsymbol{\Gamma}) - T\widehat{s}(T, \boldsymbol{\Gamma})$. Show that $s = -\partial\widehat{\psi}/\partial T|_{\boldsymbol{\Gamma}}$ and $\boldsymbol{\gamma} = \partial\widehat{\psi}/\partial\boldsymbol{\Gamma}|_T$.
3. The specific *enthalpy* is defined as $h = \widehat{h}(s, \boldsymbol{\gamma}) = \widehat{u}(s, \widehat{\boldsymbol{\Gamma}}(s, \boldsymbol{\gamma})) - \boldsymbol{\gamma} \cdot \widehat{\boldsymbol{\Gamma}}(s, \boldsymbol{\gamma})$. Show that $T = \partial\widehat{h}/\partial s|_{\boldsymbol{\gamma}}$ and $\boldsymbol{\Gamma} = -\partial\widehat{h}/\partial\boldsymbol{\gamma}|_s$.
4. The specific *Gibbs free energy* is defined as $g = \widehat{g}(T, \boldsymbol{\gamma}) = \widehat{u}(\widehat{s}(T, \boldsymbol{\gamma}), \widetilde{\boldsymbol{\Gamma}}(T, \boldsymbol{\gamma})) - T\widehat{s}(T, \boldsymbol{\gamma}) - \boldsymbol{\gamma} \cdot \widetilde{\boldsymbol{\Gamma}}(T, \boldsymbol{\gamma})$. Show that $s = -\partial\widehat{g}/\partial T|_{\boldsymbol{\gamma}}$ and $\boldsymbol{\Gamma} = -\partial\widehat{g}/\partial\boldsymbol{\gamma}|_T$.

6.2 [SECTION 6.2] A tensile test is a one-dimensional experiment where a material sample is stretched in a controlled manner to measure its response. The loading machine can control either the displacement, u, applied to the end of the sample (displacement control) or the force, f, applied to its end (load control). If displacement is controlled, the output is f/A_0, where A_0 is the reference cross-sectional area. If load is controlled, the output is $L/L_0 = (L_0 + u)/L_0$, where L_0 and L are the reference and deformed lengths of the sample. The mass of the sample is m. Describe different experiments where the relevant thermodynamic potentials are:

1. the internal energy density, u;
2. the Helmholtz free energy density, ψ;
3. the enthalpy density, h;
4. the Gibbs free energy density, g.

In each case indicate what quantity is measured in the experiment (i.e. force or length) and provide an explicit expression for it in terms of m and the appropriate potential. **Hint:** You will need to consider thermal conditions when setting up your experiments.

6.3 [SECTION 6.2] A material undergoes a homogeneous, time-dependent, simple shear motion with deformation gradient:

$$[\boldsymbol{F}] = \begin{bmatrix} 1 & \gamma(t) & 0 \\ 0 & 1 & 0 \\ 0 & 0 & 1 \end{bmatrix},$$

where $\gamma(t) = \dot{\gamma}t$ is the shear parameter and the shear rate $\dot{\gamma}$ is constant. Consider the following two cases:

1. The material is elastic, incompressible and rubber-like with a Helmholtz free energy density given by $\Psi = c_1(\operatorname{tr}\boldsymbol{B} - 3)$, where $\boldsymbol{B} = \boldsymbol{F}\boldsymbol{F}^T$ is the left Cauchy–Green deformation tensor, and c_1 is a material constant. A material of this type is called *neo-Hookean*.
   a. For constant temperature conditions, show that the Cauchy stress for a neo-Hookean material is given by $\boldsymbol{\sigma} = -p\boldsymbol{I} + \mu\boldsymbol{B}$, where p is the pressure, $\boldsymbol{I}$ is the identity tensor, $\mu = 2\rho_0 c_1$ is the shear modulus and ρ_0 is the reference mass density.
   b. Compute the Cauchy stress due to the imposed simple shear. Present your results as a 3×3 matrix of the components of $\boldsymbol{\sigma}$. Explicitly show the time dependence.
2. The material is a Newtonian fluid for which the Cauchy stress is given by $\boldsymbol{\sigma} = -p\boldsymbol{I} + 2\mu\boldsymbol{d}$, where μ is the shear viscosity and $\boldsymbol{d} = \frac{1}{2}(\nabla\boldsymbol{v} + \nabla\boldsymbol{v}^T)$ is the rate of deformation tensor.
   a. Compute the Cauchy stress due to the imposed simple shear motion.
   b. How can the pressure $p(t)$ be determined?

6.4 [SECTION 6.3] The dyad $\boldsymbol{T} = \boldsymbol{a} \otimes \boldsymbol{b}$, where $\boldsymbol{a}$ and $\boldsymbol{b}$ are vectors, is a second-order tensor. Assuming that $\boldsymbol{a}$ and $\boldsymbol{b}$ are objective, show that $\boldsymbol{T}$ satisfies the objectivity condition in Eqn. (6.66). (This can also be viewed as a way to obtain the objectivity condition.) How can this approach be extended to establish the objectivity conditions for nth-order tensors with $n \geq 3$?

6.5 [SECTION 6.3] The transformation relation between two frames of reference for the rate of deformation tensor $\boldsymbol{d} = \frac{1}{2}(\boldsymbol{l} + \boldsymbol{l}^T)$ is given in Eqn. (6.56). The spin tensor is defined as $\boldsymbol{w} = \frac{1}{2}(\boldsymbol{l} - \boldsymbol{l}^T)$. Find the relation between $\boldsymbol{w}$ and $\boldsymbol{w}^+$. Is $\boldsymbol{w}$ an objective tensor?

6.6 [SECTION 6.3] Let $\dot{\boldsymbol{\sigma}}$ stand for the material time derivative of $\boldsymbol{\sigma}$,

$$\dot{\boldsymbol{\sigma}}(\boldsymbol{x}, t) \equiv \frac{\partial}{\partial t}\boldsymbol{\sigma}(\boldsymbol{x}(\boldsymbol{X}, t), t)\bigg|_{\boldsymbol{X}}.$$

This is the rate-of-change of the Cauchy stress experienced by a fixed material particle.

1. Find a relation between $\dot{\boldsymbol{\sigma}}$ and $\dot{\boldsymbol{\sigma}}^+$, and show that this time derivative is not an objective quantity.
2. The *Jaumann stress rate* (or *corotational stress rate*) is defined by

$$\overset{\nabla}{\boldsymbol{\sigma}} \equiv \dot{\boldsymbol{\sigma}} + \boldsymbol{\sigma}\boldsymbol{w} - \boldsymbol{w}\boldsymbol{\sigma},$$

where $\boldsymbol{w}$ is the spin tensor. Show that the Jaumann stress rate is an objective tensor. **Hint:** You will need to use the result from Exercise 6.5.
3. Another example of an objective stress rate is the *Truesdell stress rate* defined by

$$\mathring{\boldsymbol{\sigma}} \equiv \dot{\boldsymbol{\sigma}} - \boldsymbol{l}\boldsymbol{\sigma} - \boldsymbol{\sigma}\boldsymbol{l}^T + \boldsymbol{\sigma}\,\mathrm{tr}\,\boldsymbol{l},$$

where $\boldsymbol{l}$ is the velocity gradient. Show that the Truesdell stress rate is an objective tensor.

6.7 [SECTION 6.3] The material frame-indifference conditions in Eqn. (6.79) involve terms of the form $\mathcal{L}_0^{-1}\mathcal{L}_t\gamma$, where γ represents a variable dependence of the constitutive relation. Show that for $\gamma = \rho$ (mass density), $\gamma = \boldsymbol{v}$ (velocity vector), and $\gamma = \boldsymbol{l} = \nabla\boldsymbol{v}$ (velocity gradient tensor), $\mathcal{L}_0^{-1}\mathcal{L}_t\gamma$ is given by

1. $\mathcal{L}_0^{-1}\mathcal{L}_t\rho = \rho$,
2. $\mathcal{L}_0^{-1}\mathcal{L}_t\boldsymbol{v} = \mathcal{Q}_0^T\dot{\boldsymbol{c}}^+ + \dot{\boldsymbol{Q}}\boldsymbol{x} + \boldsymbol{Q}\boldsymbol{v}$,
3. $\mathcal{L}_0^{-1}\mathcal{L}_t\boldsymbol{l} = \dot{\boldsymbol{Q}}\boldsymbol{Q} + \boldsymbol{Q}\boldsymbol{l}\boldsymbol{Q}^T$,

where $\mathcal{Q}_t$ is an orthogonal linear transformation between the frames of reference, $\boldsymbol{c}^+$ is the relative translation between the frames and $\boldsymbol{Q} = \mathcal{Q}_0^T\mathcal{Q}_t$ is a proper orthogonal, second-order tensor.

6.8 [SECTION 6.3] Consider a constitutive equation for the Cauchy stress that is linear in the velocity, $\boldsymbol{v}$, and the velocity gradient, $\boldsymbol{l} = \nabla\boldsymbol{v}$, namely

$$\sigma_{ij} = A_{ij} + B_{ijm}v_m + C_{ijmn}l_{mn} \quad \Leftrightarrow \quad \boldsymbol{\sigma} = \boldsymbol{A} + \boldsymbol{B}\boldsymbol{v} + \boldsymbol{C} : \boldsymbol{l},$$

where $A_{ij}, B_{ijm}, C_{ijmn}$ are tensor-valued functions of the density, ρ, and each are symmetric in the indices i and j. Our objective is to obtain constraints on the tensor functions, $\boldsymbol{A}(\rho)$, $\boldsymbol{B}(\rho)$ and $\boldsymbol{C}(\rho)$, due to material frame-indifference. Recall that the material frame-indifference condition for the stress tensor is $\widehat{\boldsymbol{\sigma}}(\mathcal{L}_0^{-1}\mathcal{L}_t\gamma) = \boldsymbol{Q}\widehat{\boldsymbol{\sigma}}(\gamma)\boldsymbol{Q}^T$, where γ represents the arguments of the stress function $\widehat{\boldsymbol{\sigma}}$ (see Eqn. (6.79)). **Hint:** To do the following you will need to use $\mathcal{L}_0^{-1}\mathcal{L}_t\rho$, $\mathcal{L}_0^{-1}\mathcal{L}_t\boldsymbol{v}$ and $\mathcal{L}_0^{-1}\mathcal{L}_t\boldsymbol{l}$ given in Exercise 6.7 and the properties of isotropic tensors in Section 2.5.6.

1. Consider a deformation for which $\boldsymbol{v} = \boldsymbol{0}$. In this case only the $\boldsymbol{A}(\rho)$ term exists. Show that material frame-indifference implies that $\boldsymbol{A} = \alpha(\rho)\boldsymbol{I}$, where $\alpha(\rho)$ is a real-valued function of the density and $\boldsymbol{I}$ is the identity tensor.

2. Consider a motion for which $\boldsymbol{v}$ is constant. Show that material frame-indifference implies that $\boldsymbol{B} = \boldsymbol{0}$.
3. Show that material frame-indifference implies that $C_{ijkl}(\rho)$ must have the following form:

$$C_{ijkl} = \beta(\rho)\delta_{ij}\delta_{kl} + \mu(\rho)(\delta_{ik}\delta_{jl} + \delta_{il}\delta_{jk}),$$

where $\beta(\rho)$ and $\mu(\rho)$ are real-valued functions of the density, and δ_{ij} is the Kronecker delta.
4. Based on the results in the previous three parts, show that after accounting for the constraints due to material frame-indifference, the most general allowable form for Eqn. (6.8) is

$$\boldsymbol{\sigma} = \alpha(\rho)\boldsymbol{I} + \beta(\rho)(\operatorname{tr}\boldsymbol{d})\boldsymbol{I} + 2\mu(\rho)\boldsymbol{d},$$

where $\boldsymbol{d}$ is the rate of deformation tensor.

6.9 [SECTION 6.4] The constitutive relation for a Reiner–Rivlin fluid is given in Eqn. (6.124).
1. Show that by only retaining terms that are linear in the rate of deformation tensor, $\boldsymbol{d}$, the Reiner–Rivlin constitutive relation reduces to that of a Newtonian fluid in Eqn. (6.125). Find expressions for the bulk viscosity, κ, and the shear viscosity, μ, in terms of functions appearing in the Reiner–Rivlin form.
2. Consider the motion, $\boldsymbol{x} = \alpha(t)\boldsymbol{X}$, where $\alpha(t)$ is a differentiable function of time, $\boldsymbol{X}$ are coordinates in the referential description and $\boldsymbol{x}$ are coordinates in the spatial description. Assuming that the fluid is Newtonian, compute the stress in the fluid. This result demonstrates why κ is called the *bulk* viscosity. Explain.
3. Consider the motion, $x_1 = X_1 + \gamma(t)X_2, x_2 = X_2, x_3 = X_3$, where $\gamma(t)$ is a differentiable function of time. Assuming that the fluid is Newtonian, compute the stress in the fluid. This result demonstrates why μ is called the *shear* viscosity. Explain.

6.10 [SECTION 6.4] It can be shown that the most general form for the internal energy density function, $\widehat{u}(\boldsymbol{C})$, for an isotropic incompressible material is

$$\widehat{u}(\boldsymbol{C}) = \psi(I_1^{\boldsymbol{C}}, I_2^{\boldsymbol{C}}), \qquad (*)$$

where $\boldsymbol{C}$ is the right Cauchy–Green deformation tensor, $I_i^{\boldsymbol{C}}$ are the principal invariants of $\boldsymbol{C}$, $\psi(\cdot,\cdot)$ is an arbitrary function of its arguments and the domain of $\widehat{u}$ is restricted to values of $\boldsymbol{C}$ for which $I_3^{\boldsymbol{C}} = 1$. It is convenient to employ the method of Lagrange multipliers which allows us to work with functions on unrestricted domains. In this case, we introduce the augmented energy function

$$\bar{u}(\boldsymbol{C}) = \widehat{u}(\boldsymbol{C}) - p(I_3^{\boldsymbol{C}} - 1), \qquad (**)$$

where p is the undetermined pressure. Note that the augmented energy function $\bar{u}$ does not have a physical meaning for values of $I_3^{\boldsymbol{C}}$ other than 1, but it is equal to $\widehat{u}$ when $I_3^{\boldsymbol{C}} = 1$. Show that the second Piola–Kirchhoff stress corresponding to Eqn. (**) is

$$\boldsymbol{S} = 2\rho_0\left[\left(\frac{\partial\psi}{\partial I_1^{\boldsymbol{C}}} + I_1^{\boldsymbol{C}}\frac{\partial\psi}{\partial I_2^{\boldsymbol{C}}}\right)\boldsymbol{I} - \frac{\partial\psi}{\partial I_2^{\boldsymbol{C}}}\boldsymbol{C}\right] - \rho_0 p\boldsymbol{C}^{-1},$$

when the incompressibility constraint is enforced by setting $I_3^{\boldsymbol{C}} = 1$. Here ρ_0 is the reference mass density, and $\boldsymbol{I}$ is the identity tensor.

6.11 [SECTION 6.5] Derive Eqn. (6.157) following the procedure outlined in the text.

6.12 [SECTION 6.5] Show that linearizing the general stress function for isotropic materials in Eqn. (6.130) gives Hooke's law in Eqn. (6.175). **Hint:** Replace the left Cauchy–Green deformation tensor $\boldsymbol{B}$ in Eqn. (6.130) by the appropriate small strain measure (see Example 3.8) and retain only linear terms.

6.13 [SECTION 6.5] Show that for a one-dimensional problem (where the only nonzero stress component is $\sigma_{11} = \sigma$), Hooke's law for an isotropic solid in Eqn. (6.175) reduces to $\sigma = E\epsilon$, where σ and ϵ are the stress and strain in the direction of loading and $E = \mu(3\lambda+2\mu)/(\lambda+\mu)$ is Young's modulus.

6.14 [SECTION 6.5] Under conditions of hydrostatic loading, $\boldsymbol{\sigma} = -p\boldsymbol{I}$, where p is the pressure, the *bulk modulus* B is defined as the negative ratio of the pressure and dilatation, $e = \operatorname{tr} \boldsymbol{\epsilon}$, so that $p = -Be$. Starting with the generalized Hooke's law in Voigt notation in Eqn. (6.172), obtain expressions for the bulk modulus of the eight crystal symmetry classes presented in Fig. 6.4. In particular, show that for tetragonal, trigonal and hexagonal symmetry the bulk modulus is given by

$$B = \frac{(\mathsf{c}_{11} + \mathsf{c}_{12})\mathsf{c}_{33} - 2\mathsf{c}_{13}^2}{\mathsf{c}_{11} + \mathsf{c}_{12} - 4\mathsf{c}_{13} + 2\mathsf{c}_{33}},$$

and for cubic and isotropic symmetry the bulk modulus is

$$B = \frac{\mathsf{c}_{11} + 2\mathsf{c}_{12}}{3}.$$

Also, show that for isotropic symmetry, the bulk modulus can also be expressed in terms of the Lamé constants as $B = \lambda + 2\mu/3$. **Hint:** This exercise is best performed on a computer using a symbolic mathematics package.

7 Boundary-value problems, energy principles and stability

In this final chapter of Part I, we discuss the formulation and specification of well-defined problems in continuum mechanics. For simplicity, we restrict our attention to the purely mechanical behavior of materials. This means that, unless otherwise explicitly stated, in this chapter we will ignore thermodynamics. The resulting theory provides a reasonable approximation of real material behavior in two extreme conditions. The first scenario is that of *isentropic processes* (see Section 6.2.5), where the motion and deformation occurs at such a high temporal rate that essentially no flow of heat occurs. In this scenario the strain energy density function should be associated with the internal energy density at constant entropy. The second scenario is that of *isothermal processes* (see Section 6.2.5), where the motion and deformation occurs at such a low temporal rate that the temperature is essentially uniform and constant. In this scenario the strain energy density function should be associated with the Helmholtz free energy density at constant temperature.

We start by discussing the specification of initial boundary-value problems in Section 7.1. Then, in Section 7.2 we develop the principle of stationary potential energy. Finally, in Section 7.3 we introduce the idea of stability and ultimately derive the principle of minimum potential energy.

7.1 Initial boundary-value problems

So far we have laid out an extensive set of concepts and derived the local balance laws to which continuous physical systems (which satisfy the various assumptions we have made along the way) must conform. Now we pull these together into a formal *problem statement* which consists of three distinct parts: (1) the partial differential *field equations* to be satisfied; (2) the *unknown fields* that constitute the sought solution of the problem and the relations between them; and (3) the *prescribed data*, which include everything else that is required to turn the problem into one that *can* be solved. If we are interested in the dynamic response of a system, then the problem is referred to as an *initial boundary-value problem* and its three parts will all have a temporal component. If we are only interested in the static equilibrium state of our system, then the term *boundary-value problem* is used.

In addition to the above considerations, continuum mechanics problems naturally divide into two further categories: those which are formulated within the spatial description and those that are formulated within the material description. The former category is most useful for fluid mechanics problems and the latter for solid mechanics problems. However, solids or fluids problems can, in principle, be solved with either description.

7.1.1 Problems in the spatial description

We first describe the initial boundary-value problem in the spatial description: the so-called Eulerian approach. The first part of a problem is the field equations which, in this case, are the continuity equation (conservation of mass, Eqn. (4.3)) and the balance of linear momentum (Eqn. (4.25)). The balance of angular momentum leads to the symmetry of the Cauchy stress tensor ($\boldsymbol{\sigma} = \boldsymbol{\sigma}^T$) that can be directly imposed. The resulting set of equations for a system occupying spatial domain B is

$$\left.\begin{aligned} &\frac{\partial \rho}{\partial t} + \operatorname{div}(\rho \boldsymbol{v}) = 0, \\ &\operatorname{div} \boldsymbol{\sigma} + \rho \boldsymbol{b} = \rho \frac{\partial \boldsymbol{v}}{\partial t} + \rho (\nabla \boldsymbol{v}) \boldsymbol{v}, \end{aligned}\right\} \quad \boldsymbol{x} \in B,\ t > 0, \tag{7.1}$$

where we have used Eqn. (3.34) to write the balance of linear momentum in terms of the velocity field. Typically (in fluids problems), B is a constant *control volume*.[1]

The second part of a problem is the set of unknown fields. Here, the unknowns are taken to be the fields $\rho(\boldsymbol{x}, t)$ and $\boldsymbol{v}(\boldsymbol{x}, t)$. The aim of the problem is to determine these fields such that they simultaneously satisfy Eqn. (7.1) subject to the conditions specified below.

The final part of a problem are the prescribed data. In this case the data include, in addition to the (initial) domain B, the *initial conditions* and *boundary conditions* for the unknown fields and the specification of functions that provide the body forces and the Cauchy stress. The partial differential equations in Eqn. (7.1) are of first order in time. Therefore, we will need to specify the initial velocity and density fields:

$$\rho(\boldsymbol{x}, 0) = \rho_{\text{init}}(\boldsymbol{x}), \quad \boldsymbol{v}(\boldsymbol{x}, 0) = \boldsymbol{v}_{\text{init}}(\boldsymbol{x}), \quad \boldsymbol{x} \in B \cup \partial B.$$

For boundary conditions we must specify, at *each* point on the boundary of B, *one* quantity for *each unknown field component*. Since the velocity field is a vector quantity we must specify three values associated with the motion at each boundary point: one value for each spatial direction. These values can correspond to either a velocity or a traction component. If only velocities are prescribed, the problem is said to have *velocity boundary conditions*:

$$\boldsymbol{v}(\boldsymbol{x}, t) = \bar{\boldsymbol{v}}(\boldsymbol{x}, t), \quad \boldsymbol{x} \in \partial B,\ t > 0,$$

where $\bar{\boldsymbol{v}}(\boldsymbol{x}, t)$ is a specified velocity field imposed at the surfaces of the body. Another possibility is to impose *traction boundary conditions* where only tractions are applied:

$$\boldsymbol{\sigma n}(\boldsymbol{x}, t) = \bar{\boldsymbol{t}}(\boldsymbol{x}, t), \quad \boldsymbol{x} \in \partial B,\ t > 0.$$

Here $\boldsymbol{n}(\boldsymbol{x}, t)$ is the outward unit normal to ∂B (which may, in fact, be constant) and $\bar{\boldsymbol{t}}(\boldsymbol{x}, t)$ is a specified field of external tractions applied to the surfaces of the body. It is also possible to combine traction and velocity boundary conditions. In this case the boundary is divided

[1] There are many situations, such as when free surfaces exist, where B will be time dependent. However, we do not consider the spatial formulation of such initial boundary-value problems in this book.

into a part ∂B_t where traction boundary conditions are applied and a part ∂B_v where velocity boundary conditions are applied, such that $\partial B_t \cup \partial B_v = \partial B$ and $\partial B_t \cap \partial B_v = \emptyset$. The resulting *mixed boundary conditions* are

$$\begin{aligned} \boldsymbol{\sigma n}(\boldsymbol{x}, t) &= \bar{\boldsymbol{t}}(\boldsymbol{x}, t), \qquad \boldsymbol{x} \in \partial B_t,\ t > 0, \\ \boldsymbol{v}(\boldsymbol{x}, t) &= \bar{\boldsymbol{v}}(\boldsymbol{x}, t), \qquad \boldsymbol{x} \in \partial B_v,\ t > 0. \end{aligned}$$

In particular, it is worth noting that "free surfaces," i.e. parts of the body where no forces and no velocities are specified, are described as traction boundary conditions with $\bar{\boldsymbol{t}} = \mathbf{0}$. These three cases do not, however, exhaust the list of possibilities. Another case is *mixed–mixed boundary conditions*, where traction and velocity boundary conditions are individually applied to different spatial directions at a single point on the surface. Thus, a point on the surface may have a velocity boundary condition along some directions and traction boundary conditions along the others. An example of a physical situation that corresponds to such a boundary condition is a frictionless piston in a cylindrical container with an external pressure (normal component of traction) p. The fluid in the container can move the piston along the cylinder's axis (if it generates a traction in that direction that is larger than the constraining pressure p), but the fluid cannot move along the piston where *no-slip*, zero velocity, conditions are assumed to hold. Assuming the axis of the cylinder is in the direction $\boldsymbol{n}$, the boundary conditions for the fluid at the piston would be: $(\boldsymbol{\sigma n}) \cdot \boldsymbol{n} = -p$, and $\boldsymbol{v} - (\boldsymbol{v} \cdot \boldsymbol{n})\boldsymbol{n} = \mathbf{0}$. An important point regarding mixed–mixed conditions that deserves reiteration is that it is not possible to apply both traction and velocity boundary conditions along the same direction at the same point. Doing so will generally result in an ill-posed boundary-value problem for which no solution exists.

We have identified three boundary conditions associated with the three components of the velocity field, but we still require one more condition associated with the density field. However, in this case no further data need to be supplied. Instead, the appropriate boundary condition takes the form of a consistency equation between the two unknown fields. This condition ensures that the mass flux across the boundary of the body is equal to the density at the boundary times the velocity, i.e.

$$(\nabla \rho - \rho \boldsymbol{v}) \cdot \boldsymbol{n} = 0, \quad \boldsymbol{x} \in \partial B,\ t > 0.$$

The final pieces of data required are the functions that determine how the body forces $\boldsymbol{b}$ (three functions) and the stresses $\boldsymbol{\sigma}$ (six functions) depend on the unknown fields and, in general, time. The body forces are governed by well-characterized physical principles and are usually given by simple functions. As we saw in Chapter 6, the relations that describe the stresses associated with a given state and history of a body – the *constitutive relations* – have very few constraints on their functional form and commonly are given by complicated nonlinear functions.

If the problem is one of steady state, this means that the time derivatives in Eqn. (7.1) are zero and so the set of differential equations reduces to

$$\left.\begin{aligned} \operatorname{div} \rho \boldsymbol{v} &= 0, \\ \operatorname{div} \boldsymbol{\sigma} + \rho \boldsymbol{b} &= \rho (\nabla \boldsymbol{v}) \boldsymbol{v}, \end{aligned}\right\} \quad \boldsymbol{x} \in B. \tag{7.2}$$

We call these the *steady-state stress equations*. In this case, we are not interested in how a system reaches steady state (the transient behavior that would be captured by the initial boundary-value problem), and thus, initial conditions are not needed. Instead, everything is independent of time and only the (constant) boundary conditions must be specified.

7.1.2 Problems in the material description

The continuum mechanics initial boundary-value problem can also be formulated in the material description. This is referred to as a Lagrangian description. The first part of a problem is the field equations. In the material description, the balance of linear momentum is given by Eqn. (4.39) in terms of the first Piola–Kirchhoff stress or by Eqn. (4.43) in terms of the second Piola–Kirchhoff stress. If we use the latter equation, we also enforce the balance of angular momentum by requiring that the second Piola–Kirchhoff stress is symmetric ($\boldsymbol{S} = \boldsymbol{S}^T$). Further, in deriving Eqn. (4.43) the conservation of mass (Eqn. (4.1)) has been used. Since the resulting equation depends only on the reference mass density ρ_0, which in the material description does not depend on time (and will be specified as part of the problem data), it is not necessary to include the continuity equation. Thus, the field equations for a problem in the material description are

$$\operatorname{Div}\left[(\nabla_0 \boldsymbol{\varphi})\boldsymbol{S}\right] + \rho_0 \breve{\boldsymbol{b}} = \rho_0 \frac{\partial^2 \boldsymbol{\varphi}}{\partial t^2}, \qquad \boldsymbol{X} \in \mathcal{B}_0,\ t > 0, \tag{7.3}$$

where we have explicitly indicated that the body force is expressed in its material form.

The second part of a problem is the set of unknown fields, which in this case consists of the deformation mapping $\boldsymbol{\varphi}(\boldsymbol{X},t)$. As shown in Chapter 3, knowledge of the deformation mapping allows for the computation of all other kinematic quantities of interest.

Just as in the spatial description, the final part of a problem is the prescribed data. For a problem in the material description, the data include the *initial conditions* and *boundary conditions* for the unknown deformation mapping field and the specification of functions that provide the body forces and the second Piola–Kirchhoff stress. The partial differential equations in Eqn. (7.3) are of second order in time. Therefore, we will need to specify the initial reference configuration $\mathcal{B}_0$, and the initial velocity fields:

$$\boldsymbol{\varphi}(\boldsymbol{X},0) = \boldsymbol{X}, \quad \breve{\boldsymbol{v}}(\boldsymbol{X},0) = \frac{\partial \boldsymbol{\varphi}(\boldsymbol{X},0)}{\partial t} = \breve{\boldsymbol{v}}_{\text{init}}(\boldsymbol{X}), \quad \boldsymbol{X} \in \mathcal{B}_0 \cup \partial\mathcal{B}_0.$$

Once again, as in the spatial description, we must specify as many boundary conditions at *each* point on the boundary of $\mathcal{B}_0$ as there are *unknown field components*. Since the deformation mapping is a vector quantity we must specify three values associated with the motion at each boundary point, one value for each spatial direction. These values can correspond to either the components of the traction or the position:[2]

$$SN(X,t) = F^{-1}\bar{T}(X,t), \qquad X \in \partial B_0,\ t > 0,$$

or

$$\varphi(X,t) = \bar{x}(X,t), \qquad X \in \partial B_0,\ t > 0,$$

where $N(X,t)$ is the outward unit normal of ∂B_0 and $\bar{T}(X,t)$ and $\bar{x}(X,t)$ are specified fields of external reference tractions and positions applied to the surfaces of the body, respectively. Often position boundary conditions are provided in terms of displacements from the reference configuration. In this case, the boundary condition reads

$$\varphi(X,t) = X + \bar{u}(X,t), \qquad X \in \partial B_0,\ t > 0,$$

where $\bar{u}(X,t)$ is the specified boundary displacement field. Clearly the two forms are related by $\bar{x}(X,t) \equiv X + \bar{u}(X,t)$. The *mixed boundary conditions* are

$$\begin{aligned} SN(X,t) &= F^{-1}\bar{T}(X,t), \qquad X \in \partial B_{0t},\ t > 0, \\ \varphi(X,t) &= \bar{x}(X,t), \qquad X \in \partial B_{0u},\ t > 0. \end{aligned}$$

As for the spatial description, in the case of mixed boundary conditions, the boundary is divided into a part ∂B_{0t} where traction boundary conditions are applied and a part ∂B_{0u} where displacement boundary conditions are applied, such that $\partial B_{0t} \cup \partial B_0 = \partial B_0$ and $\partial B_{0t} \cap \partial B_{0u} = \emptyset$. Again, it is worth emphasizing that "free surfaces," i.e. parts of the body where no forces and no positions are applied, are described as traction boundary conditions with $\bar{T} = \mathbf{0}$. *Mixed–mixed boundary conditions* can also be defined for problems in the material description. Here, a point on the surface may have a position boundary condition along some directions and traction boundary conditions along the others. A pin sliding in a frictionless rigid slot is an example of this. The pin can move freely along the slot direction (the traction component is zero), but cannot move perpendicular to the slot (displacement components are zero). Similarly to spatial description problems, it is not possible to apply both traction and position (displacement) boundary conditions along the same direction at the same point. Doing so will generally result in an ill-posed boundary-value problem for which no solution exists. The final pieces of data required are the functions that determine

[2] This is actually more complicated than it sounds when considered from a microscopic perspective. Since all physical matter interacts through forces between atoms it is actually not possible to apply "position" boundary conditions. This is clearly an approximation reflecting the relative rigidity of one material compared with another. Consider, for example, a box placed on a floor. We may choose to model this with a position boundary condition applied to the bottom of the box. In reality, though, the box will sink somewhat into the floor – a fact that is neglected by the position boundary condition. This issue is part of a larger problem associated with the application of boundary conditions at finite deformation. Since bodies always change their shape as a result of applied loading, how can traction boundary conditions be specified? Tractions are defined as the force per *deformed* surface area, but the deformed surface area is unknown before the force is applied and the body deforms. This creates a difficult problem for experimentalists attempting to design experiments with well-characterized boundary conditions at large deformation (see, for example, the discussions in [Tre48, RS51, CJ93]). This is also the essential difficulty in applying accurate stress boundary conditions in atomistic simulations (see, for example, Sections 6.4.3 and 9.5 of [TM11]).

the body forces $\boldsymbol{b}$ (three functions) and the constitutive relations for the second Piola–Kirchhoff stresses $\boldsymbol{S}$ (six functions) which were discussed in Chapter 6.

If the problem is static, the differential equation in Eqn. (7.3) reduces to the stress equilibrium equation

$$\operatorname{Div}[(\nabla_0 \boldsymbol{\varphi})\boldsymbol{S}] + \rho_0 \breve{\boldsymbol{b}} = \boldsymbol{0}, \qquad \boldsymbol{X} \in \mathcal{B}_0, \tag{7.4}$$

where we have explicitly indicated that the body force is expressed in its material form.

7.2 Equilibrium and the principle of stationary potential energy (PSPE)

In this section, we reformulate the thermomechanical equilibrium (static) boundary-value problem discussed above as a *variational* problem. This means that we seek to write the problem in such a way that its solution is a stationary point (maximum, minimum or saddle point) of some energy functional. In the next section, we will see that *stable* equilibrium solutions correspond to minima of this functional. The reformulation we seek can be performed for problems involving hyperelastic materials, i.e. the stress in the material is given by the derivative of a strain energy density function with respect to strain (Eqn. (6.43)). In this context we treat the strain energy density function as a purely mechanical quantity and ignore its connection to thermodynamics (which is discussed in Section 6.2.5).

The appropriate energy functional for the continuum mechanics boundary-value problem is the *total potential energy* Π. The total potential energy is defined as the strain energy stored in the body together with the potential of the applied loads:[3]

$$\Pi = \int_{\mathcal{B}_0} W(\boldsymbol{F})\, dV_0 - \int_{\mathcal{B}_0} \rho_0 \breve{\boldsymbol{b}} \cdot \boldsymbol{\varphi}\, dV_0 - \int_{\partial \mathcal{B}_{0t}} \bar{\boldsymbol{T}} \cdot \boldsymbol{\varphi}\, dA_0, \tag{7.5}$$

where $\boldsymbol{\varphi}$ is the deformation mapping, $\boldsymbol{F} = \nabla_0 \boldsymbol{\varphi}$ is the deformation gradient and $W(\boldsymbol{F})$ is the strain energy density (Eqn. (6.42)). Here, we are considering a body force field $\breve{\boldsymbol{b}}$ (expressed in its material form) and reference traction field $\boldsymbol{T}$, corresponding to *dead-loading*, i.e. fields whose magnitude and direction are constant and independent of the deformation $\boldsymbol{\varphi}$. The boundary conditions for a problem in the Lagrangian description are

$$\boldsymbol{P}(\boldsymbol{F})\boldsymbol{N} = \bar{\boldsymbol{T}} \qquad \text{on } \partial\mathcal{B}_{0t},$$
$$\boldsymbol{\varphi} = \bar{\boldsymbol{x}} \qquad \text{on } \partial\mathcal{B}_{0u},$$

where $\boldsymbol{P}$ is the first Piola–Kirchhoff stress, and $\bar{\boldsymbol{T}}$ and $\bar{\boldsymbol{x}}$ can depend on $\boldsymbol{X}$. For convenience, we have expressed the displacement boundary condition directly in terms of the deformation mapping $\boldsymbol{\varphi}$, instead of the displacement field $\boldsymbol{u} = \boldsymbol{\varphi} - \boldsymbol{X}$. However, since $\boldsymbol{X}$ is constant the two approaches are equivalent. A displacement field (or deformation mapping field)

[3] Notice that not all loads have a potential function. Thus, we are further restricted to considering problems where conservative body forces and traction fields are applied.

that satisfies the position boundary conditions is called *admissible*. The solution to the boundary-value problem must be drawn from the set of admissible displacement fields.

We postulate the following variational principle:

Principle of stationary potential energy (PSPE) Given the set of admissible displacement fields for a conservative system, an equilibrium state will correspond to one for which the total potential energy is stationary.

Proof Assume that the potential energy Π is stationary at φ. This means that

$$\langle \mathcal{D}_{\varphi}\Pi; \delta\boldsymbol{u}\rangle = \frac{d}{d\eta}\Pi[\varphi + \eta\delta\boldsymbol{u}]\bigg|_{\eta=0} = 0, \qquad \forall\delta\boldsymbol{u}, \tag{7.6}$$

where $\langle \mathcal{D}_{\varphi}\Pi; \delta\boldsymbol{u}\rangle$ is the functional variation of Π defined in Eqn. (3.26) and $\delta\boldsymbol{u}$ is a small displacement field with $\delta\boldsymbol{u} = \boldsymbol{0}$ on ∂B_{0u}, so that $\varphi + \eta\delta\boldsymbol{u}$ is kinematically admissible. Substituting Eqn. (7.5) into Eqn. (7.6), we have

$$\langle \mathcal{D}_{\varphi}\Pi; \delta\boldsymbol{u}\rangle = \int_{B_0} \langle \mathcal{D}_{\varphi}W(\boldsymbol{F}); \delta\boldsymbol{u}\rangle\, dV_0 - \int_{B_0} \rho_0\breve{\boldsymbol{b}}\cdot\delta\boldsymbol{u}\, dV_0 - \int_{\partial B_{0t}} \bar{\boldsymbol{T}}\cdot\delta\boldsymbol{u}\, dA_0 = 0, \tag{7.7}$$

which must be true for all admissible displacement perturbation fields $\delta\boldsymbol{u}$. Now, focus on the integrand of the first integral in Eqn. (7.7):

$$\begin{aligned}
\langle \mathcal{D}_{\varphi}W(\boldsymbol{F}); \delta\boldsymbol{u}\rangle &= \frac{\partial W}{\partial F_{iJ}}\langle \mathcal{D}_{\varphi}F_{iJ}; \delta\boldsymbol{u}\rangle \\
&= P_{iJ}(\boldsymbol{F})\,\frac{d}{d\eta}\left[\frac{\partial(\varphi_i + \eta\delta u_i)}{\partial X_J}\right]\bigg|_{\eta=0} \\
&= P_{iJ}(\boldsymbol{F})\frac{\partial\delta u_i}{\partial X_J} = \boldsymbol{P}(\boldsymbol{F}) : \nabla_0\delta\boldsymbol{u},
\end{aligned} \tag{7.8}$$

where we have used Eqn. (6.43) and set $\boldsymbol{P}(\boldsymbol{F}) = \boldsymbol{P}^{(e)}(\boldsymbol{F})$, since the material is hyperelastic and we are only considering static configurations. Substituting Eqn. (7.8) into Eqn. (7.7) we have

$$\int_{B_0} \boldsymbol{P} : \nabla_0\delta\boldsymbol{u}\, dV_0 - \int_{B_0} \rho_0\breve{\boldsymbol{b}}\cdot\delta\boldsymbol{u}\, dV_0 - \int_{\partial B_{0t}} \bar{\boldsymbol{T}}\cdot\delta\boldsymbol{u}\, dA_0 = 0, \qquad \forall\delta\boldsymbol{u}, \tag{7.9}$$

which is a special case of the *principle of virtual work*.[4] To continue, we focus on the first term in Eqn. (7.9) and integrate it by parts:

$$\int_{B_0} \boldsymbol{P} : \nabla_0\delta\boldsymbol{u}\, dV_0 = \int_{\partial B_0} (\boldsymbol{P}\boldsymbol{N})\cdot\delta\boldsymbol{u}\, dA_0 - \int_{B_0} (\operatorname{Div}\boldsymbol{P})\cdot\delta\boldsymbol{u}\, dV_0. \tag{7.10}$$

[4] The principle of virtual work is actually far more general than it appears here. It is not limited to conservative systems and the stress and displacement fields appearing in it can be completely arbitrary as long as they satisfy the balance laws and boundary conditions, respectively. In its general form, it is an important principle that is broadly used both in theoretical and computational applications of continuum mechanics. See [Mal69, Section 5.5] for a detailed explanation.

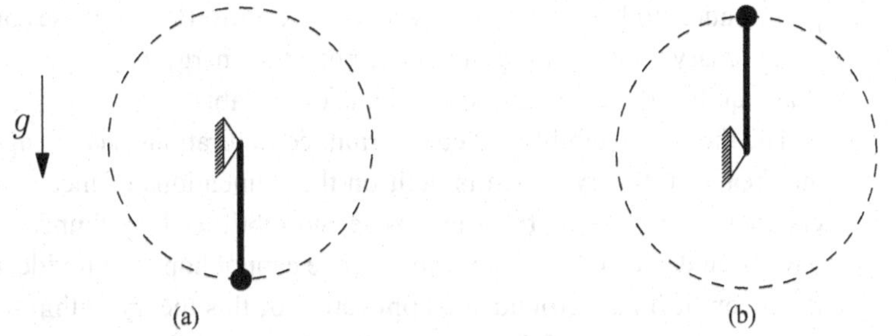

Fig. 7.1 Schematic diagram showing a pendulum consisting of a rigid rod and spherical mass connected to a fixed pin in (a) stable and (b) unstable states of equilibrium. The direction of gravity is indicated by g.

Substituting the material Cauchy relation, $\boldsymbol{PN} = \boldsymbol{T}$, into Eqn. (7.10) and then substituting this back into Eqn. (7.9), we have after rearranging terms:

$$-\int_{\mathcal{B}_0} \left(\operatorname{Div} \boldsymbol{P} + \rho_0 \breve{\boldsymbol{b}}\right) \cdot \delta \boldsymbol{u}\, dV_0 + \int_{\partial \mathcal{B}_0} \boldsymbol{T} \cdot \delta \boldsymbol{u}\, dA_0 - \int_{\partial \mathcal{B}_{0t}} \bar{\boldsymbol{T}} \cdot \delta \boldsymbol{u}\, dA_0 = 0. \quad (7.11)$$

Although the integration bounds of the last two terms are not the same, they cancel since for the middle integral $\delta \boldsymbol{u} = \boldsymbol{0}$ on $\partial \mathcal{B}_{0u}$ (as previously mentioned, $\delta \boldsymbol{u}$ must be zero wherever displacements are prescribed for $\boldsymbol{\varphi} + \delta \boldsymbol{u}$ to be kinematically admissible) and $\boldsymbol{T} = \bar{\boldsymbol{T}}$ on $\partial \mathcal{B}_{0t}$. Therefore Eqn. (7.11) reduces to

$$\int_{\mathcal{B}_0} \left(\operatorname{Div} \boldsymbol{P} + \rho_0 \breve{\boldsymbol{b}}\right) \cdot \delta \boldsymbol{u}\, dV_0 = 0. \quad (7.12)$$

This equation must be satisfied for all admissible $\delta \boldsymbol{u}$, which implies that

$$\operatorname{Div} \boldsymbol{P} + \rho_0 \breve{\boldsymbol{b}} = \boldsymbol{0}, \quad (7.13)$$

but this is exactly the static equilibrium equation (Eqn. (4.39)) and hence the principle of stationary potential energy is proved. □

7.3 Stability of equilibrium configurations

In the previous section we discovered that finding a deformed configuration that satisfies the PSPE is equivalent to finding a solution to the corresponding equilibrium boundary-value problem. However, simply finding an equilibrium configuration is insufficient to gain a clear understanding of the problem. In particular, at this point in the book, we are not able to distinguish between stable and unstable forms of equilibrium. These concepts are schematically illustrated in Fig. 7.1, which shows a pendulum consisting of a mass and a rigid bar attached by a fixed pin-joint in two equilibrium configurations. Configuration (a) corresponds to a state of *stable* equilibrium; the system will remain "close" to this configuration following small perturbations about it. Configuration (b) corresponds to a state of *unstable* equilibrium; the system will remain in this state if placed there, but any perturbation will diverge and cause it to move away. Both of these configurations correspond to stationary points of the pendulum's potential energy. Clearly, it is very important to know if an equilibrium configuration is stable or unstable.

The theory of stability for equilibrium configurations has a long history. It is a complex and beautiful theory which is built on the foundations of mechanics and continuum mechanics. It is also every bit as extensive and subtle as these foundations, and whole volumes have been dedicated to its description. We cannot hope to provide a deep understanding of the theoretical background and application of this theory within these few pages. Instead, we will present two of the most commonly used techniques for investigating the stability of an equilibrium configuration and show how these may be used to derive certain constraints on the constitutive relations for simple elastic materials. The reader interested in gaining a more complete understanding should start with the theory of stability for *finite-dimensional systems* (we recommend [Mei03] and [Kha02]). A good familiarity with the finite-dimensional theory is necessary before tackling the extensive and rigorous mathematical presentation of the modern theory of stability for *infinite-dimensional* structural and continuum mechanics. For this, we highly recommend [CG95].[5]

7.3.1 Definition of a stable equilibrium configuration

Although we are interested in a static equilibrium configuration, stability is inherently a concept related to dynamics. Its aim is to describe how a system dynamically evolves when it is subjected to perturbations of its equilibrium configuration. Accordingly, the definition of stability is phrased in terms of the solutions of initial boundary-value problems for the system of interest. Suppose $\varphi_{\text{eq}}(\boldsymbol{X})$ is the deformation mapping of an equilibrium configuration, i.e φ_{eq} satisfies the static equilibrium equations of Section 7.1 for given fixed values of the body forces, boundary displacements and tractions. Then, we say that φ_{eq} represents a (Lyapunov) stable equilibrium configuration if for every $\epsilon > 0$ there exists a $\delta = \delta(\epsilon) > 0$ such that if

$$\|\varphi(\boldsymbol{X},0) - \varphi_{\text{eq}}(\boldsymbol{X})\| < \delta \quad \text{and} \quad \|\dot{\varphi}(\boldsymbol{X},0)\| < \delta, \tag{7.14}$$

then

$$\|\varphi(\boldsymbol{X},t) - \varphi_{\text{eq}}(\boldsymbol{X})\| < \epsilon \quad \text{and} \quad \|\dot{\varphi}(\boldsymbol{X},t)\| < \epsilon, \quad \forall\, t, \tag{7.15}$$

where $\varphi(\boldsymbol{X},t)$ is the solution[6] to the initial boundary-value problem with the body forces, boundary displacements and tractions associated with the equilibrium solution φ_{eq}. The initial configuration is $\varphi(\boldsymbol{X},0) = \varphi_{\text{init}}(\boldsymbol{X})$ and the initial velocities are $\dot{\varphi}(\boldsymbol{X},0) = \breve{\boldsymbol{v}}_{\text{init}}(\boldsymbol{X})$. In words, the equilibrium configuration is stable if all small disturbances (perturbations of both configuration and velocities) lead to small responses. In Eqns. (7.14) and (7.15) the norms are associated with the function spaces of admissible deformations and

[5] Unfortunately, [CG95] is riddled with typesetting errors that make it difficult reading for the casual or less mathematically inclined reader. However, this is essentially the only book we are aware of that treats the subject with enough mathematical depth to obtain rigorous results.

[6] In the dynamical systems literature, this would be called a *trajectory* of the system.

velocity fields. For example, one possible choice of norm for the deformation map is

$$\|\varphi\| \equiv \left[\int_{B_0} \varphi \cdot \varphi \, dV_0\right]^{1/2}. \tag{7.16}$$

Other norms are possible, and in general, the concept of stability depends on the particular norm that is used. That is, an equilibrium configuration may be stable when the above norm is used but unstable when a different norm is considered.[7]

7.3.2 Lyapunov's indirect method and the linearized equations of motion

A straightforward approach to investigating the stability of an equilibrium configuration is to consider the equations of motion for the system, linearized about the equilibrium configuration. This technique is sometimes known as *Lyapunov's indirect method* because it works with the stability of the linearized equations of motion instead of directly with the equations of motion themselves. The method is also known as *Lyapunov's first method.*

The first step is to linearize the equations of motion in Eqn. (7.4) by applying to both sides of the equations the first variation in terms of the displacements $\delta \boldsymbol{u}$ relative to the deformed configuration:

$$\langle \mathcal{D}_{\varphi} \left(\text{Div}\left[(\nabla_0 \varphi) \boldsymbol{S}\right] + \rho_0 \breve{\boldsymbol{b}}\right); \delta \boldsymbol{u}\rangle = \left\langle \mathcal{D}_{\varphi} \rho_0 \frac{\partial^2 \varphi}{\partial t^2}; \delta \boldsymbol{u}\right\rangle.$$

This leads to

$$\left[\left(\delta_{ik} S_{JL} + F_{iP} F_{kQ} \frac{\partial S_{PJ}}{\partial E_{QL}}\right) \delta u_{k,L}\right]_{,J} = \rho_0 \frac{\partial^2 \delta u_i}{\partial t^2}, \tag{7.17}$$

where we have assumed that the material is hyperelastic and the symmetry of $\boldsymbol{S}$ and $\boldsymbol{E}$ has been used. Referring to Eqns. (6.151) and (6.157), we see that the partial derivative of $\boldsymbol{S}$

[7] The distinction between stability with respect to different norms occurs only in continuum systems. This is because, unlike finite-dimensional systems (see footnote 20 on page 26), not all norms are equivalent in infinite-dimensional spaces. For example, consider the norm

$$\|\varphi\|_{\infty} \equiv \max_{\boldsymbol{X} \in B_0} \left[\varphi(\boldsymbol{X}) \cdot \varphi(\boldsymbol{X})\right]^{1/2}.$$

This norm appears to be the most natural choice to make when generalizing the finite-dimensional theory of stability to continuous systems. In 1963, Shield and Green [SG63] showed that if one uses this norm, the undeformed unloaded reference configuration of a solid sphere, made of an innocuous material, is unstable. They proved that an arbitrarily small (in the sense of the $\|\cdot\|_{\infty}$ norm) spherically symmetric initial perturbation will result in a short-term concentration of energy in an infinitesimal region near the center of the sphere. The implication is that finite values (as opposed to infinitesimal values) of energy density, strain and most importantly velocity occur within the sphere. The occurrence of finite velocities near the center of the sphere violates the stability condition Eqn. $(7.15)_2$ that requires the velocity to remain small everywhere in the sphere. This result was controversial at the time of its publication; however, no one could refute its correctness. Almost immediately, Koiter [Koi63] resolved the matter with the recommendation that it is more appropriate for continuous systems to require average values to be small, instead of requiring small point-wise values. That is, it is more appropriate to use the norm in Eqn. (7.16) than it is to use the $\|\cdot\|_{\infty}$ norm for continuous systems. With this norm, Shield and Green's example no longer poses a problem. The undeformed configuration is stable with respect to the norm in Eqn. (7.16). For a more complete discussion of this subtle aspect of stability theory see [CG95].

on the left-hand side is the material elasticity tensor $\boldsymbol{C}$ (relating $d\boldsymbol{S}$ to $d\boldsymbol{E}$) and further that the entire term in parentheses is the mixed elasticity tensor $\boldsymbol{D}$ (relating $d\boldsymbol{P}$ to $d\boldsymbol{F}$). Thus,

$$[D_{iJkL}\delta u_{k,L}]_{,J} = \rho_0 \frac{\partial^2 \delta u_i}{\partial t^2}. \tag{7.18}$$

This is a linear, second-order partial differential equation with nonconstant coefficients, called the "wave equation," which describes the small-amplitude motion of a system about an equilibrium configuration φ_{eq}. The corresponding linearized boundary conditions are

$$\begin{aligned} D_{iJkL}\delta u_{k,L}N_J &= 0, & \boldsymbol{X} &\in \partial B_{0t}, t > 0, \\ \delta u_i &= 0, & \boldsymbol{X} &\in \partial B_{0u}, t > 0. \end{aligned}$$

The zero function, $\delta\boldsymbol{u}(\boldsymbol{X}, t) = \boldsymbol{0}$, is an equilibrium solution for this system and corresponds to the equilibrium φ_{eq} of the nonlinear system.

Stability of φ_{eq} It turns out that for conservative systems, such as those considered here, stability of the trivial solution $\delta\boldsymbol{u}(\boldsymbol{X}, t) = \boldsymbol{0}$ for the linearized system implies stability of the equilibrium configuration φ_{eq} for the nonlinear system.[8] Thus once we establish the conditions for the stability of the linearized system, those of the nonlinear system will follow.

Equation (7.18) is separable and admits solutions of the form $\delta\boldsymbol{u}(\boldsymbol{X}, t) = \boldsymbol{y}(\boldsymbol{X})\sin(\omega t)$. Each such solution is an "eigenfunction" of the system and is associated with the eigenvalue ω^2, where each ω is a *natural cyclic frequency* of the system. A solution of this form remains bounded for all time if its natural frequency ω is a nonzero real number.[9] If ω is imaginary, i.e. $\omega = \bar{\omega}i$ (where $\bar{\omega}$ is real), then $\sin(\omega t)$ becomes $\sinh(\bar{\omega}t)$, which diverges as t increases. Thus, a *sufficient condition* for stability of the equilibrium configuration is for all of the system's natural frequencies to be real and nonzero. Equivalently, all of the eigenvalues ω^2 must be positive. It can be shown (see, for example, [TN65]) that a *necessary condition* for all natural frequencies to be real is

$$a_i b_J a_k b_L D_{iJkL}(\boldsymbol{F}(\boldsymbol{X})) > 0, \quad \forall \boldsymbol{a}, \boldsymbol{b}, \text{ and } \forall \boldsymbol{X} \in B_0, \tag{7.19}$$

which can be rewritten as

$$(\boldsymbol{D} : \boldsymbol{A}) : \boldsymbol{A} > 0, \quad \forall \boldsymbol{A} = \boldsymbol{a} \otimes \boldsymbol{b} \neq \boldsymbol{0}. \tag{7.20}$$

A tensor $\boldsymbol{D}$ that satisfies this inequality is called a (strictly) "rank-one convex tensor".[10] Thus, $\boldsymbol{D}$ must be a rank-one convex tensor for each value of the deformation gradient $\boldsymbol{F}$, obtained from φ_{eq}, that occurs within B_0. Further, when $\boldsymbol{D}$ is constant (for example, when

[8] More generally, one must confirm that the nonlinear terms in the original system are small, in an appropriate sense, in order to ensure that stability for the linearized system implies stability for the nonlinear system. (See [CG95] for more details.)

[9] In fact, if $\omega = 0$ the solution is also bounded. However, here we avoid a number of technicalities by requiring strictly nonzero real frequencies.

[10] The term "rank-one" comes from the fact that $\boldsymbol{A} = \boldsymbol{a} \otimes \boldsymbol{b}$, viewed as a linear operator on vectors, is of rank one. That is, the matrix of components of $\boldsymbol{A}$ has only one linearly independent row. This should not be confused with the concept of a tensor's rank which is equal to the number of vector arguments it takes (or equivalently, the number of indices its component form has). The two ideas are completely distinct.

the equilibrium configuration corresponds to a uniform configuration, i.e. $\varphi_0 = F_0 X$) and the position of all boundary points is specified (i.e. $\partial B_{0u} = \partial B_0$ and $\partial B_{0t} = \emptyset$) it is found that the rank-one convexity condition in Eqn. (7.19) represents a *necessary and sufficient* condition for stability of the equilibrium configuration. In the theory of partial differential equations, the condition in Eqn. (7.19) is known as the "strong ellipticity" condition.

We can now concisely state the stability results we have just discussed.

- If $\partial B_{0u} = \partial B_0$ and D is constant, then (strict) rank-one convexity of the elasticity tensor $D(F)$ for every value of $F \in \{F \mid F = \nabla_0 \varphi_{eq}(X) \text{ for some } X \in B_0\}$ is a *necessary and sufficient* condition for the equilibrium configuration φ_{eq} to be stable.[11]
- If $\partial B_{0u} \neq \partial B_0$, then (strict) rank-one convexity of the elasticity tensor $D(F)$ for every value of $F \in \{F \mid F = \nabla_0 \varphi_{eq}(X) \text{ for some } X \in B_0\}$ is only a *necessary* condition for the equilibrium configuration φ_{eq} to be stable.

Stability of all equilibrium configurations for a hyperelastic simple material Since the above results depend only on the local properties of the material, they may be used to obtain a stability condition that applies to entire classes of equilibrium configurations for any body composed of a given hyperelastic simple material. For this purpose, the concept of rank-one convexity is generalized for the strain energy density function. The strain energy density function $W(F)$ is said to be a "rank-one convex function" if its second derivative is rank-one convex for all values of the deformation gradient.[12] That is, $W(F)$ is a rank-one convex function if $D(F)$ is a rank-one convex tensor for all values of F. With this definition, two results follow immediately from the above stability results.

- If $\partial B_{0u} = \partial B_0$, then (strict) rank-one convexity of a material's strain energy density function is a *necessary and sufficient* condition for stability of *every uniform* (spatially homogeneous) equilibrium configuration (see footnote 11).
- If $\partial B_{0u} \neq \partial B_0$, then (strict) rank-one convexity of a material's strain energy density function is only a *necessary* condition for stability of *every* equilibrium configuration.

In many instances, only one (stable) configuration of a hyperelastic material body is observed in experiments whenever the body is deformed by specifying, entirely, the deformation of its boundary. Rubber is the most common example of this type of material.[13] In other words, for many hyperelastic materials it is appropriate to require the strain energy density to be a rank-one convex function.

[11] Here we consider only displacements δu that are sufficiently smooth to ensure that δF is a continuous function of X. If more general perturbations of the equilibrium configuration must be considered, then the rank-one convexity condition is only necessary.

[12] More mathematically precise definitions of a rank-one convex function may be formulated, but these will not be necessary for the current discussion.

[13] Important counterexamples include materials that exhibit phase transformations and metals and polymers that exhibit shear-banding behavior (such as necking) which can be represented at the continuum level as a softening of the constitutive response.

Example 7.1 (Rank-one convexity of neo-Hookean models) Consider the neo-Hookean material model in Eqn. (6.139), $\overline{W}(I_1^B) = c_1(I_1^B - 3)$. The tensor $\boldsymbol{D}$ for the neo-Hookean material is obtained by taking the second derivative of this relation with respect to $\boldsymbol{F}$, to obtain

$$\boldsymbol{D} = 2c_1\boldsymbol{I},$$

where $\boldsymbol{I}$ is the fourth-order identity tensor. Applying Eqn. (7.20), we find the condition

$$(\boldsymbol{D} : \boldsymbol{a} \otimes \boldsymbol{b}) : \boldsymbol{a} \otimes \boldsymbol{b} = 2c_1 \, \|\boldsymbol{a}\|^2 \, \|\boldsymbol{b}\|^2 > 0.$$

This is satisfied for all $\boldsymbol{a}$ and $\boldsymbol{b}$ whenever $c_1 > 0$. Thus, when $c_1 > 0$ the neo-Hookean strain energy density is rank-one convex.

Example 7.2 (Rank-one convexity of Saint Venant–Kirchhoff models) The Saint Venant–Kirchhoff strain energy density function in Eqn. (6.147), $\widetilde{W}(\boldsymbol{E}) = [(\boldsymbol{C} : \boldsymbol{E}) : \boldsymbol{E}]/2$, is never rank-one convex, even when $\boldsymbol{C}$ is positive definite. (That is, when $(\boldsymbol{C} : \boldsymbol{E}) : \boldsymbol{E} > 0$ for all $\boldsymbol{E} = \boldsymbol{E}^T \neq \boldsymbol{0}$.)

To show this, we consider the homogeneous deformation given by $\varphi = \boldsymbol{F}\boldsymbol{X}$, where

$$\boldsymbol{F} = \alpha(\boldsymbol{e}_1 \otimes \boldsymbol{e}_1) + 1(\boldsymbol{e}_2 \otimes \boldsymbol{e}_2) + 1(\boldsymbol{e}_3 \otimes \boldsymbol{e}_3),$$

and α is the stretch parameter. For this deformation the Lagrangian strain tensor is

$$\boldsymbol{E} = [(\alpha^2 - 1)/2]\boldsymbol{e}_1 \otimes \boldsymbol{e}_1$$

and the second Piola–Kirchhoff stress tensor is

$$\boldsymbol{S} = [(\alpha^2 - 1)/2]\boldsymbol{C} : \boldsymbol{e}_1 \otimes \boldsymbol{e}_1.$$

Now, to establish that $\widetilde{W}$ is not rank-one convex, we need to find values of $\boldsymbol{a}$, $\boldsymbol{b}$ and α for which Eqn. (7.20) is not satisfied. Some reflection and intuition leads us to choose $\boldsymbol{a} = \boldsymbol{b} = \boldsymbol{e}_1$. Using Eqn. (6.157), we have

$$(\boldsymbol{D} : \boldsymbol{e}_1 \otimes \boldsymbol{e}_1) : \boldsymbol{e}_1 \otimes \boldsymbol{e}_1 = D_{1111} = \frac{1}{2}(3\alpha^2 - 1)C_{1111}.$$

Now, since $\boldsymbol{C}$ is positive definite, $C_{1111} > 0$. Thus, for $0 < \alpha < 1/\sqrt{3}$, the inequality Eqn. (7.20) is violated, regardless of the value of C_{1111}.

This result tells us that the Saint Venant–Kirchhoff material becomes unstable when subjected to sufficient compression. This may be surprising, since these materials are, in a sense, the most natural extension of a stable linear elastic material to the nonlinear regime. They provide a unique and invertible mapping between the second Piola–Kirchhoff stress and the Lagrangian strain. This would seem to suggest that, under all around displacement boundary conditions, there is a unique solution to the equilibrium problem. However, it is easy to see that this is not the case if one considers the Cauchy stress for these materials. The Cauchy stress for a Saint Venant–Kirchhoff material is easily found to be a nonlinear function of the deformation gradient. Thus, we should generally not expect these materials to have a relation mapping each value of *Cauchy stress* to a unique value of the deformation gradient.

In the above examples it was straightforward to determine that the neo-Hookean strain energy density function is rank-one convex and that the Saint Venant–Kirchhoff strain energy density function is not. However, it is generally difficult to directly establish the rank-one convexity of a given material model. For this, and other technical reasons, a number of additional convexity conditions have been introduced in the literature. The two most commonly encountered are called *quasiconvexity* and *polyconvexity*. The interested reader is referred to [Bal76] for further details on these concepts.

7.3.3 Lyapunov's direct method and the principle of minimum potential energy (PMPE)

In contrast to the indirect approach described above, *Lyapunov's direct method* makes straightforward use of the nonlinear equations of motion, but its success often requires considerable cleverness. Fortunately, for conservative systems a general solution is available, and Lyapunov's direct method leads immediately to the most commonly encountered criterion for evaluating the stability of an equilibrium configuration. This criterion is known as the *principle of minimum potential energy* (PMPE).

Lyapunov's direct method is a general approach for demonstrating *sufficient* conditions for φ_{eq} to be a stable equilibrium configuration. The method hinges on finding a special functional, called a *Lyapunov functional* $\mathfrak{L}(\varphi, \dot{\varphi})$, that satisfies the following conditions in the neighborhood of the equilibrium solution:[14]

1. The Lyapunov functional is positive definite. That is, $\mathfrak{L}(\varphi, \dot{\varphi}) > 0$ for all φ and $\dot{\varphi}$ such that $\|\varphi - \varphi_{\text{eq}}\| < \epsilon$ and $\|\dot{\varphi}\| < \epsilon$ for some $\epsilon > 0$. Here, φ and $\dot{\varphi}$ are taken as independent functions of $\boldsymbol{X}$, and represent the set of all possible configurations and velocity fields in the neighborhood of the equilibrium configuration. In other words, the equilibrium configuration φ_{eq} (and its trivial velocity field $\dot{\varphi}_e = \mathbf{0}$) correspond to an isolated (local) minimum of $\mathfrak{L}$.
2. The Lyapunov functional monotonically decreases along *every* solution of the equations of motion. That is, $(d/dt)[\mathfrak{L}(\varphi(t), \dot{\varphi}(t))] \leq 0$ for all t and all solutions of the equations of motion $\varphi(t)$, with initial conditions satisfying $\|\varphi(0) - \varphi_{\text{eq}}\| < \epsilon$ and $\|\dot{\varphi}(0)\| < \epsilon$.

The existence of one such Lyapunov functional is *sufficient* to guarantee the stability of φ_{eq}. In general, finding a Lyapunov functional for a system is difficult and requires creativity on the part of the investigator. However, for conservative systems a natural candidate for a Lyapunov functional is readily available. It is the system's total energy $\mathcal{E}$, which consists of kinetic energy $\mathcal{K}$ and its total potential energy Π, i.e.[15] $\mathcal{E} = \mathcal{K} + \Pi$. Here, we will assume that the datum of the total potential energy is taken at the equilibrium configuration of interest. That is, $\Pi(\varphi_{\text{eq}}) = 0$. Since the system is conservative, its total energy is constant, i.e. $d\mathcal{E}/dt = 0$, for all solutions of its equations of motion, by definition. Thus, item 2 above is satisfied for all such systems. It only remains to determine if item 1 is satisfied. For this, we first note that the kinetic energy for conservative systems is only a functional of the velocity field, i.e. $\mathcal{K} = \mathcal{K}(\dot{\varphi})$. Further, it is a positive-definite quadratic form. It follows immediately that if the potential energy is a positive-definite functional in the neighborhood of the equilibrium configuration, then the total energy is also a positive-definite functional

[14] For notational simplicity, the dependence of φ on $\boldsymbol{X}$ is suppressed.
[15] In the current purely mechanical setting, the total potential energy consists of the total strain energy plus the potential of any applied loads and is given by Eqn. (7.5).

in this neighborhood. This condition would then satisfy item 1 above and guarantee the stability of φ_{eq}. Recalling that for conservative systems the equilibrium configuration φ_{eq} is a stationary point of the potential energy, we have just proved the following principle.[16]

Principle of minimum potential energy (PMPE) If a stationary point of the potential energy corresponds to a (local) isolated minimum, then the equilibrium is stable.

Lyapunov's direct method provides *sufficient* conditions for stability of an equilibrium configuration. However, for conservative systems, it is straightforward to show that the PMPE is also a *necessary* condition for stability.[17] For additional details and discussion of the theory of stability and Lyapunov's two methods see [CG95, KW73].

Extension of the PMPE to thermomechanical systems It can be shown that the PMPE is valid for equilibrium configurations of thermomechanical systems [Koi71]. Such configurations must have uniform temperature fields in order to ensure that the heat flux is zero. If this were not the case, the configuration would not be in static equilibrium. Although the argument is complicated (and even includes one or two additional, but reasonable, assumptions), the result is the same as long as one associates the strain energy density W with the Helmholtz free energy density. That is, if we take $\widehat{W}(T, \boldsymbol{F}) = \rho_0 \widehat{\psi}(T, \boldsymbol{F})$, then for a uniform temperature field, the PMPE states that a minimum of the total Helmholtz free energy (Eqn. (7.5)) provides a *stable* equilibrium configuration.

Example 7.3 (Rivlin's cube[18]) Consider a unit cube made of an isotropic Saint Venant–Kirchhoff material (see Eqn. (6.147)) and subjected to nominal traction vectors $\boldsymbol{T}$ on each face. These tractions all have the same magnitude p and their directions are opposite to their respective face normals. That is, the nominal traction vector on a face with normal $\boldsymbol{N}$ is $\boldsymbol{T} = -p\boldsymbol{N}$. This state of loading corresponds to a compressive (for positive p) dead-loading. It is similar to hydrostatic loading since the associated second Piola–Kirchhoff stress is spherical, however, it is not the same. For hydrostatic pressure loading the traction vector is always normal to the deformed surface, whereas for dead-loading the direction of the traction vector does not change, even when the surface normal does. The

[16] This principle may remind you of the second law of thermodynamics. In fact, it is straightforward to show ([Cal85, Chapter 5]) that maximizing the entropy is equivalent to minimizing the internal energy of homogeneous systems. One should not confuse this with the PMPE; The principles have many similarities, but they are separate and distinct ideas. The second law is concerned with homogeneous systems that are in "true thermodynamic equilibrium," and therefore, does not consider dynamics of any kind. In contrast, the PMPE is applicable to all equilibrium configurations (homogeneous and nonhomogeneous) and is fundamentally based on the dynamical behavior of the system. However, in the limit where the equilibrium configuration is homogeneous, the two principles are equivalent.

[17] Certain technicalities associated with the case of a positive semi-definite potential energy are ignored here. See [Koi65a, Koi65b, Koi65c] for further details.

[18] Ronald Rivlin investigated the problem of homogeneous deformations of an elastic cube under dead-loading in 1948 [Riv48]. He revisited the problem in 1974 [Riv74]. The problem has since become known as the "Rivlin cube" problem and is a standard example for illustrating the ideas of stability, bifurcation and nonuniqueness of equilibrium solutions in continuum mechanics problems. For example, it is used by [MH94] and [Gur95], to name just two popular expositions on continuum mechanics. For more on Rivlin's pivotal role in the development of the modern theory of continuum mechanics, see Chapter 8.

unstressed reference configuration of the body is a unit cube centered at the origin. We will assume the center of the cube is fixed (to eliminate rigid-body translations) and restrict attention (for simplicity) to homogeneous deformations of the form $\varphi = \boldsymbol{F}\boldsymbol{X}$ with $\boldsymbol{F} = \sum_{i=1}^{3} \alpha_i \boldsymbol{e}_i \otimes \boldsymbol{e}_i$ (this eliminates, among other things, the rigid-body rotations).

We will explore the equilibrium solutions to this boundary-value problem. We start with the total potential energy, Eqn. (7.5) and substitute Eqn. (6.147) for the strain energy density to obtain

$$\begin{aligned}
\Pi = &\int_{-1/2}^{1/2}\int_{-1/2}^{1/2}\int_{-1/2}^{1/2} \left[\frac{1}{2}(\boldsymbol{C}:\boldsymbol{E}):\boldsymbol{E}\right] dX_1 dX_2 dX_3 \\
&- \int_{-1/2}^{1/2}\int_{-1/2}^{1/2} p(\boldsymbol{e}_3)\cdot\left(\alpha_1 X_1 \boldsymbol{e}_1 + \alpha_2 X_2 \boldsymbol{e}_2 + \alpha_3\left(-\frac{1}{2}\right)\boldsymbol{e}_3\right) dX_1 dX_2 \\
&- \int_{-1/2}^{1/2}\int_{-1/2}^{1/2} p(-\boldsymbol{e}_3)\cdot\left(\alpha_1 X_1 \boldsymbol{e}_1 + \alpha_2 X_2 \boldsymbol{e}_2 + \alpha_3\left(\frac{1}{2}\right)\boldsymbol{e}_3\right) dX_1 dX_2 \\
&- \int_{-1/2}^{1/2}\int_{-1/2}^{1/2} p(\boldsymbol{e}_1)\cdot\left(\alpha_1\left(-\frac{1}{2}\right)\boldsymbol{e}_1 + \alpha_2 X_2 \boldsymbol{e}_2 + \alpha_3 X_3 \boldsymbol{e}_3\right) dX_2 dX_3 \\
&- \int_{-1/2}^{1/2}\int_{-1/2}^{1/2} p(-\boldsymbol{e}_1)\cdot\left(\alpha_1\left(\frac{1}{2}\right)\boldsymbol{e}_1 + \alpha_2 X_2 \boldsymbol{e}_2 + \alpha_3 X_3 \boldsymbol{e}_3\right) dX_1 dX_3 \\
&- \int_{-1/2}^{1/2}\int_{-1/2}^{1/2} p(\boldsymbol{e}_2)\cdot\left(\alpha_1 X_1 \boldsymbol{e}_1 + \alpha_2\left(-\frac{1}{2}\right)\boldsymbol{e}_2 + \alpha_3 X_3 \boldsymbol{e}_3\right) dX_3 dX_1 \\
&- \int_{-1/2}^{1/2}\int_{-1/2}^{1/2} p(-\boldsymbol{e}_2)\cdot\left(\alpha_1 X_1 \boldsymbol{e}_1 + \alpha_2\left(\frac{1}{2}\right)\boldsymbol{e}_2 + \alpha_3 X_3 \boldsymbol{e}_3\right) dX_3 dX_1.
\end{aligned}$$

Taking $\boldsymbol{E} = \frac{1}{2}\sum_{i=1}^{3}(\alpha_i^2 - 1)\boldsymbol{e}_i \otimes \boldsymbol{e}_i$ and Eqn. (6.174) for the isotropic form of $\boldsymbol{C}$ and simplifying the integrals, the total potential energy becomes

$$\begin{aligned}
&\Pi(\alpha_1,\alpha_2,\alpha_3;p) \\
&= \frac{1}{8}[(\lambda+2\mu)\alpha_1^4 + (\lambda[\alpha_2^2+\alpha_3^2] - 2(3\lambda+2\mu))\alpha_1^2 + 8p\alpha_1 + (3\lambda+2\mu) \\
&\quad + (\lambda+2\mu)\alpha_2^4 + (\lambda[\alpha_3^2+\alpha_1^2] - 2(3\lambda+2\mu))\alpha_2^2 + 8p\alpha_2 + (3\lambda+2\mu) \\
&\quad + (\lambda+2\mu)\alpha_3^4 + (\lambda[\alpha_1^2+\alpha_2^2] - 2(3\lambda+2\mu))\alpha_3^2 + 8p\alpha_3 + (3\lambda+2\mu)].
\end{aligned}$$

We see that Π is a simple fourth-order polynomial function of the three stretches. It is also interesting to note that the potential energy is invariant with respect to all permutations of the stretches. This symmetry property can be useful for finding solutions to the equilibrium equations because it implies that if $\Pi(\alpha_1,\alpha_2,\alpha_3;p)$ is a stationary value of the potential energy for some particular p, then so are $\Pi(\alpha_2,\alpha_1,\alpha_3;p)$, $\Pi(\alpha_2,\alpha_3,\alpha_1;p)$ and so on.

Next, we obtain the equilibrium equations by applying the PSPE. Thus, the first partial derivatives of the total potential energy with respect to the three stretches must be zero:

$$\begin{aligned}
\frac{1}{2}[(\lambda+2\mu)\alpha_1^3 + (\lambda[\alpha_2^2+\alpha_3^2] - (3\lambda+2\mu))\alpha_1 + 2p] &= 0, \\
\frac{1}{2}[(\lambda+2\mu)\alpha_2^3 + (\lambda[\alpha_3^2+\alpha_1^2] - (3\lambda+2\mu))\alpha_2 + 2p] &= 0, \\
\frac{1}{2}[(\lambda+2\mu)\alpha_3^3 + (\lambda[\alpha_1^2+\alpha_2^2] - (3\lambda+2\mu))\alpha_3 + 2p] &= 0.
\end{aligned}$$

This is a set of three cubic polynomial equations in the three unknowns α_i with one parameter p. In general we can expect to find $3^3 = 27$ roots to these equations, however, some of these roots may be complex. These solutions are best obtained using numerical methods.

For $p = 0$ the only physical solution for these equilibrium equations is the reference configuration $\alpha_i = 1$. It is instructive to use the PMPE to show that this is a stable equilibrium configuration provided that the following assumptions are satisfied:

$$\mu > 0 \quad \text{and} \quad \lambda > -2\mu/3. \tag{7.21}$$

The necessary and sufficient conditions for the function $\Pi(\alpha_1, \alpha_2, \alpha_3, 0)$ to have an isolated local minimum at point $\alpha_i = 1$ are that this point corresponds to a stationary point and that the matrix of second derivatives of Π is positive definite. The value of this matrix of derivatives, evaluated for the general case of $\alpha_i = \alpha$, is found to be

$$\frac{1}{2}\begin{bmatrix} (5\lambda + 6\mu)\alpha^2 - (3\lambda + 2\mu) & 2\lambda\alpha^2 & 2\lambda\alpha^2 \\ 2\lambda\alpha^2 & (5\lambda + 6\mu)\alpha^2 - (3\lambda + 2\mu) & 2\lambda\alpha^2 \\ 2\lambda\alpha^2 & 2\lambda\alpha^2 & (5\lambda + 6\mu)\alpha^2 - (3\lambda + 2\mu) \end{bmatrix}.$$

The eigenvalues of this matrix are

$$(9\lambda/2 + 3\mu)\alpha^2 - (3\lambda/2 + \mu) \quad \text{and} \quad (3\lambda/2 + 3\mu)\alpha^2 - (3\lambda/2 + \mu). \tag{7.22}$$

The eigenvalue in Eqn. $(7.22)_2$ occurs twice. In the reference configuration $\alpha = 1$ and the eigenvalues reduce to $(3\lambda + 2\mu)$ and 2μ. When the assumptions of Eqn. (7.21) hold, they ensure that the matrix is positive definite and therefore the reference equilibrium configuration is stable.

Next, we explore equilibrium solutions for nonzero values of the applied load. We continue to consider the case where all stretches are equal $\alpha_i = \alpha$. The equilibrium equations degenerate to a single equation in this case and it is easy to show that there is a continuous branch of equilibrium solutions passing through the reference configuration. Solving the equilibrium equation one finds the traction–stretch relation for this branch to be

$$p(\alpha) = -\frac{3\lambda + 2\mu}{2}(\alpha^2 - 1)\alpha.$$

Thus, there is a unique equilibrium configuration, with all equal stretches, associated with each stretch value $\alpha > 0$. In order for this equilibrium configuration to be stable, the eigenvalues of the matrix of second derivatives, obtained above, must be positive. By simple algebraic manipulations it is easy to show that the first eigenvalue Eqn. $(7.22)_1$ is negative for $0 < \alpha < 1/\sqrt{3}$ and positive for $\alpha > 1/\sqrt{3}$. The second eigenvalue Eqn. $(7.22)_2$ is negative for $0 < \alpha < \sqrt{1/2 + 3\lambda/4\mu}$ and positive for $\alpha > \sqrt{1/2 + 3\lambda/4\mu}$. Thus, the matrix will be positive definite if the stretch is bigger than the critical stretch

$$\alpha_{\text{crit}} = \max(1/\sqrt{3}, \sqrt{1/2 + 3\lambda/4\mu}).$$

That is, the cubic configuration is unstable for $\alpha_i = \alpha < \alpha_{\text{crit}}$ and stable for $\alpha > \alpha_{\text{crit}}$. With the assumed inequalities in Eqn. (7.21) for μ and λ, we have

$$\alpha_{\text{crit}} = \begin{cases} \sqrt{1/2 + 3\lambda/4\mu}, & \text{if } \lambda > -2\mu/9, \\ 1/\sqrt{3}, & \text{if } -2\mu/3 < \lambda < -2\mu/9. \end{cases}$$

Finally, it is interesting to consider the (algebraically) maximum value of p for which a stable cubic equilibrium configuration is possible. From the $p(\alpha)$ relation given above it is clear that p is positive when $0 < \alpha < 1$ and negative when $\alpha > 1$. Thus, the maximum value of p is dictated by α_{crit}. The desired relation is

$$p_{\max} = \begin{cases} \dfrac{(3\lambda + 2\mu)(2\mu - 3\lambda)}{16\mu}\sqrt{1/2 + 3\lambda/\mu}, & \text{if } \alpha_{\text{crit}} = \sqrt{1/2 + 3\lambda/4\mu}, \\ (3\lambda + 2\mu)/\sqrt{3}, & \text{if } \alpha_{\text{crit}} = 1/\sqrt{3}. \end{cases}$$

In this example we have studied the equilibrium and stability properties of cubic configurations of the Rivlin cube. However, this is only part of the story. As hinted at above, the cubic configurations are not the only possible equilibrium states for this system. In fact, there exist tetragonal (two equal stretches not equal to the third) and orthorhombic (all stretches distinct) equilibrium *branches*.[19] Many of these branches actually intersect with the cubic branch at the critical stretch values (where one or more of the eigenvalues of the stability matrix become zero). Thus, in any neighborhood (no matter how small) of the critical stretch there are multiple distinct equilibrium configurations. Equilibrium configurations with this property are known as *bifurcation points* because they represent the points where the number of (real) solutions to the equilibrium equations changes. The theory of bifurcation is intimately connected to the theory of stability, but clearly, it is concerned with ideas that are distinct from those of stability. The interested reader is encouraged to consult [Tho82], [IM02] and [BT03] for more information on bifurcation theory.

This completes our brief introduction to energy principles and stability. Although we have dedicated only a few pages to these topics, we would like to emphasize their importance. These ideas are central players in essentially every modern science and engineering investigation. Thus, we encourage the reader to seek out a more complete understanding of these issues. The books listed in Chapter 11 are a good place to start.

Exercises

7.1 [SECTION 7.1] Consider a right circular cylinder of unit radius and length L. The front and back ends of the cylinder are represented by the surfaces $X_1^2 + X_2^2 \leq 1; X_3 = \pm L/2$. The lateral surface of the cylinder is given by $X_1^2 + X_2^2 = 1, -L/2 \leq X_3 \leq L/2$.

1. The cylinder is subjected to tensile dead-loads of magnitude F which are uniformly distributed over its ends. Express the complete set of boundary conditions for this equilibrium boundary-value problem in the material description in terms of the reference traction vector.
2. Using the reference traction vector values from the previous part, write these boundary conditions in terms of the Cartesian components of the first Piola–Kirchhoff stress components.
3. Explain why the imposition of two different traction vectors on the circular edges of the cylinder ($X_1^2 + X_2^2 = 1$ and $X_3 = \pm L/2$) does not create incompatible relationships between the values of the first Piola–Kirchhoff stress components.

[19] If we relaxed our assumptions to allow for shear deformation in addition to axial stretching, then even more equilibrium branches are possible.

7.2 [Section 7.1] Consider the thin plate subjected to axial compression shown in the figure below. The plate is bounded above and below by the planes $x_3 = \pm h$, respectively, and $2h \ll (b - a)$.

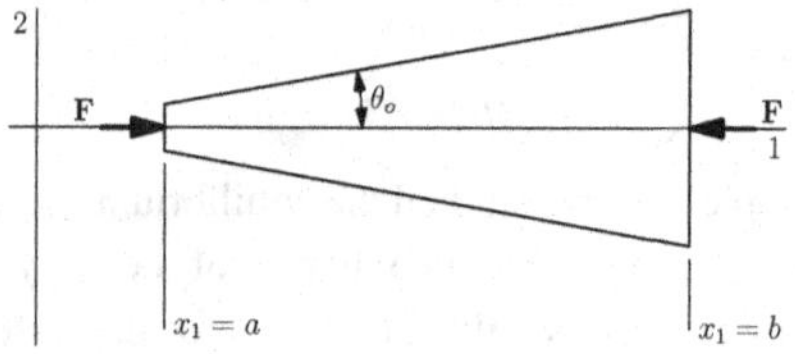

1. Using the traction-free conditions, write an expression for the relationship between the Cartesian components of the Cauchy stress tensor along the tapered edges of the plate.
2. Using the traction-free conditions, write an expression for the relationship between the polar cylindrical components of the Cauchy stress tensor along the tapered edges of the plate.
3. Assuming no body forces are acting on the plate, find a simple equilibrium stress field for this problem. You may choose any distribution of the force $\boldsymbol{F}$ over the vertical ends of the plate, so long as the resultant has magnitude F and is aligned along the 1-axis. **Hint:** Focus on finding a divergence-free stress field that satisfies the traction free boundary conditions.
4. We can find an approximation to the displacements in this plate by making the following series of approximations. First, we reduce the problem to one dimension and treat it as a bar with varying cross-sectional area $A(x)$. Second, we assume the displacement gradients are small so that we can use the small strain equation $\epsilon = u_{,x}$, where u is the displacement in the axial direction. Finally, we assume a linear one-dimensional force–strain constitutive law $F = EA\epsilon$, where E is Young's modulus and A is the cross-sectional area. These considerations lead to the following displacement-based equilibrium equation

$$(EAu_{,x})_{,x} = 0.$$

Assuming the plate is fixed at its left-hand edge ($u(a) = 0$) find the equilibrium displacement field $u(x)$.

7.3 [Section 7.2] For the one-dimensional model of the tapered plate in Exercise 7.2 the potential energy is given by

$$\Pi = \frac{1}{2}\int_a^b EA[u_{,x}]^2\, dx + Fu(b),$$

where the last term accounts for the potential energy of the applied compressive load (of magnitude F) and we recall the displacement boundary condition $u(a) = 0$. Show by explicit calculation that the PSPE is satisfied when the equilibrium equation quoted in Exercise 7.2 and the appropriate boundary conditions are used. **Hint:** Apply the fundamental theorem of calculus to the loading term.

7.4 [Section 7.3] Show, using the definition of stability, that the equilibrium solution $x(t) = 0$ for the one-dimensional second-order dynamical system

$$\ddot{x} - 2C\dot{x} + Bx = 0$$

is unstable. Assume B and C are constants and that $B > C^2 > 0$. **Hint:** Find the general solution for $x(t)$, then choose $|x(0)| < \epsilon$ and $|\dot{x}(0)| < \epsilon$ such that $|x(t)| > \epsilon$ or $|\dot{x}(t)| > \epsilon$ for some $t > 0$.

7.5 [SECTION 7.3] Starting from Eqn. (7.3) and the mixed boundary conditions given on page 246, derive the linearized field Eqn. (7.18) and the corresponding boundary conditions.

7.6 [SECTION 7.3] Consider the Saint Venant–Kirchhoff material and deformation mapping given in Example 7.2. Assume the tensor $\boldsymbol{C}$ takes the form of Eqn. (6.174) with $\lambda = \mu$.

1. Find the components of the first Piola–Kirchhoff stress tensor $\boldsymbol{P}$. Now, plot the nonzero components of $\boldsymbol{P}$ normalized by μ (i.e. P_{iJ}/μ) as a function of the stretch parameter α. In Example 7.2 we found that the Saint Venant–Kirchhoff material loses rank-one convexity for this deformation as α approaches $1/\sqrt{3}$ from the undeformed reference value of $\alpha = 1$. Using your plots explain what is special, if anything, about the value $\alpha = 1/\sqrt{3}$. Explain the physical importance of any identified special property.
2. Find the components of the Cauchy stress tensor $\boldsymbol{\sigma}$. Now, plot the nonzero components of $\boldsymbol{\sigma}$ normalized by μ (i.e. σ_{ij}/μ) as a function of the stretch parameter α. Compare the two sets of plots you have created as part of this problem. Explain the differences between them using physical terms.

7.7 [SECTION 7.3] Repeat the calculations of Example 7.3 for the case where the cube is subjected to a true hydrostatic pressure p. That is, where the potential energy of the applied pressure is given by $-pV$, so that the total potential energy has the form

$$\Pi = \int_{-1/2}^{1/2}\int_{-1/2}^{1/2}\int_{-1/2}^{1/2} \left[\frac{1}{2}(\boldsymbol{C} : \boldsymbol{E}) : \boldsymbol{E} + p\alpha_1\alpha_2\alpha_3\right] dX_1 dX_2 dX_3.$$

PART II

SOLUTIONS

8 Universal equilibrium solutions

In this chapter we study solutions to the equations of continuum mechanics instead of the equations themselves. In particular, our aim will be to obtain general equilibrium solutions to the field equations of continuum mechanics that are independent, in a specific sense, of the material from which a body is composed.[1] Such solutions are of fundamental importance to the practical application of the theory of continuum mechanics. This is because they provide valuable guidance to the experimentalist who would like to design experiments for the determination of a particular material's constitutive relations. Generally, in an experiment it is only possible to control and measure (to a greater or lesser extent) the tractions and displacements associated with the *boundary* of the body being studied. From this information one would like to infer the stress and deformation fields within the body and ultimately extract the functional form of the material's constitutive relations and the values of any coefficients belonging to this functional form. However, if the interior stress and deformation fields explicitly depend on the functional form of the constitutive relations, then it is essentially impossible to infer this information from a practical experiment.

According to Saccomandi [Sac01], a deformation which satisfies the equilibrium equations with zero body forces and is supported by suitable surface tractions alone is called a *controllable solution*. A controllable solution that is the same for all materials in a given class is a *universal solution*.[2] This chapter is devoted to a brief discussion of the best-known universal solutions. The reader interested in some examples of controllable solutions that are not universal is referred to the discussion in [Ogd84, Section 5.2]. Throughout this chapter, except where explicitly indicated, we restrict our attention to the purely mechanical formulation of continuum mechanics.

8.1 Universal equilibrium solutions for homogeneous simple elastic bodies

Universal solutions were first systematically investigated by Jerald Ericksen [Eri54, Eri55]. In fact, the problem of determining *all* universal equilibrium solutions for a given class

[1] The content of this chapter is largely based on the highly recommended review article [Sac01].

[2] Unfortunately there is not a consensus in the literature on this nomenclature. Many authors (including [SP65], [Car67] and [PC68]) do not consider the concept of a controllable solution as defined by Saccomandi. Rather they use the term "controllable solution" to refer to Saccomandi's universal solution. We prefer, and have adopted, Saccomandi's nomenclature because of its more evocative nature and its more finely grained classification of solutions.

of materials is now commonly referred to as *Ericksen's problem*. For the class of all homogeneous simple elastic bodies, *Ericksen's theorem* [Eri55] states that the only universal equilibrium solutions are the homogeneous deformations. This is a remarkable result that in many ways explains the extensive coverage of homogeneous deformations that is commonly found in books on continuum mechanics. In fact, it is quite easy to demonstrate Ericksen's Theorem.[3]

Proof It is trivial to see that the homogeneous deformations are always solutions to the equations of equilibrium. The fact that they are also the *only* solutions that satisfy the equilibrium equations *for all* simple elastic materials is much less obvious. To prove that this is true, we show that the homogeneous deformations are the only universal solutions for members of the class of simple isotropic, hyperelastic materials. Then, since this is a subclass of all simple elastic bodies, it follows that the homogeneous deformations are also the only universal solutions of all simple elastic bodies. That is, if homogeneous deformations are the only universal solutions for simple isotropic hyperelastic bodies, then it is not possible for the class of simple elastic bodies (which include simple isotropic hyperelastic bodies as a subset) to have more universal solutions.

As explained in Section 6.4.2, a general isotropic hyperelastic material can be represented by a strain energy density function that depends only on the three principal invariants of the left Cauchy–Green deformation tensor, i.e. $W(I_1^{\boldsymbol{B}}, I_2^{\boldsymbol{B}}, I_3^{\boldsymbol{B}})$. In the absence of body forces and under equilibrium conditions, the local material form of the balance of linear momentum in Eqn. (4.39) becomes

$$\operatorname{Div} \boldsymbol{P} = \boldsymbol{0}, \tag{8.1}$$

where the first Piola–Kirchhoff stress for the hyperelastic material is given by (see Eqn. (6.43))

$$\boldsymbol{P} = \sum_{i=1}^{3} \frac{\partial W}{\partial I_i^{\boldsymbol{B}}} \frac{\partial I_i^{\boldsymbol{B}}}{\partial \boldsymbol{F}}. \tag{8.2}$$

Now, consider the special case $W = \sum_i \mu_i I_i^{\boldsymbol{B}}$, where the μ_i are arbitrary constants. Substituting this into the expression for $\boldsymbol{P}$, and the result into the equilibrium equation, we find that for Eqn. (8.1) to be satisfied for all values of μ_i, we must have

$$\operatorname{Div} \frac{\partial I_i^{\boldsymbol{B}}}{\partial \boldsymbol{F}} = \boldsymbol{0}, \qquad \text{for } i = 1, 2, 3. \tag{8.3}$$

Similarly, if we consider a material with strain energy density given by $W = \sum_i \bar{\mu}_i (I_i^{\boldsymbol{B}})^2$, where again $\bar{\mu}_i$ are arbitrary constants, generally unrelated to μ_i, we obtain

$$\nabla_0 I_i^{\boldsymbol{B}} = \boldsymbol{0}, \qquad \text{for } i = 1, 2, 3. \tag{8.4}$$

Here we have used Eqn. (8.3). This is appropriate because we are searching for deformations that satisfy the equilibrium equations for all strain energy density functions *simultaneously*.

[3] The proof that follows is adapted from [Sac01] who attributes it to R. T. Shield [Shi71].

The gradient (with respect to the reference coordinates) of Eqn. (8.4) with $i = 1$ is found using Eqn. (6.134) to be

$$\varphi_{i,J}\varphi_{i,JK} = 0, \tag{8.5}$$

where φ is the deformation mapping. Taking the divergence of this expression, we obtain

$$\varphi_{i,JK}\varphi_{i,JK} + \varphi_{i,J}\varphi_{i,JKK} = 0,$$

which we can simplify by recognizing that Eqn. (8.3) for $i = 1$ gives $\text{Div}\, \boldsymbol{F} = \boldsymbol{0}$ or $\varphi_{i,JJ} = 0$ in indicial notation. Thus, the second term on the left-hand side above drops out and we are left with the expression

$$\varphi_{i,JK}\varphi_{i,JK} = 0.$$

This is a sum of squares which allows us to directly infer that

$$\varphi_{i,JK} = 0,$$

and conclude that the deformation map must be affine (see footnote 8 on page 83). This proves that the only universal solutions for the class of simple isotropic hyperelastic materials are the homogeneous deformations. Further, as noted above, it is also sufficient to prove that they are also the only universal solutions for the class of simple elastic materials. □

Finally, we note that in [PC68] it is shown that the analogous result for simple thermomechanical elastic materials (for which $W = \rho_0\psi$, where ψ is the Helmholtz free energy density) is that only constant temperature fields with homogeneous deformation are universal solutions.

Simple shear of isotropic elastic materials One of the most interesting general results that can be obtained from the homogeneous deformation universal solutions is known as the *Poynting effect*, named after the British physicist John Henry Poynting who first reported on it in 1909 [Poy09]. The Poynting effect refers to the observation that wires subjected to torsion in the elastic range exhibit an increase in their length by an amount proportional to the square of the twist [Bil86]. A similar effect occurs for materials undergoing finite simple shear where, since displacement perpendicular to the direction of shearing is precluded, the material develops normal stresses. We illustrate this case below and show that it leads to a remarkable universal relationship between the normal stresses and shear stress under simple shear conditions. To do so, we refer back to the discussion of simple shear in Example 3.2. Using Eqn. (6.130), we find the Cauchy stress to be

$$[\sigma] = \begin{bmatrix} \eta_0 + \eta_1(1+\gamma^2) + \eta_2(1+3\gamma^2+\gamma^4) & \eta_1\gamma + \eta_2(2\gamma+\gamma^3) & 0 \\ \eta_1\gamma + \eta_2(2\gamma+\gamma^3) & \eta_0 + \eta_1 + \eta_2(1+\gamma^2) & 0 \\ 0 & 0 & \eta_1 + \eta_2 + \eta_3 \end{bmatrix},$$

where we recall that γ is the shear parameter, and η_i are functions of the principal invariants of the left Cauchy–Green deformation tensor. The presence of normal stresses is clear. From this expression, a little algebra reveals the relation

$$\sigma_{12} = \frac{\sigma_{11} - \sigma_{22}}{\gamma}. \tag{8.6}$$

This is an excellent example of the type of amazing results that can be obtained from universal solutions. We see that regardless of the form of the functions η_i, the Cauchy stress components for a simple isotropic elastic material, subjected to homogeneous simple shear, must satisfy Eqn. (8.6)! (For more on this relation see Exercise 8.1.)

8.2 Universal solutions for isotropic and incompressible hyperelastic materials

Ericksen's theorem provides a complete set of universal solutions for simple elastic materials. However, if we restrict our attention to more specialized classes of materials, it becomes possible for additional (more interesting) universal solutions to exist. The most famous and fruitful results of this type are for the class of simple incompressible hyperelastic materials. Again, Ericksen [Eri54] was the first to consider the problem systematically. Ericksen identified four families of universal solutions in addition to the homogeneous deformations.[4] All of these solutions had been previously discovered by Ronald Rivlin during the late 1940s and early 1950s (see Rivlin's collected works [BJ96]).[5] Ericksen's systematic approach showed that the existence of more universal solutions was possible, but he was unable to discover any such solutions and the problem of identifying all universal solutions for this class of materials was left unanswered. In the intervening years significant progress toward the final answer to Ericksen's problem has been made and a fifth family of universal solutions has been discovered. However, it is still unknown whether this list of universal solutions is complete. Many researchers consider this one of the major open theoretical questions in the theory of elasticity.

In the remainder of this section we present the six known families of universal equilibrium solutions for simple isotropic hyperelastic materials.[6] However, two comments are in order

[4] Of course, only the volume-preserving homogeneous deformations are valid universal solutions for this class of materials.

[5] Technological advances during World War II led to a general interest by researchers in the nonlinear finite deformation behavior of rubber materials which are well approximated as simple isotropic and incompressible hyperelastic materials. In fact, it was Rivlin's discovery of the universal solutions for these materials that rekindled a theoretical interest in the general nonlinear theory of continua. Indeed, in the two decades following Rivlin's breakthrough paper of 1947 there was an explosion of new development and a general effort (lead by Clifford Truesdell and his collaborators [TT60, TN65]) to formalize and consolidate all that was known at the time in regard to the mechanics of continua. The result of these efforts is what we now know as "continuum mechanics." In this sense, Rivlin can fairly be called the father of modern continuum mechanics theory.

[6] The homogeneous family plus the five families mentioned above.

before we begin. First, the names and numbering of these families have become standard in the literature and often authors simply refer to "Family 0," "Family 1" and so on. The families are also given (standard) descriptive names. These descriptions use material bodies of specific shape that are convenient for visualizing the associated deformations. However, it is emphasized that all of the deformations discussed below satisfy the field equations of equilibrium identically. Thus, they are applicable to bodies of *any shape*.

Second, we remind the reader that we have restricted consideration to a purely mechanical setting. However, the results given below are also valid for steady-state heat conduction, when supplemented with a *constant* temperature field, within a thermomechanical setting. The only known nontrivial universal solutions (i.e. having nonconstant temperature field) for the thermomechanical formulation are associated with homogeneous deformations (see [PC68]).[7] These are discussed as part of Family 0.

For brevity, below we only describe the known families without proving that they identically satisfy the equilibrium equations, except for one example (in Family 3) to give the reader a taste of how such proofs are carried out. For help visualizing the nature of each family of deformation, the reader is directed to Exercises 8.2–8.7.

8.2.1 Family 0: homogeneous deformations

Any homogeneous deformation with $\det \nabla_0 \varphi = 1$ is a universal solution. For steady-state heat conduction there are three cases. First, we have the "trivial" case which consists of a constant temperature field and a homogeneous deformation. The second case, shown in [PC68], consists of a temperature field of the form $T = k + qx_3$ coupled with a homogeneous deformation of the form (ignoring an arbitrary rigid motion)

$$\begin{aligned} x_1 &= CA^{-1/2}X_1 - DB^{-1/2}X_2, \\ x_2 &= DA^{-1/2}X_1 + CB^{-1/2}X_2, \\ x_3 &= (AB)^{1/2}X_3, \end{aligned} \tag{8.7}$$

where $k, q, A > 0, B > 0, C$ and D are constants and $C^2 + D^2 = 1$. The third case consists of a temperature field of the form $T = k + p\theta$ coupled with homogeneous deformations of the form

$$r = C^{-1/2}R, \qquad \theta = \Theta, \qquad z = CZ, \tag{8.8}$$

where k, p and $C > 0$ are constants and where (r, θ, z) and (R, Θ, Z) are deformed and reference polar cylindrical coordinates, respectively. Notice that this last scenario must be restricted to cases for which the shape of the body ensures that the temperature field is single-valued.

[7] It is known, however, that no such nontrivial universal solutions exist with deformations found in Families 1–5 [PC68].

8.2.2 Family 1: bending, stretching and shearing of a rectangular block

The deformation mapping for Family 1 is

$$r = \sqrt{2AX_1}, \qquad \theta = BX_2, \qquad z = \frac{1}{AB}X_3 - BCX_2, \tag{8.9}$$

where A, B and C are constants and where (r, θ, z) and (X_1, X_2, X_3) are deformed polar cylindrical and reference Cartesian coordinates, respectively. This deformation is most simply described using these two coordinate systems. To show that this deformation satisfies the equilibrium equations in Eqn. (8.1), it is best to convert the reference Cartesian coordinates to polar cylindrical, assume a strain energy density function of the form $W = W(I_1^B, I_2^B)$ (since $I_3^B = 1$ due to incompressibility) and then use Eqns. (2.99) and (2.100) to obtain the equilibrium equations in polar cylindrical coordinates.[8]

8.2.3 Family 2: straightening, stretching and shearing of a sector of a hollow cylinder

The deformation mapping for Family 2 is

$$x_1 = \frac{1}{2}AB^2R^2, \qquad x_2 = \frac{1}{AB}\Theta, \qquad x_3 = \frac{1}{B}Z + \frac{C}{AB}\Theta, \tag{8.10}$$

where A, B and C are constants and where (x_1, x_2, x_3) and (R, Θ, Z) are deformed Cartesian and reference polar cylindrical coordinates, respectively. Again, this family of universal solutions is most simply described using these two coordinate systems, but a transition from polar cylindrical to Cartesian reference coordinates will expedite an effort to verify that these deformations satisfy the equilibrium equations.

8.2.4 Family 3: inflation, bending, torsion, extension and shearing of an annular wedge

The deformation mapping for Family 3 is

$$r = \sqrt{AR^2 + B}, \qquad \theta = C\Theta + DZ, \qquad z = E\Theta + FZ, \tag{8.11}$$

where A, B, C and D are constants and $A(CF - DE) = 1$. Here (r, θ, z) and (R, Θ, Z) are deformed and reference polar cylindrical coordinates, respectively. Although the name of this family refers to an annular wedge, it is also applicable to cylindrical bodies that surround the origin as long as $E = 0$ and $C = 1$, which ensure that the displacements are single-valued. Thus, solutions such as the eversion of a circular tube are also contained within this family of universal solutions.

[8] The conversion to polar cylindrical coordinates for both deformed and reference coordinates is necessary since we have not presented the general theory for dealing with arbitrary coordinate systems in this book. However, the reader should note that most discussions of this topic in other books and the technical literature take advantage of the general theory of tensor fields using curvilinear coordinates. For more on this theory, consult the books listed in Chapter 11.

Example 8.1 (Extension and torsion of a solid circular cylinder) Limiting ourselves to the values $B = 0$, $C = 1$, $E = 0$ and $AF = 1$ leads to a deformation that corresponds to the extension and torsion of a solid circular cylinder. Substituting these values along with the following change of notation $D = \Psi$ and $F = \alpha$ gives

$$r = \frac{R}{\sqrt{\alpha}}, \qquad \theta = \Theta + \Psi Z, \qquad z = \alpha Z, \tag{8.12}$$

where the relation $A\alpha = 1$ has been used to obtain the equation for r. Here α is referred to as the material stretch ratio and Ψ is referred to as the material twist rate.

We will demonstrate that this deformation is, in fact, a universal solution. The simple isotropic, incompressible, hyperelastic materials have strain energy density functions of the form $W = W(I_1^{\boldsymbol{B}}, I_2^{\boldsymbol{B}})$. Therefore, the first Piola–Kirchhoff and Cauchy stresses are, respectively, given by

$$\boldsymbol{P} = 2W_{,I_1^{\boldsymbol{B}}}\boldsymbol{F} + 2W_{,I_2^{\boldsymbol{B}}}(I_1^{\boldsymbol{B}}\boldsymbol{F} - \boldsymbol{B}\boldsymbol{F}) - c_0\boldsymbol{F}^{-T},$$

$$\boldsymbol{\sigma} = 2W_{,I_1^{\boldsymbol{B}}}\boldsymbol{B} + 2W_{,I_2^{\boldsymbol{B}}}(I_1^{\boldsymbol{B}}\boldsymbol{B} - \boldsymbol{B}^2) - c_0\boldsymbol{I},$$

where $W_{,I_k^{\boldsymbol{B}}} = \partial W/\partial I_k^{\boldsymbol{B}}$ and the constant c_0 accounts for the undetermined part of the hydrostatic pressure. The deformation mapping is $\breve{\varphi} = r\boldsymbol{e}_r(\theta) + z\boldsymbol{e}_z$ and we would like to start by computing the deformation gradient. However, there is a problem: we have not discussed how to take the material gradient of a spatial vector field when curvilinear coordinate systems are employed. Instead of using Eqn. (2.99) directly, we will return to the general definition of the gradient in Eqn. $(2.94)_1$. First, we write the deformation mapping in the material description by substituting in Eqn. (8.12):

$$\boldsymbol{\varphi} = \frac{R}{\sqrt{\alpha}}\boldsymbol{e}_r(\Theta + \Psi Z) + \alpha Z\boldsymbol{e}_z.$$

Second, Eqn. $(2.94)_1$ states that

$$\boldsymbol{F} = \nabla_0\boldsymbol{\varphi} = \frac{\partial\boldsymbol{\varphi}}{\partial R} \otimes \boldsymbol{e}_r(\Theta) + \frac{\partial\boldsymbol{\varphi}}{\partial\Theta} \otimes \left(\frac{1}{R}\boldsymbol{e}_\theta(\Theta)\right) + \frac{\partial\boldsymbol{\varphi}}{\partial Z} \otimes \boldsymbol{e}_z,$$

where we have used Eqn. (2.98). Finally, expanding the partial derivatives (and recalling that $\partial\boldsymbol{e}_r/\partial\theta = \boldsymbol{e}_\theta$) gives the following tensor form for the deformation gradient:

$$\begin{aligned}\boldsymbol{F} &= \left[\frac{1}{\sqrt{\alpha}}\boldsymbol{e}_r(\Theta + \Psi Z)\right] \otimes \boldsymbol{e}_r(\Theta) + \frac{1}{R}\left[\frac{R}{\sqrt{\alpha}}\boldsymbol{e}_\theta(\Theta + \Psi Z)\right] \otimes \boldsymbol{e}_\theta(\Theta) \\ &\quad + \left[\frac{R}{\sqrt{\alpha}}\Psi\boldsymbol{e}_\theta(\Theta + \Psi Z) + \alpha\boldsymbol{e}_z\right] \otimes \boldsymbol{e}_z \\ &= \frac{1}{\sqrt{\alpha}}\boldsymbol{e}_r(\theta) \otimes \boldsymbol{e}_r(\Theta) + \frac{1}{\sqrt{\alpha}}\boldsymbol{e}_\theta(\theta) \otimes \boldsymbol{e}_\theta(\Theta) + \frac{\Psi R}{\sqrt{\alpha}}\boldsymbol{e}_\theta(\theta) \otimes \boldsymbol{e}_z + \alpha\boldsymbol{e}_z \otimes \boldsymbol{e}_z.\end{aligned} \tag{8.13}$$

This form explicitly displays the two-point nature of the deformation tensor. The basis vectors on the left in each tensor product are evaluated at the point (r, θ, z), whereas the basis vectors on the right in each tensor product are evaluated at the point (R, Θ, Z). In matrix form (with respect to the above two-point basis) we have

$$[\boldsymbol{F}] = \begin{bmatrix} 1/\sqrt{\alpha} & 0 & 0 \\ 0 & 1/\sqrt{\alpha} & \Psi R/\sqrt{\alpha} \\ 0 & 0 & \alpha \end{bmatrix}. \tag{8.14}$$

The left Cauchy–Green tensor is $\boldsymbol{B} = \boldsymbol{F}\boldsymbol{F}^T$. Since this is a spatial tensor, it is most natural to write it in the spatial description. Thus, we form the matrix product[9] $[\boldsymbol{F}]\,[\boldsymbol{F}]^T$ using Eqn. (8.14) and then substitute for R and Z with the expressions $R = r\sqrt{\alpha}$ and $Z = z/\alpha$, respectively, to obtain

$$[\boldsymbol{B}] = \begin{bmatrix} 1/\alpha & 0 & 0 \\ 0 & 1/\alpha + \Psi^2 r^2 & \Psi\alpha r \\ 0 & \Psi\alpha r & \alpha^2 \end{bmatrix}. \tag{8.15}$$

The basis is now evaluated only at (r, θ, z). It is interesting and important to note that the polar cylindrical components of $\boldsymbol{B}$ depend only on r and are independent of θ and z. The principal invariants are

$$I_1^{\boldsymbol{B}} = \alpha^2 + 2/\alpha + \Psi^2 r^2, \qquad I_2^{\boldsymbol{B}} = 2\alpha + 1/\alpha^2 + \Psi^2 r^2/\alpha, \qquad I_3^{\boldsymbol{B}} = 1.$$

Notice that $I_3^{\boldsymbol{B}}$ is equal to 1 which confirms that this is a volume preserving deformation.

We continue by considering the equations of equilibrium in the deformed configuration (Eqn. (4.27)) with zero body forces. Solving the equations in the deformed configuration is the simplest way to make progress, since we can simply use Eqn. (2.100) to take the divergence of a spatial (or material) tensor.[10] Substituting $\boldsymbol{B}$ into the above expression for the Cauchy stress and simplifying results in the following (polar cylindrical) components:

$$\begin{aligned}
\sigma_{rr} &= 2[\alpha W_{,I_1^{\boldsymbol{B}}} + (1 + \alpha^3 + \Psi^2\alpha r^2)W_{,I_2^{\boldsymbol{B}}}]/\alpha^2 - c_0, \\
\sigma_{\theta\theta} &= 2[\alpha(1 + \Psi^2\alpha r^2)W_{,I_1^{\boldsymbol{B}}} + (1 + \alpha^3 + \Psi^2\alpha r^2)W_{,I_2^{\boldsymbol{B}}}]/\alpha^2 - c_0, \\
\sigma_{zz} &= 2\alpha^2(W_{,I_1^{\boldsymbol{B}}} + 2W_{,I_2^{\boldsymbol{B}}}/\alpha) - c_0, \\
\sigma_{\theta z} &= 2\Psi\alpha(W_{,I_1^{\boldsymbol{B}}} + W_{,I_2^{\boldsymbol{B}}}/\alpha)r, \\
\sigma_{rz} &= 0, \\
\sigma_{r\theta} &= 0.
\end{aligned} \tag{8.16}$$

Taking the divergence with respect to the deformed polar cylindrical coordinates (using Eqn. (2.100)) we obtain

$$\begin{aligned}
\boldsymbol{0} &= \operatorname{div}\boldsymbol{\sigma} \\
&= \Bigg(4\Psi^2 r\Bigg[-\frac{1}{2}W_{,I_1^{\boldsymbol{B}}} + \frac{1}{\alpha}W_{,I_2^{\boldsymbol{B}}} + \frac{1}{\alpha}W_{,I_1^{\boldsymbol{B}}I_1^{\boldsymbol{B}}} + \frac{2 + \alpha^3 + \Psi^2\alpha r^2}{\alpha^2}W_{,I_1^{\boldsymbol{B}}I_2^{\boldsymbol{B}}} \\
&\quad + \frac{1 + \alpha^3 + \Psi^2\alpha r^2}{\alpha^3}W_{,I_2^{\boldsymbol{B}}I_2^{\boldsymbol{B}}}\Bigg] - \frac{\partial c_0}{\partial r}\Bigg)\boldsymbol{e}_r - \frac{1}{r}\frac{\partial c_0}{\partial\theta}\boldsymbol{e}_\theta - \frac{\partial c_0}{\partial z}\boldsymbol{e}_z.
\end{aligned} \tag{8.17}$$

This is a set of uncoupled, linear, first-order differential equations for the undetermined part of the hydrostatic pressure $c_0(r, \theta, z)$. These equations can be easily integrated, and thus we are ensured that there exists a function c_0 for which $\operatorname{div}\boldsymbol{\sigma} = \boldsymbol{0}$. Thus, we have finally arrived at the conclusion that the deformation given by Eqn. (8.12) is, in fact, a universal solution for the class of simple, isotropic, incompressible hyperelastic materials.

It is instructive to take this process one step further and write down the explicit solution of a well-defined boundary-value problem. However, to do this we must choose a particular strain energy density function. Here we will consider the neo-Hookean material discussed previously in Section 6.4.2. In

[9] This is equivalent to the corresponding tensor product $\boldsymbol{F}\boldsymbol{F}^T$ in this case because we are using an *orthonormal* basis. If nonorthonormal basis vectors are employed this is generally not true.

[10] To obtain the divergence of a mixed tensor (such as the first Piola–Kirchhoff tensor) we would have to perform a procedure similar to the one we just used to compute $\boldsymbol{F}$.

this case the equilibrium equation reduces to

$$\mathbf{0} = -\left[2c_1\Psi^2 r + \frac{\partial c_0}{\partial r}\right]\mathbf{e}_r - \frac{1}{r}\frac{\partial c_0}{\partial \theta}\mathbf{e}_\theta - \frac{\partial c_0}{\partial z}\mathbf{e}_z,$$

which has the general solution $c_0(r, \theta, z) = -c_1\Psi^2 r^2 + d$, where d is a constant. With this solution for the undetermined part of the hydrostatic pressure, the Cauchy stress components reduce to

$$\begin{aligned}
\sigma_{rr} &= c_1(2/\alpha + \Psi^2 r^2) - d, \\
\sigma_{\theta\theta} &= c_1(1/\alpha - \Psi^2 r^2) - d, \\
\sigma_{zz} &= c_1(2/\alpha^2 + \Psi^2 r^2) - d, \\
\sigma_{\theta z} &= 2c_1\Psi\alpha r, \\
\sigma_{rz} &= 0, \\
\sigma_{r\theta} &= 0,
\end{aligned} \tag{8.18}$$

and the hydrostatic pressure (given by Eqn. (4.23)) is

$$p = \left(d - c_1\left[\frac{1}{\alpha} + \frac{2}{3\alpha^2}\right]\right) - \frac{c_1\Psi^2 r^2}{3}.$$

Now, suppose the undeformed cylinder has radius R_1 and length L. We must consider the three surfaces $r = R_1/\sqrt{\alpha}$, $z = 0$ and $z = \alpha L$, with outward unit normals $\mathbf{e}_r$, $-\mathbf{e}_z$ and $\mathbf{e}_z$, respectively. According to Eqn. (4.18), and using Eqn. (8.18), the traction vectors on these three surfaces are

$$\begin{aligned}
\boldsymbol{t}(R_1/\sqrt{\alpha}, \theta, z) &= \left[\frac{c_1}{\alpha}(2 + \Psi^2 R_1^2) - d\right]\mathbf{e}_r, \\
\boldsymbol{t}(r, \theta, 0) &= -2c_1\Psi\alpha r\mathbf{e}_\theta + \left[d - c_1\left(\frac{2}{\alpha^2} + \Psi^2 r^2\right)\right]\mathbf{e}_z, \\
\boldsymbol{t}(r, \theta, \alpha L) &= \quad 2c_1\Psi\alpha r\mathbf{e}_\theta - \left[d - c_1\left(\frac{2}{\alpha^2} + \Psi^2 r^2\right)\right]\mathbf{e}_z.
\end{aligned}$$

Thus, we see that, for given values of α and Ψ, the traction on the lateral surface of the cylinder has a magnitude that is always a constant and a direction perpendicular to the surface. The tractions on the ends of the cylinder include an axial component whose magnitude varies with the square of the distance from the center of the cylinder. There is also a shear component that varies linearly with the distance from the cylinder's center.

We now return to the general solution to derive a curious universal property of isotropic, incompressible, solid circular cylinders with traction free lateral surfaces. We will compare the axial force required to stretch the cylinder by α without torsion to the torsional stiffness associated with an infinitesimal spatial twist rate applied to the stretched cylinder. Thus, we will need expressions for the total axial force without torsion (for arbitrary α and $\Psi = 0$) and the total moment applied to the ends of the cylinder (for arbitrary values of α and Ψ).

Consider a cylinder with a traction-free lateral surface and suppose it is initially of length L and radius R_1. When $\Psi = 0$ the equation for c_0 indicates that it is a constant. Using the traction-free lateral surface condition we obtain the particular value

$$c_0 = 2[\alpha W^*_{,I_1^{\boldsymbol{B}}} + (1 + \alpha^3)W^*_{,I_2^{\boldsymbol{B}}}]/\alpha^2.$$

Here the superscript $*$ indicates that the derivatives of the strain energy density are evaluated at the values $I_1^{\boldsymbol{B}} = \alpha^2 + 2/\alpha$ and $I_2^{\boldsymbol{B}} = 2\alpha + 1/\alpha^2$. The traction vector at the end $z = \alpha L$ is $\boldsymbol{t} = \boldsymbol{\sigma n}$, where $\boldsymbol{n} = \boldsymbol{e}_z$. The axial force applied to this end of the cylinder is

$$N(\alpha) = \int_0^{\frac{R_1}{\sqrt{\alpha}}} \int_0^{2\pi} \boldsymbol{t} \cdot \boldsymbol{e}_z r dr d\theta = \int_0^{\frac{R_1}{\sqrt{\alpha}}} \int_0^{2\pi} (\boldsymbol{\sigma} \boldsymbol{e}_z) \cdot \boldsymbol{e}_z r dr d\theta = \int_0^{\frac{R_1}{\sqrt{\alpha}}} \int_0^{2\pi} \sigma_{zz} r dr d\theta.$$

Substituting for σ_{zz} from Eqn. (8.16), using the above expression for c_0 and noting that σ_{zz} is constant (in space) when $\Psi = 0$ results in

$$N(\alpha) = 2\pi R_1^2 (\alpha - 1/\alpha^2)[W^*_{,I_1^{\boldsymbol{B}}} + W^*_{,I_2^{\boldsymbol{B}}}/\alpha]. \tag{8.19}$$

We proceed similarly to obtain the applied moment (about the center of the cylinder at $z = \alpha L$). The expression for arbitrary values of α and Ψ is

$$\begin{aligned} M(\alpha, \Psi) &= \int_0^{\frac{R_1}{\sqrt{\alpha}}} \int_0^{2\pi} [\boldsymbol{t} \times (r\boldsymbol{e}_r)] \cdot \boldsymbol{e}_z r dr d\theta \\ &= \int_0^{\frac{R_1}{\sqrt{\alpha}}} \int_0^{2\pi} [(\boldsymbol{\sigma} \boldsymbol{e}_z) \times (r\boldsymbol{e}_r)] \cdot \boldsymbol{e}_z r dr d\theta \\ &= \int_0^{\frac{R_1}{\sqrt{\alpha}}} \int_0^{2\pi} \sigma_{\theta z} r^2 dr d\theta. \end{aligned}$$

Substituting for $\sigma_{\theta z}$ we obtain

$$M(\alpha, \Psi) = 4\pi\Psi\alpha \int_0^{\frac{R_1}{\sqrt{\alpha}}} [W_{,I_1^{\boldsymbol{B}}} + W_{,I_2^{\boldsymbol{B}}}/\alpha] r^3 dr.$$

Next, we define the moment, $m(\alpha, \psi) \equiv M(\alpha, \psi\alpha)$, as a function of the material stretch ratio and the spatial twist rate, $\psi = \Psi/\alpha$, and take the derivative with respect to ψ:

$$m_{,\psi} = 4\pi\alpha^2 \int_0^{\frac{R_1}{\sqrt{\alpha}}} [W_{,I_1^{\boldsymbol{B}}} + W_{,I_2^{\boldsymbol{B}}}/\alpha] r^3 dr + 4\pi\psi\alpha^2 \int_0^{\frac{R_1}{\sqrt{\alpha}}} \frac{\partial}{\partial \psi} [W_{,I_1^{\boldsymbol{B}}} + W_{,I_2^{\boldsymbol{B}}}/\alpha] r^3 dr.$$

Evaluating this at $\psi = 0$ results in

$$m_{,\psi}(\alpha, 0) = 4\pi\alpha^2 \int_0^{\frac{R_1}{\sqrt{\alpha}}} [W^*_{,I_1^{\boldsymbol{B}}} + W^*_{,I_2^{\boldsymbol{B}}}/\alpha] r^3 dr = \pi R_1^4 [W^*_{,I_1^{\boldsymbol{B}}} + W^*_{,I_2^{\boldsymbol{B}}}/\alpha], \tag{8.20}$$

which can be identified as the (initial) torsional stiffness of the stretched cylinder. Finally, we form the ratio of the axial force to the torsional stiffness

$$\left(\frac{N(\alpha) R_1^2}{m_{,\psi}(\alpha, 0)} \right) = 2 \left(\alpha - \frac{1}{\alpha^2} \right), \tag{8.21}$$

where we have included the factor of R_1^2 in order to obtain a dimensionless ratio. It is remarkable that all terms related to the material's constitutive relations cancel out of this ratio. This means that the ratio of the axial force to the torsional stiffness for *arbitrary* extension of a cylinder with traction-free lateral surface is independent of the material model!

8.2.5 Family 4: inflation or eversion of a sector of a spherical shell

The deformation mapping for Family 4 is

$$r = (\pm R^3 + A)^{1/3}, \qquad \theta = \Theta \quad \text{or} \quad \theta = \pi - \Theta, \qquad \phi = \Phi, \tag{8.22}$$

where A is a constant and where (r, θ, ϕ) and (R, Θ, Φ) are deformed and reference spherical coordinates, respectively.

8.2.6 Family 5: inflation, bending, extension and azimuthal shearing of an annular wedge

The deformation mapping for Family 5 is

$$r = \sqrt{A}R, \qquad \theta = D\ln(BR) + C\Theta, \qquad z = FZ, \tag{8.23}$$

where A, B, C, D and F are constants and $ACF = 1$. Here (r, θ, z) and (R, Θ, Z) are deformed and reference polar cylindrical coordinates, respectively.

8.3 Summary and the need for numerical solutions

In this chapter we took the following approach to finding solutions of continuum mechanics theory. We first looked for exact solutions to the equilibrium field equations for various classes of material constitutive relations and then considered what boundary-value problems are solved by the obtained solutions. It is clear that there exists an extremely limited (but important) set of problems for which analytical solutions are available. This set includes the universal solutions discussed above and controllable solutions such as those found in [Ogd84, Section 5.2]. In addition, if one considers the approximate theory of small strain linear elasticity (see Section 10.4), then a wide range of analytical solutions for important problems becomes available. However, in almost all other cases we must resort to numerical methods whenever we are interested in a problem that is not included in the set just listed. In the next chapter we start by identifying a particular boundary value problem of interest and then seek an accurate approximate solution for this problem. To this end, we develop the *finite element method* which is a general methodology for numerically computing approximate solutions to a given, well-defined, boundary-value problem.

Exercises

8.1 [SECTION 8.1] Consider a unit cube of material (whose sides are aligned with the coordinate axes in the reference configuration) subjected to the simple shear

$$x_1 = X_1 + \gamma X_2, \quad x_2 = X_2, \quad x_3 = X_3,$$

with Cauchy stress components given by

$$[\sigma] = \begin{bmatrix} \sigma_{11} & \sigma_{12} & 0 \\ \sigma_{12} & \sigma_{22} & 0 \\ 0 & 0 & \sigma_{33} \end{bmatrix}.$$

1. Compute the traction components on the six faces of the deformed cube. Since nothing changes in the 3-direction, for the rest of the problem we will treat the body as a two-dimensional object. In the 1–2 plane, draw the deformed geometry, a rhombus, and the tractions that act on the edges of this rhombus. These tractions are independent of the X_3 component and can, therefore, be plotted as a (two-dimensional) vector at each point on these edges.
2. Show by writing the sum of forces in the 1- and 2-directions and the sum of moments in the 3-direction that the deformed rhombus is in static equilibrium for any value of γ and that these equations provide no restrictions on the Cauchy stress components σ_{11}, σ_{12} and σ_{22}. This shows that for a stress tensor with the above form, the equilibrium conditions place no further restrictions on the nonzero stress components. Thus, Eqn. (8.6) is a result based purely on the isotropy properties of the constitutive relation and does not depend explicitly on the particular (simple shear) geometric configuration of the body.

8.2 [SECTION 8.2] Consider a unit cube of an isotropic incompressible hyperelastic material.
1. Plot the reference and deformed configurations associated with Eqn. (8.7) and shade the deformed configuration to indicate the variation of temperature throughout the body. Experiment with the free parameters k, q, A, B and C to show the variety of deformed configurations described by this deformation mapping.
2. Repeat part 1 for Eqn. (8.8), experimenting with the free parameters k, p and C.

8.3 [SECTION 8.2] Consider a unit cube of an isotropic incompressible hyperelastic material. Plot the reference and deformed configurations associated with Eqn. (8.9). Experiment with the free parameters A, B and C to show the variety of deformed configurations described by this deformation mapping.

8.4 [SECTION 8.2] Repeat Exercise 8.3 for Eqn. (8.10).

8.5 [SECTION 8.2] Repeat Exercise 8.3 for Eqn. (8.11), experimenting with the free parameters A, B, C and D.

8.6 [SECTION 8.2] Repeat Exercise 8.3 for Eqn. (8.22), experimenting with the free parameter A.

8.7 [SECTION 8.2] Repeat Exercise 8.3 for Eqn. (8.23), experimenting with the free parameters A, B, C and D.

8.8 [SECTION 8.2] Consider a brick-shaped body, composed of a neo-Hookean material, bounded by the following planes: $X_1 = X \pm W$ (where $X > W$), $X_2 = \pm L$, and $X_3 = \pm H$. Using the deformation mapping for Family 1 in Eqn. (8.9):
1. Show that the deformed configuration satisfies the equilibrium field equations and find the explicit form of the (polar cylindrical) stress components, including the integrated form for the undetermined part of the hydrostatic pressure c_0.
2. Write expressions for the nominal tractions that are required to act on each of the six surfaces of the body in order to bring about this deformation, and plot the deformed configuration projected on the 1–2 plane with the traction vectors shown.

8.9 [SECTION 8.2] Consider a spherical shell, composed of a neo-Hookean material, with inner radius $R = R_\text{i}$ and outer radius $R = R_\text{o}$. Use the special case of the deformation mapping for Family 4 in Eqn. (8.22):

$$r = (R_\text{o}^3 + R_\text{i}^3 - R^3)^{1/3}, \qquad \theta = \Theta, \qquad \phi = \Phi,$$

which corresponds to the eversion of the sphere.
1. Show that the deformed configuration satisfies the equilibrium field equations and find the explicit form of the (spherical) stress components, including the integrated form for the undetermined part of the hydrostatic pressure c_0.
2. Write expressions for the nominal tractions that are required to act on each surface of the body in order to bring about this deformation.

9 Numerical solutions: the finite element method

The rapid growth of computer power since the 1960s has been accompanied by a similarly rapid growth and development of computational methods, to the point where the stress analysis of complex components is a routine part of almost any engineering design. To demonstrate how continuum mechanics problems can be accurately and efficiently solved by an approximate numerical representation on a computer, we will focus on the solution of static problems in solid mechanics, and we will not consider the effects of temperature. While there is certainly no shortage of numerical techniques to solve fluid mechanics, heat transfer or other continuum problems, our focus on solids reflects the emphasis of this book in general. And while we will start out on a relatively general footing applicable to many of the computational techniques available for solid mechanics, our focus will be on the finite element method (FEM). This is because the FEM has clearly emerged as the most common and powerful approach for solid mechanics and materials science. Further, we view the FEM as a natural bridge between continuum mechanics and atomistic methods. In Part IV of the companion book to this one [TM11], we explicitly use it as a way to build multiscale models combining atomistic and continuum frameworks.

A perusal of Chapter 11 on Further Reading makes it clear that the FEM is a subject that can easily fill an entire book on its own. Here, we provide a very brief introduction to the FEM which explains how this method works and why it is useful. Our development is somewhat nonconventional when compared with the usual approach taken in the FEM literature, reflecting our particular interest in making the connection to the atomic scale in Chapters 12 and 13 of the companion volume [TM11]. It also has the advantage of connecting to the variational approaches described in Chapter 7. In Section 9.3.7, we make the connection between our description and the more common approaches taken in other FEM introductions.

9.1 Discretization and interpolation

The problem we wish to solve is the static boundary-value problem of Fig. 9.1 subject to mixed boundary conditions as described in Section 7.1.2. A body B_0 in the reference configuration has surface ∂B_0 with surface normal $\boldsymbol{N}$. This surface is divided into a portion ∂B_{0u} over which the displacements are prescribed as $\bar{\boldsymbol{u}}$ and the remainder (∂B_{0t}) which is either free or subject to a prescribed traction, $\bar{\boldsymbol{T}}$. Our goal is to determine the stress, strain and displacement fields throughout the body due to the applied loads.

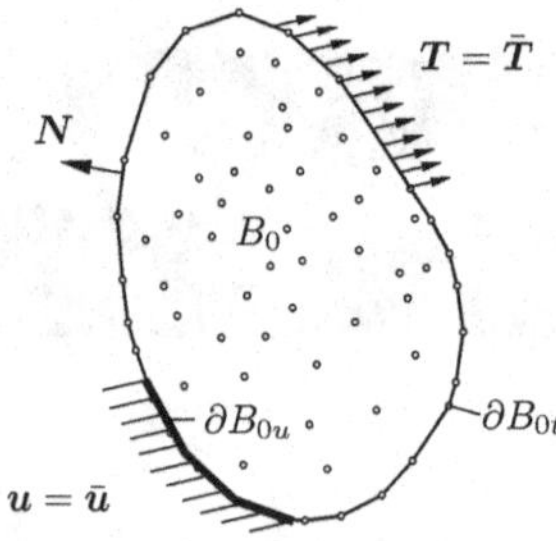

Fig. 9.1 A general continuum mechanics boundary-value problem and an arbitrary set of nodes selected to discretize it for solution by the FEM.

We adopt a Lagrangian, finite deformation framework for hyperelastic materials. This has several implications. First, the choice of a Lagrangian framework means that we will write all quantities in terms of the undeformed, reference configuration of the body, B_0. Second, finite strain implies that we expect, in general, that the gradients of the displacements in the body will be too large for the simplifications of small strain elasticity (described in Section 10.4) to be accurate. Finally, the hyperelasticity assumption restricts our attention to materials for which we can write a strain energy density function, W, in terms of some suitable strain measure (see Section 6.2.5).

The static boundary-value problem is conveniently posed using the *principle of stationary potential energy* of Section 7.2. The total potential energy, Π, given in Eqn. (7.5), in the absence of body forces takes the following form:

$$\Pi = \int_{B_0} W(\boldsymbol{F}(\boldsymbol{X}))dV_0 - \int_{\partial B_{0t}} \bar{\boldsymbol{T}} \cdot \boldsymbol{u}\, dA_0. \tag{9.1}$$

We seek a displacement field $\boldsymbol{u}(\boldsymbol{X})$ for which Π is stationary subject to the constraint that $\boldsymbol{u}(\boldsymbol{X}) = \bar{\boldsymbol{u}}$ for $\boldsymbol{X} \in \partial B_{0u}$. Our first step is to replace the continuous variable $\boldsymbol{u}(\boldsymbol{X})$ with a discrete variable,[1] $\mathsf{u}(\mathsf{X})$, stored at a finite set of points in the body, called *nodes*, as shown schematically in Fig. 9.1. The goal will be to approximate the continuous displacements from these discrete values using interpolation. For efficient computer implementation, we write u and X as column matrices with $n_{\mathrm{dof}} = n_{\mathrm{nodes}} \times n_{\mathrm{d}}$ entries (where n_{nodes} is the number of nodes and n_{d} is the number of spatial dimensions):

$$\mathsf{X} = \begin{bmatrix} \boldsymbol{X}^1 \\ \boldsymbol{X}^2 \\ \vdots \\ \boldsymbol{X}^{n_{\mathrm{nodes}}} \end{bmatrix}, \qquad \mathsf{u} = \begin{bmatrix} \mathsf{u}^1 \\ \mathsf{u}^2 \\ \vdots \\ \mathsf{u}^{n_{\mathrm{nodes}}} \end{bmatrix} = \begin{bmatrix} \mathsf{u}_1^1 \\ \mathsf{u}_2^1 \\ \mathsf{u}_3^1 \\ \vdots \\ \mathsf{u}_1^{n_{\mathrm{nodes}}} \\ \mathsf{u}_2^{n_{\mathrm{nodes}}} \\ \mathsf{u}_3^{n_{\mathrm{nodes}}} \end{bmatrix}, \tag{9.2}$$

[1] We adopt the sans serif font in this chapter in a manner that is essentially consistent with the convention described in Section 2.2.4, in that $\boldsymbol{u}$ and $\boldsymbol{X}$ are vectors whereas u and X are column matrices. In some instances, these column matrices behave as first-order tensors (e.g. u behaves as a tensor in the $\mathbb{R}^{n_{\mathrm{dof}}}$ space on which it is defined), but we do not make use of tensorial properties here. As such, we retain the notation mainly as a way to differentiate between the continuum variables and their discrete representation as column matrices.

where $\boldsymbol{X}^\alpha$ and $\mathbf{u}^\alpha$ are the *nodal* coordinate and displacement vectors associated with node α. The latter will be obtained as part of the solution process.

A brief comment about notation Before we get too far along, it is worth warning the reader that this chapter is going to be notationally challenging. We have tried to use a notation that is accurate and detailed, but still sufficiently clear to follow. A brief description of the notation now will help us as we proceed. It is sometimes convenient to refer to the nodal displacements in either invariant form (as simply a bold-faced $\mathbf{u}$), or alternatively in indicial form as $\mathsf{u}_{\bar{\beta}}$ ($\bar{\beta} = 1, \dots, n_{\text{dof}}$). The overbar on the subscript reminds us that this is an index spanning nodes and coordinates (i.e. $1, \dots, n_{\text{dof}}$). It is sometimes convenient to split up the index implied by $\bar{\beta}$ and write u_i^α to indicate the ith component of displacement associated with node α, and thus $i = 1, \dots, n_{\text{d}}$ and $\alpha = 1, \dots, n_{\text{nodes}}$. At other times, the displacements of only a specific node α are required, again in either invariant form as $\mathbf{u}^\alpha$ or indicial form as u_i^α. All of these are essentially the same quantities (or subsets of each other) presented in different forms. We adopt the convention that Latin subscripts refer to spatial components ($i = 1, \dots, n_{\text{d}}$), and Greek indices will serve double duty. When they appear as *subscripts* with an overbar, they refer to the index embodied in Eqn. (9.2), ranging over $\bar{\alpha} = 1, \dots, n_{\text{dof}}$. Greek *superscripts* refer to node numbering and thus range over $\alpha = 1, \dots, n_{\text{nodes}}$. The Einstein summation convention will be applied to repeated *subscript* indices as usual.

Despite the discrete representation of $\boldsymbol{u}$, Eqn. (9.1) will still require a continuous displacement field defined throughout the body in order to be evaluated. This is achieved through a set of so-called *shape functions* that interpolate the discrete displacements to all points between the nodes to yield an approximate displacement field, $\tilde{\boldsymbol{u}}$. (Throughout this chapter, we use the $\tilde{\cdot}$ notation to indicate FEM approximations to continuous quantities.) In terms of the reference position vectors, $\boldsymbol{X}$, this can be written generally as

$$\boldsymbol{u}(\boldsymbol{X}) \approx \tilde{\boldsymbol{u}}(\boldsymbol{X}) = \mathbf{S}\mathbf{u} = \begin{bmatrix} \mathbf{S}^1(\boldsymbol{X}) & \mathbf{S}^2(\boldsymbol{X}) & \dots & \mathbf{S}^{n_{\text{nodes}}}(\boldsymbol{X}) \end{bmatrix} \begin{bmatrix} \mathbf{u}^1 \\ \mathbf{u}^2 \\ \vdots \\ \mathbf{u}^{n_{\text{nodes}}} \end{bmatrix}, \tag{9.3}$$

where $\mathbf{S}$ is a $3 \times 3n_{\text{nodes}}$ matrix in three dimensions. In indicial notation this is $\tilde{u}_i = \mathsf{S}_{i\bar{\alpha}} \mathsf{u}_{\bar{\alpha}}$, where $\mathsf{S}_{i\bar{\alpha}}$ refers to the components of the $3 \times 3n_{\text{nodes}}$ matrix of shape functions defined in Eqn. (9.3). Each entry in this matrix is a scalar shape function related to a specific node, interpolating one entry in the displacement matrix onto one component of the continuous displacement field. Normally, the same functional form is used for interpolating all three components of $\tilde{\boldsymbol{u}}$. Also, it is not physically sensible to use information from one degree of freedom to interpolate the other (e.g. displacements in the X_1-direction should not depend on displacements in the X_2-direction.) Thus, for the case of a three-dimensional displacement field, Eqn. (9.3) becomes

$$\tilde{u}_i(\boldsymbol{X}) = \mathsf{S}_{i\bar{\alpha}} \mathsf{u}_{\bar{\alpha}} = \begin{bmatrix} S^1 & 0 & 0 & & S^{n_{\text{nodes}}} & 0 & 0 \\ 0 & S^1 & 0 & \dots & 0 & S^{n_{\text{nodes}}} & 0 \\ 0 & 0 & S^1 & & 0 & 0 & S^{n_{\text{nodes}}} \end{bmatrix} \begin{bmatrix} \mathsf{u}_1^1 \\ \mathsf{u}_2^1 \\ \mathsf{u}_3^1 \\ \vdots \\ \mathsf{u}_1^{n_{\text{nodes}}} \\ \mathsf{u}_2^{n_{\text{nodes}}} \\ \mathsf{u}_3^{n_{\text{nodes}}} \end{bmatrix},$$

where $S^\alpha(\boldsymbol{X})$ is a scalar shape function associated with node α. An equivalent, but often more transparent way of writing Eqn. (9.3) is

$$\widetilde{u}_i(\boldsymbol{X}) = \sum_{\alpha=1}^{n_{\text{nodes}}} S^\alpha(\boldsymbol{X})\mathsf{u}_i^\alpha. \tag{9.4}$$

Finite element formulations have the desirable property that

$$\mathbf{u}^\alpha = \widetilde{\boldsymbol{u}}(\boldsymbol{X}^\alpha) = \sum_{\beta=1}^{n_{\text{nodes}}} S^\beta(\boldsymbol{X}^\alpha)\mathbf{u}^\beta, \tag{9.5}$$

meaning that the approximate displacement field exactly interpolates the displacement, $\mathbf{u}^\alpha$, stored at each nodal position, $\boldsymbol{X}^\alpha$. This implies that $S^\beta(\boldsymbol{X}^\alpha) = \delta_{\alpha\beta}$, where $\delta_{\alpha\beta}$ is the Kronecker delta, as we shall see later in Eqn. (9.22).

Equations (9.3) and (9.4) represent two alternative notations that we will employ in this chapter. Analogous to how we used indicial and invariant notation in continuum mechanics, these two notations are equivalent, but for certain purposes one or the other is more convenient for illustrating a particular derivation or expression. We will often present key expressions in both forms for this reason.

Given a solution vector $\mathbf{u}$, one can now obtain an approximation to the displacements everywhere in the body. Since the energy of Eqn. (9.1) depends on the displacements through the deformation gradient, we require that the interpolation be suitably smooth to provide piecewise bounded first derivatives of the displacements everywhere. Recalling from Eqn. (3.29) that the deformation gradient is defined as $\boldsymbol{F} = \boldsymbol{I} + \partial \boldsymbol{u}/\partial \boldsymbol{X}$, we can find an approximate deformation gradient $\widetilde{\boldsymbol{F}}$ as

$$\widetilde{F}_{iJ} = \delta_{iJ} + \frac{\partial \widetilde{u}_i}{\partial X_J} = \delta_{iJ} + \frac{\partial \mathsf{S}_{i\bar{\alpha}}}{\partial X_J}\mathsf{u}_{\bar{\alpha}}, \tag{9.6}$$

where δ_{iJ} is the Kronecker delta. Alternatively if we start from Eqn. (9.4)

$$\widetilde{F}_{iJ} = \delta_{iJ} + \sum_{\alpha=1}^{n_{\text{nodes}}} \frac{\partial S^\alpha}{\partial X_J}\mathsf{u}_i^\alpha. \tag{9.7}$$

The strain energy density can now be written as a function of the nodal displacement through the deformation gradient, $W(\widetilde{\boldsymbol{F}}(\mathbf{u}))$, and the approximate potential energy becomes

$$\widetilde{\Pi} = \int_{B_0} W(\widetilde{\boldsymbol{F}}(\mathbf{u}))\, dV_0 - \int_{\partial B_{0t}} \bar{\boldsymbol{T}} \cdot \mathbf{S}\mathbf{u}\, dA_0. \tag{9.8}$$

Let us assume for the moment that a suitable set of shape functions has been chosen for Eqn. (9.3). The solution procedure is then to determine a stationary point of Eqn. (9.8) with respect to the nodal displacements, $\mathbf{u}$, subject to appropriate boundary conditions. Stationary points of the energy functional satisfy

$$-\frac{\partial \widetilde{\Pi}}{\partial \mathsf{u}_{\bar{\alpha}}} = 0.$$

Note that the negative sign is added only for convenience, so that we can refer to forces instead of gradients. For an out-of-equilibrium displacement field we have a residual force vector, $\mathsf{f} \in \mathbb{R}^{n_{\rm dof}}$, such that

$$\mathsf{f}_{\bar{\alpha}}(\mathbf{u}) \equiv -\frac{\partial \widetilde{\Pi}}{\partial \mathsf{u}_{\bar{\alpha}}} = -\int_{B_0} \frac{\partial W}{\partial \widetilde{F}_{iJ}} \frac{\partial \widetilde{F}_{iJ}}{\partial \mathsf{u}_{\bar{\alpha}}} \, dV_0 + \int_{\partial B_{0t}} \bar{T}_i \mathsf{S}_{i\bar{\alpha}} \, dA_0.$$

We recognize the derivative $\partial W / \partial \boldsymbol{F}$ in the first integral as the first Piola–Kirchhoff stress, $\boldsymbol{P}$, and note that from Eqn. (9.6)

$$\frac{\partial \widetilde{F}_{iJ}}{\partial \mathsf{u}_{\bar{\alpha}}} = \frac{\partial \mathsf{S}_{i\bar{\alpha}}}{\partial X_J}. \tag{9.9}$$

This leads to the following expression for the residual force vector:

$$\mathsf{f}_{\bar{\alpha}}(\mathbf{u}) = -\int_{B_0} P_{iJ}(\widetilde{\boldsymbol{F}}(\mathbf{u})) \frac{\partial \mathsf{S}_{i\bar{\alpha}}}{\partial X_J} \, dV_0 + \int_{\partial B_{0t}} \bar{T}_i \mathsf{S}_{i\bar{\alpha}} \, dA_0. \tag{9.10}$$

These are the out-of-balance forces on the nodes for a given displacement vector. For an *equilibrium displacement vector*, these out-of-balance forces must be zero.

Finding the displacement vector that renders $\widetilde{\Pi}$ stationary will generally require an iterative solution since $\boldsymbol{P}$ is a nonlinear function of the displacements for a hyperelastic material. In many cases we are only interested in *stable* equilibrium configurations. For this reason, in the rest of this chapter we will employ the *principle of minimum potential energy* (PMPE) and focus on energy *minimization*. Thus, in the next section we elaborate on the details of nonlinear energy minimization (or more generally "optimization").

9.2 Energy minimization

The search for minima of a nonconvex, multidimensional function is one of the great challenges of computational mathematics, and in nonlinear finite elements it is the principal computational effort associated with finding static solutions. It is interesting that the human eye and brain can look at a hilly landscape and almost immediately find the point of lowest elevation, in addition to establishing where most of the other local minima lie. To do the same with a high-dimensional mathematical function on a computer is much more difficult. Although finding *any minimum* is not too difficult, finding it *quickly* is a bit more of a challenge, and finding the *global* minimum confidently and quickly is still an open field of research. The entire branch of numerical mathematics known as "optimization" is essentially dedicated to this goal. It is not our intention to exhaustively discuss the latest in nonlinear optimization. Rather, we present the theory and implementation of the most common workhorse used in finite elements, the Newton–Raphson (NR) method. The core ideas of the NR algorithm serve as the basis for more advanced optimization approaches, which form the subject of entire optimization textbooks (see, for example, [Pol71, Rus06]). Strictly speaking, what we describe herein is a *modified NR* approach since standard NR is a method for finding roots (i.e. stationary points of a function) and not a minimization approach. However, for simplicity we refer to it as "NR."

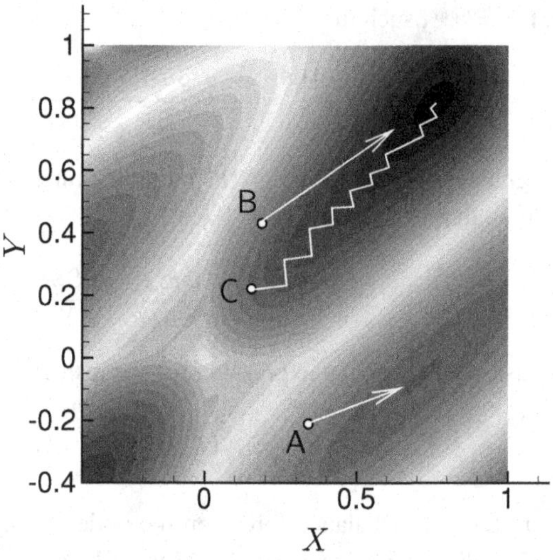

Fig. 9.2 An energy landscape in two dimensions. Points A–C represent different initial guesses, and the path from C to the minimum for the steepest descent method is shown.

9.2.1 Solving nonlinear problems: initial guesses

Our goal is to find a minimum of a generic energy function $\widetilde{\Pi}(\mathbf{u})$, where $\mathbf{u}$ (referred to as the "configuration") represents a n_{dof}-dimensional column matrix of variables upon which the energy depends. In the context of finite elements, this function is given by Eqn. (9.8) and $\mathbf{u}$ is the nodal displacement vector, but the minimization process is, of course, perfectly general. The method proceeds by evolving the system from some initial configuration $\mathbf{u}^{(0)}$ (with energy $\widetilde{\Pi}(\mathbf{u}^{(0)})$) to the configuration $\mathbf{u}_{\text{min}}$ that locally minimizes $\widetilde{\Pi}$.

Figure 9.2 illustrates an energy landscape for a system with $n_{\text{dof}} = 2$, $[\mathbf{u}]^T = [X, Y]^T$. There are several local minima, with the global minimum occurring at about $[X, Y]^T = [0.75, 0.80]^T$. A minimization method will invariably converge to different minima depending on the initial guess one makes for the configuration. In Fig. 9.2, for instance, starting from point "A" is likely to take the system to the minimum near $[X, Y]^T = [0.70, -0.05]^T$, while starting from point "B" will converge to the global minimum.

In physical terms, the need for an initial guess, and the dependence of the solution on that guess may or may not be problematic. Sometimes, we may not even be interested in the true global minimum, but rather a local minimum that is nearby some physically motivated starting point for the system. In plasticity models, for example, loading is typically applied incrementally from zero, such that an equilibrium solution is found for each quasistatic[2] load step along the way and used as the initial guess for the next load step. However, the plasticity formulation is path dependent, meaning it depends on such details as the size of the load steps and whether multiple loads are incremented simultaneously or in series. As

[2] See Section 5.4 for the definition of quasistatic processes.

Algorithm 9.1 A generic minimization algorithm

$n := 0$
$\mathbf{f}^{(0)} := -\nabla_{\mathbf{u}}\widetilde{\Pi}(\mathbf{u}^{(0)})$
while $\left\|\mathbf{f}^{(n)}\right\| >$ tol **do**
  find the search direction $\mathbf{d}^{(n)}$
  find step size $\alpha^{(n)} > 0$
  $\mathbf{u}^{(n+1)} := \mathbf{u}^{(n)} + \alpha^{(n)}\mathbf{d}^{(n)}$
  $\mathbf{f}^{(n+1)} := -\nabla_{\mathbf{u}}\widetilde{\Pi}(\mathbf{u}^{(n+1)})$
  $n := n + 1$
end while
$\mathbf{u}_{\min} := \mathbf{u}^{(n)}$

such, the question of what is the appropriate initial guess is replaced by the question of whether the loading program is physically realistic and of sufficient numerical accuracy. For other problems, however, it is not clear what the initial guess should be, and the dependence of the solution on this arbitrary choice is disconcerting. In the absence of good physical grounds for a particular initial guess, it may be necessary to run multiple simulations from different starting points to assess the sensitivity of the solution. In the companion book to this one [TM11, Chapter 6], we talk about this in more detail in the context of atomistic systems, where the true global minimum is usually a perfect crystal but we are interested in more complex configurations containing such defects as dislocations and grain boundaries.

While energy minimization can be achieved without directly computing the forces (energy gradients), gradient methods are almost always more efficient for problems of interest to us here. We have already seen the forces on the finite element nodes in Eqn. (9.10), and later we will develop efficient ways to evaluate them. Let us begin by considering the generic approach one takes given an expression for the energy and forces.

9.2.2 The generic nonlinear minimization algorithm

Given an energy function $\widetilde{\Pi}(\mathbf{u})$ and its gradient (forces) $\mathbf{f}(\mathbf{u}) = -\nabla_{\mathbf{u}}\widetilde{\Pi}(\mathbf{u})$, we seek the configuration $\mathbf{u}_{\min}$ such that $\widetilde{\Pi}(\mathbf{u}_{\min})$ is a local minimum. This corresponds to a point where the forces on all the degrees of freedom are zero, so we typically test for convergence using some prescribed tolerance on the force norm:

$$\text{if } \|\mathbf{f}(\mathbf{u})\| < \text{tol}, \quad \mathbf{u}_{\min} := \mathbf{u}. \tag{9.11}$$

We adopt the notation := to denote "is assigned," to distinguish it from an equality. In other words, the above statement takes the current value of **u** and "overwrites" it into $\mathbf{u}_{\min}$.

If the forces are nonzero, we can lower the energy by iteratively moving the system along some *search direction*, **d**, that is determined from the energy and forces at the current and possibly past configurations visited during the minimization. As such, all minimization methods are based on the simple steps presented in Algorithm 9.1.

The methods differ principally in how they determine the search direction, $\mathbf{d}$ at line 4 and the step size α at line 5 of Algorithm 9.1. Generally, the more local information one has about the function being minimized, the more intelligently these things can be chosen. Higher derivatives of the energy, if they are not too onerous to compute, are usually a good source of such information. For example the NR method requires the *stiffness* or *Hessian* matrix of the system (the second derivative of the energy), which can greatly improve convergence rates.

9.2.3 The steepest descent method

The steepest descent method is generally an inefficient approach to finding a local minimum, but it has several advantages. First and foremost, it is a very simple algorithm to code without error. If one is more concerned with reliability than speed (or if one wants to do as little coding and debugging as possible) it is a good choice. Second, the steepest descent trajectory followed in going from the initial configuration to the minimized state has a clear physical interpretation as an overdamped dynamical system. This can be important when one is actually interested in entire pathways in configuration space, as opposed to just the minimum. Third, the steepest descent method is robust. It may be slow, but it almost always works. Finally, the steepest descent method is pedagogically useful as an introduction to energy minimization. Once you understand the steepest descent algorithm, you are equipped to understand the more complicated methods discussed later.

As the name "steepest descent" suggests, the idea is simply to choose the search direction at each iteration to be along the direction of the forces. This corresponds to the steepest "downhill" direction at that particular point in the energy landscape. Referring to the generic minimization method in Algorithm 9.1, line 4 becomes

$$\mathbf{d} \equiv \mathbf{f} \quad \text{for steepest descent.}$$

In the absolutely simplest implementation of the steepest descent method, the step size α may be prescribed to be some fixed, small value, although a check needs to be made to ensure that taking the full step does not lead to an increase in energy (something that could happen if $\mathbf{u}$ is already near the minimum and the full step $\alpha\mathbf{f}$ overshoots it). In more sophisticated implementations, the system is moved some variable amount α along the direction of the forces until the one-dimensional minimum along that direction is found. In other words, the multi-dimensional minimization problem is replaced by a series of constrained one-dimensional minimizations. Details of this *line minimization* process are discussed below, but for now we note the essential idea: for a fixed $\mathbf{u}$ and $\mathbf{d}$ we seek a positive real number α such that $\widetilde{\Pi}(\mathbf{u} + \alpha\mathbf{d})$ is minimized with respect to α, and then update the system as

$$\mathbf{u} := \mathbf{u} + \alpha\mathbf{d}.$$

The new $\mathbf{u}$ is used to compute a new force, and the process repeats until the force norm is below the set tolerance. The steepest descent method is summarized in Algorithm 9.2.

The steepest descent algorithm is an "intuitive" one: from where you are, move in the local direction of steepest descent, determine the new direction of steepest descent and repeat. It is not especially fast, however, because for many landscapes the most direct route

Algorithm 9.2 The steepest descent algorithm

1: $n := 0$
2: $\mathbf{f}^{(0)} := -\nabla_{\mathbf{u}}\widetilde{\Pi}(\mathbf{u}^{(0)})$
3: **while** $\left\|\mathbf{f}^{(n)}\right\| > \text{tol}$ **do**
4: $\quad \mathbf{d}^{(n)} = \mathbf{f}^{(n)}$
5: $\quad$ find $\alpha^{(n)} > 0$ such that $\phi(\alpha^{(n)}) \equiv \widetilde{\Pi}(\mathbf{u}^{(n)} + \alpha^{(n)}\mathbf{d}^{(n)})$ is minimized.
6: $\quad \mathbf{u}^{(n+1)} := \mathbf{u}^{(n)} + \alpha^{(n)}\mathbf{d}^{(n)}$
7: $\quad \mathbf{f}^{(n+1)} := -\nabla_{\mathbf{u}}\widetilde{\Pi}(\mathbf{u}^{(n+1)})$
8: $\quad n := n + 1$
9: **end while**
10: $\mathbf{u}_{\min} := \mathbf{u}^{(n)}$

to the minimum is not in the direction of steepest descent. Consider a long narrow trench dug straight down the side of a mountain. At a short distance up either side of the trench, the steepest descent direction is back into the trench bottom, which is almost at right angles to the "global" downhill direction taking us down the mountain. Taking the steepest descent path results in many short hops back and forth across the trench floor, gradually moving us down the mountain. This is illustrated by the jagged line in Fig. 9.2.

9.2.4 Line minimization

Most multi-dimensional minimization algorithms are carried out by a series of one-dimensional constrained minimizations (for example, see line 5 of Algorithm 9.2). Since it is used many times, the efficiency of the *line minimization* (or *line search*) is important.

Line minimization is an interesting area of computational mathematics because we can actually gain overall efficiency in this part of the algorithm through sloppiness; it is not necessary to find the line minimum exactly, so long as each line minimization does a reasonable job of lowering the energy of the system. In other words, we would like to replace line 5 of Algorithm 9.2 with

find $\alpha^{(n)} > 0$ such that $\phi(\alpha^{(n)}) \equiv \widetilde{\Pi}(\mathbf{u}^{(n)} + \alpha^{(n)}\mathbf{d}^{(n)})$ is *sufficiently reduced.*

If we can quantify "sufficiently reduced" we can avoid wasting time unnecessarily polishing our effort to minimize ϕ when starting along a new search direction would be more efficient. One approach is a combination of backtracking and the so-called "sufficient decrease" condition,[3] as follows. First, we must choose some sensible initial guess for α. This can be tricky in some methods, since $\mathbf{d}$ need not have the same units as $\mathbf{u}$. However, this is not the case in the NR method, where it is most efficient[4] to start with $\alpha = 1$. We then march along

[3] The sufficient decrease condition makes up part of the so-called "Wolfe Conditions" described in more detail in [NW99].

[4] As we shall see in Section 9.2.5, the NR method converges exactly in one step if the function to be minimized is quadratic. For this reason, $\alpha = 1$ should be tried as the initial step size in case the system is sufficiently close to the minimum that the quadratic approximation will move it directly to within tolerance of the minimum. For other minimization algorithms, more sophisticated and robust methods of choosing the step size α are described in [NW99, Chapter 3].

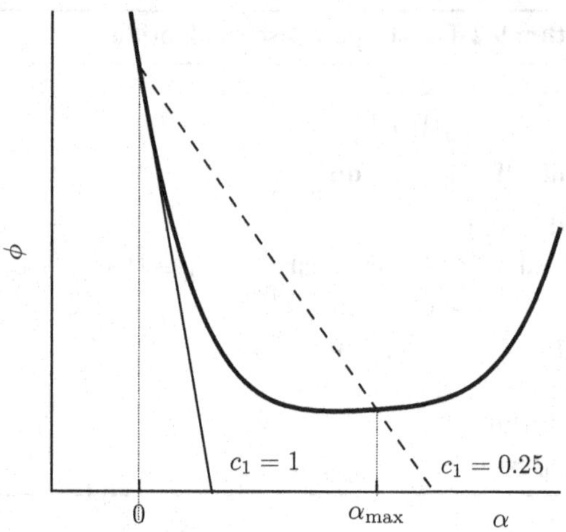

Fig. 9.3 The sufficient reduction condition determines a maximum value for α. In this case, it is pictured for a value of $c_1 = 0.25$. The region where the function ϕ is less than the dashed line is the range of acceptable values for α.

d until we find two points such that $0 < \alpha_1 < \alpha_2$ and

$$\phi(0) > \phi(\alpha_1) < \phi(\alpha_2), \tag{9.12}$$

so that there must be a minimum in the interval $(0, \alpha_2)$. Now, we can approximate the function ϕ as a parabola passing through $\phi(0)$, $\phi(\alpha_1)$ and $\phi(\alpha_2)$, and through simple algebra arrive at the minimum of the parabola at α_p:

$$\alpha_p = \frac{\phi(0)[\alpha_2^2 - \alpha_1^2] - \phi(\alpha_1)\alpha_2^2 + \phi(\alpha_2)\alpha_1^2}{2\,(\phi(0)[\alpha_2 - \alpha_1] - \phi(\alpha_1)\alpha_2 + \phi(\alpha_2)\alpha_1)}. \tag{9.13}$$

Now, we can make $\alpha := \alpha_p$ our initial guess and determine whether $\phi(\alpha)$ is sufficiently decreased compared to $\phi(0)$. This condition requires α to satisfy

$$\phi(\alpha) \leq \phi(0) - c_1 \alpha \mathbf{f}(\mathbf{u}^{(n)}) \cdot \mathbf{d}(\mathbf{u}^{(n)}),$$

for some value of $c_1 \in (0, 1)$. Note that $-\mathbf{f}(\mathbf{u}^{(n)}) \cdot \mathbf{d}(\mathbf{u}^{(n)}) = \phi'(0)$, i.e. this is equivalent to

$$\phi(\alpha) \leq \phi(0) + c_1 \alpha \phi'(0),$$

and as such it is just a way to estimate the expected decrease in ϕ based on the slope at $\alpha = 0$. When $c_1 = 1$, the last term is exactly the expected decrease in the energy based on a linear interpolation from the point $\mathbf{u}^{(n)}$, and this will limit α to very small values as shown in Fig. 9.3. Typically, a value of c_1 on the order of 10^{-4} is chosen. Figure 9.3 shows how this condition imposes a *maximum* value on α, and provides a way to decide when to quit searching along a particular direction **d**, as outlined in Algorithm 9.3. There, ρ is a scaling factor $\rho \in (0, 1)$ typically chosen on the order of $\rho = 0.5$. This algorithm would replace, for example, line 5 in Algorithm 9.2.

Algorithm 9.3 Line minimization using quadratic interpolation

choose ρ, such that $0 < \rho < 1$, c_1 and a tolerance tol
find $0 < \alpha_1 < \alpha_2$, such that $\phi(0) > \phi(\alpha_1) < \phi(\alpha_2)$
compute α_p using Eqn. (9.13).
$\alpha := \alpha_p$
while $\phi(\alpha) > \phi(0) - c_1 \alpha \mathbf{f}(\mathbf{u}) \cdot \mathbf{d}(\mathbf{u})$ **do**
  $\alpha := \rho\alpha$
  if $\alpha \leq$ tol **then**
    exit with error code {Line minimization has failed.}
  end if
end while

9.2.5 The Newton–Raphson (NR) method

Suppose that the strain energy is a simple quadratic function of the configuration

$$\widetilde{\Pi}(\mathbf{u}) = \frac{1}{2}\mathbf{u}^T\mathbf{K}\mathbf{u} - \mathbf{f}^T\mathbf{u}, \tag{9.14}$$

where $\mathbf{K}$ is a constant, positive-definite matrix and $\mathbf{f}$ is a constant vector. The condition for a stationary point of this function amounts to finding $\mathbf{u}$ such that

$$\nabla_{\mathbf{u}}\widetilde{\Pi} = \mathbf{K}\mathbf{u} - \mathbf{f} = \mathbf{0}.$$

If we are willing to invert the stiffness matrix, we can solve this directly:

$$\mathbf{u} = \mathbf{K}^{-1}\mathbf{f}.$$

The NR method applies this same approach iteratively to more general functions. We start from the Taylor expansion of the energy about the current guess at the configuration, $\mathbf{u}^{(n)}$:

$$\widetilde{\Pi}(\mathbf{u}) \approx \frac{1}{2}(\mathbf{u} - \mathbf{u}^{(n)})^T\mathbf{K}^{(n)}(\mathbf{u} - \mathbf{u}^{(n)}) - (\mathbf{f}^{(n)})^T(\mathbf{u} - \mathbf{u}^{(n)}) + \widetilde{\Pi}(\mathbf{u}^{(n)}),$$

where as usual

$$\mathbf{f}^{(n)} = -\left.\frac{\partial\widetilde{\Pi}(\mathbf{u})}{\partial\mathbf{u}}\right|_{\mathbf{u}=\mathbf{u}^{(n)}}, \qquad \mathbf{K}^{(n)} = \left.\frac{\partial^2\widetilde{\Pi}(\mathbf{u})}{\partial\mathbf{u}\partial\mathbf{u}}\right|_{\mathbf{u}=\mathbf{u}^{(n)}},$$

so that

$$\nabla_{\mathbf{u}}\widetilde{\Pi}(\mathbf{u}) \approx \mathbf{K}^{(n)}(\mathbf{u} - \mathbf{u}^{(n)}) - \mathbf{f}^{(n)}. \tag{9.15}$$

Now instead of solving for the next approximation to $\mathbf{u}$ by setting the above expression to zero (which could result in convergence to a maximum or saddle point rather than a minimum), we search for a solution by heading in the direction of the minimum of this quadratic approximation to the real energy. Thus we only use Eqn. (9.15) to obtain the search direction. Setting Eqn. (9.15) to zero gives us

$$\mathbf{d}^{(n)} \equiv \mathbf{u} - \mathbf{u}^{(n)} = (\mathbf{K}^{(n)})^{-1}\mathbf{f}^{(n)}, \tag{9.16}$$

Algorithm 9.4 The NR algorithm

$n := 0$
$\mathbf{f}^{(0)} := -\nabla_{\mathbf{u}}\widetilde{\Pi}(\mathbf{u}^{(0)})$
$\mathbf{K}^{(0)} := \partial^2\widetilde{\Pi}(\mathbf{u}^{(0)})/\partial\mathbf{u}\partial\mathbf{u}$
while $\left\|\mathbf{f}^{(n)}\right\| >$ tol **do**
  $\mathbf{d}^{(n)} := (\mathbf{K}^{(n)})^{-1}\mathbf{f}^{(n)}$
  find $\alpha^{(n)} > 0$ using line minimization (Algorithm 9.3). If this fails, set $\mathbf{d}^{(n)} := \mathbf{f}^{(n)}$ and retry the line minimization.
  $\mathbf{u}^{(n+1)} := \mathbf{u}^{(n)} + \alpha^{(n)}\mathbf{d}^{(n)}$
  $\mathbf{f}^{(n+1)} := -\nabla_{\mathbf{u}}\widetilde{\Pi}(\mathbf{u}^{(n+1)})$
  $\mathbf{K}^{(n+1)} := \partial^2\widetilde{\Pi}(\mathbf{u}^{(n+1)})/\partial\mathbf{u}\partial\mathbf{u}$
  $n := n + 1$
end while

and then we move the system by a line minimization in the usual way:

$$\mathbf{u}^{(n+1)} = \mathbf{u}^{(n)} + \alpha^{(n)}\mathbf{d}^{(n)}, \tag{9.17}$$

where $\alpha^{(n)} > 0$ is obtained from line minimization (Algorithm 9.3). This approach can fail when $\mathbf{K}^{(n)}$ is not positive definite, in which case $\mathbf{d}^{(n)}$ may not be a descent direction. In this case, one option is to abandon NR for the current step and set $\mathbf{d}^{(n)}$ to the steepest descent direction. Alternatively, $\mathbf{K}^{(n)}$ can be modified in some way to force it to be positive definite (see, for example, [FF77]). The NR method is summarized in Algorithm 9.4.

The FEM is particularly well suited to the NR method, since the stiffness matrix takes on a relatively simple form that permits efficient storage and inversion. Essentially, the foundations of the FEM to be presented in Section 9.3 revolve around developing an efficient way to compute the stiffness matrix.

9.2.6 Quasi-Newton methods

Often, one may want to use Eqn. (9.17), but it is too expensive or difficult to obtain and invert the Hessian matrix. There are several methods to produce approximations to $\mathbf{K}^{-1}$, or more generally to provide an algorithm for generating search directions of the form of a matrix multiplying the force vector. These methods are broadly classified as "quasi-Newton methods," and they can be advantageous for problems where the second derivatives required for the Hessian are sufficiently complex to make the code either tedious to implement or slow to execute.[5] For more details, the interested reader may try [Pol71, Rus06, PTVF08].

[5] The more sophisticated of the quasi-Newton methods are amongst the fastest algorithms for finding stationary points. Wales [Wal03] argues that one such method in particular, Nocedal's limited memory Broyden–Fletcher–Goldfarb–Shanno (L-BFGS) method, is in fact currently the fastest method that can be applied to relatively large systems.

9.2.7 The finite element tangent stiffness matrix

In order to implement the NR method within finite elements, it is necessary to compute the *tangent stiffness matrix* (or Hessian), K

$$\mathsf{K}_{\bar{\alpha}\bar{\beta}}(\mathbf{u}) = -\left.\frac{\partial \mathsf{f}_{\bar{\alpha}}}{\partial \mathsf{u}_{\bar{\beta}}}\right|_{\mathbf{u}} = \left.\frac{\partial^2 \widetilde{\Pi}}{\partial \mathsf{u}_{\bar{\alpha}} \partial \mathsf{u}_{\bar{\beta}}}\right|_{\mathbf{u}}. \tag{9.18}$$

This can be obtained from Eqn. (9.10) as

$$\mathsf{K}_{\bar{\alpha}\bar{\beta}} = \int_{\mathcal{B}_0} \frac{\partial P_{iJ}}{\partial F_{mN}} \frac{\partial F_{mN}}{\partial \mathsf{u}_{\bar{\beta}}} \frac{\partial \mathsf{S}_{i\bar{\alpha}}}{\partial X_J} \, dV_0 = \int_{\mathcal{B}_0} D_{iJmN} \frac{\partial \mathsf{S}_{m\bar{\beta}}}{\partial X_N} \frac{\partial \mathsf{S}_{i\bar{\alpha}}}{\partial X_J} \, dV_0, \tag{9.19}$$

where the last expression makes use of Eqn. (9.9) and the definition of the mixed elasticity tensor $\boldsymbol{D}$ from Eqn. (6.155).

If the strain energy were a quadratic function of $\mathbf{u}$, as would be the case for a linear elastic material subjected to small strains, the solution would be directly obtained from inverting the stiffness matrix. For nonlinear problems, we can use the NR method as we just described, and iteratively update the displacements according to Eqn. (9.17).

9.3 Elements and shape functions

To summarize up to this point, the approach of the FEM is as follows. Starting from a suitably accurate approximation to the potential energy, $\widetilde{\Pi}$, achieved through a discretization of the displacement variables, the minimization of the energy proceeds once we have an efficient scheme for computing the energy $\widetilde{\Pi}$, the residual $\mathbf{f}$ and the tangent stiffness K. In terms of the discretized displacement variables, these quantities are:

$$\widetilde{\Pi} = \int_{\mathcal{B}_0} W(\widetilde{\boldsymbol{F}}(\mathbf{u})) \, dV_0 - \mathsf{f}^{\text{ext}}_{\bar{\alpha}} \mathsf{u}_{\bar{\alpha}}, \tag{9.20a}$$

$$\mathsf{f}_{\bar{\alpha}} = \mathsf{f}^{\text{int}}_{\bar{\alpha}} + \mathsf{f}^{\text{ext}}_{\bar{\alpha}}, \tag{9.20b}$$

$$\mathsf{K}_{\bar{\alpha}\bar{\beta}} = \int_{\mathcal{B}_0} D_{iJmN}(\widetilde{\boldsymbol{F}}(\mathbf{u})) \frac{\partial \mathsf{S}_{m\bar{\beta}}}{\partial X_N} \frac{\partial \mathsf{S}_{i\bar{\alpha}}}{\partial X_J} \, dV_0, \tag{9.20c}$$

where we have defined the *internal nodal force vector*, $\mathbf{f}^{\text{int}}$ and *external nodal force vector*, $\mathbf{f}^{\text{ext}}$ with components

$$\mathsf{f}^{\text{int}}_{\bar{\alpha}} = -\int_{\mathcal{B}_0} P_{iJ}(\widetilde{\boldsymbol{F}}(\mathbf{u})) \frac{\partial \mathsf{S}_{i\bar{\alpha}}}{\partial X_J} \, dV_0, \qquad \mathsf{f}^{\text{ext}}_{\bar{\alpha}} = \int_{\partial \mathcal{B}_{0t}} \bar{T}_i \mathsf{S}_{i\bar{\alpha}} \, dA_0. \tag{9.21}$$

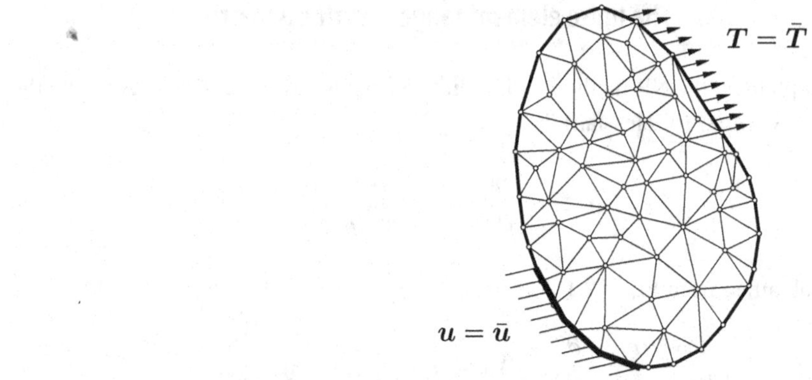

Fig. 9.4 Elements connecting the nodes to form a finite element mesh.

We will discuss the external nodal forces in Section 9.3.5. For now, we simply note that they are independent of the solution variable, $\mathbf{u}$. Thus, we confine our attention to techniques for the calculation of the internal force vector and stiffness matrix.

As written, Eqns. (9.20) do not permit a computationally efficient implementation without further consideration. We mention in passing an active area of research in "meshless methods" [BKO$^+$96, OIZT96, BM97, AZ00, AS02, SA04], whereby continuum mechanics is discretized using an unstructured array of points and shape functions that are not dependent on a finite element mesh. While these methods are beyond the scope of this work, we note here that meshless methods also start from Eqns. (9.20).

The key to the efficiency of FEM lies in the restrictions imposed on the shape functions. These functions are defined with respect to a very general tessellation (mesh) of the body into *elements* as shown in Fig. 9.4. These elements need not be triangular (or tetrahedral in three dimensions), although triangles represent the simplest geometry that can be used to fill the space between the nodes. Also, the elements need not all be the same type or size, but they must not overlap nor leave any gaps.[6] Each element is associated with a strict number of nodes, and conversely each node is associated only with the elements that it touches. The shape function for node α, S^α, is defined to have the following properties:

- *C^0 continuity* The shape function should be continuous across element boundaries but can have a discontinuous first derivative: such a function is referred to as having C^0 continuity. The continuity demanded of the shape functions is dictated by the shape function derivatives that appear in Eqns. (9.20), which we see are first derivatives. Since these equations must be integrable, the integrands can be discontinuous across element boundaries but they must be finite within each element.
- *The Kronecker delta property* S^α must satisfy

$$S^\alpha(\boldsymbol{X}^\beta) = \delta_{\alpha\beta} = \begin{cases} 1, & \text{when } \alpha = \beta, \\ 0, & \text{when } \alpha \neq \beta. \end{cases} \tag{9.22}$$

[6] There are variations of the FEM that do allow elements to overlap or to leave gaps. Examples include the so-called "natural element method" [SMB98] and others that share features with both finite elements and meshless techniques.

This ensures that the value of the interpolated displacement field at the position of node α is equal to the nodal value, since

$$\widetilde{u}_i(\boldsymbol{X}^\alpha) = \sum_{\beta=1}^{n_{\text{nodes}}} S^\beta(\boldsymbol{X}^\alpha)\mathsf{u}_i^\beta = \sum_{\beta=1}^{n_{\text{nodes}}} \delta_{\alpha\beta}\mathsf{u}_i^\beta = \mathsf{u}_i^\alpha. \tag{9.23}$$

This permits a direct physical interpretation of the values in $\mathbf{u}$.

- *The interpolation property* For the special case when the displacements are equal at every node in the mesh, the interpolated field should be exactly uniform. This property ensures the physically sensible behavior that a uniform displacement of all the nodes (which corresponds to a rigid translation of the body) produces a uniform interpolated displacement field and thus no strain in the body. Given a constant displacement vector, $\bar{u}_i$, we see that if all nodal displacements are the same, $\mathsf{u}_i^\alpha = \bar{u}_i$, we have

$$\widetilde{u}_i(\boldsymbol{X}) = \sum_{\alpha=1}^{n_{\text{nodes}}} S^\alpha(\boldsymbol{X})\mathsf{u}_i^\alpha = \bar{u}_i \sum_{\alpha=1}^{n_{\text{nodes}}} S^\alpha(\boldsymbol{X}),$$

so we require that the shape functions satisfy

$$\sum_{\alpha=1}^{n_{\text{nodes}}} S^\alpha(\boldsymbol{X}) = 1, \tag{9.24}$$

for all $\boldsymbol{X}$. For this reason, shape functions are sometimes referred to as a *partition of unity*.

- *Compact support* S^α is defined to be identically zero in any element not touching node α. It is this feature of the FEM shape functions that makes the method computationally very attractive, as we shall see next.

Without loss of generality, the integrals in Eqns. (9.20) can be treated as sums over integrals on each individual element, i.e.

$$\widetilde{\Pi} = \sum_{e=1}^{n_{\text{elem}}} \int_{B_0^e} W(\widetilde{\boldsymbol{F}}(\mathbf{u}))\,dV_0 - \mathsf{f}_{\bar{\alpha}}^{\text{ext}}\mathsf{u}_{\bar{\alpha}}, \tag{9.25a}$$

$$\mathsf{f}_{\bar{\alpha}} = -\sum_{e=1}^{n_{\text{elem}}} \int_{B_0^e} P_{iJ}(\widetilde{\boldsymbol{F}}(\mathbf{u}))\frac{\partial \mathsf{S}_{i\bar{\alpha}}}{\partial X_J}\,dV_0 + \mathsf{f}_{\bar{\alpha}}^{\text{ext}}, \tag{9.25b}$$

$$\mathsf{K}_{\bar{\alpha}\bar{\beta}} = \sum_{e=1}^{n_{\text{elem}}} \int_{B_0^e} D_{iJmN}(\widetilde{\boldsymbol{F}}(\mathbf{u}))\frac{\partial \mathsf{S}_{m\bar{\beta}}}{\partial X_N}\frac{\partial \mathsf{S}_{i\bar{\alpha}}}{\partial X_J}\,dV_0, \tag{9.25c}$$

where B_0^e is the domain of element e. This element-by-element parceling of the integrals is only useful, however, if the compact support property of the shape functions is exploited, as demonstrated by a simple one-dimensional example.

Example 9.1 (One-dimensional shape functions) Figure 9.5 shows a one-dimensional region discretized by seven nodes between $X = a$ and $X = b$. The simplest possible choice of mesh is to define each element by two nodes (elements labeled A–F in Fig. 9.5(b)). Imposing the restrictions

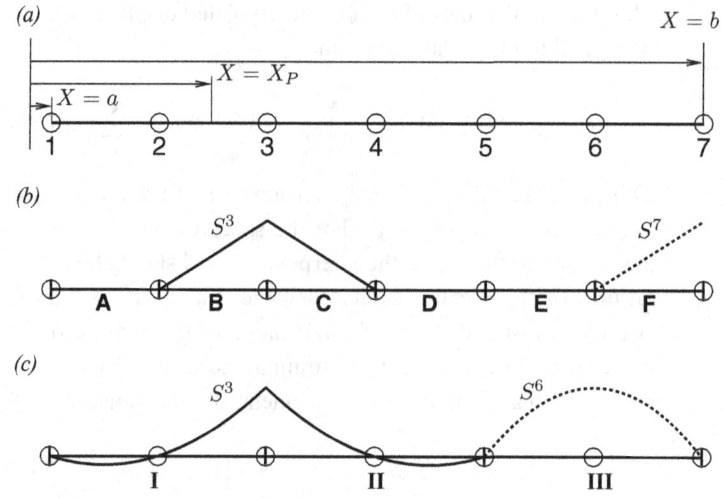

Fig. 9.5 Shape functions for a one-dimensional domain: (a) discretized domain; (b) linear elements A–F; (c) quadratic elements I–III.

listed previously leads to shape functions as shown for node 3 and node 7. These will necessarily be linear functions within each element, since Eqns. (9.22) and (9.24) effectively require the lowest-order function that can satisfy $S = 1$ at the node of interest and $S = 0$ at the other nodes in the element. The shape functions within a single element are shown in Tab. 9.1. The effect of this choice of shape functions is a piecewise linear interpolation of the displacements, as illustrated in Fig. 9.6(b).

Due to the compact support property, the interpolated displacements $\tilde{u}$ within each element depend only on the nodes connected to the element. For example, the displacement at position X_P in Fig. 9.5(a) is completely determined from the displacement of nodes 2 and 3. Precisely, we have

$$\tilde{u}(X_P) = S^2(X_P)\mathrm{u}^2 + S^3(X_P)\mathrm{u}^3 = \frac{\mathrm{X}^3 - X_P}{\mathrm{X}^3 - \mathrm{X}^2}\mathrm{u}^2 + \frac{X_P - \mathrm{X}^2}{\mathrm{X}^3 - \mathrm{X}^2}\mathrm{u}^3.$$

Note that the superscripts are node numbers, not exponents. Derivatives of this displacement field, which determine the strains and stresses at each point, are found from this equation and therefore they also depend only on the nodes connected to the element. Thus, an integral over any one element can be completely determined by considering only the displacements of these nodes.

Alternatively, we could divide the domain in Fig. 9.5 into three elements, each containing a node at each end and one in the center. These choices lead to quadratic shape functions, as illustrated for nodes 3 and 6 in Fig. 9.5(c) and given in detail in Tab. 9.1. This produces a piecewise quadratic interpolation of the displacements, as shown in Fig. 9.6(c). Quadratic interpolation generally improves the accuracy of the results for a fixed number of nodes, but also increases the computational effort by making the integration more difficult.

At this stage, we have an element-by-element description of the energy, residual forces and stiffness. Further, the compact support of the shape functions has ensured that the integration within an element is dependent only on the displacements of nodes connected to it. This paves the way for rapid and efficient computation of Eqns. (9.25) once a suitable numerical integration scheme is chosen.

Table 9.1. Linear and quadratic elemental shape functions for a one-dimensional domain. Superscripts denote node numbers

Linear elements	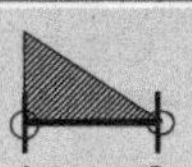 1 2	$S^1(X) = \dfrac{\mathsf{X}^2 - X}{\mathsf{X}^2 - \mathsf{X}^1}$
	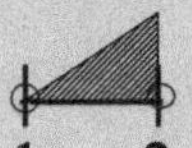 1 2	$S^2(X) = \dfrac{X - \mathsf{X}^1}{\mathsf{X}^2 - \mathsf{X}^1}$
Quadratic elements	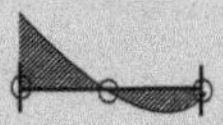 1 3 2	$S^1(X) = \dfrac{X \cdot X - X(\mathsf{X}^2 + \mathsf{X}^3) + \mathsf{X}^2\mathsf{X}^3}{\mathsf{X}^1 \cdot \mathsf{X}^1 - \mathsf{X}^1(\mathsf{X}^2 + \mathsf{X}^3) + \mathsf{X}^2\mathsf{X}^3}$
	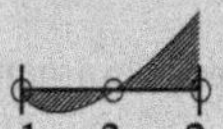 1 3 2	$S^2(X) = \dfrac{X \cdot X - X(\mathsf{X}^1 + \mathsf{X}^3) + \mathsf{X}^1\mathsf{X}^3}{\mathsf{X}^2 \cdot \mathsf{X}^2 - \mathsf{X}^2(\mathsf{X}^1 + \mathsf{X}^3) + \mathsf{X}^1\mathsf{X}^3}$
	 1 3 2	$S^3(X) = \dfrac{X \cdot X - X(\mathsf{X}^1 + \mathsf{X}^2) + \mathsf{X}^1\mathsf{X}^2}{\mathsf{X}^3 \cdot \mathsf{X}^3 - \mathsf{X}^3(\mathsf{X}^1 + \mathsf{X}^2) + \mathsf{X}^1\mathsf{X}^2}$

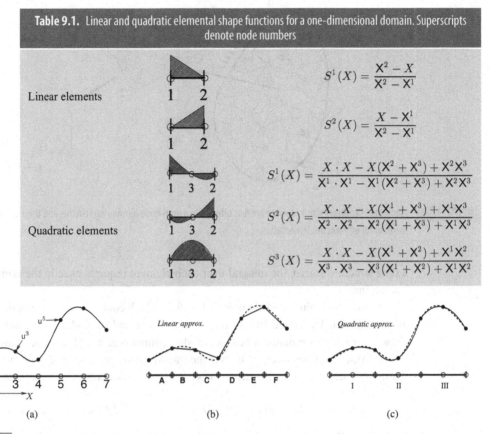

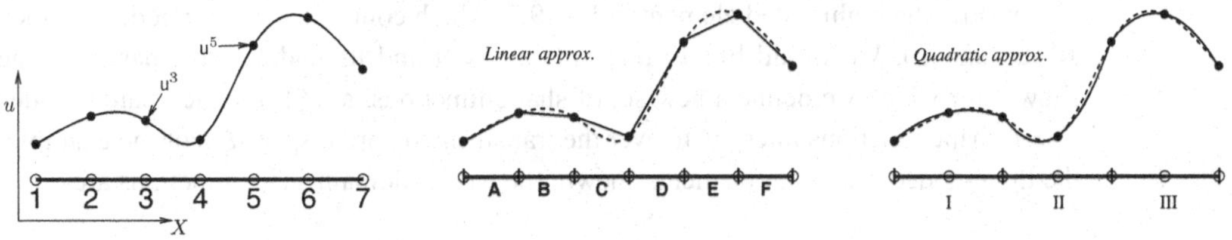

(a) (b) (c)

Fig. 9.6 One-dimensional example of interpolation with linear elements. In (a), the exact function is shown, whereas (b) and (c) show the interpolated functions given the exact values u^α at each node. Note that the nodal values will not generally be a perfect match to the exact function as shown in (a); the point of this figure is only to illustrate the different interpolations in (b) and (c).

9.3.1 Element mapping and the isoparametric formulation

In the simple one-dimensional case outlined in Example 9.1, one may envision some computational scheme by which to evaluate each integral in the sums of Eqns. (9.25). The domain in one dimension is always a simple line and the functions to be integrated are generally polynomials, so a straightforward scheme like Simpson's rule may be used. However, in higher dimensions the problem becomes considerably more complex, as illustrated in Fig. 9.4. In this case of two-dimensional triangular elements, the domain of integration differs for each element, and setting up a general, efficient and accurate routine to perform these integrals is not straightforward. But the compact support of the shape functions allows us to perform the integrations, not over the physical space B_0^e, but over the space of a so-called *parent* element Ω into which each element is mapped. The idea is to interpolate *both* the displacements *and* the reference configuration of the body itself. The advantage is that every integral is over the same domain, and once some preprocessing is completed and

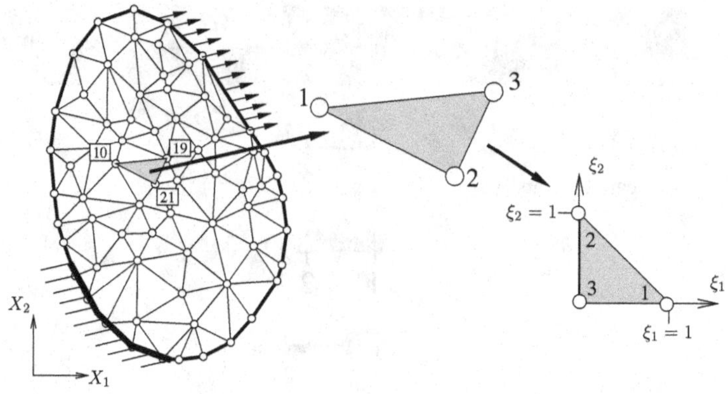

Fig. 9.7 Elements of arbitrary size and shape are first referred to a local node numbering scheme, and then mapped to a parent element for efficient implementation.

a data structure stored, the integral over each element requires exactly the same computer operations.

Consider the highlighted element in Fig. 9.7, which connects the three nodes numbered 10, 21 and 19. We would like to map this element and its nodes to the parent element shown, for which we define a new set of shape functions, $\mathfrak{s}^\alpha(\boldsymbol{\xi})$, for each parent node α. These shape functions interpolate over the transformed parent space $\boldsymbol{\xi}$. For the example of the three-noded triangular elements shown, the parent-domain shape functions are

$$\mathfrak{s}^1(\boldsymbol{\xi}) = \xi_1, \qquad \mathfrak{s}^2(\boldsymbol{\xi}) = \xi_2, \qquad \mathfrak{s}^3(\boldsymbol{\xi}) = 1 - \xi_1 - \xi_2. \tag{9.26}$$

We can readily verify that the interpolation and Kronecker delta properties hold for these shape functions. The numbering refers to the numbering on the parent element shown in Fig. 9.7, and so each element mapping must be accompanied by a mapping from the global node numbers (in this example, 10, 21 and 19) to the local parent node numbers.

We redefine the shape function matrix $\mathsf{S}_{I\bar{\alpha}}$ introduced in Eqn. (9.3) as follows:

$$S_{I\bar{\alpha}}(\boldsymbol{\xi}, e) = \begin{cases} \mathfrak{s}^\beta(\boldsymbol{\xi}) & \text{if } \bar{\alpha} \text{ maps to node } \beta \text{ of element } e, \\ 0 & \text{otherwise.} \end{cases} \tag{9.27}$$

Now, the mapping between physical coordinates and the parent coordinates within each element is obtained using these shape functions:

$$\widetilde{X}^e_I(\boldsymbol{\xi}) = \mathsf{S}_{I\bar{\alpha}}(\boldsymbol{\xi}, e)\mathsf{X}_{\bar{\alpha}}. \tag{9.28}$$

To write this in the alternative notation introduced in Eqn. (9.4) requires the introduction of a new symbol to map between the global node numbers and the local node numbers of the parent element, which we indicate with $\vec{\alpha}_e$. If a global node numbered α is attached to the element e in which the interpolation is being performed, then $\vec{\alpha}_e$ is the local node number in the parent element and the appropriate shape function is $\mathfrak{s}^{\vec{\alpha}_e}$ (In the example of Fig. 9.7, $\alpha = 10$ maps to $\vec{\alpha}_e = 1$, $\alpha = 21$ maps to $\vec{\alpha}_e = 2$ and $\alpha = 19$ maps to $\vec{\alpha}_e = 3$). Otherwise, the shape function is zero by compact support. Thus we write

$$\widetilde{X}_I^e(\boldsymbol{\xi}) = \sum_{\alpha=1}^{n_{\text{nodes}}} \mathfrak{s}^{\vec{\alpha}_e}(\boldsymbol{\xi}) \mathsf{X}_I^\alpha, \tag{9.29}$$

Similarly to the interpolation of the physical coordinates, the nodal displacements can be interpolated inside the parent element as

$$\widetilde{\boldsymbol{u}}^e(\boldsymbol{\xi}) = \mathbf{S}(\boldsymbol{\xi}, e)\mathbf{u} \quad \text{or} \quad \widetilde{u}_i^e(\boldsymbol{\xi}) = \sum_{\alpha=1}^{n_{\text{nodes}}} \mathfrak{s}^{\vec{\alpha}_e}(\boldsymbol{\xi}) \mathsf{u}_i^\alpha. \tag{9.30}$$

When the displacements and the coordinates are interpolated using the same shape functions in this way, it is referred to as the *isoparametric* formulation. This is the most common formulation of the FEM, but it is by no means the only one. For example, one could interpolate the displacements using shape functions of a lower order than those used for the coordinates, in what is referred to as the *subparametric* formulation. Conversely, if the displacements are interpolated using higher-order shape functions than those used for the coordinates, it is referred to as a *superparametric* formulation. Such formulations are useful when it is known, for instance, that the displacement field is likely to be much more (or less) difficult to interpolate than the geometry of the body.

Tables 9.2 and 9.3 show a sampling of isoparametric elements in two and three dimensions, respectively, together with their shape functions. In Fig. 9.8, we illustrate how such elements can be mixed within a mesh, although it is common for a single element type to be used throughout a model. Since the number of nodes determines the polynomial order of the interpolation (as evidenced by the shape functions), mixing of element types is usually limited to those of the same polynomial order, to ensure that continuity of the interpolated displacements across the element boundaries is satisfied.

The integrals in Eqns. (9.25) now require a change of variables from the physical to the parent coordinates, which depends on the Jacobian determinant, $\hat{J}$, of the mapping[7] from $\widetilde{X}$ to $\boldsymbol{\xi}$:

$$\hat{J} = \det \mathbf{J} = \det \nabla_{\xi} \widetilde{X},$$

where we adopt the notation $\hat{J}$ to distinguish this Jacobian from the one defined in Eqn. (3.7). We obtain $\mathbf{J}^e$ for element e from the interpolation in Eqn. (9.28)

$$\mathsf{J}_{IJ}^e = \frac{\partial \widetilde{X}_I^e}{\partial \xi_J} = \frac{\partial \mathsf{S}_{I\bar{\alpha}}}{\partial \xi_J} \mathsf{X}_{\bar{\alpha}},$$

or equally well from Eqn. (9.29) as

$$\mathsf{J}_{IJ}^e = \sum_{\alpha=1}^{n_{\text{nodes}}} \frac{\partial \mathfrak{s}^{\vec{\alpha}_e}}{\partial \xi_J} \mathsf{X}_I^\alpha. \tag{9.31}$$

[7] This mapping is precisely the same, mathematically, as a mapping between a reference and a deformed configuration as discussed in Chapter 3. The symbol $\mathbf{J}$ plays the same role in the element mapping as the deformation gradient plays in a deformation mapping (see Section 3.4, Eqn. (3.4)). $\mathbf{J}$ and $\boldsymbol{F}$ have all the same properties and must obey the same rules to be physically sensible. In particular, the requirement $\hat{J} > 0$ implies that the mapping of nodes from physical space to the parent space must not turn the element "inside-out."

Table 9.2. Geometry, shape functions and Gauss point information for some common isoparametric parent elements in two dimensions

		$\int_\Omega h(\boldsymbol{\xi})\, d\Omega = \sum_{g=1}^{n_q} w_g h(\boldsymbol{\xi}^g)$	
Element	Shape functions	$\boldsymbol{\xi}^g$	w_g
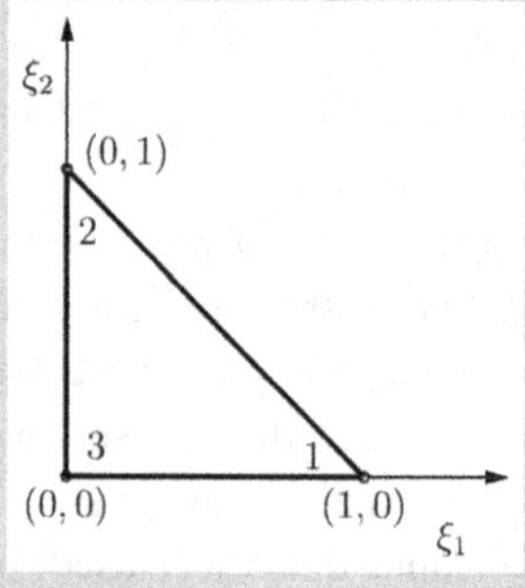	$\mathfrak{s}^1 = \xi_1$	$(\frac{1}{3}, \frac{1}{3})$	$\frac{1}{2}$
	$\mathfrak{s}^2 = \xi_2$		
	$\mathfrak{s}^3 = 1 - \xi_1 - \xi_2$		
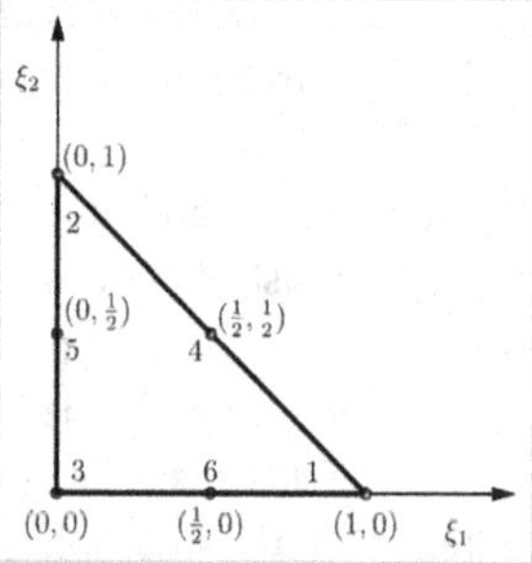	$\mathfrak{s}^1 = \xi_1(2\xi_1 - 1)$	$(\frac{1}{2}, \frac{1}{2})$	$\frac{1}{6}$
	$\mathfrak{s}^2 = \xi_2(2\xi_2 - 1)$	$(0, \frac{1}{2})$	$\frac{1}{6}$
	$\mathfrak{s}^3 = (1 - \xi_1 - \xi_2)[1 - 2(\xi_1 + \xi_2)]$	$(\frac{1}{2}, 0)$	$\frac{1}{6}$
	$\mathfrak{s}^4 = 4\xi_1\xi_2$		
	$\mathfrak{s}^5 = 4\xi_2(1 - \xi_1 - \xi_2)$		
	$\mathfrak{s}^6 = 4\xi_1(1 - \xi_1 - \xi_2)$		
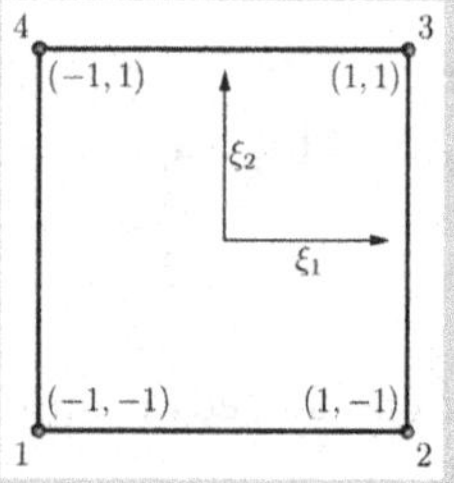	$\mathfrak{s}^1 = (1 - \xi_1)(1 - \xi_2)/4$	$(+\frac{1}{\sqrt{3}}, +\frac{1}{\sqrt{3}})$	1
	$\mathfrak{s}^2 = (1 + \xi_1)(1 - \xi_2)/4$	$(+\frac{1}{\sqrt{3}}, -\frac{1}{\sqrt{3}})$	1
	$\mathfrak{s}^3 = (1 + \xi_1)(1 + \xi_2)/4$	$(-\frac{1}{\sqrt{3}}, +\frac{1}{\sqrt{3}})$	1
	$\mathfrak{s}^4 = (1 - \xi_1)(1 + \xi_2)/4$	$(-\frac{1}{\sqrt{3}}, -\frac{1}{\sqrt{3}})$	1
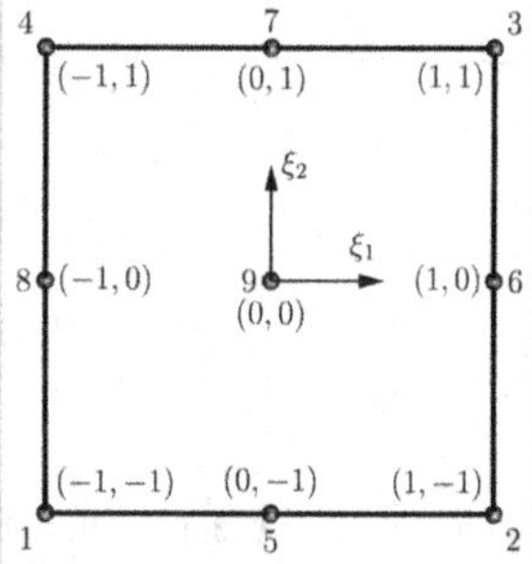	$\mathfrak{s}^1 = (-\xi_1 + \xi_1^2)(-\xi_2 + \xi_2^2)/4$	$(-\sqrt{\frac{3}{5}}, -\sqrt{\frac{3}{5}})$	$\frac{25}{81}$
	$\mathfrak{s}^2 = (\xi_1 + \xi_1^2)(-\xi_2 + \xi_2^2)/4$	$(0, -\sqrt{\frac{3}{5}})$	$\frac{40}{81}$
	$\mathfrak{s}^3 = (\xi_1 + \xi_1^2)(\xi_2 + \xi_2^2)/4$	$(\sqrt{\frac{3}{5}}, -\sqrt{\frac{3}{5}})$	$\frac{25}{81}$
	$\mathfrak{s}^4 = (-\xi_1 + \xi_1^2)(\xi_2 + \xi_2^2)/4$	$(-\sqrt{\frac{3}{5}}, 0)$	$\frac{40}{81}$
	$\mathfrak{s}^5 = (1 - \xi_1^2)(\xi_2^2 - \xi_2)/2$	$(0, 0)$	$\frac{64}{81}$
	$\mathfrak{s}^6 = (\xi_1^2 + \xi_1)(1 - \xi_2^2)/2$	$(\sqrt{\frac{3}{5}}, 0)$	$\frac{40}{81}$
	$\mathfrak{s}^7 = (1 - \xi_1^2)(\xi_2^2 + \xi_2)/2$	$(-\sqrt{\frac{3}{5}}, \sqrt{\frac{3}{5}})$	$\frac{25}{81}$
	$\mathfrak{s}^8 = (\xi_1^2 - \xi_1)(1 - \xi_2^2)/2$	$(0, \sqrt{\frac{3}{5}})$	$\frac{40}{81}$
	$\mathfrak{s}^9 = (1 - \xi_1^2)(1 - \xi_2^2)$	$(\sqrt{\frac{3}{5}}, \sqrt{\frac{3}{5}})$	$\frac{25}{81}$

Table 9.3. Geometry, shape functions and Gauss point information for some common isoparametric parent elements in three dimensions

Element	Shape functions	$\int_\Omega h(\boldsymbol{\xi})\,d\Omega = \sum_{g=1}^{n_q} w_g h(\boldsymbol{\xi}^g)$ $\boldsymbol{\xi}^g$	w_g
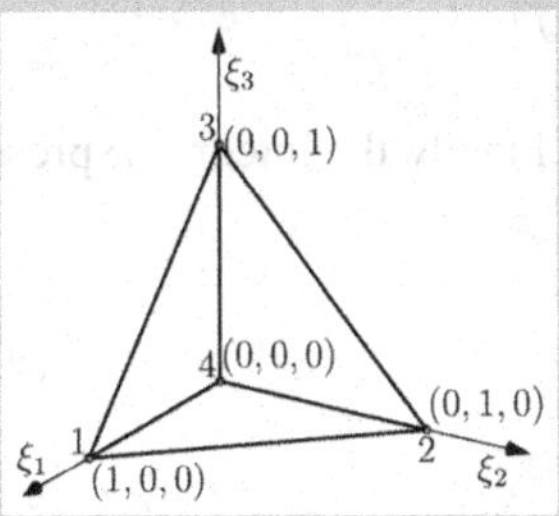	$\mathfrak{s}^1 = \xi_1$ $\mathfrak{s}^2 = \xi_2$ $\mathfrak{s}^3 = \xi_3$ $\mathfrak{s}^4 = 1 - \xi_1 - \xi_2 - \xi_3$	$(\frac{1}{3}, \frac{1}{3}, \frac{1}{3})$	$\frac{1}{6}$
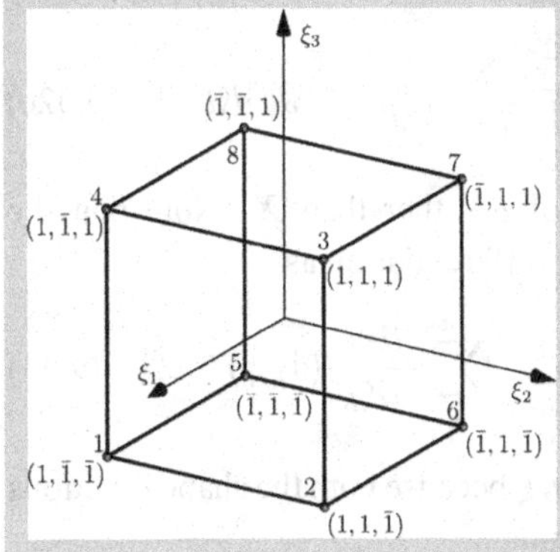	$\mathfrak{s}^1 = \frac{1}{8}(1+\xi_1)(1-\xi_2)(1-\xi_3)$ $\mathfrak{s}^2 = \frac{1}{8}(1+\xi_1)(1+\xi_2)(1-\xi_3)$ $\mathfrak{s}^3 = \frac{1}{8}(1+\xi_1)(1+\xi_2)(1+\xi_3)$ $\mathfrak{s}^4 = \frac{1}{8}(1+\xi_1)(1-\xi_2)(1+\xi_3)$ $\mathfrak{s}^5 = \frac{1}{8}(1-\xi_1)(1-\xi_2)(1-\xi_3)$ $\mathfrak{s}^6 = \frac{1}{8}(1-\xi_1)(1+\xi_2)(1-\xi_3)$ $\mathfrak{s}^7 = \frac{1}{8}(1-\xi_1)(1+\xi_2)(1+\xi_3)$ $\mathfrak{s}^8 = \frac{1}{8}(1-\xi_1)(1-\xi_2)(1+\xi_3)$	$(+\frac{1}{\sqrt{3}}, +\frac{1}{\sqrt{3}}, +\frac{1}{\sqrt{3}})$ $(+\frac{1}{\sqrt{3}}, +\frac{1}{\sqrt{3}}, -\frac{1}{\sqrt{3}})$ $(+\frac{1}{\sqrt{3}}, -\frac{1}{\sqrt{3}}, +\frac{1}{\sqrt{3}})$ $(+\frac{1}{\sqrt{3}}, -\frac{1}{\sqrt{3}}, -\frac{1}{\sqrt{3}})$ $(-\frac{1}{\sqrt{3}}, +\frac{1}{\sqrt{3}}, +\frac{1}{\sqrt{3}})$ $(-\frac{1}{\sqrt{3}}, +\frac{1}{\sqrt{3}}, -\frac{1}{\sqrt{3}})$ $(-\frac{1}{\sqrt{3}}, -\frac{1}{\sqrt{3}}, +\frac{1}{\sqrt{3}})$ $(-\frac{1}{\sqrt{3}}, -\frac{1}{\sqrt{3}}, -\frac{1}{\sqrt{3}})$	1 1 1 1 1 1 1 1

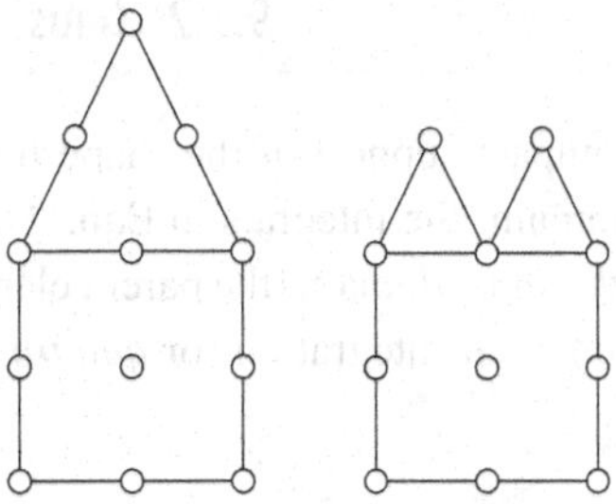

(a) (b)

Fig. 9.8 Examples of mixing element types in the same mesh. In (a), continuity of the shape functions across the element boundaries is preserved since both elements are quadratic. Strictly speaking, the mixing of linear triangles with quadratic rectangles in (b) is not permitted, but such combinations are sometimes used in special circumstances.

Infinitesimal volume elements then transform (cf. Eqn. (3.7)) as

$$dV_0 = \hat{J}\,d\Omega,$$

where $d\Omega$ is an infinitesimal volume in the parent space, while the volume in the physical space is dV_0. Derivatives of the shape functions appearing in Eqns. (9.25) are evaluated using the chain rule

$$\frac{\partial \mathsf{S}_{I\bar{\alpha}}}{\partial \widetilde{X}^e_J} = \frac{\partial \mathsf{S}_{I\bar{\alpha}}}{\partial \xi_K} \frac{\partial \xi_K}{\partial \widetilde{X}^e_J} = \frac{\partial \mathsf{S}_{I\bar{\alpha}}}{\partial \xi_K} (\mathsf{J}^e_{KJ})^{-1},$$

or alternatively we can use

$$\frac{\partial \mathfrak{s}^{\alpha}}{\partial \widetilde{X}^e_J} = \frac{\partial \mathfrak{s}^{\alpha}}{\partial \xi_K} \frac{\partial \xi_K}{\partial \widetilde{X}^e_J} = \frac{\partial \mathfrak{s}^{\alpha}}{\partial \xi_K} (\mathsf{J}^e_{KJ})^{-1},$$

when considering the scalar shape function at each node. Finally the integral expressions in Eqn. (9.25) become

$$\widetilde{\Pi} = \sum_{e=1}^{n_{\text{elem}}} \int_{\Omega} W(\widetilde{\boldsymbol{F}}(\mathbf{u})) \hat{\jmath}^e \, d\Omega - \mathsf{f}^{\text{ext}}_{\bar{\alpha}} \mathsf{u}_{\bar{\alpha}}, \tag{9.32a}$$

$$\mathsf{f}_{\bar{\alpha}} = - \sum_{e=1}^{n_{\text{elem}}} \int_{\Omega} P_{iJ}(\widetilde{\boldsymbol{F}}(\mathbf{u})) \frac{\partial \mathsf{S}_{i\bar{\alpha}}}{\partial \xi_R} (\mathsf{J}^e_{RJ})^{-1} \hat{\jmath}^e \, d\Omega + \mathsf{f}^{\text{ext}}_{\bar{\alpha}}, \tag{9.32b}$$

$$\mathsf{K}_{\bar{\alpha}\bar{\beta}} = \sum_{e=1}^{n_{\text{elem}}} \int_{\Omega} D_{iJmN}(\widetilde{\boldsymbol{F}}(\mathbf{u})) \frac{\partial \mathsf{S}_{m\bar{\beta}}}{\partial \xi_S} (\mathsf{J}^e_{SN})^{-1} \frac{\partial \mathsf{S}_{i\bar{\alpha}}}{\partial \xi_R} (\mathsf{J}^e_{RJ})^{-1} \hat{\jmath}^e \, d\Omega, \tag{9.32c}$$

where the shape functions $\mathbf{S}$ are now all functions of $\boldsymbol{\xi}$ rather than $\boldsymbol{X}$. Note that the deformation gradient must be found using a chain rule differentiation as

$$\widetilde{F}_{iJ} = \delta_{iJ} + \frac{\partial \mathsf{S}_{i\bar{\alpha}}}{\partial \xi_K} (\mathsf{J}^e_{KJ})^{-1} \mathsf{u}_{\bar{\alpha}} \quad \text{or} \quad \widetilde{F}_{iJ} = \delta_{iJ} + \sum_{\alpha=1}^{n_{\text{nodes}}} \frac{\partial \mathfrak{s}^{\vec{\alpha}_e}}{\partial \xi_K} (\mathsf{J}^e_{KJ})^{-1} \mathsf{u}^{\alpha}_i. \tag{9.33}$$

Further, it is important to remember that $\widetilde{F}_{iJ}$ depends on $\boldsymbol{\xi}$ because both the shape functions and the Jacobian are functions of $\boldsymbol{\xi}$ in the above equations.

9.3.2 Gauss quadrature

Through the compact support of the shape functions and the mapping from the reference to the parent domain, the integrals in Eqn. (9.32) have been reduced to a sum of different integrals over *the same* domain (the parent element). These integrals can be efficiently evaluated using numerical integration (or *quadrature*). Consider the general one-dimensional integral

$$H = \int_{-1}^{1} h(x) \, dx.$$

Any quadrature scheme to evaluate H can be expressed by the general formula

$$H \approx \sum_{g=1}^{n_{\text{q}}} w_g h(x_g), \tag{9.34}$$

where the function $h(x)$ is evaluated at n_{q} distinct quadrature points, x_g, $g = 1, \ldots, n_{\text{q}}$. Each $h(x_g)$ is then multiplied by an appropriate weight w_g and the sum is computed. If we know nothing about the nature of the function $h(x)$, it is natural to choose the points x_g to be equally spaced by a distance h, and choose the weights based on an assumed interpolation

Table 9.4. Gaussian integration points and weights in one dimension

$\int_{-1}^{1} h(x)\, dx = \sum_{g=1}^{n_q} w_g h(x_g)$

Polynomial order of $h(x)$, m	n_q	x_g	w_g
1	1	0	2
3	2	$\pm\sqrt{1/3}$	1
5	3	0	8/9
		$\pm\sqrt{3/5}$	5/9
7	4	±0.33998104	0.65214515
		±0.86113631	0.34785485
9	5	0	0.56888889
		±0.53846931	0.47862867
		±0.90617985	0.23692689
15	8	±0.18343464	0.36268378
		±0.52553241	0.31370665
		±0.79666648	0.22238103
		±0.96028986	0.10122854

scheme between the points that approximates the function. The well-known Simpson's rule, for instance, quadratically interpolates between the points to lead to weights $h/3$, $2h/3$ or $4h/3$ depending on the location of the point along the line.

Gauss recognized that the positions x_g of the points represented unused degrees of freedom that could improve the accuracy of the integration. Specifically, it is possible to show that an integrand $h(x)$ of known polynomial order m can be *exactly* integrated with only $(m+1)/2$ points, provided the positions of these points are optimal.[8] This optimized quadrature scheme is known as *Gaussian quadrature*. The optimization is achieved if the integrand is represented in terms of a set of orthogonal polynomials, such as the Legendre polynomials [AW95], and the points x_g chosen at the polynomial roots. Table 9.4 shows the optimal choice of the quadrature points and weights for polynomials of different order. Extension of this approach to two- and three-dimensional parent domains is conceptually straightforward but mathematically cumbersome. Therefore, we simply include some typical examples of Gauss points and weights in Tabs. 9.2 and 9.3.

If we closely consider the integrals in Eqn. (9.32), we see that while we can determine the polynomial order of most terms, the quantities W, $\boldsymbol{P}$ and $\boldsymbol{D}$ are general functions of the deformation gradient $\boldsymbol{F}$. If the polynomial order of this relationship is known, such as in the special case of linear elements (in which $\boldsymbol{F}$ is constant within the element) we can, in principle, determine the number of Gauss points necessary to integrate the functions exactly. In general though, this is not the case. However, it has been shown that number of

[8] It is also known that it is generally impossible to obtain the exact result using less than $(m+1)/2$ points, regardless of where these points are located. Thus, Gauss quadrature is optimal in the sense that it obtains the exact value using the minimal amount of computational effort possible.

Gauss points should *not* be chosen for exact integration. Rather, the most effective choice is the minimum number of points required to ensure that the same *rate of convergence with decreasing element size* is preserved as when exact integration is used. This is due to a very interesting curiosity of FEM. In essence, there is an advantageous cancellation of errors that occurs between discretization errors on the one hand and integration accuracy on the other. More discussion of this can be found in [ZT05].

Using the convergence rate as the criterion for the required accuracy of integration makes it possible to determine the number of Gauss points independently of the functional form of W, $\boldsymbol{P}$ and $\boldsymbol{D}$. This is because the convergence rate is dominated by the fact that at a sufficiently small element size, the displacement variation becomes linear and the deformation gradient is uniform within each element. As such, the number and location of the Gauss points is strictly determined by the element type, as shown in Tabs. 9.2 and 9.3.

We are now in a position to apply Gaussian quadrature to each of the integrals in Eqns. (9.32). First, we note that each quantity in Eqns. (9.32) is a sum over contributions independently obtained from each element:

$$\widetilde{\Pi} = \sum_{e=1}^{n_{\mathrm{elem}}} \mathcal{U}^e - \mathsf{f}_{\bar{\alpha}}^{\mathrm{ext}} \mathsf{u}_{\bar{\alpha}}, \qquad \mathsf{f}_{\bar{\alpha}} = \sum_{e=1}^{n_{\mathrm{elem}}} \mathsf{f}_{\bar{\alpha}}^{\mathrm{int},e} + \mathsf{f}_{\bar{\alpha}}^{\mathrm{ext}}, \qquad \mathsf{K}_{\bar{\alpha}\bar{\beta}} = \sum_{e=1}^{n_{\mathrm{elem}}} \mathsf{K}_{\bar{\alpha}\bar{\beta}}^e, \tag{9.35}$$

where the elemental quantities (denoted by the superscript e) are the Gauss quadrature expressions for each of the integrals in the equations:

$$\mathcal{U}^e \equiv \sum_{g=1}^{n_q} w_g W(\widetilde{\boldsymbol{F}}^g(\mathbf{u}))\hat{\jmath}^e, \tag{9.36a}$$

$$\mathsf{f}_{\bar{\alpha}}^{\mathrm{int},e} \equiv -\sum_{g=1}^{n_q} w_g P_{iJ}(\widetilde{\boldsymbol{F}}^g(\mathbf{u}))\frac{\partial \mathsf{S}_{i\bar{\alpha}}}{\partial \xi_R}(\mathsf{J}_{RJ}^e)^{-1}\hat{\jmath}^e, \tag{9.36b}$$

$$\mathsf{K}_{\bar{\alpha}\bar{\beta}}^e \equiv \sum_{g=1}^{n_q} w_g D_{iJmN}(\widetilde{\boldsymbol{F}}^g(\mathbf{u}))\frac{\partial \mathsf{S}_{m\bar{\beta}}}{\partial \xi_S}(\mathsf{J}_{SN}^e)^{-1}\frac{\partial \mathsf{S}_{i\bar{\alpha}}}{\partial \xi_R}(\mathsf{J}_{RJ}^e)^{-1}\hat{\jmath}^e. \tag{9.36c}$$

These are the sums to be evaluated for each element.

Note that for all but linear shape functions, the shape function derivatives and the Jacobian vary through the element, and therefore take a different value at each Gauss point. This is to say that even though we only explicitly show a dependence on g for the deformation gradient and the Gauss weight, it is tacitly contained in the other factors as well. The deformation gradient at each Gauss point, $\widetilde{\boldsymbol{F}}^g(\mathbf{u})$, depends on the current displacement vector and therefore needs to be evaluated during each iteration as outlined in Section 9.2, but the remaining quantities, i.e. the shape function derivatives and the Jacobian matrix, need to be computed only once and stored when the initial mesh is set up.

It is sometimes convenient to rewrite the residual in a form that explicitly separates the node number and the components. If we start from Eqn. (9.4) to define the displacements and Eqn. (9.7) for the deformation gradient in Eqn. (9.1) we obtain the form

$$\mathsf{f}_i^{\alpha} = \mathsf{f}_i^{\mathrm{ext},\alpha} - \sum_{e=1}^{n_{\mathrm{elem}}}\sum_{g=1}^{n_q} w_g P_{iJ}^g \frac{\partial \mathsf{s}^{\vec{\alpha}_e}}{\partial \xi_K}(\mathsf{J}_{KJ}^e)^{-1}\hat{\jmath}^e. \tag{9.37}$$

9.3.3 Practical issues of implementation

It is worth spending some time looking at the practical implementation of the method just outlined. For an NR minimization approach, this amounts to the iterative solution of Eqn. (9.16) until $\|\mathbf{f}\|$ is less than some tolerance. Therefore, we expect to have to evaluate Eqns. (9.35) multiple times during the solution, building the residual vector and constructing and inverting the stiffness matrix.

For the sake of concise notation, we have indicated that the elemental quantities in Eqns. (9.36) depend on the entire array of shape functions and displacements, but we know that the compact support of the shape functions will make most of these contributions zero. Thus, in practical implementation, local elemental arrays of displacements are extracted from the global vector, and these are used to produce small elemental vectors and matrices which are then added, one component at a time, to their global counterparts. A specific example helps to demonstrate this. We will consider a mesh of three-dimensional, four-node tetrahedral elements. Thus the number of dimensions is $n_\mathrm{d} = 3$ and the number of nodes per element $n_\mathrm{en} = 4$. The shape functions for this element are shown in Tab. 9.3.

The terms in Eqns. (9.36) arise from the formal differentiation of the approximate energy functional, but they are not in a form that is especially amenable to efficient computer implementation. Specifically, we would like to recast our tensor quantities in matrix form (similar to the Voigt notation of Section 6.5.1) in order to avoid contractions over tensors of higher order. With this goal in mind, it is possible to rearrange terms as follows. We start by treating the quantities P_{iJ} and F_{iJ} as column matrices, defining

$$\mathbf{P} = \begin{bmatrix} P_{11} \\ P_{21} \\ P_{31} \\ P_{12} \\ P_{22} \\ P_{32} \\ P_{13} \\ P_{23} \\ P_{33} \end{bmatrix}, \qquad \mathbf{F} = \begin{bmatrix} F_{11} \\ F_{21} \\ F_{31} \\ F_{12} \\ F_{22} \\ F_{32} \\ F_{13} \\ F_{23} \\ F_{33} \end{bmatrix} = \begin{bmatrix} 1 \\ 0 \\ 0 \\ 0 \\ 1 \\ 0 \\ 0 \\ 0 \\ 1 \end{bmatrix} + \begin{bmatrix} \partial u_1/\partial X_1 \\ \partial u_2/\partial X_1 \\ \partial u_3/\partial X_1 \\ \partial u_1/\partial X_2 \\ \partial u_2/\partial X_2 \\ \partial u_3/\partial X_2 \\ \partial u_1/\partial X_3 \\ \partial u_2/\partial X_3 \\ \partial u_3/\partial X_3 \end{bmatrix}. \tag{9.38}$$

We identify the first term in $\mathbf{F}$ as the identity tensor $\boldsymbol{I}$ written as a column matrix. The second term can be written as

$$\begin{bmatrix} \partial u_1/\partial X_1 \\ \partial u_2/\partial X_1 \\ \partial u_3/\partial X_1 \\ \partial u_1/\partial X_2 \\ \partial u_2/\partial X_2 \\ \partial u_3/\partial X_2 \\ \partial u_1/\partial X_3 \\ \partial u_2/\partial X_3 \\ \partial u_3/\partial X_3 \end{bmatrix} = \mathbf{E}\boldsymbol{u} = \begin{bmatrix} \partial/\partial X_1 & 0 & 0 \\ 0 & \partial/\partial X_1 & 0 \\ 0 & 0 & \partial/\partial X_1 \\ \partial/\partial X_2 & 0 & 0 \\ 0 & \partial/\partial X_2 & 0 \\ 0 & 0 & \partial/\partial X_2 \\ \partial/\partial X_3 & 0 & 0 \\ 0 & \partial/\partial X_3 & 0 \\ 0 & 0 & \partial/\partial X_3 \end{bmatrix} \begin{bmatrix} u_1 \\ u_2 \\ u_3 \end{bmatrix}, \tag{9.39}$$

which defines the "strain operator" $\mathbf{E}$. Within an element, u can be approximated by the interpolated finite element displacements, $\tilde{u}$ using Eqn. (9.3). However, the compact support we have introduced for the shape functions means that we can limit the extent of the matrices to only shape functions that are nonzero within the element in question. Specifically we write for each element e

$$\tilde{u} = \mathbf{S}^e \mathbf{u}^e = \begin{bmatrix} \mathfrak{s}^1 & 0 & 0 & \mathfrak{s}^2 & 0 & 0 & \mathfrak{s}^3 & 0 & 0 & \mathfrak{s}^4 & 0 & 0 \\ 0 & \mathfrak{s}^1 & 0 & 0 & \mathfrak{s}^2 & 0 & 0 & \mathfrak{s}^3 & 0 & 0 & \mathfrak{s}^4 & 0 \\ 0 & 0 & \mathfrak{s}^1 & 0 & 0 & \mathfrak{s}^2 & 0 & 0 & \mathfrak{s}^3 & 0 & 0 & \mathfrak{s}^4 \end{bmatrix} \begin{bmatrix} \mathsf{u}_1^1 \\ \mathsf{u}_2^1 \\ \mathsf{u}_3^1 \\ \mathsf{u}_1^2 \\ \mathsf{u}_2^2 \\ \mathsf{u}_3^2 \\ \mathsf{u}_1^3 \\ \mathsf{u}_2^3 \\ \mathsf{u}_3^3 \\ \mathsf{u}_1^4 \\ \mathsf{u}_2^4 \\ \mathsf{u}_3^4 \end{bmatrix}, \tag{9.40}$$

where the numbering now refers to the $n_{\text{en}} = 4$ nodes of the tetrahedron in Tab. 9.3, rather then the totality of n_{nodes} nodes in the mesh. Combining Eqns. (9.40), (9.39) and (9.38) we define a matrix operator $\mathbf{B}^e$ as

$$\mathbf{F}^e = \boldsymbol{I} + \mathbf{E}\mathbf{S}^e\mathbf{u}^e = \boldsymbol{I} + \mathbf{B}^e\mathbf{u}^e, \tag{9.41}$$

where

$$\mathbf{B}^e \equiv \mathbf{E}\mathbf{S}^e \tag{9.42}$$

is a $9 \times 3n_{\text{en}}$ matrix in three dimensions that will play the role of the shape function derivatives in our computer implementation. Roughly speaking,

$$\frac{\partial \mathsf{S}_{i\bar{a}}}{\partial \xi_R}(\mathsf{J}^e_{RJ})^{-1} \rightarrow \mathbf{B}^e$$

in our implementation-friendly formulation.

Now we consider, for example, the elemental internal force vector in Eqn. (9.36b). This can be written in terms of the local elemental matrices as

$$\mathbf{f}^{\text{int},e} = -\frac{1}{6}\hat{J}^e(\mathbf{B}^e)^T\mathbf{P}, \tag{9.43}$$

where the $1/6$ is the Gauss weight for the single Gauss point of a tetrahedral element and the matrix $\mathbf{B}^e$ takes the place of the quantity $(\partial \mathsf{S}_{i\bar{a}}/\partial \xi_R)(\mathsf{J}^e_{RJ})^{-1}$. Note that $\mathbf{f}^e$ is a $3n_{\text{en}} \times 1$ $(=12 \times 1)$ vector, $(\mathbf{B}^e)^T$ is $3n_{\text{en}} \times 9$ $(=12 \times 9)$ and $\mathbf{P}$ is 9×1.

Similarly, the elemental stiffness matrix from Eqn. (9.36d) is a 12×12 matrix that can be found from the multiplication

$$\mathbf{K}^e = \frac{1}{6}\hat{J}^e(\mathbf{B}^e)^T\mathbf{D}\mathbf{B}^e, \tag{9.44}$$

where $\mathbf{D}$ is a 9×9 symmetric matrix containing the unique components of D_{iJkL} (there are 45 unique entries due to symmetries). Specifically,

$$\mathbf{D} = \begin{bmatrix} D_{1111} & D_{1121} & D_{1131} & D_{1112} & D_{1122} & D_{1132} & D_{1113} & D_{1123} & D_{1133} \\ D_{1121} & D_{2121} & D_{2131} & D_{2112} & D_{2122} & D_{2132} & D_{2113} & D_{2123} & D_{2133} \\ D_{1131} & D_{2131} & D_{3131} & D_{3112} & D_{3122} & D_{3132} & D_{3113} & D_{3123} & D_{3133} \\ D_{1112} & D_{2112} & D_{3112} & D_{1212} & D_{1222} & D_{1232} & D_{1213} & D_{1223} & D_{1233} \\ D_{1122} & D_{2122} & D_{3122} & D_{1222} & D_{2222} & D_{2232} & D_{2213} & D_{2223} & D_{2233} \\ D_{1132} & D_{2132} & D_{3132} & D_{1232} & D_{2232} & D_{3232} & D_{3213} & D_{3223} & D_{3233} \\ D_{1113} & D_{2113} & D_{3113} & D_{1213} & D_{2213} & D_{3213} & D_{1313} & D_{1323} & D_{1333} \\ D_{1123} & D_{2123} & D_{3123} & D_{1223} & D_{2223} & D_{3223} & D_{1323} & D_{2323} & D_{2333} \\ D_{1133} & D_{2133} & D_{3133} & D_{1233} & D_{2233} & D_{3233} & D_{1333} & D_{2333} & D_{3333} \end{bmatrix}.$$

It is possible, but tedious, to show that summing Eqns. (9.43) and (9.44) over the Gauss points in the element is equivalent to Eqns. (9.36b) and (9.36c). The advantage is the elimination of the higher-order tensors and the many zeroes and symmetries that they concealed. As a result, the equations can be much more rapidly evaluated on a computer.

Spatial forms; material and geometric stiffness matrices Often, a constitutive law is given entirely in terms of the spatial quantities, i.e. $\boldsymbol{c}$ and $\boldsymbol{\tau}$ (recall that $\boldsymbol{\tau} = J\boldsymbol{\sigma}$). It is therefore convenient to transform Eqns. (9.43) and (9.44) to a form that operates directly on these quantities. We start by defining a matrix form of Eqn. (4.35) such that

$$\mathbf{P} = \mathbf{V}^T \mathbf{t},$$

where

$$\mathbf{V}^T = \begin{bmatrix} \widetilde{F}^{-1}_{11} & 0 & 0 & 0 & \widetilde{F}^{-1}_{13} & \widetilde{F}^{-1}_{12} \\ 0 & \widetilde{F}^{-1}_{12} & 0 & \widetilde{F}^{-1}_{13} & 0 & \widetilde{F}^{-1}_{11} \\ 0 & 0 & \widetilde{F}^{-1}_{13} & \widetilde{F}^{-1}_{12} & \widetilde{F}^{-1}_{11} & 0 \\ \widetilde{F}^{-1}_{21} & 0 & 0 & 0 & \widetilde{F}^{-1}_{23} & \widetilde{F}^{-1}_{22} \\ 0 & \widetilde{F}^{-1}_{22} & 0 & \widetilde{F}^{-1}_{23} & 0 & \widetilde{F}^{-1}_{21} \\ 0 & 0 & \widetilde{F}^{-1}_{23} & \widetilde{F}^{-1}_{22} & \widetilde{F}^{-1}_{21} & 0 \\ \widetilde{F}^{-1}_{31} & 0 & 0 & 0 & \widetilde{F}^{-1}_{33} & \widetilde{F}^{-1}_{32} \\ 0 & \widetilde{F}^{-1}_{32} & 0 & \widetilde{F}^{-1}_{33} & 0 & \widetilde{F}^{-1}_{31} \\ 0 & 0 & \widetilde{F}^{-1}_{33} & \widetilde{F}^{-1}_{32} & \widetilde{F}^{-1}_{31} & 0 \end{bmatrix}, \quad \mathbf{t} = \begin{bmatrix} \tau_{11} \\ \tau_{22} \\ \tau_{33} \\ \tau_{23} \\ \tau_{13} \\ \tau_{12} \end{bmatrix}, \tag{9.45}$$

from which we can rewrite Eqn. (9.43) as

$$\mathbf{f}^{\text{int},e} = -\frac{1}{6}\hat{J}^e(\mathbf{B}^e)^T\mathbf{P} = -\frac{1}{6}\hat{J}^e(\mathbf{B}^e)^T\mathbf{V}^T\mathbf{t} = -\frac{1}{6}\hat{J}^e(\mathbf{B}^e_{\text{c}})^T\mathbf{t}.$$

We have defined a new matrix

$$\mathbf{B}^e_{\text{c}} = \mathbf{V}\mathbf{B}^e \tag{9.46}$$

that transforms the Kirchhoff stress column matrix directly to the nodal forces.

The stiffness matrix calculation is not quite as simple, but it is still possible to write it in terms of the spatial quantities. Recalling Eqn. (6.166) we have

$$D_{iJmN} = J\left(c_{ijmn} + \delta_{im}\sigma_{jn}\right)F^{-1}_{Jj}F^{-1}_{Nn},$$

which we can insert into Eqn. (9.36c) to obtain two distinct terms:

$$\mathsf{K}^e_{\bar{\alpha}\bar{\beta}} = \sum_{g=1}^{n_q} w_g J c_{ijmn} \widetilde{F}^{-1}_{Jj} \widetilde{F}^{-1}_{Nn} \frac{\partial \mathsf{S}_{m\bar{\beta}}}{\partial \xi_S} (\mathsf{J}^e_{SN})^{-1} \frac{\partial \mathsf{S}_{i\bar{\alpha}}}{\partial \xi_R} (\mathsf{J}^e_{RJ})^{-1} \hat{\jmath}^e$$
$$+ \sum_{g=1}^{n_q} w_g \delta_{im} \tau_{jn} \widetilde{F}^{-1}_{Jj} \widetilde{F}^{-1}_{Nn} \frac{\partial \mathsf{S}_{m\bar{\beta}}}{\partial \xi_S} (\mathsf{J}^e_{SN})^{-1} \frac{\partial \mathsf{S}_{i\bar{\alpha}}}{\partial \xi_R} (\mathsf{J}^e_{RJ})^{-1} \hat{\jmath}^e.$$

We have dropped the explicit dependence on the Gauss points to streamline the notation, but recall that c, τ and F all depend on the deformation and the location in the element at which we are evaluating the terms. These two sums are referred to as the *material stiffness* and the *geometric stiffness*, respectively, to emphasize the dependence of the former on the material property c and the latter on the current state of stress and deformation. Analogous to Eqn. (9.44), this can be written in a compact matrix form as

$$\mathbf{K}^e = \mathbf{K}^e_{\mathrm{mat}} + \mathbf{K}^e_{\mathrm{geo}}, \tag{9.47}$$

where

$$\mathbf{K}^e_{\mathrm{mat}} = \frac{1}{6} \hat{\jmath}^e J (\mathbf{B}^e_{\mathrm{c}})^T \mathbf{c} \mathbf{B}^e_{\mathrm{c}}, \qquad \mathbf{K}^e_{\mathrm{geo}} = \frac{1}{6} \hat{\jmath}^e (\mathbf{B}^e_{\mathrm{T}})^T \mathbf{T} \mathbf{B}^e_{\mathrm{T}}. \tag{9.48}$$

In these expressions, $\mathbf{c}$ is the 6×6 form of the spatial stiffness in Voigt notation (see Eqn. (6.171)) and $\mathbf{T}$ is a 9×9 matrix that represents $\delta_{im}\tau_{jn}$ analogous to how $\mathbf{D}$ represents D_{iJmN}. Specifically, $\mathbf{T}$ is the symmetric matrix

$$\mathbf{T} = \begin{bmatrix} \tau_{11}\boldsymbol{I} & \tau_{12}\boldsymbol{I} & \tau_{13}\boldsymbol{I} \\ \tau_{12}\boldsymbol{I} & \tau_{22}\boldsymbol{I} & \tau_{23}\boldsymbol{I} \\ \tau_{13}\boldsymbol{I} & \tau_{23}\boldsymbol{I} & \tau_{33}\boldsymbol{I} \end{bmatrix},$$

where $\boldsymbol{I}$ is the 3×3 identity matrix.

The matrix $\mathbf{B}^e_{\mathrm{c}}$ in Eqn. $(9.48)_1$ has already been defined in Eqn. (9.46), whereas $\mathbf{B}^e_{\mathrm{T}}$ plays a similar role in Eqn. $(9.48)_2$, but has different dimensions due to the difference in the symmetries of $\delta_{im}\tau_{jn}$ versus c_{ijmn}. In analogy with Eqn. $(9.45)_1$, we define a matrix $\mathbf{U}$ such that

$$\mathbf{U}^T = \begin{bmatrix}
\widetilde{F}^{-1}_{11} & 0 & 0 & \widetilde{F}^{-1}_{12} & 0 & 0 & \widetilde{F}^{-1}_{13} & 0 & 0 \\
0 & \widetilde{F}^{-1}_{11} & 0 & 0 & \widetilde{F}^{-1}_{12} & 0 & 0 & \widetilde{F}^{-1}_{13} & 0 \\
0 & 0 & \widetilde{F}^{-1}_{11} & 0 & 0 & \widetilde{F}^{-1}_{12} & 0 & 0 & \widetilde{F}^{-1}_{13} \\
\widetilde{F}^{-1}_{21} & 0 & 0 & \widetilde{F}^{-1}_{22} & 0 & 0 & \widetilde{F}^{-1}_{23} & 0 & 0 \\
0 & \widetilde{F}^{-1}_{21} & 0 & 0 & \widetilde{F}^{-1}_{22} & 0 & 0 & \widetilde{F}^{-1}_{23} & 0 \\
0 & 0 & \widetilde{F}^{-1}_{21} & 0 & 0 & \widetilde{F}^{-1}_{22} & 0 & 0 & \widetilde{F}^{-1}_{23} \\
\widetilde{F}^{-1}_{31} & 0 & 0 & \widetilde{F}^{-1}_{32} & 0 & 0 & \widetilde{F}^{-1}_{33} & 0 & 0 \\
0 & \widetilde{F}^{-1}_{31} & 0 & 0 & \widetilde{F}^{-1}_{32} & 0 & 0 & \widetilde{F}^{-1}_{33} & 0 \\
0 & 0 & \widetilde{F}^{-1}_{31} & 0 & 0 & \widetilde{F}^{-1}_{32} & 0 & 0 & \widetilde{F}^{-1}_{33}
\end{bmatrix}$$

from which we build

$$\mathbf{B}^e_{\mathrm{T}} = \mathbf{U}\mathbf{B}^e.$$

Note that in the spatial form the matrices $\mathbf{B}^e_{\mathrm{T}}$ and $\mathbf{B}^e_{\mathrm{c}}$ depend directly on the state of deformation through F^{-1}. This means that they must be recomputed at each step in the iterative solution. The matrix $\mathbf{B}^e$, on the other hand, is constant.

Data stored prior to NR iteration Further steps to an efficient implementation can now be made apparent. For example, $\mathbf{B}^e$ comprises terms which do not depend on the solution vector $\mathbf{u}$, but only on the nodal coordinates and the location of the Gauss point. Thus, when each element is initially defined, the following steps can be taken:

- For each Gauss point, a matrix of shape function derivatives with respect to the parent coordinates, evaluated at the Gauss point, is loaded into memory. In this example of a three-dimensional tetrahedral element, we require a 3×4 matrix and there is only one Gauss point:

$$\nabla_\xi \mathbf{S}^e = \begin{bmatrix} \frac{\partial \mathfrak{s}^1}{\partial \xi_1} & \frac{\partial \mathfrak{s}^2}{\partial \xi_1} & \frac{\partial \mathfrak{s}^3}{\partial \xi_1} & \frac{\partial \mathfrak{s}^4}{\partial \xi_1} \\ \frac{\partial \mathfrak{s}^1}{\partial \xi_2} & \frac{\partial \mathfrak{s}^2}{\partial \xi_2} & \frac{\partial \mathfrak{s}^3}{\partial \xi_2} & \frac{\partial \mathfrak{s}^4}{\partial \xi_2} \\ \frac{\partial \mathfrak{s}^1}{\partial \xi_3} & \frac{\partial \mathfrak{s}^2}{\partial \xi_3} & \frac{\partial \mathfrak{s}^3}{\partial \xi_3} & \frac{\partial \mathfrak{s}^4}{\partial \xi_3} \end{bmatrix} = \begin{bmatrix} 1 & 0 & 0 & -1 \\ 0 & 1 & 0 & -1 \\ 0 & 0 & 1 & -1 \end{bmatrix}.$$

  Note that this matrix is the same for every element of the same (linear tetrahedral) type.
- A matrix of the coordinates of the nodes defining the element is extracted from the global coordinate array. In this case, we have the 4×3 matrix:

$$\mathbf{X}^e = \begin{bmatrix} X_1^1 & X_2^1 & X_3^1 \\ X_1^2 & X_2^2 & X_3^2 \\ X_1^3 & X_2^3 & X_3^3 \\ X_1^4 & X_2^4 & X_3^4 \end{bmatrix},$$

  which permits the calculation of the 3×3 Jacobian matrix, $\mathbf{J}^e$, from Eqn. (9.31), but with n_{nodes} replaced with the number of nodes on the element, $n_{\mathrm{en}} = 4$. The determinant of $\mathbf{J}^e$ is stored as $\hat{J}^e$ for each element, e. The Jacobian matrix and its determinant are different for every element, but they does not change during the solution iterations since they are independent of the displacement vector. Thus, they can be computed and stored for each element once as a preprocessing step.
- The inverse of $\mathbf{J}^e$ is computed and stored to be used in subsequent computations of $\widetilde{F}$, which is dependent on the displacements and is computed during each minimization step.
- The inverse of $\mathbf{J}^e$ is used to find the components of $\mathbf{B}^e$ which are stored for the element. While it is tempting, because of the simple code which would result, to directly implement Eqns. (9.43) and (9.44) exactly as they appear as matrix multiplications, this approach would not be especially efficient due to many multiplications by zero. Alternatively and more efficiently, the gradient of the shape functions with respect to the global coordinate, $\nabla_0 \mathbf{S}^e$, can be stored as a 3×4 matrix analogous to $\nabla_\xi \mathbf{S}$, but now containing unique

values for each element:

$$\nabla_0 \mathbf{S}^e = \begin{bmatrix} \dfrac{\partial \mathfrak{s}^1}{\partial X_1} & \dfrac{\partial \mathfrak{s}^2}{\partial X_1} & \dfrac{\partial \mathfrak{s}^3}{\partial X_1} & \dfrac{\partial \mathfrak{s}^4}{\partial X_1} \\ \dfrac{\partial \mathfrak{s}^1}{\partial X_2} & \dfrac{\partial \mathfrak{s}^2}{\partial X_2} & \dfrac{\partial \mathfrak{s}^3}{\partial X_2} & \dfrac{\partial \mathfrak{s}^4}{\partial X_2} \\ \dfrac{\partial \mathfrak{s}^1}{\partial X_3} & \dfrac{\partial \mathfrak{s}^2}{\partial X_3} & \dfrac{\partial \mathfrak{s}^3}{\partial X_3} & \dfrac{\partial \mathfrak{s}^4}{\partial X_3} \end{bmatrix}, \tag{9.49}$$

from which Eqns. (9.43) and (9.44) can be more carefully coded. It is the use of this kind of optimization that typically makes FEM code tedious to write and hard to read.

The quantities **P** and **D** are dependent on the displacements through $\widetilde{F}$, and as such must be computed at each iteration during the solution. Rapid computation of the deformation gradient at a Gauss point is achieved by extracting a local elemental displacement vector and evaluating Eqn. (9.41). The deformation gradient can then be passed to an independent routine that returns **P** and **D**, which are used to compute the elemental internal force (Eqn. (9.43)) and elemental stiffness (Eqn. (9.44)). The entries of these matrices can be added to the global force and stiffness through the mapping of the local node numbering within the parent element and the global node numbers. This will be illustrated for a simple one-dimensional example in the next section.

The FEM solution algorithm Figure 9.9 is a sketch of the flow of the FEM solution process, and helps to illustrate the benefits of the rearrangement of terms and element-by-element treatment. Primarily, it illustrates the modularity of the FEM. For example, we see that the constitutive model is completely contained in **D** and **P** for a given deformation gradient, and it is therefore entirely independent of the type of element used and the dimensionality of the problem. Also, the elemental data stored in $\mathbf{B}^e$ can be computed once at the start of the process, and a carefully written code can easily swap between element types (since this only changes the size of the $\mathbf{B}^e$ matrices and the number of Gauss points). Every element is independent from every other element in the sense that a problem can contain elements with different constitutive responses and different shapes. Well-written FEM code can be used for multiple element types and multiple materials, without loss of efficiency.

The main iterative loop of the algorithm shown in Fig. 9.9 is essentially the NR process explained in Section 9.2.5, containing three main processes (as well as the simple convergence test). The first is the element-by-element construction of the forces, indicated by the first loop over the elements. If the forces have not converged, we start the second main process, which is to build the stiffness matrix, **K**, again element-by-element.[9] The third and final process, comprising the "solve" and "update" steps, is to invert **K** and take an NR step to update the displacement field. In Section 9.3.5, we discuss the application of the boundary conditions required within the "solve" process.

[9] The process of efficiently assembling **K** is described in Section 9.3.4. A similar process is used in assembling **f**. The details are obvious from the discussion of the stiffness matrix assembly.

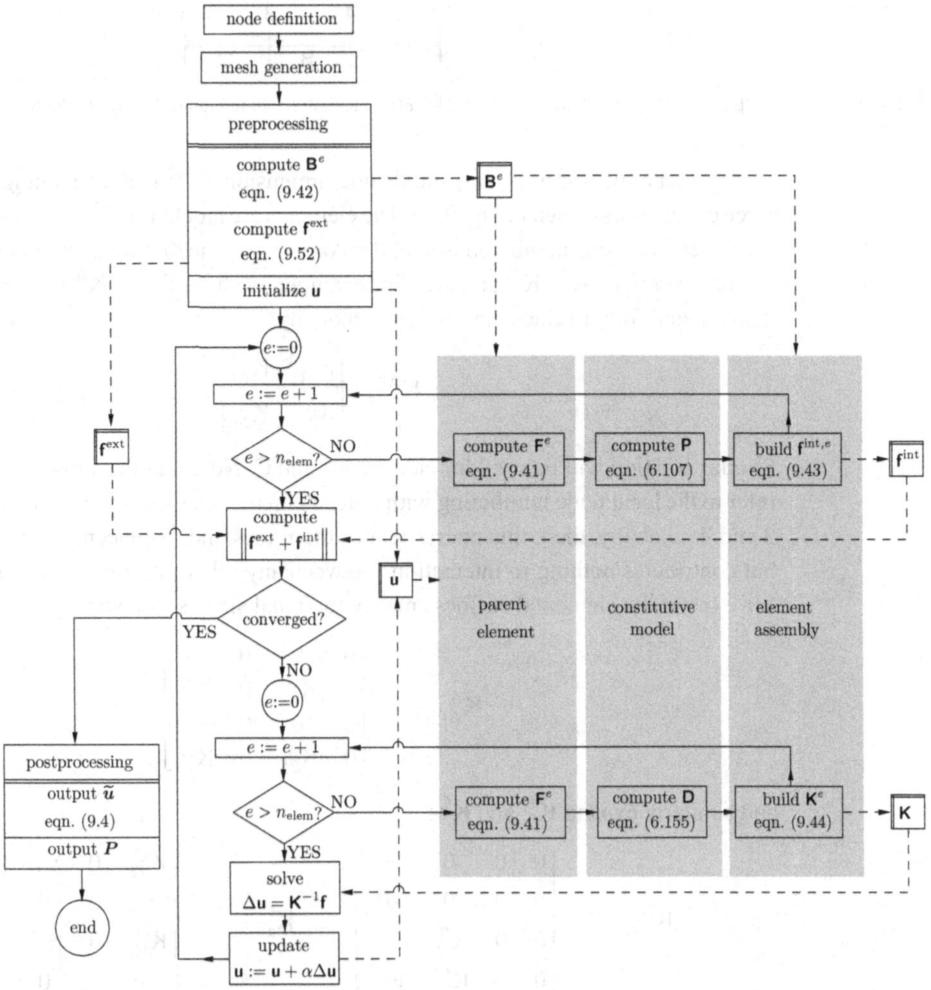

Fig. 9.9 Flow chart of the FEM solution process. Solid lines indicate the flow of the algorithm, while dashed lines indicate the flow of data as they are computed or used by various processes. The counter e refers to the element number. The chart highlights the modularity of the elements, constitutive law, and pre- and post- processing aspects of the FEM.

9.3.4 Stiffness matrix assembly

In the previous section, we computed the elemental stiffness matrix, $\mathbf{K}^e$. Note that this matrix contains $(n_d \cdot n_{en}) \times (n_d \cdot n_{en})$ entries, where n_{en} is the number of nodes per element and n_d is the number of dimensions of the problem. The final step before solving the matrix equation is to assemble these elemental matrices into the global stiffness matrix.

The elemental stiffness matrix was computed with reference to a local numbering scheme for the nodes, but the numbering of the nodes in the global displacements $\mathbf{u}$ must be followed in the final equation. However, a straightforward mapping can be used to insert the elemental stiffness entries into the global stiffness matrix.

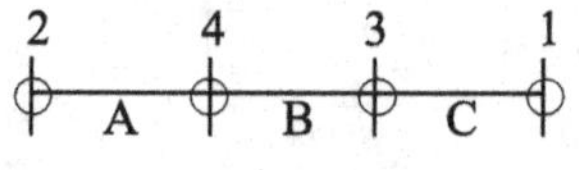

Fig. 9.10 Simple one-dimensional mesh with three linear elements and four randomly numbered nodes.

Consider a specific mesh in a simple one-dimensional domain, containing four nodes and three elements as shown in Fig. 9.10. The elements are labeled A, B and C, but for generality the nodes have been numbered in a random order. Assume that we have computed elemental stiffness matrices $\mathbf{K}^{\mathrm{A}}$, $\mathbf{K}^{\mathrm{B}}$ and $\mathbf{K}^{\mathrm{C}}$. For example, we have found $\mathbf{K}^{\mathrm{A}}$ by considering nodes 2 and 4, and found values that we will denote by[10]

$$\mathbf{K}^{\mathrm{A}} = \begin{bmatrix} \mathsf{K}^{\mathrm{A}}_{11} & \mathsf{K}^{\mathrm{A}}_{12} \\ \mathsf{K}^{\mathrm{A}}_{21} & \mathsf{K}^{\mathrm{A}}_{22} \end{bmatrix}.$$

Similar notation will be used for elements B and C. Note that the subscripts 1 and 2 in $\mathbf{K}^{\mathrm{A}}$ refer to the local node numbering within the element, and globally these nodes are numbers 2 and 4. Globally, then, this matrix relates the forces and displacements of nodes 2 and 4, but contributes nothing to interactions between any other pair of nodes. *Conceptually*, we can expand the elemental stiffness matrix to global size as follows:

$$\mathbf{K}^{\mathrm{A}}_{\mathrm{global}} = \begin{bmatrix} 0 & 0 & 0 & 0 \\ 0 & \mathsf{K}^{\mathrm{A}}_{11} & 0 & \mathsf{K}^{\mathrm{A}}_{12} \\ 0 & 0 & 0 & 0 \\ 0 & \mathsf{K}^{\mathrm{A}}_{21} & 0 & \mathsf{K}^{\mathrm{A}}_{22} \end{bmatrix},$$

and similarly expand $\mathbf{K}^{\mathrm{B}}$ and $\mathbf{K}^{\mathrm{C}}$:

$$\mathbf{K}^{\mathrm{B}}_{\mathrm{global}} = \begin{bmatrix} 0 & 0 & 0 & 0 \\ 0 & 0 & 0 & 0 \\ 0 & 0 & \mathsf{K}^{\mathrm{B}}_{22} & \mathsf{K}^{\mathrm{B}}_{21} \\ 0 & 0 & \mathsf{K}^{\mathrm{B}}_{12} & \mathsf{K}^{\mathrm{B}}_{11} \end{bmatrix}, \qquad \mathbf{K}^{\mathrm{C}}_{\mathrm{global}} = \begin{bmatrix} \mathsf{K}^{\mathrm{C}}_{22} & 0 & \mathsf{K}^{\mathrm{C}}_{21} & 0 \\ 0 & 0 & 0 & 0 \\ \mathsf{K}^{\mathrm{C}}_{12} & 0 & \mathsf{K}^{\mathrm{C}}_{11} & 0 \\ 0 & 0 & 0 & 0 \end{bmatrix},$$

since element B connects nodes 3 and 4, while element C joints nodes 1 and 3. The global stiffness matrix from Eqn. $(9.35)_3$ is then

$$\mathbf{K} = \sum_{e=1}^{n_{\mathrm{elem}}} \mathbf{K}^{e}_{\mathrm{global}} = \mathbf{K}^{\mathrm{A}}_{\mathrm{global}} + \mathbf{K}^{\mathrm{B}}_{\mathrm{global}} + \mathbf{K}^{\mathrm{C}}_{\mathrm{global}}$$

and therefore

$$\mathbf{K} = \begin{bmatrix} \mathsf{K}^{\mathrm{C}}_{22} & 0 & \mathsf{K}^{\mathrm{C}}_{21} & 0 \\ 0 & \mathsf{K}^{\mathrm{A}}_{11} & 0 & \mathsf{K}^{\mathrm{A}}_{12} \\ \mathsf{K}^{\mathrm{C}}_{12} & 0 & \mathsf{K}^{\mathrm{B}}_{22} + \mathsf{K}^{\mathrm{C}}_{11} & \mathsf{K}^{\mathrm{B}}_{21} \\ 0 & \mathsf{K}^{\mathrm{A}}_{21} & \mathsf{K}^{\mathrm{B}}_{12} & \mathsf{K}^{\mathrm{A}}_{22} + \mathsf{K}^{\mathrm{B}}_{11} \end{bmatrix}. \tag{9.50}$$

We emphasize that this is a *conceptual* process only. It would be extremely wasteful to build the $\mathbf{K}_{\mathrm{global}}$ matrices on the computer, since they would be mostly filled with zeroes.

[10] Often, and certainly for a hyperelastic material, $\mathbf{K}^{e}$ is symmetric and therefore $\mathsf{K}^{e}_{12} = \mathsf{K}^{e}_{21}$.

This expansion and summation can be efficiently carried out through the storage of a book-keeping array that maps each element to its place in the global problem (see the discussion on sparse matrix storage and inversion in [PTVF92, Saa03]).

9.3.5 Boundary conditions

In Section 7.1, we discussed the nature of boundary conditions for continuum mechanics problems. Here, we see how those boundary conditions translate into constraints on the solution to an FEM problem. Boundary conditions for static problems consist of two types:[11] the so-called "natural" (or traction) boundary condition and the "essential" (or displacement) boundary condition. As the name suggests, the traction boundary condition arises "naturally" from the potential energy due to the applied loads in Eqn. (9.1), and manifests itself as a constant external force vector (Eqn. $(9.21)_2$) applied to the nodes. Later, we will discuss why displacement boundary conditions are indeed "essential" to the solution process as the name suggests, but we first look at the external nodal forces more closely.

Traction boundary conditions We specified the boundary-value problem with a general traction applied to part of the body's surface. Recall that this gives rise to a contribution to the nodal forces (Eqn. $(9.21)_2$):

$$\mathsf{f}^{\mathrm{ext}}_{\tilde{\alpha}} = \int_{\partial B_{0t}} \bar{T}_i \mathsf{S}_{i\tilde{\alpha}} \, dA_0 .$$

The tractions are assumed to be prescribed independently from the solution variable, and therefore this integral needs to be evaluated only once at the time that the model is initialized.

Rigorous treatment of this term is often glossed over in the finite element literature because of its complexity and also because of the difficulty of exactly prescribing a traction boundary condition in the first place (see footnote 2 on page 246). In practice, a traction boundary condition is often either relatively simple and treatable as a special case (e.g. constant pressure over a surface), too complex to know exactly (e.g. contact forces), or more easily represented as a displacement boundary condition (e.g. the end conditions in a fixed-grip tensile experiment). If, at the end of the day, one still wants to apply a traction, the exact traction needed to mimic the experiment is probably sufficiently vague that any reasonable approximation to the equivalent nodal forces will be good enough.[12]

To rigorously evaluate $\mathsf{f}^{\mathrm{ext}}$ in three dimensions for a general case is tricky since we must carry out an integration of an arbitrary function (the traction vector as a function of position on the surface) over an irregularly-shaped surface. However, it is possible to write the applied traction in terms of the reference surface normal as

$$\bar{T} = \bar{P} N, \tag{9.51}$$

[11] There can also be "mixed" boundary conditions, as discussed in Section 7.1. Their application in FEM is a straightforward extension of the discussion herein.

[12] The fact that the nodal forces need not be exactly derived from the surface tractions is related to *Saint Venant's principle*, which states that the stresses, strains and displacements "far" from the location of the applied traction do not depend explicitly on the details of the traction distribution. Rather, they depend only on the resultant force (and moment) that the traction creates. For a rigorous statement of this principle see, for example, [Ste54].

where $\bar{\boldsymbol{P}}$ is an applied stress that gives rise to the correct tractions. If the loading and geometry are relatively simple, it is not difficult to work out the functional form of $\bar{\boldsymbol{P}}$. When this is the case, we can use Nanson's formula (Eqn. (3.9)) to carry out a mapping into the parent space[13]

$$\mathsf{f}^{\mathrm{ext}}_{\bar{\alpha}} = \sum_{e=1}^{n_{\mathrm{elem}}} \int_{\partial\Omega_t} (\mathsf{J}^e_{RJ})^{-1} \widehat{N}_R \mathsf{S}_{i\bar{\alpha}} \bar{P}_{iJ} \hat{J}^e \, da, \tag{9.52}$$

where $\widehat{N}$ is the normal to the surface in the parent space (which is a constant on each facet of the parent element) and da is an element of area in the parent space. It is now possible to carry out this integration using appropriately located Gauss points for the reduced-dimensional facet of the parent element. Of course, this need only be evaluated for the subset of element facets upon which nonzero tractions act.

Displacement boundary conditions Displacement boundary conditions are called "essential" because they take the form of constraints that serve to make the stiffness matrix invertible. In three dimensions, any finite element mesh has six degrees of freedom that do not change the energy of the system (three translations and three rotations). Mathematically, these *zero-energy eigenmodes* of the stiffness matrix render it uninvertible. We must constrain enough nodes to make rigid rotations and translations impossible, with the mathematical effect of building a reduced stiffness matrix that will be invertible. This is achieved by constraining nodes on ∂B_{0u} to the prescribed displacements there. This means that we no longer want to "solve" for the displacement of these nodes but rather use them to eliminate some of the equations governing an NR iteration.

The process is best illustrated by a simple rearrangement of the order of the scalar equations in Eqn. (9.16). Practically speaking, this amounts to a renumbering of the nodes, although efficient FEM implementations can perform this operation through appropriate book-keeping without actual renumbering. Imagine we renumber so that all the nodes which have fixed displacement appear first in the vector $\mathbf{u}$. Then we can partition our matrix equation as

$$\begin{bmatrix} \mathbf{K}_{\mathrm{CC}} & \mathbf{K}_{\mathrm{CF}} \\ \mathbf{K}_{\mathrm{FC}} & \mathbf{K}_{\mathrm{FF}} \end{bmatrix} \begin{bmatrix} \Delta\mathbf{u}_{\mathrm{C}} \\ \Delta\mathbf{u}_{\mathrm{F}} \end{bmatrix} = \begin{bmatrix} \mathbf{f}_{\mathrm{C}} \\ \mathbf{f}_{\mathrm{F}} \end{bmatrix}.$$

Here, the subscript C refers to the "constrained" degrees of freedom where the displacement is prescribed and the subscript F means "free." Assuming that the displacement of the constrained nodes is already imposed, then $\Delta\mathbf{u}_{\mathrm{C}} = \mathbf{0}$. This set of equations can now be written as two separate equations:

$$\mathbf{K}_{\mathrm{CF}} \Delta\mathbf{u}_{\mathrm{F}} = \mathbf{f}_{\mathrm{C}}, \qquad \mathbf{K}_{\mathrm{FF}} \Delta\mathbf{u}_{\mathrm{F}} = \mathbf{f}_{\mathrm{F}}. \tag{9.53}$$

The second of these can be inverted to find displacement increments of the free nodes:

$$\Delta\mathbf{u}_{\mathrm{F}} = \mathbf{K}_{\mathrm{FF}}^{-1} \mathbf{f}_{\mathrm{F}}. \tag{9.54}$$

13 Note by comparing the definitions of $\mathbf{J}$ and $\boldsymbol{F}$ that the parent space here plays the role of the *reference* configuration in the derivation of Nanson's formula.

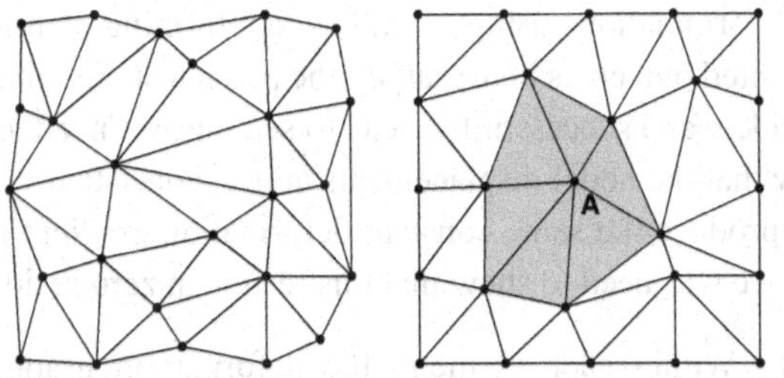

Fig. 9.11 Example meshes for the patch test. Node A is an example of an interior node on which the forces must be identically zero under uniform deformation. It is an interior node because it is surrounded by elements on all sides.

Generally, forces will arise on the constrained nodes due to the fact that they are held fixed. These forces can now be computed directly from Eqn. $(9.53)_1$ if they are desired.

9.3.6 The patch test

In order to be useful, the FEM should converge in the limit of high nodal density. Once elements are small enough in this limit, it is reasonable to expect that all fields can be approximated as uniform within an element, and thus we should require that the FEM reproduces uniform fields exactly. The test of this convergence property is the so-called patch test. It derives its name from an arbitrary patchwork of elements like those illustrated in Fig. 9.11, and is succinctly stated as follows.

Patch test A method passes the patch test if, for any arbitrary arrangement of nodes with nodal displacements consistent with a uniform deformation, the residual force on internal nodes is identically zero.[14]

In other words, we take any of the patches shown in Fig. 9.11 and apply displacements to all of the nodes of the form

$$\mathsf{u}_i^\alpha = (F_{iJ}^{\mathrm{app}} - \delta_{iJ})\mathsf{X}_J^\alpha, \tag{9.55}$$

where $\boldsymbol{F}^{\mathrm{app}}$ is a constant deformation gradient. The resulting residual force on any *internal* node must be exactly zero. Note that we do not require the residual on the *boundary* nodes to be zero. We think of this as the physical problem of applying displacement boundary conditions consistent with a uniform deformation gradient. In Section 8.1 we saw that homogeneous deformation of uniform material is a universal equilibrium solution that can be sustained by application of appropriate boundary tractions. Thus, we should expect that, no matter what simple elastic constitutive relation is used, the uniformly deformed FEM mesh for the patch test will be in equilibrium away from the boundary nodes. In other words, the internal nodes will be at equilibrium positions with zero out-of-balance forces.

[14] In the computational literature a weaker form of the patch test is often invoked. Instead of requiring the residual to be identically zero, a method must only satisfy this condition to a specified numerical tolerance in order to pass the test. We prefer the strict definition used here.

FEM formulations using so-called "conforming elements" (the type with which we have contented ourselves here) satisfy the patch test, and this is one of the reasons why they are so widely and successfully used. To see this, we need to prove two things. First, we need to show that the nodal displacements above, consistent with an applied $\boldsymbol{F}^{\text{app}}$ that is constant, also produce the same constant deformation gradient inside each element. Once we have that, we will need to show that this results in zero residual on the internal nodes.

Proof Within each element, the deformation gradient is given by the expression in Eqn. $(9.33)_2$. Inserting the prescribed displacement field from Eqn. (9.55) gives us

$$\widetilde{F}_{iJ} = \delta_{iJ} + F^{\text{app}}_{iM} \sum_{\alpha=1}^{n_{\text{nodes}}} \frac{\partial \mathfrak{s}^{\vec{\alpha}_e}}{\partial \xi_K} (\mathsf{J}^e_{KJ})^{-1} \mathsf{X}^{\alpha}_M - \delta_{iM} \sum_{\alpha=1}^{n_{\text{nodes}}} \frac{\partial \mathfrak{s}^{\vec{\alpha}_e}}{\partial \xi_K} (\mathsf{J}^e_{KJ})^{-1} \mathsf{X}^{\alpha}_M .$$

Note that by the definition of $\mathbf{J}^e$ in Eqn. (9.31), this becomes

$$\widetilde{F}_{iJ} = \delta_{iJ} + F^{\text{app}}_{iM} \mathsf{J}^e_{MK} (\mathsf{J}^e_{KJ})^{-1} - \delta_{iM} \mathsf{J}^e_{MK} (\mathsf{J}^e_{KJ})^{-1}.$$

The summation convention on the repeated indices allows us to cancel the first and third terms while simplifying the second to give

$$\widetilde{F}_{iJ} = F^{\text{app}}_{iJ}.$$

Thus the deformation gradient is equal to the constant applied value in every element. Since we assume that the constitutive law is the same in each element and a function only of $\boldsymbol{F}$, this further implies that the stress $\boldsymbol{P}$ and stiffness $\boldsymbol{D}$ are also constant everywhere.

Now consider the residual, as defined in Eqn. (9.37), for the special case of no externally applied forces. Since $\boldsymbol{P}$ is constant we can take it outside the sums to yield

$$\mathsf{f}^{\alpha}_i = -P_{iJ} \sum_{e=1}^{n_{\text{elem}}} \sum_{g=1}^{n_{\text{q}}} w_g \frac{\partial \mathfrak{s}^{\vec{\alpha}_e}}{\partial \xi_K} (\mathsf{J}^e_{KJ})^{-1} \hat{\jmath}^e. \tag{9.56}$$

We know the polynomial order of all terms within the sums, so we can choose the quadrature points and weights such that the integral is evaluated exactly. Next, we remind ourselves what this integral is by returning to the analytical integration over the *real* space instead of the *mapped parent* space

$$\mathsf{f}^{\alpha}_i = -P_{iJ} \sum_{e=1}^{n_{\text{elem}}} \int_{B^e_0} \frac{\partial \mathfrak{s}^{\vec{\alpha}_e}}{\partial X_J} \, dV_0. \tag{9.57}$$

By the compact support of the shape functions, this is an integral over the elements touching the node α, since the shape functions are identically zero outside this support. In the example patch of Fig. 9.11, a typical interior node A is shown along with the shaded region over which this integral needs to be considered. Using the divergence theorem (Eqn. (2.106)) allows us to transform this integral over the volume of each element to an integral only over the element surfaces, and the residual becomes

$$\mathsf{f}^{\alpha}_i = -P_{iJ} \sum_{e=1}^{n_{\text{elem}}} \int_{\partial B^e_0} \mathfrak{s}^{\vec{\alpha}_e} N_J \, dA_0, \tag{9.58}$$

where $\boldsymbol{N}$ is the outward normal to the element surface. If we perform this integration element by element, going around each element facet-by-facet, we see that along the facets that do not touch node α, the contribution is zero since the shape function must be zero on this facet. On the other hand, contributions from facets which include node α may be nonzero, but they will always be canceled by the contribution from a neighboring element. This follows from the assumed continuity of the shape functions across element boundaries, and from the fact that $\boldsymbol{N}$ on a face of one element is $-\boldsymbol{N}$ for the same face of a neighboring element. Thus, as long as node α is completely surrounded by elements (as it must be on any interior node), this evaluates to zero. The patch test is therefore identically satisfied. □

Advanced modifications to the FEM include types of elements for which it is not possible to show that the patch test is generally satisfied as we have here. In some instances, one can show a numerical patch test is satisfied. In other cases, care must be taken as to how the elements are used, and such FEM formulations are best left to the FEM experts. On the other hand, the relatively simple FEM approach outlined in this book can be implemented and used by FEM novices with confidence that the results will generally be reliable, stable and accurate. An essential reason for this reliability is the satisfaction of the patch test.

9.3.7 The linear elastic limit with small and finite strains

An important limit of continuum mechanics and finite element solutions is the case of linear elastic, small strain (see Sections 3.5, 6.5 and 10.4). In this limit, the gradients of the displacement, $u_{i,j}$, are small and the strain energy density function becomes (see Eqn. (6.170))

$$W = \frac{1}{2} c_{ijkl} u_{i,j} u_{k,l}, \tag{9.59}$$

from which the Cauchy stress follows as

$$\sigma_{ij} = c_{ijkl} u_{k,l}. \tag{9.60}$$

In effect, the small-strain assumption is that all components of $\nabla_0 \boldsymbol{u}$ are small compared with unity. From this, we can say that for small strains

$$\boldsymbol{F} = \boldsymbol{I} + \nabla_0 \boldsymbol{u} \approx \boldsymbol{I} \tag{9.61}$$

and

$$J = \det \boldsymbol{F} \approx 1. \tag{9.62}$$

We can now use this to simplify the relations between the various stress measures and elastic moduli. Equations (4.35) and (4.41) clearly lead to

$$\boldsymbol{\sigma} \approx \boldsymbol{P} \approx \boldsymbol{S}, \quad \text{for small strains.}$$

For the moduli, we start from Eqn. (6.166) and insert Eqn. (9.60) to eliminate the stress from the equation. Using Eqns. (9.61) and (9.62) this becomes

$$D_{iJkL} \approx \delta_{Jj} \delta_{Ll} \left(c_{ijkl} + \delta_{ik} u_{m,n} c_{jlmn} \right) \quad \text{for small strains.}$$

We note that by the assumption of small $\nabla_0 \boldsymbol{u}$, the second term in the parentheses is much smaller than the first, and we can therefore simply write

$$D_{iJkL} \approx \delta_{Jj}\delta_{Ll}c_{ijkl} \quad \text{for small strains.} \tag{9.63}$$

Finally, we can insert Eqns. (9.59), (9.60) and (9.63) into Eqns. (9.35) and (9.36) to get the small-strain, linear elastic form of the governing equations:

$$\widetilde{\Pi} = \frac{1}{2}\mathsf{K}_{\bar{\alpha}\bar{\beta}}\mathsf{u}_{\bar{\alpha}}\mathsf{u}_{\bar{\beta}}, \tag{9.64a}$$

$$\mathsf{f}_{\bar{\alpha}} = \mathsf{K}_{\bar{\alpha}\bar{\beta}}\mathsf{u}_{\bar{\beta}} + \mathsf{f}^{\text{ext}}_{\bar{\alpha}}, \tag{9.64b}$$

$$\mathsf{K}_{\bar{\alpha}\bar{\beta}} = \sum_{e=1}^{n_{\text{elem}}}\sum_{g=1}^{n_q} w_g c_{ijmn}\frac{\partial \mathsf{S}_{m\bar{\beta}}}{\partial \xi_S}(\mathsf{J}^e_{Sn})^{-1}\frac{\partial \mathsf{S}_{i\bar{\alpha}}}{\partial \xi_R}(\mathsf{J}^e_{Rj})^{-1}\hat{\jmath}^e. \tag{9.64c}$$

These equations can be implemented in a more compact form than our previous version, due to the symmetries of c_{ijkl} and σ_{ij}. We can define a small-strain version of the strain operator as (cf. Eqn. (9.39)):

$$\begin{bmatrix} u_{1,1} \\ u_{2,2} \\ u_{3,3} \\ u_{2,3}+u_{3,2} \\ u_{1,3}+u_{3,1} \\ u_{1,2}+u_{2,1} \end{bmatrix} = \mathsf{E}_{\text{ss}}\boldsymbol{u} = \begin{bmatrix} \partial/\partial X_1 & 0 & 0 \\ 0 & \partial/\partial X_2 & 0 \\ 0 & 0 & \partial/\partial X_3 \\ 0 & \partial/\partial X_3 & \partial/\partial X_2 \\ \partial/\partial X_3 & 0 & \partial/\partial X_1 \\ \partial/\partial X_2 & \partial/\partial X_1 & 0 \end{bmatrix} \begin{bmatrix} u_1 \\ u_2 \\ u_3 \end{bmatrix}, \tag{9.65}$$

from which $\mathsf{B}^e_{\text{ss}} = \mathsf{E}_{\text{ss}}\mathsf{S}^e$, allowing a compact expression for implementation of the stiffness matrix:

$$\mathsf{K} = \sum_{e=1}^{n_{\text{elem}}}\sum_{g=1}^{n_q} \hat{\jmath}^e w_g (\mathsf{B}^e_{\text{ss}})^T \mathsf{c}\mathsf{B}^e_{\text{ss}},$$

where c is the 6×6 form of the spatial stiffness in Voigt notation (see Eqn. (6.171)), and the stiffness matrix assembly process of Section 9.3.4 is implied. The column matrix form of the stress, obtained from $\mathsf{c}\mathsf{B}^e_{\text{ss}}\mathsf{u}$, can be stored as a 6×1 matrix instead of a 9×1 one thanks to the symmetry of the Cauchy stress.

In this small-strain limit, the equations become completely linear in the solution variable, u, and therefore the solution is exactly obtained in a single iteration of the NR process. However, this formulation does not take into account the effects of geometric nonlinearity, and must therefore be used carefully. The most striking manifestation of this is the dependence of the energy on rigid-body rotations (see Exercise 3.12), which can lead to considerable error in the results. Take, for example, a slender beam in bending. Although the strains may be small everywhere, the rotations of many elements are large and the resulting errors in the finite element solution will be substantial. For this reason, finite elements are normally formulated in terms of the Lagrangian strain tensor even when the material is linear elastic. In this case the constitutive law becomes

$$W = \frac{1}{2}C_{IJKL}E_{IJ}E_{KL}, \tag{9.66}$$

where $\boldsymbol{C}$ is the Lagrangian elasticity tensor. This is precisely the Saint Venant–Kirchhoff material discussed in more detail in Section 6.4.2. This strain measure is nonlinear in the displacement, and so the overall formulation is nonlinear even though the stress is linear in the strain. This is clear from the term $\boldsymbol{F}^T\boldsymbol{F}$ in $\boldsymbol{E}$, which is quadratic in the displacement from Eqn. (9.41), and explicitly highlights the role of the geometric nonlinearity.

Exercises

9.1 [SECTION 9.2] Write a program that implements the steepest descent method in Algorithm 9.2 to minimize a real, multivariate scalar function. Instead of line 5, use a fixed value $\alpha^{(n)} = \alpha_0$. Note that α_0 is a dimensional constant in this algorithm, and therefore must be chosen carefully. Since the units of α_0 are the same as the units of $r = \|\mathbf{u}\| \,/\, \|\mathbf{f}\|$, one approach is to set α_0 to be some small fraction of the initial value of r. Explore the effect of changing the magnitude of α_0.

9.2 [SECTION 9.2] Write a subroutine that implements the line minimization method of Algorithm 9.3. Incorporate this subroutine into the steepest descent code from the previous exercise by using it to find $\alpha^{(n)}$ for each step. Explore how well this improves the rate of convergence to the solution. Explore the effects of varying ρ and c_1.

9.3 [SECTION 9.3] Verify that Eqn. (9.24) is satisfied for the shape functions shown in Tab. 9.1.

9.4 [SECTION 9.3] Verify that the interpolation and Kronecker delta properties hold for the three-noded triangular element of Eqn. (9.26).

9.5 [SECTION 9.3] Use Gaussian quadrature to integrate the quadratic function $h(x) = Ax^2 + Bx + C$ on the domain $-1 \leq x \leq 1$.
1. Verify that using two Gauss points at $x_g = \pm 1/\sqrt{3}$ and $w_g = 1$ yields the exact integral.
2. Compute the error if $x_g = \pm 1/2$ and $w_g = 1$ are used instead.

9.6 [SECTION 9.3] Using the shape functions from Tab. 9.2, verify that the configuration of elements in Fig. 9.8(a) satisfies continuity of the interpolated displacements across the element boundaries. Similarly, show that this continuity is lost in Fig. 9.8(b).

9.7 [SECTION 9.3] Reproduce the derivation of Section 9.3.3 for the simpler case of a two-dimensional, three-node triangular element. Assume plane strain (i.e. $F_{13} = F_{23} = F_{31} = F_{32} = 0$ and $F_{33} = 1$), and optimize all matrices for the two-dimensional case (eliminate unnecessary storage of zeroes). Be sure to note the size of each matrix if it is not explicitly written out in one of the steps. Note that plane strain does not imply plane stress, but out-of-plane stress components can be treated separately and computed, if desired, as a postprocessing step.

9.8 [SECTION 9.3] Analogously to Eqn. (9.46), derive a matrix $\mathbf{B}_s^e$ that relates the second Piola–Kirchhoff stress to the internal nodal force vector. In other words, find $\mathbf{B}_s^e$ such that

$$\mathbf{f}^{\text{int},e} = -\frac{1}{6}\hat{J}^e(\mathbf{B}_s^e)^T\mathbf{z},$$

where $\mathbf{z}$ is a column vector of the six independent components of the second Piola–Kirchhoff stress tensor, $\mathbf{z} = [S_{11}, S_{22}, S_{33}, S_{23}, S_{13}, S_{12}]^T$.

9.9 [SECTION 9.3] Consider the case of a square body spanning the domain from $(X_1, X_2) = (-10\text{ m}, -10\text{ m})$ to $(X_1, X_2) = (10\text{ m}, 10\text{ m})$. The top face of the body ($X_2 = 10$ m) experiences a compressive dead-load traction of 100 MPa, i.e. $\bar{P}_{iJ} = -p\delta_{iJ}$, where

$p = 100$ MPa in Eqn. (9.51). (Note that this is similar to hydrostatic loading, but not exactly the same since the surface normal may not remain parallel to the traction vector, see Example 7.3.) Verify that if the body is represented by a single four-noded square element, then Eqn. (9.52) leads to an external force of 1000 MN/m in the downward direction on each of the top corner nodes.

10 Approximate solutions: reduction to the engineering theories

Continuum mechanics is in many ways the "grand unified theory" of engineering science. As long as the fundamental continuum assumptions are valid and relativistic effects are negligible, the governing equations of continuum mechanics[1] provide the most general description of the behavior of materials (solid and fluid) under arbitrary loading.[2] Any such engineering problem can therefore be described as a solution to the following coupled system of equations (balance of mass, linear momentum, angular momentum and energy):

$$\frac{\partial \rho}{\partial t} + \operatorname{div}(\rho \boldsymbol{v}) = 0, \tag{10.1}$$

$$\operatorname{div}\boldsymbol{\sigma} + \rho \boldsymbol{b} = \rho\left[\frac{\partial \boldsymbol{v}}{\partial t} + (\nabla \boldsymbol{v})\boldsymbol{v}\right], \tag{10.2}$$

$$\boldsymbol{\sigma} = \boldsymbol{\sigma}^T, \tag{10.3}$$

$$\boldsymbol{\sigma} : \boldsymbol{d} + \rho r - \operatorname{div}\boldsymbol{q} = \rho\left[\frac{\partial u}{\partial t} + \boldsymbol{v}\cdot\nabla u\right], \tag{10.4}$$

together with the appropriate constitutive relations and initial and/or boundary conditions.[3] As discussed in Chapter 8, the difficulty is that due to the nonlinearity (material and geometric) of the resulting initial/boundary-value problem, analytical solutions are unavailable except in very few cases. This leaves two options. Either a numerical solution must be pursued or the governing equations and/or constitutive relations must be simplified, usually through linearization. We discussed numerical solutions of the continuum boundary-value problem using the finite element method in Chapter 9. In this chapter, we discuss various simplifications of the continuum equations that lead to more approximate theories that nevertheless provide great insight into physical behavior.

The fact that most of the courses taught in an engineering curriculum are closely related to and derive from the common source of continuum mechanics is lost on most undergraduate and even some graduate engineering students.[4] Figure 10.1 illustrates the connections

[1] We include thermodynamics under this heading.

[2] It is also possible to include electromagnetic effects in the theory. However, we have not pursued this here.

[3] Recall that we have required the constitutive relations to satisfy the Clausius–Duhem inequality a priori, and therefore this inequality does not enter into the formulation explicitly.

[4] This state of affairs is not universal to all the engineering disciplines. In chemical engineering, for example, undergraduate students enjoy a more sophisticated view of engineering science due to the groundbreaking book by Bird, Stewart and Lightfoot on *Transport Phenomena* [BSL60], which was first published in 1960 and which presents a unified view of momentum, energy and mass transport. There are other examples of similar books, but generally the typical undergraduate engineering education remains fragmented.

This comment should not be understood as a call to restructure all engineering education by beginning with continuum mechanics and then specializing to the various engineering subjects. We believe that the current approach, which begins with simpler subjects like *statics* and gradually builds up to more sophisticated theories,

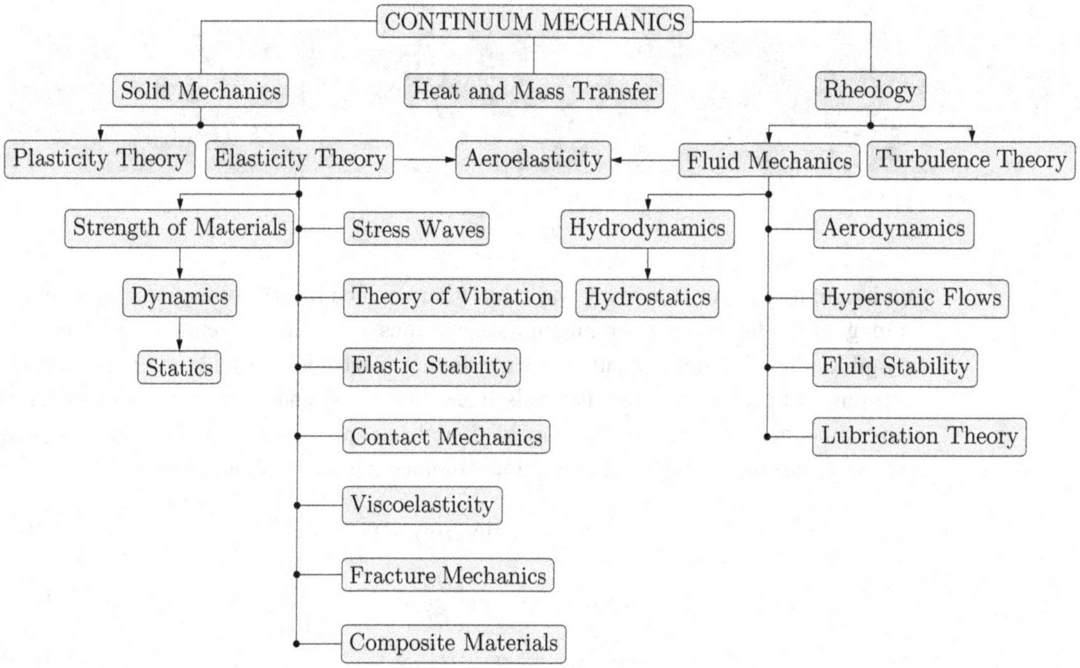

Fig. 10.1 Continuum mechanics as the "grand unified theory" of engineering science. Many of the courses taught in an engineering curriculum can be obtained as special cases of the general framework of continuum mechanics. Lines without arrows indicate that the lower course is a subset of the course it is connected with above. Lines with an arrow indicate that some sort of approximation is associated with the lower course relative to the one it comes from (typically linearization of the governing equations and/or the constitutive relations).

between continuum mechanics and engineering courses in the form of a flow chart. The names in the boxes are to be understood as titles of courses in an undergraduate/graduate engineering curriculum. At the very top of the figure is *Continuum Mechanics* where the most general coupled nonlinear governing equations (balance of mass, momentum and energy) are solved for general nonlinear constitutive relations. Under this we have *Solid Mechanics*, *Heat and Mass Transfer* and *Rheology*. These courses involve the application of the continuum mechanics framework to a particular type of problem (deformation of solids, transfer of heat or mass in rigid materials, flow of complex fluids). Although these courses do not normally involve simplification of the equations, they do compartmentalize the different subjects. For example, most engineering students have no idea that heat transfer is intimately coupled with deformation.

is both more in tune with the historical development of these theories and provides greater physical understanding to the students. It is at the *graduate* level that engineering students should begin to perceive the connections between the different subjects that they have been taught. For these students, a course in continuum mechanics is the ideal mechanism for demonstrating the unified framework for engineering science. Having said that, it also would not hurt to educate undergraduate students as much as possible during their standard educational curriculum by repeatedly pointing out the relationship between the different subjects as they are developed.

At the next level, we have courses that involve some level of simplification. Courses on *Elasticity Theory* usually involve linearization of the governing equations and the constitutive relations. Courses on *Fluid Mechanics* normally focus on Newtonian fluids which leads to the Navier–Stokes equations. A course on *Aeroelasticity*, normally taught in aerospace departments, stands between these two courses and deals with solid–fluid interactions. We have also placed courses on *Plasticity Theory* and *Turbulence Theory* at this level. Both involve "failure" at some level (either within the material or in the nature of the flow). Both courses also involve additional phenomenological assumptions absent from the continuum mechanics framework.

Below this level are specialized courses that emerge from *Elasticity Theory* and *Fluid Mechanics*. These are divided into two major categories. The branches heading left involve additional simplifications and are often encountered in undergraduate curricula. *Elasticity Theory* simplifies to *Strength of Materials*, by introducing additional approximations due to specialized geometries (two-dimensional plate and shell structures and one-dimensional beam structures). *Dynamics* adds on the additional constraint of rigid bodies and the most basic course on *Statics* also assumes equilibrium. On the fluids side, *Hydrodynamics* deals with the flow of a particular fluid, water, and *Hydrostatics* deals with its equilibrium states.

The branches heading down under *Elasticity Theory* and *Fluid Mechanics* are specialized courses where the governing equations of the parent subject are applied to particular applications. On the solids side, we have courses from *Stress Waves* to *Composite Materials* and on the fluids side, courses from *Aerodynamics* to *Lubrication Theory*. There are some parallels between the solids and fluids courses. *Aerodynamics* and *Hypersonic Flows* deal with dynamic phenomena as do *Stress Waves* and the *Theory of Vibration*. The courses on *Fluid Stability* and *Elastic Stability* deal with similar issues as do *Lubrication Theory* and *Contact Mechanics*, which are both important in the science of tribology.

Missing from the diagram are electromagnetic courses since these topics are not covered in this book. It is possible, however, to formulate a complete continuum theory that includes electromagnetic phenomena. See for examples the books by Eringen and Maugin [EM90a, EM90b], Kovetz [Kov00], and Hehl and Obukhov [HO03]. The diagram could then be expanded to include many of the courses in an electrical engineering curriculum as well.

Below we show how four main engineering theories, *Mass Transfer*, *Heat Transfer*, *Fluid Mechanics* and *Elasticity Theory* are derived as special cases of the continuum mechanics equations.

10.1 Mass transfer theory

The theory of mass transfer begins with the continuity equation (Eqn. (10.1)):

$$\frac{\partial \rho}{\partial t} + \operatorname{div}(\rho \boldsymbol{v}) = 0.$$

Define $\boldsymbol{j} \equiv \rho \boldsymbol{v}$ as the *mass flux vector* and make the constitutive assumption, referred to as *Fick's law*, that

$$\boldsymbol{j} = -\widehat{D}(\rho)\nabla\rho(\boldsymbol{x}, t), \tag{10.5}$$

where $D = \widehat{D}(\rho)$ is the *diffusion coefficient*, which can in general depend on the density ρ. Substituting Fick's law into the continuity equation, we obtain the *nonlinear diffusion equation*:

$$\frac{\partial \rho}{\partial t} - \operatorname{div}\left[\widehat{D}(\rho)\nabla\rho\right] = 0. \tag{10.6}$$

If $\widehat{D} = D$ is a constant, the result is the *linear diffusion equation*:

$$\frac{\partial \rho}{\partial t} = D\rho_{,kk} \quad \Leftrightarrow \quad \frac{\partial \rho}{\partial t} = D\nabla^2\rho, \tag{10.7}$$

where ∇^2 is the Laplacian. See [TT60, Sect. 295] for a more in-depth discussion of the diffusion equation.

10.2 Heat transfer theory

As the name suggests, the theory of heat transfer focuses entirely on the transfer of energy via heat. Energy flux due to mechanical work (which couples with the balance of linear momentum) is neglected. The energy equation is then an independent equation. Formally, this is achieved by assuming a rigid material so that Eqn. (10.4) reduces to[5]

$$\rho r - \operatorname{div}\boldsymbol{q} = \rho\frac{\partial u}{\partial t}. \tag{10.8}$$

We add to this two constitutive postulates:

1. The local form of Joule's law (Eqn. (5.7)),

$$u = u_0 + c_v T, \tag{10.9}$$

   where u_0 is a reference internal energy density and $c_v = \partial u/\partial T|_V$ is the *specific heat capacity at constant volume*, which is the amount of heat required to change the temperature of a unit mass of material by one degree. The specific heat capacity c_v is related to the molar heat capacity C_v, defined earlier in Eqn. (5.5), through

$$c_v = \frac{C_v}{M}, \tag{10.10}$$

   where M is the molar mass (the mass of one mole of the substance).

[5] Strictly, the variables appearing in this equation should be replaced with their reference counterparts. We retain the spatial notation to be consistent with the notation used in the engineering literature.

2. Fourier's law (Eqn. (6.99)),

$$q = -k\nabla T, \tag{10.11}$$

where k is the *thermal conductivity* of the material.

Substituting the two constitutive laws into the energy equation, we obtain

$$\rho r + k\nabla^2 T = \rho c_v \frac{\partial T}{\partial t}. \tag{10.12}$$

In the absence of internal heat sources ($r = 0$), Eqn. (10.12) reduces to

$$kT_{,kk} = \rho c_v \frac{\partial T}{\partial t} \quad \Leftrightarrow \quad k\nabla^2 T = \rho c_v \frac{\partial T}{\partial t}, \tag{10.13}$$

which is called the *heat equation*. Note that it has the same mathematical form as the diffusion equation in Eqn. (10.7), although physically the equations describe different phenomena.

10.3 Fluid mechanics theory

The basic theory of fluid mechanics deals with the flow of Newtonian fluids for which, as we showed earlier (Eqn. (6.125)), the constitutive relation is

$$\boldsymbol{\sigma} = -p(\rho)\boldsymbol{I} + \left[\kappa(\rho) - \frac{2}{3}\mu(\rho)\right](\operatorname{tr}\boldsymbol{d})\boldsymbol{I} + 2\mu(\rho)\boldsymbol{d},$$

where $p(\rho)$ is the elastic pressure response and κ and μ are the bulk and shear viscosities. Substituting this relation into the balance of linear momentum (Eqn. (10.2)) gives

$$\underbrace{-\nabla p}_{\text{pressure gradient force}} \underbrace{+\nabla\left[\left(\kappa - \frac{2}{3}\mu\right)\operatorname{tr}\boldsymbol{d}\right] + 2\operatorname{div}(\mu\boldsymbol{d})}_{\text{viscous forces}} + \underbrace{\rho\boldsymbol{b}}_{\text{body forces}} = \rho\left[\frac{\partial \boldsymbol{v}}{\partial t} + (\nabla\boldsymbol{v})\boldsymbol{v}\right].$$

These equations are called the *Navier–Stokes equations*. The terms on the left represent the forces acting on a volume element of fluid as indicated by the descriptions under the braces. Together, the terms on the right make up the acceleration of the fluid element. Note that the equations are nonlinear due to the convective part of the acceleration, $(\nabla\boldsymbol{v})\boldsymbol{v}$. The generalized Navier–Stokes equations can describe the most general kinds of *laminar* flows, i.e. flows in which the fluid elements move in parallel layers, that Newtonian fluids can undergo. The application to *turbulent* flows that involve both chaotic and regular motion over a broad range of temporal and spatial scales constitutes a separate area of research (see, for example, [MM98]). The transition from laminar flow to turbulent flow is reminiscent of the phenomenon of yielding in solids in which plastic flow associated with the motion of microstructural defects is initiated.

The Navier–Stokes equations are often simplified by making some additional approximations. If κ and μ are assumed to be material constants that do not depend on the density or position, then the Navier–Stokes equations become

$$-\nabla p + \left(\kappa + \frac{1}{3}\mu\right)\nabla(\operatorname{div} \boldsymbol{v}) + \mu\nabla^2\boldsymbol{v} + \rho\boldsymbol{b} = \rho\left[\frac{\partial \boldsymbol{v}}{\partial t} + (\nabla\boldsymbol{v})\boldsymbol{v}\right], \tag{10.14}$$

where some differential identities were used. Further simplification is obtained by assuming incompressible flow for which $\operatorname{div} \boldsymbol{v} = 0$ (Eqn. (3.58)):

$$-\nabla p + \mu\nabla^2\boldsymbol{v} + \rho\boldsymbol{b} = \rho\left[\frac{\partial \boldsymbol{v}}{\partial t} + (\nabla\boldsymbol{v})\boldsymbol{v}\right]. \tag{10.15}$$

This is the form of the Navier–Stokes equations that is most familiar to engineers and is used most often in practical applications. For an ideal nonviscous fluid, $\mu = 0$, and we obtain the *Euler equation*,

$$-\nabla p + \rho\boldsymbol{b} = \rho\left[\frac{\partial \boldsymbol{v}}{\partial t} + (\nabla\boldsymbol{v})\boldsymbol{v}\right], \tag{10.16}$$

which represents the flow of frictionless incompressible fluids. Finally, in the static case ($\boldsymbol{v} = \boldsymbol{0}$), we obtain the hydrostatic equations:

$$\nabla p = \rho\boldsymbol{b}, \tag{10.17}$$

which describe the behavior of a stationary fluid subjected to body forces.

10.4 Elasticity theory

In elasticity theory attention is restricted to linear elastic materials. Most materials only exhibit a linear response for small perturbations about the reference state. For this reason, a further simplification introduced in the theory is to assume that the displacement gradients are small relative to unity, so that the Lagrangian strain tensor,

$$\boldsymbol{E} = \frac{1}{2}\left[\nabla\boldsymbol{u} + (\nabla\boldsymbol{u})^T + (\nabla\boldsymbol{u})^T\nabla\boldsymbol{u}\right],$$

where $\boldsymbol{u}$ is the displacement field, can be approximated by the small-strain tensor,

$$\boldsymbol{\epsilon} = \frac{1}{2}\left[\nabla\boldsymbol{u} + (\nabla\boldsymbol{u})^T\right].$$

The appropriate constitutive relation for this case is the generalized Hooke's law (given in Eqn. (6.167)),

$$\sigma_{ij} = c_{ijkl}\epsilon_{kl} = c_{ijkl}u_{k,l},$$

where c_{ijkl} is the elasticity tensor representing the elastic stiffness of the material and where in the last term we have used the symmetry of c_{ijkl} with respect to the indices k and l.

Substituting Hooke's law into the balance of linear momentum (Eqn. (10.2)) and assuming small perturbations, we obtain

$$(c_{ijkl}u_{k,l})_{,j} + \rho b_i = \rho \frac{\partial^2 u_i}{\partial t^2}, \tag{10.18}$$

which are called the *Navier equations* for a linear elastic solid. The form of the elasticity tensor for different forms of symmetry was discussed in Section 6.4. For the simplest case of a homogeneous isotropic material, the elasticity tensor is given in Eqn. (6.174), and the Navier equations take the form:

$$\mu u_{i,kk} + (\lambda + \mu)u_{k,ki} + \rho b_i = \rho \frac{\partial^2 u_i}{\partial t^2} \quad \Leftrightarrow \quad \mu \nabla^2 \boldsymbol{u} + (\lambda + \mu)\nabla(\operatorname{div} \boldsymbol{u}) + \rho \boldsymbol{b} = \rho \frac{\partial^2 \boldsymbol{u}}{\partial t^2}. \tag{10.19}$$

Unlike the Navier–Stokes equations for a fluid (Eqn. (10.15)), the Navier equations are linear and for this reason closed-form solutions for elasticity problems are much easier to find than those for fluid mechanics. In fact, much of the work in elasticity theory focuses on obtaining such solutions, for special cases. This is particularly true for the special case of static boundary-value problems for which the Navier equations reduce to

$$\mu \nabla^2 \boldsymbol{u} + (\lambda + \mu)\nabla(\operatorname{div} \boldsymbol{u}) + \rho \boldsymbol{b} = \boldsymbol{0}. \tag{10.20}$$

Further simplification is possible by restricting the equations to two dimensions and making certain kinematic assumptions about the response of the material. If the body is very thin in the third direction, *plane stress* conditions are assumed to hold. Conversely, if the body is "infinite" in the third direction, *plane strain* conditions are assumed. Under these conditions (which are surprisingly useful, both because of the significant mathematical simplifications they produce and because of their applicability to a wide range of real engineering problems), powerful techniques exist for obtaining accurate closed-form approximations and exact closed-form solutions, respectively. It is beyond the scope of this book to go into such methods. See, for example, the classic texts by Timoshenko and Goodier [TG51] and Sokolnikoff [Sok56].

Afterword

We have endeavored herein to lay out the full story of continuum mechanics, starting with very general mathematical ideas and ending with the practical engineering approximations outlined above. We hope that you found the book as interesting to read as we found it to write, and that you can appreciate that continuum mechanics is a rich and extensive subject. Since we have tried to keep this book relatively concise we were not able to cover all topics in full detail. If you are interested in learning more on any of these topics, we direct you to the suggestions for further reading provided in the next chapter.

11 Further reading

The suggestions for further reading given below are divided according to the two parts of the book: theory and solutions.

11.1 Books related to Part I on theory

There exists an impressive assortment of books addressing the topics contained in the first part of this book. Here we list either those books that have become standard references in the field, or titles that focus on specific aspects of the theory and therefore provide a deeper presentation than the relatively few pages of this book will permit.

- Readers interested in the connection between continuum mechanics and more fundamental microscopic theories of material behavior are referred to the companion book to this one, written by two of the authors, called *Modeling Materials: Continuum, Atomistic and Multiscale Techniques* and also published by Cambridge University Press [TM11]. That book includes a concise summary of the continuum theory presented in this book (which serves as a good abbreviated reference to the subject), followed by a discussion of atomistics (quantum mechanics, atomistic models of materials and molecular statics), atomistic foundations of continuum concepts (statistical mechanics, microscopic expressions for continuum fields and molecular dynamics) and multiscale methods (atomistic constitutive relations and computational techniques for coupling continuum and atomistics). [TM11] is consistent in spirit and notation with this book and is likewise targeted at a broad readership including chemists, engineers, materials scientists and physicists.
- Although published in 1969, Malvern's book [Mal69] continues to be considered the classic text in the field. It is not the best organized of books, but it is thorough and correct. It will be found on most continuum mechanicians' book shelves.
- A mathematically rigorous presentation is provided by Truesdell and Toupin's volume in the *Handbuch der Physik* [TT60]. This authoritative and comprehensive book presents the foundations of continuum mechanics in a deep and readable way. The companion book [TN65] (currently available as [TN04]) continues where [TT60] left off and discusses everything known (up to the original date of publication) regarding all manner of constitutive laws. Surprisingly approachable and in-depth, both of these books are a must read for those interested in the foundations of continuum mechanics and constitutive theory, respectively.

- Ogden's book [Ogd84] has long been considered to be an important classic text on the subject of nonlinear elastic materials. Mathematical in nature, it provides a high-level authoritative discussion of many topics not covered in other books.
- A *very* concise and yet complete introduction to continuum mechanics is given by Chadwick [Cha99]. This excellent book takes a self-work approach, where many details and derivations are left to the reader as exercises along the way.
- A mathematically concise presentation of the subject, aimed at the advanced reader, is that of Gurtin [Gur95]. More recently, Gurtin, Fried and Anand have published a much larger book [GFA10] covering many advanced topics, which can serve as a reference for the advanced practitioner.
- Holzapfel's book [Hol00] presents a clear derivation of equations and provides a good review of tensor algebra. It also has a good presentation of constitutive relations used in different applications.
- Salençon's book [Sal01] provides a complete introduction from the viewpoint of the French school. The interested reader will find a number of differences in the philosophical approach to developing the basic theory. In this sense, the book complements the above treatments well.
- Truesdell's *A First Course in Rational Continuum Mechanics* [Tru77] is a highly mathematical treatment of the most basic foundational ideas and concepts on which the theory is based. This title is for the more mathematically inclined and/or advanced reader.
- Marsden and Hughes' book [MH94] is a modern, authoritative and highly mathematical presentation of the subject.
- We would also like to mention a book by Jaunzemis [Jau67] that is not well known in the continuum mechanics community.[1] Published at about the same time as Malvern's book, Jaunzemis takes a completely different tack. Written with humor (a rare quality in a continuum text) it is a pleasure to read. Since the terminology and some of the principles are inconsistent with modern theory, it is not recommended for the beginner, but a more advanced reader will find it a refreshing read.
- Lanczos's classic *The Variational Principles of Mechanics* [Lan70] provides an accessible discussion and exploration of variational principles including the principle of virtual work and the principles of stationary and minimum potential energy.
- Timoshenko and Gere's *Theory of Elastic Stability* [TG61] is a classic that takes a practical engineering approach to the study of the stability of continuous structures.
- Thompson's book [Tho82] provides a very readable introduction to the ideas of bifurcation, instabilities and catastrophes from an engineering perspective.
- Como and Grimaldi's book [CG95] was extensively cited in Section 7.3 and provides a rigorous mathematical discussion of stability and bifurcation theory.

[1] We thank Roger Fosdick for pointing out this book to us. Professor Fosdick studied with Walter Jaunzemis as an undergraduate. He still has the original draft of the book that also included a discussion of the electrodynamics of continuous media which was dropped from the final book due to length constraints.

11.2 Books related to Part II on solutions

Universal solutions of some type are discussed in every book on continuum mechanics. However, few volumes, if any, have been devoted entirely to the subject. In contrast, there are probably hundreds of books written on the finite element method (FEM), and many of them are very good. Finite elements are as often used by civil engineers as mechanical or materials engineers, and so many of the books have a slant towards "structural" elements like beams or plates. Our focus has been on solid elements that can be used for modeling materials. Here, we mention a handful of references that we like.

- The collected works of Rivlin [BJ96] contain the original groundbreaking papers in which most of the currently known universal solutions were first discovered.
- The three-volume set *The Finite Element Method* by Zienkiewicz and Taylor is currently in its sixth edition [ZT05] and has been a popular reference on the subject of FEM since the first edition was published in 1967. The book is comprehensive and clear, and in the later editions it features many interesting example problems from diverse fields. Our personal preference is for the fourth edition [ZT89, ZT91], as it is our view that some of the clarity has been lost as the length of the book has grown, but it is still an essential reference. The accompanying website for the book provides lots of useful finite element code.
- The FEM book by Hughes [Hug87] has been popular for long enough that it has been made into an inexpensive paperback by Dover. As such, it is still a great reference on the subject but it can be acquired inexpensively.
- The writing and teaching style of Ted Belytschko make his coauthored book on FEM a good introduction to the subject [BLM00]. Since it is focused on applications to continuum mechanics, it provides a refreshingly concise take on the field.
- Older FEM books, like that of Grandin [Gra91], are a little dated in terms of the computer code they provide, but many (and Grandin's in particular) do a good job of clearly laying out the fundamentals. Older books are often better than newer books at laying out clear details for someone writing their own subroutines, since they do not depend on the benefit of a web-based suite of codes.

A Heuristic microscopic derivation of the total energy

In Section 5.6.1, we stated that the internal energy accounts for the strain energy due to deformation and the microscopic vibrational kinetic energy. To motivate that this is indeed the case, we recall the concept of a continuum particle (see Fig. 3.1). The particle P represents a microscopic system with characteristic length ℓ. The volume of the particle is $dV \sim \ell^3$ and its mass is $dm = \rho dV$. The total energy of the N atoms represented by P is given by the Hamiltonian (see Section 4.3 of [TM11]),

$$\mathcal{H}(\boldsymbol{r}^1, \dots, \boldsymbol{r}^N, \dot{\boldsymbol{r}}^1, \dots, \dot{\boldsymbol{r}}^N) = \sum_{\alpha=1}^{N} \frac{1}{2} m^\alpha \left\| \dot{\boldsymbol{r}}^\alpha \right\|^2 + \mathcal{V}(\boldsymbol{r}^1, \dots, \boldsymbol{r}^N; \boldsymbol{F}),$$

where $\boldsymbol{r}^\alpha$ and $\dot{\boldsymbol{r}}^\alpha$ are, respectively, the position and velocity of atom α, and $\mathcal{V}(\boldsymbol{r}^1, \dots, \boldsymbol{r}^N; \boldsymbol{F})$ is the potential energy of the atoms constrained by the deformation gradient $\boldsymbol{F}$ at particle P in the body.[1] We associate the continuum total energy density with the temporal average of the Hamiltonian density, i.e. $d\mathcal{E}/dV = \overline{\widetilde{\mathcal{H}}}$, where

$$\widetilde{\mathcal{H}} = \frac{1}{dV}\mathcal{H}; \qquad \overline{\widetilde{\mathcal{H}}} = \frac{1}{\tau}\int_0^\tau \widetilde{\mathcal{H}}\, dt, \tag{A.1}$$

where the Hamiltonian depends on time through its arguments (the atomic positions and velocities) and τ is a time interval long enough for the microscopic system to achieve local thermodynamic equilibrium, but short relative to continuum timescales over which continuum variables vary appreciably.[2]

The velocity $\boldsymbol{v}$ of the continuum particle is identified with the time-averaged velocity of the center of mass of the microscopic system,

$$\boldsymbol{v} = \dot{\boldsymbol{x}} \equiv \frac{1}{\tau}\int_0^\tau \left[\frac{1}{dm} \sum_{\alpha=1}^{N} m^\alpha \dot{\boldsymbol{r}}^\alpha \right] dt, \tag{A.2}$$

where $dm = \sum_\alpha m^\alpha$ is the total mass of the microscopic system, which is equal to the mass of the continuum particle. Define the velocity of an atom relative to the continuum velocity as $\Delta \boldsymbol{v}^\alpha \equiv \dot{\boldsymbol{r}}^\alpha - \boldsymbol{v}$, so that

$$\dot{\boldsymbol{r}}^\alpha = \boldsymbol{v} + \Delta \boldsymbol{v}^\alpha. \tag{A.3}$$

Note that *unless* the center of mass is constant, $\Delta \boldsymbol{v}^\alpha$ is *not* the velocity of atom α relative to the instantaneous velocity of the center of mass (called the "center of mass velocity" and

[1] The calculation of the potential energy from atomistic considerations subject to the constraint of the continuum deformation gradient is described in Chapters 8 and 11 of [TM11]. Here it is treated as a known function.

[2] The derivation given here is only meant to be a heuristic exercise to gain insight into the continuum energy variable. A rigorous derivation based on nonequilibrium statistical mechanics is given in [AT11].

denoted $\boldsymbol{v}^{\alpha}_{\text{rel}}$). Substituting Eqn. (A.3) into Eqn. (A.1)$_1$ gives

$$\begin{aligned}
\widetilde{\mathcal{H}} &= \frac{1}{dV}\left[\sum_{\alpha=1}^{N}\frac{1}{2}m^{\alpha}\left\|\boldsymbol{v}+\Delta\boldsymbol{v}^{\alpha}\right\|^{2}+\mathcal{V}\right] \\
&= \frac{1}{dV}\left[\sum_{\alpha=1}^{N}\frac{1}{2}m^{\alpha}\left\|\boldsymbol{v}\right\|^{2}+\sum_{\alpha=1}^{N}m^{\alpha}\boldsymbol{v}\cdot\Delta\boldsymbol{v}^{\alpha}+\sum_{\alpha=1}^{N}\frac{1}{2}m^{\alpha}\left\|\Delta\boldsymbol{v}^{\alpha}\right\|^{2}+\mathcal{V}\right] \\
&= \frac{1}{2}\rho\left\|\boldsymbol{v}\right\|^{2}+\frac{1}{dV}\left[\boldsymbol{v}\cdot\left(\sum_{\alpha=1}^{N}m^{\alpha}\Delta\boldsymbol{v}^{\alpha}\right)+\sum_{\alpha=1}^{N}\frac{1}{2}m^{\alpha}\left\|\Delta\boldsymbol{v}^{\alpha}\right\|^{2}+\mathcal{V}\right].
\end{aligned}$$

Passing from the second to the third equation, we have used $(\sum_{\alpha} m^{\alpha})/dV = dm/dV = \rho$. The temporal average of $\widetilde{\mathcal{H}}$, defined in Eqn. (A.1)$_2$, is

$$\overline{\widetilde{\mathcal{H}}} = \frac{1}{2}\rho\left\|\boldsymbol{v}\right\|^{2}+\frac{\boldsymbol{v}}{dV}\cdot\sum_{\alpha=1}^{N}m^{\alpha}\left[\frac{1}{\tau}\int_{0}^{\tau}\Delta\boldsymbol{v}^{\alpha}\,dt\right]+\frac{1}{\tau dV}\int_{0}^{\tau}\left[\sum_{\alpha=1}^{N}\frac{1}{2}m^{\alpha}\left\|\Delta\boldsymbol{v}^{\alpha}\right\|^{2}+\mathcal{V}\right]dt. \tag{A.4}$$

The second term in this equation is identically zero as a result of the definition of $\Delta\boldsymbol{v}^{\alpha}$:

$$\begin{aligned}
\sum_{\alpha=1}^{N}m^{\alpha}\left[\frac{1}{\tau}\int_{0}^{\tau}\Delta\boldsymbol{v}^{\alpha}\,dt\right] &= \frac{1}{\tau}\int_{0}^{\tau}\sum_{\alpha=1}^{N}m^{\alpha}(\dot{\boldsymbol{r}}^{\alpha}-\boldsymbol{v})\,dt \\
&= dm\left\{\left[\frac{1}{\tau}\int_{0}^{\tau}\frac{1}{dm}\sum_{\alpha=1}^{N}m^{\alpha}\dot{\boldsymbol{r}}^{\alpha}\,dt\right]-\boldsymbol{v}\right\} = dm(\boldsymbol{v}-\boldsymbol{v}) = \boldsymbol{0}.
\end{aligned}$$

Consequently Eqn. (A.4) takes the form,

$$\overline{\widetilde{\mathcal{H}}} = \frac{1}{2}\rho\left\|\boldsymbol{v}\right\|^{2}+\rho u, \tag{A.5}$$

where u is the specific internal energy:

$$u = \frac{1}{\tau dm}\int_{0}^{\tau}\left[\sum_{\alpha=1}^{N}\frac{1}{2}m^{\alpha}\left\|\Delta\boldsymbol{v}^{\alpha}\right\|^{2}+\mathcal{V}(\boldsymbol{r}^{1},\ldots,\boldsymbol{r}^{N};\boldsymbol{F})\right]dt. \tag{A.6}$$

The total energy $\mathcal{E}$ follows as

$$\mathcal{E} = \int_{B}\overline{\widetilde{\mathcal{H}}}\,dV = \mathcal{K}+\mathcal{U}, \qquad \mathcal{K} = \int_{B}\frac{1}{2}\rho\left\|\boldsymbol{v}\right\|^{2}dV, \qquad \mathcal{U} = \int_{B}\rho u\,dV, \tag{A.7}$$

where $\mathcal{K}$ and $\mathcal{U}$ are the total kinetic and internal energies. We see that the microscopic derivation leads to the macroscopic definitions given in Eqns. (5.38), (5.39) and (5.40) with the added benefit that the significance of the internal energy is made clear by Eqn. (A.6).

Under conditions of thermodynamic equilibrium, the velocity of the center of mass is constant and so $\Delta\boldsymbol{v}^{\alpha} = \boldsymbol{v}^{\alpha}_{\text{rel}}$. It follows from Eqns. (A.6) and (A.7)$_3$ that

$$\mathcal{U} = \overline{\mathcal{H}}. \tag{A.8}$$

B Summary of key continuum mechanics equations

This appendix presents a brief summary of the main continuum mechanics and thermodynamics equations derived in Part I to serve as a quick reference. Each entry includes the relevant equation number in the main text, the equation in both indicial and invariant form (where applicable) and a brief description. The reader is referred back the text for details of the derivation and variables appearing in the equations.

Kinematic relations

(3.1)	$x_i = \varphi_i(X_1, X_2, X_3)$	$\boldsymbol{x} = \boldsymbol{\varphi}(\boldsymbol{X})$	deformation mapping
(3.4)	$F_{iJ} = \dfrac{\partial \varphi_i}{\partial X_J} = \dfrac{\partial x_i}{\partial X_J} = x_{i,J}$	$\boldsymbol{F} = \dfrac{\partial \boldsymbol{\varphi}}{\partial \boldsymbol{X}} = \dfrac{\partial \boldsymbol{x}}{\partial \boldsymbol{X}} = \nabla_0 \boldsymbol{x}$	deformation gradient
(3.10)	$F_{iJ} = R_{iI} U_{IJ} = V_{ij} R_{jJ}$	$\boldsymbol{F} = \boldsymbol{R}\boldsymbol{U} = \boldsymbol{V}\boldsymbol{R}$	polar decomposition
(3.6)	$C_{IJ} = F_{kI} F_{kJ}$	$\boldsymbol{C} = \boldsymbol{F}^T \boldsymbol{F}$	right Cauchy–Green deformation tensor
(3.15)	$B_{ij} = F_{iK} F_{jK}$	$\boldsymbol{B} = \boldsymbol{F}\boldsymbol{F}^T$	left Cauchy–Green deformation tensor
(3.7)	$J = \frac{1}{6} \epsilon_{ijk} \epsilon_{mnp} F_{mi} F_{nj} F_{pk}$	$J = \det \boldsymbol{F}$	Jacobian
(3.9)	$n_i \, dA = J F^{-1}_{Ii} N_I \, dA_0$	$\boldsymbol{n} \, dA = J \boldsymbol{F}^{-T} \boldsymbol{N} \, dA_0$	Nanson's formula
(3.23)	$E_{IJ} = \frac{1}{2}(C_{IJ} - \delta_{IJ})$	$\boldsymbol{E} = \frac{1}{2}(\boldsymbol{C} - \boldsymbol{I})$	Lagrangian strain tensor
(3.25)	$e_{ij} = \frac{1}{2}(\delta_{ij} - B^{-1}_{ij})$	$\boldsymbol{e} = \frac{1}{2}(\boldsymbol{I} - \boldsymbol{B}^{-1})$	Euler–Almansi strain tensor
(3.28)	$\epsilon_{ij} = \frac{1}{2}(u_{i,j} + u_{j,i})$	$\boldsymbol{\epsilon} = \frac{1}{2}\left[\nabla \boldsymbol{u} + (\nabla \boldsymbol{u})^T\right]$	small-strain tensor

Kinematic rates

(3.35)	$l_{ij} = v_{i,j}$	$\boldsymbol{l} = \nabla \boldsymbol{v}$	velocity gradient tensor
(3.36)	$\dot{F}_{iJ} = l_{ij} F_{jJ}$	$\dot{\boldsymbol{F}} = \boldsymbol{l}\boldsymbol{F}$	deformation gradient rate
(3.38)	$d_{ij} = \frac{1}{2}(l_{ij} + l_{ji})$	$\boldsymbol{d} = \frac{1}{2}(\boldsymbol{l} + \boldsymbol{l}^T)$	rate of deformation tensor
(3.40)	$w_{ij} = \frac{1}{2}(l_{ij} - l_{ji})$	$\boldsymbol{w} = \frac{1}{2}(\boldsymbol{l} - \boldsymbol{l}^T)$	spin tensor
(3.51)	$\dot{E}_{IJ} = F_{iI} d_{ij} F_{jJ}$	$\dot{\boldsymbol{E}} = \boldsymbol{F}^T \boldsymbol{d} \boldsymbol{F}$	Lagrangian strain rate tensor
(3.54)	$\dot{e}_{ij} = \frac{1}{2}(l_{ki} B^{-1}_{kj} + B^{-1}_{ik} l_{kj})$	$\dot{\boldsymbol{e}} = \frac{1}{2}(\boldsymbol{l}^T \boldsymbol{B}^{-1} + \boldsymbol{B}^{-1} \boldsymbol{l})$	Euler–Almansi strain rate tensor
(3.57)	$\dot{J} = J v_{k,k} = J d_{kk}$	$\dot{J} = J \operatorname{div} \boldsymbol{v} = J \operatorname{tr} \boldsymbol{d}$	Jacobian rate
(3.61)	$\dfrac{D}{Dt} \int_E g(\boldsymbol{x}, t) \, dV = \int_E [\dot{g} + g(\operatorname{div} \boldsymbol{v})] \, dV = \int_E \dfrac{\partial g}{\partial t} \, dV + \int_{\partial E} g \boldsymbol{v} \cdot \boldsymbol{n} \, dA$		Reynolds transport theorem

Conservation of mass

(4.1)	$J\rho = \rho_0$		conservation of mass (material form)
(4.2)	$\dot{\rho} + \rho v_{k,k} = 0$	$\dot{\rho} + \rho(\operatorname{div} \boldsymbol{v}) = 0$	conservation of mass (spatial form I)
(4.3)	$\dfrac{\partial \rho}{\partial t} + (\rho v_k)_{,k} = 0$	$\dfrac{\partial \rho}{\partial t} + \operatorname{div}(\rho \boldsymbol{v}) = 0$	conservation of mass (spatial form II)
(4.4)	$\rho a_i = \dfrac{\partial}{\partial t}(\rho v_i) + (\rho v_i v_j)_{,j}$	$\rho \boldsymbol{a} = \dfrac{\partial}{\partial t}(\rho \boldsymbol{v}) + \operatorname{div}(\rho \boldsymbol{v} \otimes \boldsymbol{v})$	conservation of mass (spatial form III)
(4.5)	$\dfrac{D}{Dt} \int_E \rho \psi \, dV = \int_E \rho \dot{\psi} \, dV$		Reynolds transport theorem for extensive ψ

Balance of momentum

(4.18)	$t_i(\boldsymbol{n}) = \sigma_{ij} n_j$	$\boldsymbol{t}(\boldsymbol{n}) = \boldsymbol{\sigma} \boldsymbol{n}$	Cauchy's relation (spatial form)
(4.25)	$\sigma_{ij,j} + \rho b_i = \rho \ddot{x}_i$	$\operatorname{div} \boldsymbol{\sigma} + \rho \boldsymbol{b} = \rho \ddot{\boldsymbol{x}}$	balance of linear momentum (spatial form)
(4.26)	$\sigma_{ij,j} + \rho b_i = \dfrac{\partial(\rho v_i)}{\partial t} + (\rho v_i v_j)_{,j}$	$\operatorname{div} \boldsymbol{\sigma} + \rho \boldsymbol{b} = \dfrac{\partial(\rho \boldsymbol{v})}{\partial t} + \operatorname{div}(\rho \boldsymbol{v} \otimes \boldsymbol{v})$	continuity momentum equation
(4.27)	$\sigma_{ij,j} + \rho b_i = 0$	$\operatorname{div} \boldsymbol{\sigma} + \rho \boldsymbol{b} = \boldsymbol{0}$	stress equilibrium
(4.30)	$\sigma_{ij} = \sigma_{ji}$	$\boldsymbol{\sigma} = \boldsymbol{\sigma}^T$	balance of angular momentum (spatial form)
(4.35)	$P_{iJ} = J \sigma_{ij} F^{-1}_{Jj}$	$\boldsymbol{P} = J \boldsymbol{\sigma} \boldsymbol{F}^{-T}$	first Piola–Kirchhoff stress
(4.37)	$\tau_{ij} = J \sigma_{ij}$	$\boldsymbol{\tau} = J \boldsymbol{\sigma}$	Kirchhoff stress
(4.38)	$T_i = P_{iJ} N_J$	$\boldsymbol{T} = \boldsymbol{P} \boldsymbol{N}$	Cauchy's relation (material form)
(4.39)	$P_{iJ,J} + \rho_0 \breve{b}_i = \rho_0 \breve{a}_i$	$\operatorname{Div} \boldsymbol{P} + \rho_0 \breve{\boldsymbol{b}} = \rho_0 \breve{\boldsymbol{a}}$	balance of linear momentum (material form)
(4.40)	$P_{kM} F_{jM} = P_{jM} F_{kM}$	$\boldsymbol{P} \boldsymbol{F}^T = \boldsymbol{F} \boldsymbol{P}^T$	balance of angular momentum (material form)
(4.41)	$S_{IJ} = F^{-1}_{Ii} P_{iJ}$	$\boldsymbol{S} = \boldsymbol{F}^{-1} \boldsymbol{P}$	second Piola–Kirchhoff stress
(4.43)	$(F_{iI} S_{IJ})_{,J} + \rho_0 \breve{b}_i = \rho_0 \breve{a}_i$	$\operatorname{Div}(\boldsymbol{F} \boldsymbol{S}) + \rho_0 \breve{\boldsymbol{b}} = \rho_0 \breve{\boldsymbol{a}}$	balance of linear momentum (alt. material form)

Thermodynamics

(5.24)	$T = \bar{T}(N, \mathbf{\Gamma}, \mathcal{S}) \equiv \left.\frac{\partial \bar{\mathcal{U}}}{\partial \mathcal{S}}\right\|_{N,\mathbf{\Gamma}}$		temperature
(5.25)	$\gamma_\alpha = \bar{\gamma}_\alpha(N, \mathbf{\Gamma}, \mathcal{S}) \equiv \left.\frac{\partial \bar{\mathcal{U}}}{\partial \Gamma_\alpha}\right\|_{N,\mathcal{S}}$		thermodynamic tensions
(5.26)	$\mu = \bar{\mu}(N, \mathbf{\Gamma}, \mathcal{S}) \equiv \left.\frac{\partial \bar{\mathcal{U}}}{\partial N}\right\|_{\mathbf{\Gamma},\mathcal{S}}$		chemical potential
(5.2)	$\Delta\mathcal{U} = \Delta\mathcal{W}^{\mathrm{def}} + \Delta\mathcal{Q}$		first law for internal energy
(5.3)	$\Delta\mathcal{E} = \Delta\mathcal{W}^{\mathrm{ext}} + \Delta\mathcal{Q}$		first law for total energy
(5.13)	$\Delta\mathcal{S} \geq 0$		second law for an isolated system
(5.31)	$d\mathcal{U} = \sum_\alpha \gamma_\alpha d\Gamma_\alpha + T d\mathcal{S}$		entropy form of the first law
(5.36)	$d\mathcal{S} \geq \frac{đ\mathcal{Q}}{T^{\mathrm{RHS}}}$		Clausius–Planck inequality
(5.41)	$\dot{\mathcal{K}} + \dot{\mathcal{U}} = \mathcal{P}^{\mathrm{ext}} + \mathcal{R}$		first law for continuum systems
(5.45)	$\mathcal{P}^{\mathrm{ext}} = \dot{\mathcal{K}} + \mathcal{P}^{\mathrm{def}}$		external power
(5.46)	$\mathcal{P}^{\mathrm{def}} = \int_B \sigma_{ij} d_{ij}\, dV$	$\mathcal{P}^{\mathrm{def}} = \int_B \boldsymbol{\sigma} : \boldsymbol{d}\, dV$	deformation power (spatial form)
(5.49)	$\mathcal{P}^{\mathrm{def}} = \int_{B_0} P_{iJ} \dot{F}_{iJ}\, dV_0$	$\mathcal{P}^{\mathrm{def}} = \int_{B_0} \boldsymbol{P} : \dot{\boldsymbol{F}}\, dV_0$	deformation power (material form I)
(5.50)	$\mathcal{P}^{\mathrm{def}} = \int_{B_0} S_{IJ} \dot{E}_{IJ}\, dV_0$	$\mathcal{P}^{\mathrm{def}} = \int_{B_0} \boldsymbol{S} : \dot{\boldsymbol{E}}\, dV_0.$	deformation power (material form II)
(5.56)	$h(\boldsymbol{n}) = q_i n_i$	$h(\boldsymbol{n}) = \boldsymbol{q} \cdot \boldsymbol{n}$	heat flux relation (spatial form)
(5.57)	$\sigma_{ij} d_{ij} + \rho r - q_{i,i} = \rho \dot{u}$	$\boldsymbol{\sigma} : \boldsymbol{d} + \rho r - \operatorname{div} \boldsymbol{q} = \rho \dot{u}$	energy equation (spatial form)
(5.58)	$P_{iJ} \dot{F}_{iJ} + \rho_0 r_0 - q_{0I,I} = \rho_0 \dot{u}_0$	$\boldsymbol{P} : \dot{\boldsymbol{F}} + \rho_0 r_0 - \operatorname{Div} \boldsymbol{q}_0 = \rho_0 \dot{u}_0$	energy equation (material form)
(5.63)	$\dot{s} \geq \frac{r}{T} - \frac{1}{\rho}\left(\frac{q_i}{T}\right)_{,i}$	$\dot{s} \geq \frac{r}{T} - \frac{1}{\rho} \operatorname{div} \frac{\boldsymbol{q}}{T}$	Clausius–Duhem inequality

Constitutive relations

(6.40)	$\psi = u - Ts$		Helmholtz free energy density
(6.42)	$W = \rho_0 \psi$		strain energy density
(6.44)	$h = u - \boldsymbol{\gamma} \cdot \boldsymbol{\Gamma}$		specific enthalpy
(6.46)	$g = u - Ts - \boldsymbol{\gamma} \cdot \boldsymbol{\Gamma}$		specific Gibbs free energy
(6.94)	$u = \bar{u}(s, \boldsymbol{C})$		reduced internal energy density function
(6.95)	$T = \dfrac{\partial \bar{u}(s, \boldsymbol{C})}{\partial s}$		reduced temperature function
(6.100)	$q_i = R_{iJ} \widetilde{q}_J (s, C_{KL}, R_{jM} T_{,j})$	$\boldsymbol{q} = \boldsymbol{R} \widetilde{\boldsymbol{q}}(s, \boldsymbol{C}, \boldsymbol{R}^T \nabla T)$	reduced heat flux function
(6.106)	$\sigma_{ij}^{(e)} = \dfrac{2}{J} F_{iJ} \dfrac{\partial \widetilde{W}}{\partial C_{JK}} F_{jK}$	$\boldsymbol{\sigma}^{(e)} = \dfrac{2}{J} \boldsymbol{F} \dfrac{\partial \widetilde{W}}{\partial \boldsymbol{C}} \boldsymbol{F}^T$	Cauchy stress function (elastic part)
(6.107)	$P_{iJ}^{(e)} = 2 F_{iK} \dfrac{\partial \widetilde{W}}{\partial C_{KJ}}$	$\boldsymbol{P}^{(e)} = 2 \boldsymbol{F} \dfrac{\partial \widetilde{W}}{\partial \boldsymbol{C}}$	first Piola–Kirchhoff stress (elastic part)
(6.107)	$S_{IJ}^{(e)} = 2 \dfrac{\partial \widetilde{W}}{\partial C_{IJ}}$	$\boldsymbol{S}^{(e)} = 2 \dfrac{\partial \widetilde{W}}{\partial \boldsymbol{C}}$	second Piola–Kirchhoff stress (elastic part)
(6.110)	$\sigma_{ij}^{(v)} = R_{iJ} \widetilde{\sigma}_{JK}^{(v)} (s, C_{LM}, R_{jN} d_{jk} R_{kP}) R_{jK}$	$\boldsymbol{\sigma}^{(v)} = \boldsymbol{R} \widetilde{\boldsymbol{\sigma}}^{(v)} (s, \boldsymbol{C}, \boldsymbol{R}^T \boldsymbol{dR}) \boldsymbol{R}^T$	Cauchy stress function (viscous part)
(6.150)	$dS_{IJ} = C_{IJKL} dE_{KL}$	$d\boldsymbol{S} = \boldsymbol{C} : d\boldsymbol{E}$	incremental stress–strain relation (material form)
(6.154)	$dP_{iJ} = D_{iJkL} dF_{kL}$	$d\boldsymbol{P} = \boldsymbol{D} : d\boldsymbol{F}$	incremental stress–strain relation (alt. material form)
(6.163)	$\mathring{\sigma}_{ij} = c_{ijkl} \dot{\epsilon}_{kl}$	$\mathring{\boldsymbol{\sigma}} = \boldsymbol{c} : \dot{\boldsymbol{\epsilon}}$	incremental stress–strain rate relation
(6.167)	$\sigma_{ij} = c_{ijkl} \epsilon_{kl}$	$\boldsymbol{\sigma} = \boldsymbol{c} : \boldsymbol{\epsilon}$	generalized Hooke's law
(6.151)	$C_{IJKL} = \dfrac{\partial \widetilde{S}_{IJ}(\boldsymbol{E})}{\partial E_{KL}} = \dfrac{\partial^2 \widetilde{W}(\boldsymbol{E})}{\partial E_{IJ} E_{KL}}$	$\boldsymbol{C} = \dfrac{\partial \widetilde{\boldsymbol{S}}(\boldsymbol{E})}{\partial \boldsymbol{E}} = \dfrac{\partial^2 \widetilde{W}(\boldsymbol{E})}{\partial \boldsymbol{E}^2}$	material elasticity tensor
(6.157)	$D_{iJkL} = C_{IJKL} F_{iI} F_{kK} + \delta_{ik} S_{JL}$		mixed elasticity tensor
(6.165)	$c_{ijkl} = J^{-1} F_{iI} F_{jJ} F_{kK} F_{lL} C_{IJKL}$		spatial elasticity tensor
(6.166)	$c_{ijkl} = J^{-1} (F_{jJ} F_{lL} D_{iJkL} - \delta_{ik} F_{lL} P_{jL})$		spatial elasticity tensor

References

[Adk83] C. J. Adkins. *Equilibrium Thermodynamics*. Cambridge: Cambridge University Press, third edition, 1983.

[AF73] R. J. Atkin and N. Fox. On the frame-dependence of stress and heat flux in polar fluids. *J. Appl. Math. Phys. (ZAMP)*, **24**:853–860, 1973.

[AK86] P. G. Appleby and N. Kadianakis. A frame-independent description of the principles of classical mechanics. *Arch. Ration. Mech. Anal.*, **95**:1–22, 1986.

[Ale56] H. G. Alexander. *The Leibniz–Clarke Correspondence*. Manchester: Manchester University Press, 1956.

[Art95] R. T. W. Arthur. Newton's fluxions and equably flowing time. *Stud. Hist. Phil. Sci.*, **26**:323–351, 1995.

[AS02] S. N. Atluri and S. P. Shen. The meshless local Petrov–Galerkin (MLPG) method: A simple & less-costly alternative to the finite element and boundary element methods. *Comput. Model. Eng. Sci.*, **3**(**1**):11–51, 2002.

[AT11] N. C. Admal and E. B. Tadmor. Stress and heat flux for arbitrary multi-body potentials: A unified framework. *J. Chem. Phys.*, **134**:184106, 2011.

[AW95] G. B. Arfken and H. J. Weber. *Mathematical Methods of Physicists*. San Diego: Academic Press, Inc., fourth edition, 1995.

[AZ00] S. N. Atluri and T. Zhu. New concepts in meshless methods. *Int. J. Numer. Methods Eng.*, **47**:537–556, 2000.

[Bal76] J. M. Ball. Convexity conditions and existence theorems in nonlinear elasticity. *Arch. Ration. Mech. Anal.*, **63**(**4**):337–403, 1976.

[Ban84] W. Band. Effect of rotation on radial heat flow in a gas. *Phys. Rev. A*, **29**:2139–2144, 1984.

[BBS04] A. Bóna, I. Bucataru, and M. A. Slawinski. Material symmetries of elasticity tensors. *Q. J. Mech. Appl. Math.*, **57**(**4**):583–598, 2004.

[BdGH83] R. B. Bird, P. G. de Gennes, and W. G. Hoover. Discussion. *Physica A*, **118**:43–47, 1983.

[Bea67] M. F. Beatty. On the foundation principles of general classical mechanics. *Arch. Ration. Mech. Anal.*, **24**:264–273, 1967.

[BG68] R. L. Bishop and S. I. Goldberg. *Tensor Analysis on Manifolds*. New York: Macmillan, 1968.

[Bil86] E. W. Billington. The Poynting effect. *Acta Mech.*, **58**:19–31, 1986.

[BJ96] G. I. Barenblatt and D. D. Joseph, editors. *Collected Papers of R. S. Rivlin*, volumes I & II. New York: Springer, 1996.

[BK62] P. J. Blatz and W. L. Ko. Application of finite elasticity to the deformation of rubbery materials. *Trans. Soc. Rheol.*, **6**:223–251, 1962.

[BKO$^+$96] T. Belytschko, Y. Krongauz, D. Organ, M. Fleming, and P. Krysl. Meshless methods: An overview and recent developments. *Comput. Meth. Appl. Mech. Eng.*, **139**:3–47, 1996.

[BLM00] T. Belytschko, W. K. Liu, and B. Moran. *Nonlinear Finite Elements for Continua and Structures*. Chichester: Wiley, 2000.

[BM80] F. Bampi and A. Morro. Objectivity and objective time derivatives in continuum mechanics. *Found. Phys.*, **10**:905–920, 1980.

[BM97] I. Babuska and J. M. Melenk. The partition of unity method. *Int. J. Numer. Methods Eng.*, **40(4)**:727–758, 1997.

[BSL60] R. B. Bird, W. E. Stewart, and E. N. Lightfoot. *Transport Phenomena*. New York: Wiley, 1960.

[BT03] B. Buffoni and J. Toland. *Analytic Theory of Global Bifurcation: An Introduction*. Princeton Series in Applied Mathematics. Princeton: Princeton University Press, first edition, 2003.

[Cal85] H. B. Callen. *Thermodynamics and an Introduction to Thermostatics*. New York: John Wiley and Sons, second edition, 1985.

[Car67] M. M. Carroll. Controllable deformations of incompressible simple materials. *Int. J. Eng. Sci.*, **5**:515–525, 1967.

[CG95] M. Como and A. Grimaldi. *Theory of Stability of Continuous Elastic Structures*. Boca Raton: CRC Press, 1995.

[CG01] P. Cermelli and M. E. Gurtin. On the characterization of geometrically necessary dislocations in finite plasticity. *J. Mech. Phys. Solids*, **49**:1539–1568, 2001.

[Cha99] P. Chadwick. *Continuum Mechanics: Concise Theory and Problems*. Mineola: Dover, second edition, 1999.

[CJ93] C. Chu and R. D. James. Biaxial loading experiments on Cu–Al–Ni single crystals. In K. S. Kim, editor, *Experiments in Smart Materials and Structures*, volume 181, pages 61–69. New York: ASME-AMD, 1993.

[CM87] S. C. Cowin and M. M. Mehrabadi. On the identification of material symmetry for anisotropic elastic materials. *Q. J. Mech. Appl. Math.*, **40**:451–476, 1987.

[CN63] B. D. Coleman and W. Noll. The thermodynamics of elastic materials with heat conduction and viscosity. *Arch. Ration. Mech. Anal.*, **13**:167–178, 1963.

[CR07] D. Capecchi and G. C. Ruta. Piola's contribution to continuum mechanics. *Arch. Hist. Exact Sci.*, **61**:303–342, 2007.

[CVC01] P. Chadwick, M. Vianello, and S. C. Cowin. A new proof that the number of linear elastic symmetries is eight. *J. Mech. Phys. Solids*, **49**:2471–2492, 2001.

[dGM62] S. R. de Groot and P. Mazur. *Non-Equilibrium Thermodynamics*. Amsterdam: North-Holland Publishing Company, 1962.

[DiS91] R. DiSalle. Conventionalism and the origins of the inertial frame concept. In A. Fine, M. Forbes, and L. Wessels, editors, *PSA 1990*, volume 2 of PSA - Philosophy of Science Association Proceedings Series. Biennial Meeting of the Philosophy of Science Assoc, Minneapolis, MN, 1990, pages 139–147. Chicago: University of Chicago Press, 1991.

[DiS02] R. DiSalle. Space and time: Inertial frames. In E. N. Zalta, editor, *The Stanford Encyclopedia of Philosophy*. Stanford University, Summer 2002. http://plato.stanford.edu/archives/sum2002/entries/spacetime-iframes

[DiS06] R. DiSalle. *Understanding Space-Time*. Cambridge: Cambridge University Press, 2006.

[Duf84] J. W. Dufty. Viscoelastic and non-Newtonian effects in shear-flow. *Phys. Rev. A*, **30**:622–623, 1984.

[Ear70] J. Earman. Who's afraid of absolute space? *Australasian J. Philosophy*, **48**:287–319, 1970.

[EH89] M. W. Evans and D. M. Heyes. On the material frame indifference controversy: Some results from group theory and computer simulation. *J. Mol. Liq.*, **40**:297–304, 1989.

[Ein16] A. Einstein. Die Grundlage der allgemeinen Relativitätstheorie. *Ann. der Phys.*, **49**:769–822, 1916.

[EM73] D. G. B. Edelen and J. A. McLennan. Material indifference: A principle or convenience. *Int. J. Eng. Sci.*, **11**:813–817, 1973.

[EM90a] A. C. Eringen and G. A. Maugin. *Electrodynamics of Continua I: Foundations and Solid Media*. New York: Springer, 1990.

[EM90b] A. C. Eringen and G. A. Maugin. *Electrodynamics of Continua II: Fluids and Complex Media*. New York: Springer, 1990.

[EM90c] D. J. Evans and G. P. Morriss. *Statistical Mechanics of Nonequilibrium Liquids*. London: Academic Press, 1990.

[Eri54] J. L. Ericksen. Deformations possible in every isotropic, incompressible, perfectly elastic body. *J. Appl. Math. Phys. (ZAMP)*, **5**:466–488, 1954.

[Eri55] J. L. Ericksen. Deformations possible in every isotropic compressible, perfectly elastic body. *J. Math. Phys*, **34**:126–128, 1955.

[Eri77] J. L. Ericksen. Special topics in elastostatics. In C.-S. Yih, editor, *Advances in Applied Mechanics*, volume 17, pages 189–244. New York: Academic Press, 1977.

[Eri02] A. C. Eringen. *Nonlocal Continuum Field Theories*. New York: Springer, 2002.

[Eu85] B. C. Eu. On the corotating frame and evolution equations in kinetic theory. *J. Chem. Phys.*, **82**:3773–3778, 1985.

[Eu86] B. C. Eu. Reply to "comment on 'on the corotating frame and evolution equations in kinetic theory' ". *J. Chem. Phys.*, **86**:2342–2343, 1986.

[FF77] R. Fletcher and T. L. Freeman. A modified Newton method for minimization. *J. Optimiz. Theory App.*, **23**:357–372, 1977.

[FMAH94] N. A. Fleck, G. M. Muller, M. F. Ashby, and J. W. Hutchinson. Strain gradient plasticity: Theory and experiment. *Acta Metall. Mater.*, **42**:475–487, 1994.

[Fre09] M. Frewer. More clarity on the concept of material frame-indifference in classical continuum mechanics. *Acta Mech.*, **202**:213–246, 2009.

[FV89] R. L. Fosdick and E. G. Virga. A variational proof of the stress theorem of Cauchy. *Arch. Ration. Mech. Anal.*, **105**:95–103, 1989.

[FV96] S. Forte and M. Vianello. Symmetry classes for elasticity tensors. *J. Elast.*, **43**:81–108, 1996.

[Gen96] A. N. Gent. A new constitutive relation for rubber. *Rubber Chemistry Tech.*, **69**:59–61, 1996.

[GFA10] M. E. Gurtin, E. Fried, and L. Anand. *The Mechanics and Thermodynamics of Continua*. Cambridge: Cambridge University Press, 2010.

[GR64] A. E. Green and R. S. Rivlin. On Cauchy's equations of motion. *J. Appl. Math. Phys. (ZAMP)*, **15**:290–292, 1964.

[Gra91] H. Grandin. *Fundamentals of the Finite Element Method*. Prospect Heights: Waveland Press, 1991.

[Gre04] B. Greene. *The Fabric of the Cosmos*. New York: Vintage Books, 2004.

[Gur65] M. E. Gurtin. Thermodynamics and the possibility of spatial interaction in elastic materials. *Arch. Ration. Mech. Anal.*, **19**:339–352, 1965.

[Gur81] M. E. Gurtin. *An Introduction to Continuum Mechanics*, volume 158 of Mathematics in Science and Engineering. New York: Academic Press, 1981.

[Gur95] M. E. Gurtin. The nature of configurational forces. *Arch. Ration. Mech. Anal.*, **131**:67–100, 1995.

[GW66] M. E. Gurtin and W. O. Williams. On the Clausius–Duhem inequality. *J. Appl. Math. Phys. (ZAMP)*, **17**:626–633, 1966.

[HM83] M. Heckl and I. Müller. Frame dependence, entropy, entropy flux, and wave speeds in mixtures of gases. *Acta Mech.*, **50**:71–95, 1983.

[HMML81] W. G. Hoover, B. Moran, R. M. More, and A. J. C. Ladd. Heat conduction in a rotating disk via nonequilibrium molecular dynamics. *Phys. Rev. B*, **24**:2109–2115, 1981.

[HO03] F. W. Hehl and Y. N. Obukhov. *Foundations of Classical Electrodynamics*. Boston: Birkhauser, 2003.

[Hol00] G. A. Holzapfel. *Nonlinear Solid Mechanics*. Chichester: Wiley, 2000.

[Hug87] T. J. R. Hughes. *The Finite Element Method: Linear Static and Dynamic Finite Element Analysis*. Englewood Cliffs: Prentice-Hall, 1987.

[IK50] J. H. Irving and J. G. Kirkwood. The statistical mechanical theory of transport processes. IV. the equations of hydrodynamics. *J. Chem. Phys.*, **18**:817–829, 1950.

[IM02] K. Ikeda and K. Murota. *Imperfect Bifurcation in Structures and Materials: Engineering Use of Group-Theoretic Bifurcation Theory*, volume 149 of Applied Mathematical Sciences. New York: Springer, first edition, 2002.

[Jau67] W. Jaunzemis. *Continuum Mechanics*. New York: Macmillan, 1967.

[JB67] L. Jansen and M. Boon. *Theory of Finite Groups. Applications in Physics*. Amsterdam: North Holland, 1967.

[Kem89] L. J. T. M. Kempers. The principle of material frame indifference and the covariance principle. *Il Nuovo Cimento B*, **103**:227–236, 1989.

[Kha02] H. K. Khalil. *Nonlinear Systems*. New York: Prentice-Hall, third edition, 2002.

[Kir52] G. Kirchhoff. Über die Gleichungen des Gleichgewichtes eines elastischen Körpers bei nicht unendlich kleinen Verscheibungen seiner Theile. *Sitzungsberichte der Akademie der Wissenschaften Wien*, **9**:762–773, 1852.

[Koi63] W. T. Koiter. The concept of stability of equilibrium for continuous bodies. *Proc. Koninkl. Nederl. Akademie van Wetenschappen*, **66(4)**:173–177, 1963.

[Koi65a] W. T. Koiter. The energy criterion of stability for continuous elastic bodies. – I. *Proc. of the Koninklijke Nederlandse Akademie Van Wetenschappen, Ser. B*, **68(4)**:178–189, 1965.

[Koi65b] W. T. Koiter. The energy criterion of stability for continuous elastic bodies. – II. *Proc. of the Koninklijke Nederlandse Akademie Van Wetenschappen, Ser. B*, **68(4)**:190–202, 1965.

[Koi65c] W. T. Koiter. On the instability of equilibrium in the absence of a minimum of the potential energy. *Proc. of the Koninklijke Nederlandse Akademie Van Wetenschappen, Ser. B*, **68(3)**:107–113, 1965.

[Koi71] W. T. Koiter. Thermodynamics of elastic stability. In P. G. Glockner, editor, *Proceedings [of the] Third Canadian Congress of Applied Mechanics May 17–21, 1971 at the University of Calgary*, pages 29–37, Calgary: University of Calgary.

[Kov00] A. Kovetz. *Electromagnetic Theory*. New York: Oxford University Press, 2000.

[KW73] R. Knops and W. Wilkes. Theory of elastic stability. In C. Truesdell, editor, *Handbook of Physics*, volume VIa/3, pages 125–302. Berlin: Springer-Verlag, 1973.

[Lan70] C. Lanczos. *The Variational Principles of Mechanics*. Mineola: Dover, fourth edition, 1970.

[Lap51] Pierre Simon Laplace. *A Philosophical Essay on Probabilities* [English translation by F. W. Truscott and F. L. Emery]. Dover, New York, 1951.

[LBCJ86] A. S. Lodge, R. B. Bird, C. F. Curtiss, and M. W. Johnson. A comment on "on the corotating frame and evolution equations in kinetic theory". *J. Chem. Phys.*, **85**:2341–2342, 1986.

[Lei68] D. C. Leigh. *Nonlinear Continuum Mechanics*. New York: McGraw-Hill, 1968.

[Les74] A. M. Lesk. Do particles of an ideal gas collide? *J. Chem. Educ.*, **51**:141–141, 1974.

[Liu04] I.-S. Liu. On Euclidean objectivity and the principle of material frame-indifference. *Continuum Mech. Thermodyn.*, **16**:177–183, 2004.

[Liu05] I.-S. Liu. Further remarks on Euclidean objectivity and the principle of material frame-indifference. *Continuum Mech. Thermodyn.*, **17**:125–133, 2005.

[LJCV08] G. Lebon, D. Jou, and J. Casas-Vázquez. *Understanding Non-equilitrium Thermodynamics: Foundations, Applications, Frontiers*. Berlin: Springer-Verlag, 2008.

[LL09] S. Lipschutz and M. Lipson. *Schaum's Outline for Linear Algebra*. New York: McGraw-Hill, fourth edition, 2009.

[LRK78] W. M. Lai, D. Rubin, and E. Krempl. *Introduction to Continuum Mechanics*. New York: Pergamon Press, 1978.

[Lub72] J. Lubliner. On the thermodynamic foundations of non-linear solid mechanics. *Int. J. Nonlinear Mech.*, **7**:237–254, 1972.

[Lum70] J. L. Lumley. Toward a turbulent constitutive relation. *J. Fluid Mech.*, **41**:413–434, 1970.

[Mac60] E. Mach. *The Science of Mechanics: A Critical and Historical Account of its Development*. Translated by Thomas J. McCormack. La Salle: Open Court, sixth edition, 1960.

[Mal69] L. E. Malvern. *Introduction to the Mechanics of a Continuous Medium*. Englewood Cliffs: Prentice-Hall, 1969.

[Mar90] E. Marquit. A plea for a correct translation of Newton's law of inertia. *Am. J. Phys.*, **58**:867–870, 1990.

[Mat86] T. Matolcsi. On material frame-indifference. *Arch. Ration. Mech. Anal.*, **91**:99–118, 1986.

[McW02] R. McWeeny. *Symmetry: An Introduction to Group Theory and its Applications*. Mineola: Dover, 2002.

[Mei03] L. Meirovitch. *Methods of Analytical Dynamics*. Mineola: Dover, 2003.

[MH94] J. E. Marsden and T. J. R. Hughes. *Mathematical Foundations of Elasticity*. New York: Dover, 1994.

[Mil72] W. Miller, Jr. *Symmetry Groups and Their Applications*, volume 50 of Pure and Applied Mathematics. New York: Academic Press, 1972. Available online at http://www.ima.umn.edu/~miller/.

[MM98] P. Moin and K. Mahesh. Direct numerical simulation: a tool in turbulence research. *Annu. Rev. Fluid Mech.*, **30**:539–578, 1998.

[Moo90] D. M. Moody. Unsteady expansion of an ideal gas into a vacuum. *J. Fluid Mech.*, **214**:455–468, 1990.

[MR08] W. Muschik and L. Restuccia. Systematic remarks on objectivity and frame-indifference, liquid crystal theory as an example. *Arch. Appl. Mech*, **78**:837–854, 2008.

[Mül72] I. Müller. On the frame dependence of stress and heat flux. *Arch. Ration. Mech. Anal.*, **45**:241–250, 1972.

[Mül76] I. Müller. Frame dependence of electric-current and heat flux in a metal. *Acta Mech.*, **24**:117–128, 1976.

[Mur82] A. I. Murdoch. On material frame-indifference. *Proc. R. Soc. London, Ser. A*, **380**:417–426, 1982.

[Mur83] A. I. Murdoch. On material frame-indifference, intrinsic spin, and certain constitutive relations motivated by the kinetic theory of gases. *Arch. Ration. Mech. Anal.*, **83**:185–194, 1983.

[Mur03] A. I. Murdoch. Objectivity in classical continuum physics: a rationale for discarding the principle of invariance under superposed rigid body motions in favour of purely objective considerations. *Continuum Mech. Thermodyn.*, **15**:209–320, 2003.

[Mur05] A. I. Murdoch. On criticism of the nature of objectivity in classical continuum physics. *Continuum Mech. Thermodyn.*, **17**:135–148, 2005.

[Nan74] E. J. Nanson. Note on hydrodynamics. *Messenger of Mathematics*, **3**:120–121, 1874.

[Nan78] E. J. Nanson. Note on hydrodynamics. *Messenger of Mathematics*, **7**:182–183, 1877–1878.

[New62] I. Newton. *Philosophiae Naturalis Principia Mathematica* [translated by A. Motte revised by F. Gajori], volume I. Berkeley: University of California Press, 1962.

[Nio87] E. M. S. Niou. A note on Nanson's rule. *Public Choice*, **54**:191–193, 1987.

[Nol55] W. Noll. Die Herleitung der Grundgleichungen der Thermomechanik der Kontinua aus der statischen Mechanik. *J. Ration. Mech. Anal.*, **4**:627–646, 1955.

[Nol58] W. Noll. A mathematical theory of the mechanical behaviour of continuous media. *Arch. Ration. Mech. Anal.*, **2**:197–226, 1958.

[Nol63] W. Noll. La mécanique classique, basée sur un axiome d'objectivité. In *La Méthode Axiomatique dans les Mécaniques Classiques et Nouvelles*, pages 47–56, Paris: Gauthier-Villars, 1963.

[Nol73] W. Noll. Lectures on the foundations of continuum mechanics and thermodynamics. *Arch. Ration. Mech. Anal.*, **52**:62–92, 1973.

[Nol87] W. Noll. *Finite-Dimensional Spaces: Algebra, Geometry and Analysis*, volume I. Dordrecht: Kluwer, 1987. Available online at http://www.math.cmu.edu/~wn0g/.

[Nol04] W. Noll. Five contributions to natural philosophy, 2004. Available online at http://www.math.cmu.edu/~wn0g/noll.

[Nol06] W. Noll. A frame-free formulation of elasticity. *J. Elast.*, **83**:291–307, 2006.

[NW99] J. Nocedal and S. J. Wright. *Numerical Optimization*. New York: Springer Verlag, 1999.

[Nye85] J. F. Nye. *Physical Properties of Crystals*. Oxford: Clarendon Press, 1985.

[Ogd84] R. W. Ogden. *Non-linear Elastic Deformations*. Ellis Horwood, Chichester, 1984.

[OIZT96] E. Onate, S. Idelsohn, O. C. Zienkiewicz, and R. L. Taylor. A finite point method in computational mechanics. Applications to convective transport and fluid flow. *Int. J. Numer. Methods Eng.*, **39(22)**:3839–3866, 1996.

[Old50] J. G. Oldroyd. On the formulation of rheological equations of state. *Proc. R. Soc. London, Ser. A*, **200**:523–541, 1950.

[PC68] H. J. Petroski and D. E. Carlson. Controllable states of elastic heat conductors. *Arch. Ration. Mech. Anal.*, **31(2)**:127–150, 1968.

[Pio32] G. Piola. La meccanica de' corpi naturalmente estesi trattata col calcolo delle variazioni. In *Opuscoli Matematici e Fisici di Diversi Autori*, volume 1, pages 201–236. Milano: Paolo Emilio Giusti, 1832.

[PM92] A. R. Plastino and J. C. Muzzio. On the use and abuse of Newton's second law for variable mass problems. *Celestial Mech. and Dyn. Astron.*, **53**:227–232, 1992.

[Pol71] E. Polak. *Computational Methods in Optimization: A Unified Approach*, volume 77 of Mathematics in Science and Engineering. New York: Academic Press, 1971.

[Poy09] J. H. Poynting. On pressure perpendicular to the shear-planes in finite pure shears, and on the lengthening of loaded wires when twisted. *Proc. R. Soc. London, Ser. A*, **82**:546–559, 1909.

[PTVF92] W. H. Press, S. A. Teukolsky, W. T. Vetterling, and B. P. Flannery. *Numerical Recipes in FORTRAN: The Art of Scientific Computing*. Cambridge: Cambridge University Press, second edition, 1992.

[PTVF08] W. H. Press, S. A. Teukolsky, W. T. Vetterling, and B. P. Flannery. Numerical recipes: The art of scientific computing. http://www.nr.com, 2008.

[Rei45] M. Reiner. A mathematical theory of dilatancy. *Am. J. Math.*, **67**:350–362, 1945.

[Rei59] H. Reichenbach. *Modern Philosophy of Science*. New York: Routledge & Kegan Paul, 1959.

[Riv47] R. S. Rivlin. Hydrodynamics of non-newtonian fluids. *Nature*, **160**:611–613, 1947.

[Riv48] R. S. Rivlin. Large elastic deformations of isotropic materials. II. Some uniqueness theorems for pure, homogeneous deformation. *Philos. Trans. R. Soc. London, Ser. A*, **240**:491–508, 1948.

[Riv74] R. S. Rivlin. Stability of pure homogeneous deformations of an elastic cube under dead loading. *Q. Appl. Math.*, **32**:265–271, 1974.

[Ros08] J. Rosen. *Symmetry Rules: How Science and Nature are Founded on Symmetry*. The Frontiers Collection. Berlin: Springer, 2008.

[RS51] R. S. Rivlin and D. W. Saunders. Large elastic deformations of isotropic materials VII. experiments on the deformation of rubber. *Philos. Trans. R. Soc. London, Ser. A*, **243**:251–288, 1951.

[Rub00] M. B. Rubin. *Cosserat Theories: Shells, Rods and Points*, volume 79 of Solid Mechanics and its Applications. Dordrecht: Kluwer, 2000.

[Rue99] D. Ruelle. Smooth dynamics and new theoretical ideas in nonequilibrium statistical mechanics. *J. Stat. Phys.*, **95**:393–468, 1999.

[Rus06] A. Ruszczyński. *Nonlinear Optimization*. Princeton: Princeton University Press, 2006.

[Rys85] G. Ryskin. Misconception which led to the "material frame-indifference" controversy. *Phys. Rev. A*, **32**:1239–1240, 1985.

[SA04] S. Shen and S. N. Atluri. Multiscale simulation based on the meshless local Petrov–Galerkin (MLPG) method. *Comput. Model. Eng. Sci.*, **5**:235–255, 2004.

[Saa03] Y. Saad. *Iterative Methods for Sparse Linear Systems*. Philadelphia: Society for Industrial and Applied Mathematics, 2003.

[Sac01] Giuseppe Saccomandi. Universal results in finite elasticity. In Y. B. Fu and R. W. Ogden, editors, *Nonlinear Elasticity: Theory and Applications*, number 283 in London Mathematical Society Lecture Note Series, chapter 3, pages 97–134. Cambridge: Cambridge University Press, 2001.

[Sal01] J. Salençon. *Handbook of Continuum Mechanics: General Concepts, Thermoelasticity*. Berlin: Springer, 2001.

[SB99] B. Svendsen and A. Bertram. On frame-indifference and form-invariance in constitutive theory. *Acta Mech.*, **132**:195–207, 1999.

[SG63] R. T. Shield and A. E. Green. On certain methods in the stability theory of continuous systems. *Arch. Ration. Mech. Anal.*, **12(4)**:354–360, 1963.

[SH96] A. Sadiki and K. Hutter. On the frame dependence and form invariance of the transport equations for the Reynolds stress tensor and the turbulent heat flux vector: Its consequences on closure models in turbulence modelling. *Continuum Mech. Thermodyn.*, **8**:341–349, 1996.

[Shi71] R. T. Shield. Deformations possible in every compressible isotropic perfectly elastic material. *J. Elast.*, **1**:145–161, 1971.

[Sil02] S. A. Silling. The reformulation of elasiticity theory for discontinuities and long-range forces. *J. Mech. Phys. Solids*, **48**:175–209, 2002.

[SK95] F. M. Sharipov and G. M. Kremer. On the frame dependence of constitutive equations. I. Heat transfer through a rarefied gas between two rotating cylinders. *Continuum Mech. Thermodyn.*, **7**:57–71, 1995.

[SL09] R. Soutas-Little. History of continuum mechanics. In J. Merodio and G. Saccomandi, editors, *Continuum Mechanics, EOLSS-UNESCO Encyclopedia*, chapter 2. Paris: UNSECO, 2009. Available online at http://www.eolss.net.

[SMB98] N. Sukumar, B. Moran, and T. Belytschko. The natural element method in solid mechanics. *Int. J. Numer. Methods Eng.*, **43(5)**:839+, 1998.

[Söd76] L. H. Söderholm. The principle of material frame-indifference and material equations of gases. *Int. J. Eng. Sci.*, **14**:523–528, 1976.

[Sok56] I. S. Sokolnikoff. *Mathematical Theory of Elasticity*. New York: McGraw-Hill, second edition, 1956.

[SP65] M. Singh and A. C. Pipkin. Note on Ericksen's problem. *Z. angew. Math. Phys.*, **16**:706–709, 1965.

[Spe81] C. G. Speziale. Some interesting properties of two-dimensional turbulence. *Phys. Fluids*, **24**:1425–1427, 1981.

[Spe87] C. G. Speziale. Comments on the "material frame-indifference" controversy. *Phys. Rev. A*, **36**:4522–4525, 1987.

[Ste54] E. Sternberg. On Saint-Venant's principle. *Q. Appl. Math.*, **11(4)**:393–402, 1954.

[TA86] N. Triantafyllidis and E. C. Aifantis. A gradient approach to localization of deformation. 1. Hyperelastic materials. *J. Elast.*, **16**:225–237, 1986.

[TG51] S. P. Timoshenko and J. N. Goodier. *Theory of Elasticity*. New York: McGraw-Hill, 1951.

[TG61] S. P. Timoshenko and J. Gere. *Theory of Elastic Stability*. New York: McGraw-Hill, second edition, 1961. Note: A new Dover edition came out in 2009.

[TG06] Z. Tadmor and C. G. Gogos. *Principles of Polymer Processing*. Hoboken: Wiley, second edition, 2006.

[Tho82] J. M. T. Thompson. *Instabilities and Catastrophes in Science and Engineering*. Chichester: Wiley, 1982.

[TM04] P. A. Tipler and G. Mosca. *Physics for Scientists and Engineers*, volume 2. New York: W. H. Freeman, fifth edition, 2004.

[TM11] E. B. Tadmor and R. E. Miller. *Modeling Materials: Continuum, Atomistic and Multiscale Techniques*. Cambridge: Cambridge University Press, 2011.

[TN65] C. Truesdell and W. Noll. The non-linear field theories of mechanics. In S. Flügge, editor, *Handbuch der Physik*, volume III/3, pages 1–603. Springer, 1965.

[TN04] C. Truesdell and W. Noll. In S. S. Antman, editor, *The Non-linear Field Theories of Mechanics*. Berlin: Springer-Verlag, third edition, 2004.

[Tre48] L. R. G. Treloar. Stress and birefringence in rubber subjected to general homogeneous strain. *Proc. Phys. Soc. London*, **60**:135–144, 1948.

[Tru52] C. Truesdell. The mechanical foundations of elasticity and fluid dynamics. *J. Ration. Mech. Anal.*, **1(1)**:125–300, 1952.

[Tru66a] C. Truesdell. *The Elements of Continuum Mechanics*. New York: Springer-Verlag, 1966.

[Tru66b] C. Truesdell. Thermodynamics of deformation. In S. Eskinazi, editor, *Modern Developments in the Mechanics of Continua*, pages 1–12, New York: Academic Press, 1966.

[Tru68] C. Truesdell. *Essays in the History of Mechanics*. New York: Springer-Verlag, 1968.

[Tru76] C. Truesdell. Correction of two errors in the kinetic theory of gases which have been used to cast unfounded doubt upon the principle of material frame-indifference. *Meccanica*, **11**:196–199, 1976.

[Tru77] C. Truesdell. *A First Course in Rational Continuum Mechanics*. New York: Academic Press, 1977.

[Tru84] C. Truesdell. *Rational Thermodynamics*. New York: Springer-Verlag, second edition, 1984.

[TT60] C. Truesdell and R. Toupin. The classical field theories. In S. Flügge, editor, *Handbuch der Physik*, volume III/1, pages 226–793. Berlin: Springer, 1960.

[Voi10] W. Voigt. *Lehrbuch der Kristallphysik (mit Ausschluss der Kristalloptik)*. Leipzig: Teubner, 1910.

[Wal72] D. C. Wallace. *Thermodynamics of Crystals*. Mineola: Dover, 1972.

[Wal03] D. J. Wales. *Energy Landscapes*. Cambridge: Cambridge University Press, 2003.

[Wan75] C. C. Wang. On the concept of frame-indifference in continuum mechanics and in the kinetic theory of gases. *Arch. Ration. Mech. Anal.*, **45**:381–393, 1975.

[Wei11] E. W. Weisstein. Einstein summation. http://mathworld.wolfram.com/EinsteinSummation.html. Mathworld – A Wolfram Web Resource, 2011.

[Wik10] Wikipedia. Leopold Kronecker – Wikipedia, the free encyclopedia. http://en.wikipedia.org/wiki/Leopold_kronecker, 2010. Online; accessed 30 May 2010. Based on E. T. Bell, *Men of Mathematics*. New York: Simon and Schuster, 1968, p. 477.

[Woo83] L. C. Woods. Frame-indifferent kinetic theory. *J. Fluid Mech.*, **136**:423–433, 1983.

[ZM67] H. Ziegler and D. McVean. On the notion of an elastic solids. In B. Broberg, J. Hult, and F. Niordson, editors, *Recent Progress in Applied Mechanics (The Folke Odquist Volume)*, pages 561–572. Stockholm: Almquist and Wiksell, 1967.

[ZT89] O. C. Zienkiewicz and R. L. Taylor. *The Finite Element Method*, volume I, *Basic Formulations and Linear Problems*. London: McGraw-Hill, 1989.

[ZT91] O. C. Zienkiewicz and R. L. Taylor. *The Finite Element Method*, volume II, *Solid and Fluid Mechanics: Dynamics and Non-Linearity*. London: McGraw-Hill, fourth edition, 1991.

[ZT05] O. C. Zienkiewicz and R. L. Taylor. *The Finite Element Method*. London: McGraw-Hill, sixth edition, 2005.

Index

Printed in the United States
By Bookmasters